CONVERSIONS BETWEEN U.S. CUSTOMARY UNITS AND SI UNITS (Continued)

U.S. Customary unit		Times conversion factor		Equals SI unit	
		Accurate	Practical		
Moment of inertia (area)					
inch to fourth power	in.^{4}	416,231	416,000	millimeter to fourth power	mm^4
inch to fourth power	in.^{4}	0.416231×10^{-6}	0.416×10^{-6}	meter to fourth power	m^4
Moment of inertia (mass)					
slug foot squared	slug-ft^2	1.35582	1.36	kilogram meter squared	kg·m^2
Power					
foot-pound per second	ft-lb/s	1.35582	1.36	watt (J/s or N·m/s)	W
foot-pound per minute	ft-lb/min	0.0225970	0.0226	watt	W
horsepower (550 ft-lb/s)	hp	745.701	746	watt	W
Pressure; stress					
pound per square foot	psf	47.8803	47.9	pascal (N/m^2)	Pa
pound per square inch	psi	6894.76	6890	pascal	Pa
kip per square foot	ksf	47.8803	47.9	kilopascal	kPa
kip per square inch	ksi	6.89476	6.89	megapascal	MPa
Section modulus					
inch to third power	in.^{3}	16,387.1	16,400	millimeter to third power	mm^3
inch to third power	in.^{3}	16.3871×10^{-6}	16.4×10^{-6}	meter to third power	m^3
Velocity (linear)					
foot per second	ft/s	0.3048*	0.305	meter per second	m/s
inch per second	in./s	0.0254*	0.0254	meter per second	m/s
mile per hour	mph	0.44704*	0.447	meter per second	m/s
mile per hour	mph	1.609344*	1.61	kilometer per hour	km/h
Volume					
cubic foot	ft^3	0.0283168	0.0283	cubic meter	m^3
cubic inch	in.^{3}	16.3871×10^{-6}	16.4×10^{-6}	cubic meter	m^3
cubic inch	in.^{3}	16.3871	16.4	cubic centimeter (cc)	cm^3
gallon (231 in.^{3})	gal.	3.78541	3.79	liter	L
gallon (231 in.^{3})	gal.	0.00378541	0.00379	cubic meter	m^3

*An asterisk denotes an *exact* conversion factor

Note: To convert from SI units to USCS units, *divide* by the conversion factor

Temperature Conversion Formulas

$$T(^\circ C) = \frac{5}{9}[T(^\circ F) - 32] = T(K) - 273.15$$

$$T(K) = \frac{5}{9}[T(^\circ F) - 32] + 273.15 = T(^\circ C) + 273.15$$

$$T(^\circ F) = \frac{9}{5}T(^\circ C) + 32 = \frac{9}{5}T(K) - 459.67$$

SUSTAINABLE ENERGY

SI Edition

RICHARD A. DUNLAP

Dalhousie University

CENGAGE
Learning

Australia • Brazil • Japan • Korea • Mexico • Singapore • Spain • United Kingdom • United States

![Cengage Learning logo]

Sustainable Energy, SI Edition
Richard A. Dunlap

Publisher: Timothy Anderson

Senior Developmental Editor: Mona Zeftel

Senior Editorial Assistant: Tanya Altieri

Senior Content Project Manager:
Jennifer Ziegler

Production Director: Sharon Smith

Media Assistant: Ashley Kaupert

Rights Acquisition Director:
Audrey Pettengill

Rights Acquisition Specialist, Text
and Image: Amber Hosea

Text and Image Researcher: Kristiina Paul

Manufacturing Planner: Doug Wilke

Copyeditor: Fred Dahl

Proofreader: Erin Buttner

Indexer: Shelly Gerger-Knechtl

Compositor: MPS Limited

Senior Art Director: Michelle Kunkler

Internal Designer: MPS Limited

Cover Designer: Rose Alcorn

Cover Image: Sergey Nivens/
Shutterstock.com

For product information and technology assistance, contact us at
Cengage Learning Customer & Sales Support, 1-800-354-9706.

For permission to use material from this text or product,
submit all requests online at **www.cengage.com/permissions**.
Further permissions questions can be emailed to
permissionrequest@cengage.com.

Library of Congress Control Number: 2013957163

ISBN-13: 978-1-133-10877-1

ISBN-10: 1-133-10877-6

Cengage Learning
200 First Stamford Place, Suite 400
Stamford, CT 06902
USA

Cengage Learning is a leading provider of customized learning solutions with office locations around the globe, including Singapore, the United Kingdom, Australia, Mexico, Brazil, and Japan. Locate your local office at:
international.cengage.com/region

Cengage Learning products are represented in Canada by Nelson Education Ltd.

For your course and learning solutions, visit
www.cengage.com/engineering.

Purchase any of our products at your local college store or at our preferred online store **www.cengagebrain.com**.

Unless otherwise noted, all items © Cengage Learning.

Printed in Canada
1 2 3 4 5 6 7 18 17 16 15 14

In Memory of My Father

Robert Bennett Dunlap

BRIEF CONTENTS

CONTENTS

Appendices

Our society uses substantial quantities of energy. This energy use amounts to about 5.7×10^{20} J per year worldwide, or an average of 8.1×10^{10} J per year per person. This energy economy is based on fossil fuels. More than 85% of the world's energy comes from fossil fuels, which are preferred because they are inexpensive (relatively speaking), are readily available (at least at present), and have a high energy density. As a result, an enormous infrastructure has been established for the location, production, and use of fossil fuels. The fuel of choice is oil because it is convenient, and the gasoline and diesel fuel it produces are portable and constitute our major source of fuel for transportation.

For the purpose of planning for methods to meet our future energy needs, it is important to begin by asking two questions: How long will our fossil fuel reserves last? Is it wise, from an environmental perspective, to continue to use fossil fuels?

The answers to both questions are not simple. The answer to the first question can be 40 years or 1000 years depending on the conditions that are put on our fossil fuel use. Will fossil fuels continue to supply 85% of our energy needs? Will a fossil fuel–derived product be required to fulfill our needs for a portable transportation fuel? Perhaps most importantly, How much are we willing to pay for fuel? There is certainly some limit to how much we, as individuals, are willing or able to pay for a liter of gasoline for our automobiles or for the oil to heat our homes. However, it is important to realize that the cost of fuel is not only a financial cost. Producing fossil fuels in a form that is suitable for our needs requires energy input in order to undertake exploration to locate new fuel reserves, the extraction of the fuel from those reserves, and the subsequent processing of the fuel. If the energy needed to produce a liter of fuel is greater than the energy we obtain from burning it, then the process is not only economically unattractive but is ultimately not energy productive. If only the use of oil in the traditional sense, from known and economically recoverable reserves, is considered, then the longevity of fossil fuels will certainly be at the low end of the timescale. If coal and less traditional oil reserves are also considered, then the answer can be near the upper end of the timescale. This will be especially true if alternative sources are used to supply a substantial fraction of our energy needs.

The answer to the second question is also not straightforward. There is overwhelming evidence that the emission of greenhouse gases that results from the burning of fossil fuels has a severe impact on the environment. The magnitude and the timescale of this impact are not fully understood. If the use of fossil fuels continues for an extended period of time, then our willingness or even our ability to take steps to mitigate the effects on the environment are also unclear.

To ensure an adequate supply of energy in the future and to avoid causing a negative impact on our environment, it is important to understand how energy is utilized at present, our future energy needs, and the options for fulfilling these needs.

In terms of our reliance on fossil fuels, two extreme approaches can be taken: to stop using fossil fuels now or to stop using fossil fuels when our supply is exhausted. The first approach would certainly minimize the environmental impact of fossil fuel use but would be impossible to implement because of our lack of infrastructure for the use of other energy sources. The latter approach would maximize the environmental effects and would best make use of the resources available. Whatever the final course of events, it is essential that steps toward eliminating our dependence on fossil fuel be taken immediately by developing and implementing alternative energy sources so that the environmental impact of our fossil fuel use is minimized. The latter would involve the reduction in greenhouse gas emissions by not only a reduction in fossil fuel use but also by processes such as carbon sequestration.

To put the magnitude of this task (however it is approached) into perspective, it is necessary to consider the current world power requirement of about 1.8×10^{13} W. In 50 years (a probable time scale for substantial reduction in fossil fuel use), our needs could be twice that amount. This is a rough goal that should be kept in mind when assessing the viability of any energy policy. These power requirements can be related to the output of a typical large electric generating station. These stations most commonly use fossil fuels (mostly coal and natural gas) to produce electricity and might have a typical output of about 10^9 W. The conversion to a nonfossil fuel energy economy on a timescale of about 50 years will require the construction of about $(3.6 \times 10^{13}$ W$)/(10^9$ W$) = 36,000$ large replacement facilities (or a corresponding number of smaller facilities). These might be large nuclear power plants, large hydroelectric stations, or equivalent-capacity facilities utilizing solar energy, wave energy, wind energy, and other sources. This amounts to the construction of two major nonfossil fuel power stations every day for the next five decades. Clearly, this task requires a substantial commitment.

The textbook *Sustainable Energy* considers in detail our present and future energy needs, options for continued use of fossil fuels, and options for establishing an alternative energy economy. This text was developed out of a course entitled "Energy and the Environment" that has been taught in the Department of Physics and Atmospheric Science at Dalhousie University since 2003. This one-semester introductory course is aimed at undergraduate science and engineering students and is taught at the sophomore level. Most students have taken freshman-level chemistry and physics, and most have had some introductory calculus. These prior courses make a suitable prerequisite for a course taught from the present text. The course at Dalhousie is taught as a follow-up to a course on climate change to give students the overall picture of how humanity interacts with the environment, although this previous course is not a necessary prerequisite for a course taught from *Sustainable Energy*. Although such a course is intended to be introductory, there is enough technical detail that upper-level science or engineering will find it a useful and informative elective.

The purpose of this textbook is to fill a niche between a significant number of texts on similar topics at a very descriptive level intended for a freshman survey course and a few advanced (and often fairly specialized) texts aimed at senior undergraduate or beginning graduate students in engineering. The textbook is useful for science and engineering students with an interest in energy-related matters, particularly those looking to pursue a professional career in a related field, to take an introductory energy course with some reasonable technical content. In addition to filling a mostly unfilled gap in the field, the present text also provides an up-to-date introduction to a fairly rapidly changing field.

Organization

This text begins with an overview of the basic science needed for the remainder of the book, as well as a summary of our past, present, and anticipated future energy needs. The technologies currently in use to meet our energy needs are described, and the need for the development of new energy technologies on the basis of future resource availability and environmental concerns is emphasized. The text includes a separate chapter on every future renewable energy technology that could be viewed as a viable option for the production of a significant portion of our energy needs. How these developing technologies can be integrated efficiently with existing technologies is discussed, as well as approaches to conserving available energy resources. Finally, the text considers options for perhaps our greatest energy-related challenge: transportation. The viability of any alternative energy technologies is determined by its ability to fulfill various criteria. The important criteria are described in this text by the acronym CURVE, for *c*lean, *u*nlimited, *r*enewable, *v*ersatile, and *e*conomical. This acronym makes it easy for students to appreciate how different technologies may, or may not, play an important role in our future energy production. The final chapter of the book summarizes the various alternative energy sources that have been presented and analyzes how these different technologies succeed or fail in satisfying the various CURVE criteria.

Throughout the text, the complexity of energy issues is emphasized, as is the need for a multidisciplinary approach to solving our energy problems. This approach provides students with an appreciation for the real problems that are encountered in the understanding of how we produce and use energy, as well as the realization that, while exact calculations are important and necessary, a broadly based analysis is often most appropriate. The text also stresses the fact that solutions to our energy problems, both now and in the future, are not straightforward and do not have simple, well-defined solutions and that the way ahead is far from certain. The book contains enough material for a typical one-semester (12- to 14-week) course with about 20% excess material to allow the instructor some flexibility in course design. This coverage of material allows about 2–3 hours of lecture, on average, per chapter. Instructors may also focus on specific topics to provide a more in-depth picture of certain aspects of energy. This approach may include a more detailed and probing look at some of the topics presented in the Energy Extra boxes and may require the omission of other components of the text. Some chapters from the text can be covered in less detail and/or even eliminated. Chapters 7, 12, 13, 14, and 20 can be skipped with minimal effect on continuity. Certain approaches to sustainable energy may be more or less relevant from some national and/ or regional perspectives and may warrant more or less detailed course coverage.

Finally, Chapter 21 acts as a summary of the ideas presented in the text and shows how they can be integrated into our approach to future energy production. This chapter includes a number of research and design projects that provide the student with the challenge of integrating information presented throughout the text to the solution of practical problem related to energy production and use. These projects give the student the opportunity to assess information and to make decisions about the most reasonable approach to energy production and use. Such decisions often involve a consideration of scientific, technological, environmental, and economic factors and illustrate not only the complexity but the multidisciplinary nature of sustainable energy.

Chapter Pedagogical Elements

- **Learning Objectives**. Each chapter starts with a bulleted list of learning objectives, making it very clear to both instructors and students what is covered in the chapter.

- **Examples**: The text includes numerous worked examples to provide the student with the basic approach to deal with end-of-chapter problems
- **Energy Extra Boxes**. Energy Extra boxes are included in nearly all chapters. These boxes provide insight into details of specific aspects of energy and often emphasize the complex nature of the decisions required to plan for our future energy needs. They also stress that ostensibly advantageous approaches to energy are often not as beneficial as they seem and that a critical analysis is necessary to understand all aspects of the topic.
- **End-of-chapter Problems**. The end-of-chapter problems are predominantly quantitative in nature. However, most are not straightforward calculations based on substituting values from the chapter into the appropriate formulas. The problems are designed to require the students to analyze information, to make use of material from previous chapters, to correlate data from various sources (not only from the textbook itself but from library, Internet, or other sources), and in many cases to estimate quantities based on interpretation of graphical data, interpolation of values, and sometimes just plain common sense.

Ancillaries

A variety of ancillaries are available to accompany this book to supplement your course. These supplements include:

- An *Instructor's Solution Manual*.
- *PowerPoint Slides*, which include the important figures, tables, and equations from the book.
- *Annotated Lecture Builder PowerPoint Slides*, which include suggestions for teaching the material in the book.
- Sample test items for instructors
- Additional practice problems for students.

MindTap Online Course and Reader

In addition to the print version, this textbook is also available online through MindTap, a personalized learning program. Students who purchase the MindTap version will have access to the book's MindTap Reader and will be able to complete homework and assessment material online, through their desktop, laptop, or iPad. If your class is using a Learning Management System (such as Blackboard, Moodle, or Angel) for tracking course content, assignments, and grading, you can seamlessly access the MindTap suite of content and assessments for this course.

In MindTap, instructors can:

- Personalize the Learning Path to match the course syllabus by rearranging content, hiding sections, or appending original material to the textbook content
- Connect a Learning Management System portal to the online course and Reader
- Customize online assessments and assignments
- Track student progress and comprehension with the Progress app
- Promote student engagement through interactivity and exercises

Additionally, students can listen to the text through ReadSpeaker, take notes and highlight content for easy reference, and check their understanding of the material.

Acknowledgments

I am grateful for the assistance of many individuals during the development of this text. First, I am indebted to the students who have taken courses and whom I have taught at Dalhousie University on Sustainable Energy. They have served as the inspiration for this textbook and have provided feedback on the course material. I would also like to thank Jeff Dahn for numerous discussions over the years on energy related matters and Harm Rotermund for his continued encouragement and comments during the writing of the manuscript. I am also grateful to Ewa Dunlap for assistance, support, and advice throughout this project, to German Rojas Orozco for checking the accuracy of the Examples and the Solutions to the Problems, and to Amy Hill of KHLowery Consulting for numerous comments and suggestions. Finally, I would like to thank the following reviewers who provided invaluable comments on the manuscript:

- Julie Albertson, *University of Colorado, Colorado Springs*
- Prabhakar Bandaru, *University of California, San Diego*
- Ronald Besser, *Stevens Institute of Technology*
- Christopher Bull, *Brown University*
- Larry Caretto, *California State University, Northridge*
- Gerald Cecil, *University of North Carolina at Chapel Hill*
- Timothy J. Cochran, *Alfred State College*
- Tim Healy, *Santa Clara University*
- Charles Knisely, *Bucknell University*
- David Marx, *Illinois State University*
- Chiang Shih, *Florida A&M University and Florida State University*
- Robert J. Stevens, *Rochester Institute of Technology*
- Thomas Ortmeyer, *Clarkson University*

R. A. Dunlap
Halifax, Nova Scotia

PREFACE TO THE SI EDITION

This edition of *Sustainable Energy* has been adapted to incorporate the International System of Units (*Le Système International d'Unités* or SI) throughout the book.

Le Système International d'Unités

The United States Customary System (USCS) of units uses FPS (foot–pound–second) units (also called English or Imperial units). SI units are primarily the units of the MKS (meter–kilogram–second) system. However, CGS (centimeter–gram–second) units are often accepted as SI units, especially in textbooks.

Using SI Units in this Book

In this book, we have used both MKS and CGS units. USCS units or FPS units used in the US Edition of the book have been converted to SI units throughout the text and problems. However, in case of data sourced from handbooks, government standards, and product manuals, it is not only extremely difficult to convert all values to SI, it also encroaches upon the intellectual property of the source. Some data in figures, tables, and references, therefore, remains in FPS units. For readers unfamiliar with the relationship between the FPS and the SI systems, a conversion table has been provided inside the front cover.

To solve problems that require the use of sourced data, the sourced values can be converted from FPS units to SI units just before they are to be used in a calculation. To obtain standardized quantities and manufacturers' data in SI units, the readers may contact the appropriate government agencies or authorities in their countries/regions.

Instructor Resources

The Instructors' Solution Manual in SI units is available through your Sales Representative or online through the book website at www.login.cengage.com. A digital version of the ISM and PowerPoint slides of figures, tables, and examples and equations from the SI text are available for instructors registering on the book website.

Feedback from users of this SI Edition will be greatly appreciated and will help us improve subsequent editions.

Cengage Learning

ABOUT THE AUTHOR

Richard A. Dunlap is a professor in the Department of Physics and Atmospheric Science at Dalhousie University and has a cross-appointment in the College of Sustainability. He received a B.S. in Physics from Worcester Polytechnic Institute (1974), an A.M. in Physics from Dartmouth College (1976), and a Ph.D. in Physics from Clark University (1981). Since 1981 he has been on the faculty at Dalhousie University. From 2001 to 2006 he was Killam Research Professor of Physics, and since 2009 he has been Director of the Dalhousie University Institute for Research in Materials. He currently is a member of the DREAMS Program (Dalhousie Research in Energy, Advanced Materials and Sustainability). Prof. Dunlap is author of three previous textbooks, *Experimental Physics: Modern Methods* (Oxford 1988), *The Golden Ratio and Fibonacci Numbers* (World Scientific 1997), and *An Introduction to the Physics of Nuclei and Particles* (Brooks/Cole 2004). Over the years his research interests have included critical phenomena, magnetic materials, amorphous materials, quasicrystals, hydrogen storage, and superconductivity. His current research activities are primarily in the area of materials for advanced rechargeable batteries. He has published more than 280 refereed research papers.

SUSTAINABLE
ENERGY

SI Edition

PART I

Background

Energy is an essential component of our daily lives. Throughout human history, our energy use has increased, and we now depend on a complex energy infrastructure to meet our needs for heating, lighting, transportation, and the production and distribution of all manufactured materials. Our increased energy needs have put increasing demands on the earth's resources and have had increasingly adverse effects on our environment. We are now at a stage of human development where our energy use must be critically analyzed to determine suitable future approaches to the production and use of this vital component of our lives.

Chapter 1 of this text begins with an overview of the basic scientific principles related to energy and a description of the quantitative scientific tools needed to analyze our energy use. This overview includes a summary of the various forms of energy and a quantitative description of the processes by which energy can be converted from one form to another. Also included is a survey of fundamental thermodynamics and a description of the basic principles of electricity distribution.

An overview of energy use throughout history is presented in Chapter 2. The chapter also provides the mathematical basis needed to assess future energy needs and a summary of the factors that need to be evaluated when considering possible future energy production methods.

The photograph at the beginning of this part of the text shows the Gordon Dam in Tasmania. This high head hydroelectric dam is 192 m long and 140 m high and has a maximum capacity of 432 MW_e. It became operational in 1978 and was one of the last major hydroelectric facilities to be constructed during an era of hydroelectric power development in Tasmania that began in the 1950s and continued until the 1980s. This trend in major hydroelectric development is paralleled in many other parts of the world. ■

Energy Basics

Learning Objectives: After reading the material in Chapter 1, you should understand:

- The relationship between energy and power.
- The forms of energy.
- The laws of thermodynamics.
- Heat engines and their Carnot efficiency.
- Heat pumps and their coefficient of performance.
- How electricity is generated and distributed.

1.1 Introduction

Energy may be categorized in different ways. One approach is to classify energy as either kinetic or potential. From a classical point of view, *kinetic* energy is merely the energy associated with the motion of a body. *Potential* energy may be described in terms of the nature of the interactions in a system. For example, gravitational potential energy arises from the gravitational interaction between two masses. The potential energy of water in an elevated reservoir may be converted into kinetic energy, which can be utilized to turn a turbine.

From a practical standpoint, however, it is convenient to describe the forms of energy according to how they are produced and utilized. It is also crucial to understand how one form of energy can be converted into another. In fact, when one says that energy is "produced" (e.g., by a nuclear reactor), one refers to the conversion of one form of energy that is not suitable for our needs (i.e., nuclear binding energy) into another form (e.g., electrical energy) that is more readily utilized. Energy that is extracted from our environment is *primary* energy; for example, chemical energy in the form of fossil fuels, kinetic energy in the wind, potential energy of water in a reservoir, or incident solar energy. To utilize energy, it is nearly always necessary to convert primary energy into a form that suits our needs. In this chapter, some of the basic physics of energy are explained, as well as the characteristics of some forms of energy and their conversions.

Tim Collins/Shutterstock.com

1.2 Work, Energy, and Power

Energy, *E*, is defined as the ability to do work, *W*. *Work* is the consequence of the expenditure of energy and is defined as the product of a force, *F*, acting on an object times the distance, *d*, that the object moves. This relationship can be written as

$$W = Fd. \tag{1.1}$$

This expression assumes that the force acting on the object is a constant over the time during which the object moves a distance, *d*, and that the force is acting in the same direction as the displacement. The units of work are the same as the units of energy. In the metric system, the standard unit of energy is the *joule* (J) when the force is expressed in *newtons* (N) and the displacement in *meters* (m). In terms of fundamental metric units, the *joule* is equal to $kg \cdot m^2 \cdot s^{-2}$.

It is perhaps convenient to think of the concept expressed in equation (1.1) in terms of a mechanical system. An object with a mass, *m*, lying at rest on the floor exerts a force (the gravitational force), *mg* (here *g* is the gravitational acceleration), downward on the floor. The floor exerts a force, *mg*, upward on the mass (the normal force) that cancels out the gravitational force. The net vertical force on the object is zero, and, from Newton's law,

$$F = ma, \tag{1.2}$$

the acceleration is zero, and the object does not move. The work done is zero because the distance that the object travels is zero. If an external vertical force that is equal to (or greater than) *mg* is exerted on the object, then the object can be lifted from the floor. If the object is lifted to a height *h*, then the work done, from equation (1.1), is the force times the distance, or

$$W = mgh. \tag{1.3}$$

According to the law of conservation of energy, this work is converted into gravitational potential energy, also equal to *mgh*. The work done is independent of how long the process takes or of the path taken to reach height *h*. Because of this latter property, the gravitational force is said to be *conservative*.

It is sometimes convenient to deal with power rather than with energy or work. *Power*, *P*, is the rate at which work is done (or the rate at which energy is expended). Power is measured in *watts* (W), and the watt is defined as 1 joule per second. Assuming that power is a constant in time, *t*, then the total energy utilized is

$$E = Pt. \tag{1.4}$$

This definition shows that 1 W·s = 1 J. Total energy is the power integrated over time so that producing (or using) 1000 W of power for 1 second represents the same amount of energy as producing (or using) 1 W of power for 1000 seconds. Equation (1.4) provides the basis for an alternative unit for the measurement of energy, the *kilowatt-hour* (kWh). The kWh is defined as the energy corresponding to a power of 1 kilowatt (1000 W) over a period of 1 hour (3600 s) so that 1 kWh = (1000 W) × (3600 s) = 3.6×10^6 J.

> **Example 1.1**
>
> If a system produces 10^6 J (i.e. 1 MJ) of energy every hour, what is the power produced in watts?
>
> **Solution**
>
> If 10^6 J is released over a period of 1 hour (or 3600 s), then the energy per unit time in joules per second, which is equivalent to watts, is
>
> $$P = E/t = (10^6 \text{ J})/(3600 \text{ s}) = 278 \text{ W}.$$

1.3 Forms of Energy

Energy can take on many forms:

- *Kinetic energy* (e.g., of a moving automobile)
- *Gravitational potential energy* (e.g., of water in a reservoir)
- *Thermal energy* (e.g., in a pot of boiling water)
- *Chemical energy* (e.g., stored in a liter of gasoline)
- *Nuclear energy* (e.g., stored in a gram of uranium)
- *Electrical energy* (e.g., used by a light bulb)
- *Electromagnetic energy* (e.g., that associated with a beam of sunlight)

As explained, these categories of energy are merely convenient ways of describing energy from different sources. They are not necessarily unique or mutually exclusive, nor is the list necessarily comprehensive. For example, thermal energy might be thought of as the microscopic kinetic energy of the molecules of a material. Both chemical energy and nuclear energy can be viewed as manifestations of the mass-energy associated with bonds in a material. However, these seven categories are a convenient way of defining the forms of energy from a practical standpoint.

To make use of energy, it is generally necessary to convert energy from the form in which it is obtained to a form that is compatible with our needs. For example, the stored chemical energy in a liter of gasoline can be converted to heat and then into mechanical energy to move a vehicle. Energy conversions are an important aspect of the utilization of any energy source, and the efficiency of these conversions is crucial to the viable utilization of the energy source. In any process, energy is always conserved. (In nuclear physics, the conservation of mass-energy, rather than the conservation of energy itself, is employed because there is an equivalence between these two quantities.) However, in any energy conversion process, all of the energy does not end up in the form needed. Each of these forms will now be discussed briefly.

1.3a Kinetic Energy

Kinetic energy is most obviously associated with moving objects. For an object of mass, m, moving at a velocity, v, the kinetic energy is

$$E = \frac{1}{2} mv^2, \tag{1.5}$$

In the metric system, m is in kilograms (kg), v is in meters per second (m/s), and the resulting energy is in joules (kg·m^2·s^{-2}).

Example 1.2

What is the kinetic energy associated with a 1500-kg automobile traveling at 100 km/h?

Solution

The velocity converted to m/s is $(100 \text{ km/h}) \times (1000 \text{ m/km})/(3600 \text{ s/h}) = 27.8$ m/s. Using equation (1.5), the energy is given as

$$E = \frac{1}{2}mv^2 = (0.5) \times (1500 \text{ kg}) \times (27.8 \text{ m/s})^2 = 5.8 \times 10^5 \text{ kg·m}^2/\text{s}^2 = 5.8 \times 10^5 \text{ J}.$$

Kinetic energy is also associated with the rotational motion of rotating objects. The energy is given as

$$E = \frac{1}{2}I\omega^2, \tag{1.6}$$

where I is the moment of inertia of the object, and ω is its angular velocity. (The moment of inertia of an object and further details of rotational motion will be discussed in Chapter 18.) The moment of inertia is given in units of kg·m^2, and the angular velocity is given in units of s^{-1}. As before, the energy is measured in joules. Objects that have both translational motion and rotational motion have both translational kinetic energy, as given by equation (1.5), and rotational kinetic energy, as given by equation (1.6).

Example 1.3

A wheel in the form of a solid disk with a mass of $m = 400$ kg, a diameter of $d = 0.85$ m and a moment of inertia of $I = md^2/8 = mr^2/2$ rolls without slipping. The velocity of its center of mass is 30 m/s. This is a rough approximation of a wheel on a freight train. Compare the wheel's translational kinetic energy to its rotational energy.

Solution

From equation (1.5), its translational kinetic energy is

$$E_{\text{kinetic}} = \frac{1}{2}mv^2 = (0.5) \times (400 \text{ kg}) \times (30 \text{ m/s})^2 = 1.8 \times 10^5 \text{ J}.$$

If the wheel rolls without slipping, then its angular velocity is related to the velocity of its center of mass, v, and its radius, r, because $\omega = v/r$. Substituting for ω and I in equation (1.6) gives

$$E_{\text{rotational}} = \frac{1}{2}\left(\frac{mr^2}{2}\right)\left(\frac{v}{r}\right)^2 = \frac{1}{4}mv^2$$

Substituting these values,

$$E_{\text{rotational}} = \frac{1}{4}mv^2 = (0.25) \times (400 \text{ kg}) \times (30 \text{ m/s})^2 = 9.0 \times 10^4 \text{ J}.$$

Note that the rotational energy is independent of the wheel diameter and is exactly one-half of the translational kinetic energy. These features are basic characteristics of a solid disk that rolls without slipping.

1.3b Potential Energy

Potential energy is most conveniently thought of in terms of gravitational potential, as explained. The concept of potential energy also applies to other situations, such as the energy contained in a compressed spring. In the case of gravitational potential energy, an object of mass, m, at a height h has potential energy given by

$$E = mgh. \tag{1.7}$$

This potential energy can be converted into kinetic energy by allowing the object to fall through the distance h (assuming there are no drag forces), yielding

$$E = \frac{1}{2}mv^2 = mgh. \tag{1.8}$$

The velocity of the object may thus be calculated to be

$$v = \sqrt{2gh}. \tag{1.9}$$

Example 1.4

A 75-kg person walks up a flight of stairs with a vertical height of 3 m. What is the change in that person's potential energy?

Solution
From equation (1.7),

$$E = mgh = (75 \text{ kg}) \times (9.8 \text{ m/s}^2) \times (3 \text{ m}) = 2.2 \times 10^3 \text{ J},$$

1.3c Thermal Energy

The *thermal energy* of a gas results from the kinetic energy of the microscopic movement of the molecules. Each molecule of gas has a kinetic energy associated with it that is given by equation (1.5), where m is the mass of the molecule, and v is its average velocity. It can be shown by applying ideal gas theory that the right-hand side of equation (1.5) can be expressed in terms of the temperature of the gas as

$$\frac{1}{2}mv^2 = \frac{3}{2}k_\text{B}T. \tag{1.10}$$

Here k_B is *Boltzmann's constant* with a value of 1.3806×10^{-23} J/K, and T is the absolute temperature in Kelvin (K) (more on this in Section 1.4). The total internal energy of a collection of gas molecules is obtained from equation (1.10) by multiplying the right-hand side by the number of gas molecules present. From a practical standpoint, it is convenient to deal with macroscopic quantities such as the number of moles of gas. Thus

$$E = \frac{3}{2}nRT \tag{1.11}$$

where n is the number of moles of gas, and R is the *universal gas constant*; $R = N_A k_B = 8.315$ J/(mol·K). *NA* is *Avogadro's number* (6.022×10^{23} mol^{-1}).

It is sometimes convenient (particularly for solids and liquids) to describe changes in the macroscopic thermal energy of the material in terms of the *specific heat, C*, of the material. If a quantity of energy, Q, is supplied to a piece of material of mass, m, then its temperature will increase by an amount, ΔT, given by

$$\Delta T = \frac{Q}{mC}. \tag{1.12}$$

Materials with a large specific heat require a large amount of energy per unit mass to raise their temperature by a given amount. On the other hand, these materials are able to store large amounts of thermal energy per unit mass when its temperature is raised by a relatively small amount. (The utilization of these principles is discussed in detail in Chapter 8.) If a solid is heated to its melting point, then additional energy must be provided to melt it. This energy is used to break the chemical bonds holding the solid together and is referred to as the *latent heat of fusion*. The term *latent heat* is used to distinguish it from *sensible heat* because latent heat does not change the temperature of a solid. When a liquid is heated to its boiling point, then additional energy, the *latent heat of vaporization*, is needed to cause the material to undergo a phase transition and become a gas.

Example 1.5

The specific heat of water is 4180 J/(kg·°C). Calculate the energy required to heat 500 g of water from 20°C to 80°C.

Solution
Rearranging equation (1.12) to solve for the heat gives

$$Q = mC\Delta T.$$

Using $m = 0.5$ kg, $C = 4180$ J/(kg·°C), and $\Delta T = (80°C - 20°C) = 60°C$, then

$$Q = (0.5 \text{ kg}) \times [4180 \text{ J/(kg·°C)}] \times (60°C) = 1.25 \times 10^5 \text{ J}.$$

1.3d Chemical Energy

Chemical energy is the energy associated with chemical bonds, that is, the interaction energy between atomic electrons in a material. Energy can be absorbed or released during a chemical reaction as a result of changes in the bonds between the atoms. If a process requires energy to be input for the reaction to occur, then the process is referred to as *endothermic*. In general, these types of processes are not useful in the production of energy, although they can be useful in the storage of it. The dissociation of water (into hydrogen and oxygen) is of interest in this respect and will be discussed in more detail in Chapter 20.

Processes that release energy are referred to as *exothermic* and are of interest in this discussion. In general, oxidation reactions (i.e., the burning of materials) fall into this category. Some of the most relevant for the production of energy are reactions that involve the oxidation of carbon. The simplest of these is the oxidation of pure carbon (using oxygen from the atmosphere) and the production of carbon dioxide. This process is given by the formula

$$C + O_2 \rightarrow CO_2 + 32.8 \text{ MJ/kg}. \tag{1.13}$$

The equation indicates the amount of energy released by the combustion of 1 kg of pure carbon, that is the *heat of combustion* (in MJ/kg) of carbon, which corresponds to a release of 4.09 eV (eV = 1.6×10^{-19} J) of energy from the oxidation of one atom of carbon. Note that the energy in electron volts (eV) per atom (or molecule) for a substance of *molecular mass, M* (given in g/mol), is

$$E(\text{eV}) = \frac{(10^6 \text{ J/MJ}) \cdot M \cdot (Q/m)}{(10^3 \text{ g/kg}) \cdot (6.022 \times 10^{23} \text{ mol}^{-1}) \cdot (1.602 \times 10^{-19} \text{J/eV})}, \tag{1.14}$$

where the energy released per unit mass, Q/m, is expressed in MJ/kg. The oxidation of pure carbon is sometimes a suitable approximation of the burning of coal. The burning of other fossil fuels [or other organic materials such as wood, ethanol, or municipal waste (Chapter 16)] generally involves the oxidation of hydrocarbons. Some of the simple reactions of this type are the burning of *methane* (the major component of natural gas),

$$CH_4 + 2O_2 \rightarrow CO_2 + 2H_2O + 55.5 \text{ MJ/kg}; \tag{1.15}$$

the burning of *ethanol* (a common biofuel),

$$C_2H_6O + 3O_2 \rightarrow 2CO_2 + 3H_2O + 29.8 \text{ MJ/kg}; \tag{1.16}$$

and the burning of *octane* (an important component of gasoline),

$$2C_8H_{18} + 25O_2 \rightarrow 16CO_2 + 18H_2O + 46.8 \text{ MJ/kg}. \tag{1.17}$$

Note that combustion reactions of hydrocarbons produce steam (not liquid water) as a by-product. The energies given above are referred to as higher heating values (HHV) and include the latent heat of vaporization of the steam. This energy can only be recovered if the steam produced is condensed.

An important reaction for the production of energy by animals is the combustion of *glucose*,

$$C_6H_{12}O_6 + 6O_2 \rightarrow 6CO_2 + 6H_2O + 16.0 \text{ MJ/kg}. \tag{1.18}$$

In Chapter 20, the energy released by the oxidation of hydrogen, that is,

$$2H_2 + O_2 \rightarrow 2H_2O + 142 \text{ MJ/kg}, \tag{1.19}$$

will be considered in detail.

1.3e Nuclear Energy

Nuclear energy is similar to chemical energy in that it is the energy associated with the bonds between particles. The relevant energy scale of nuclear energy relates to the bonds between the neutrons and protons within the nucleus rather than to the bonds between atoms that involve the atomic electrons. Nuclear bonds represent much larger amounts of energy than chemical bonds, typically many megaelectron volts (MeV) compared with a few electron volts. As a result, the amount of energy that can be released in nuclear reactions is many orders of magnitude larger than the energy that can be released in chemical reactions. Through the equivalence of mass and energy as given by Einstein's relation,

$$E = mc^2,$$ **(1.20)**

where c is the speed of light, nuclear energy can be related to changes in the nuclear mass (discussed in much more detail in Chapter 5). The energy released in an exothermic nuclear reaction is given in terms of the change in nuclear mass as

$$E_{exo} = \Delta mc^2.$$ **(1.21)**

Although this is also true for chemical energy, the energy associated with chemical bonds and hence the changes in mass associated with chemical reactions are very much smaller than in the nuclear case. As discussed in detail in Chapters 6 and 7, the release of nuclear energy can accompany the fission (breaking up) of heavy nuclei like those in uranium and plutonium or the fusion (bonding) of light nuclei like the isotopes of hydrogen.

1.3f **Electrical Energy**

Electrical energy is associated with electrons in a conductor. It is convenient to deal with the macroscopic representation of electrical energy in terms of *voltages* and *currents* without the need to be concerned with the microscopic description of the electrons. If a current, I, flows through a circuit with a resistance, R, then there is a voltage decrease, V, across the resistance given by

$$V = IR,$$ **(1.22)**

where V is given in *volts* (V) if I is in *amperes* (A) and R is in *ohms* (Ω). The power dissipated though the resistance is given by

$$P = VI.$$ **(1.23)**

When V and I are in volts and amperes, respectively, then P is in watts. From equation (1.22), this may be written as

$$P = I^2R = \frac{V^2}{R}.$$ **(1.24)**

From equation (1.4), electrical energy (in joules) is power (in watts) multiplied by time (in seconds). This is most often expressed in kilowatt-hours by dividing the energy in joules by the conversion factor 3.6×10^6 J/kWh. Correspondingly, the conversion factor between kilowatt-hours and megajoules (MJ) is 0.278 kWh/MJ.

1.3g **Electromagnetic Energy**

Electromagnetic radiation may be thought of in terms of associated electric and magnetic fields that form waves (such as light waves). It may also be thought of, in the quantum mechanical sense, as a collection of particles called *photons*. This radiation covers a wide range of wavelengths, as illustrated in Figure 1.1. Different wavelength regimes are, for example, X-rays, ultraviolet radiation, visible light, infrared radiation radio waves, and so on. Electromagnetic radiation from the sun (which falls largely in the visible region of the spectrum) is one of our most important sources of energy because it is the basic source responsible for most other sources of energy, such as fossil fuels, wind, solar radiation, and biomass energy. Figure 1.1 illustrates that, as the wavelength decreases, the energy content of the radiation increases. For any wave, the wavelength, λ, is related to the frequency, f, in terms of the velocity (in this case, the velocity of light, $c \approx 3 \times 10^8$ m/s) by

$$\lambda = \frac{c}{f}, \tag{1.25}$$

where the wavelength is given in meters when the velocity is given in meters per second (m/s) and the frequency in Hertz (s^{-1}).

Example 1.6

Estimate the number of wavelengths of yellow light that span the distance between a computer monitor and a user.

Solution

From Figure 1.1, the wavelength of yellow light is about 600 nm, or 6.0×10^{-7} m. If a typical user sits 0.5 m from a computer monitor, then the number of wavelengths of yellow light that fit into that distance is

$$N = \frac{0.5 \text{ m}}{6.0 \times 10^{-7} \text{m}} = 8.3 \times 10^5 \text{ wavelengths.}$$

If the electromagnetic energy is considered in terms of quanta of energy (i.e., photons), then the energy associated with each photon (E) is related to the frequency of the electromagnetic radiation (f) as

$$E = hf. \tag{1.26}$$

where h is *Planck's constant* (Appendix II). Long-wavelength radio waves may be produced artificially by electronic devices (i.e., radio transmitters). Radiation in the infrared to ultraviolet and X-ray regions of the spectrum are most commonly produced (either naturally or artificially) by electrons undergoing transitions between atomic energy levels. Short-wavelength (i.e., high energy) radiation most commonly comes from transitions involving excited states of nuclei. The energy in equation (1.26) is in joules when Planck's constant is 6.626×10^{-34} J·s. It is often convenient (as in Chapter 9) to express the energy per photon in electron volts. In this case, Planck's constant is given by 4.136×10^{-15} eV·s.

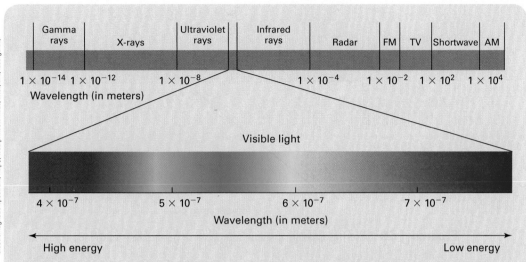

Figure 1.1: The electromagnetic spectrum. The visible portion of the spectrum is shown in the expanded insert.

1.4 Some Basic Thermodynamics

The thermodynamic behavior of systems is described by the *laws of thermodynamics*. A number of physical laws influence thermodynamic behavior, but four are of importance for the present discussion, and these are traditionally numbered 0 to 3:

0. Two systems that are both in thermodynamic equilibrium with a third system are in equilibrium with each other.
1. Energy is conserved.
2. A closed system will move toward equilibrium.
3. It is impossible to attain absolute zero temperature.

These are each discussed briefly below.

1.4a The Zeroth Law of Thermodynamics

The zeroth law is a generalization of the definition of thermal equilibrium. Two systems are in thermal equilibrium if they are able to transfer heat between each other but do not. This law implies that the thermodynamic state of a system can be defined by a single parameter, the *temperature*. Two systems in thermal equilibrium are defined to be at the same temperature.

Temperature may be defined in terms of the *ideal gas law*,

$$PV = Nk_BT, \tag{1.27}$$

where P is pressure, V is volume, N is the number of gas molecules, k_B is Boltzmann's constant, and T is the temperature in Kelvin. No matter what system is in use (Celsius, Kelvin, etc.), two parameters define a temperature scale: the location of zero and the size of the degree. Reference points that define a temperature scale are often based

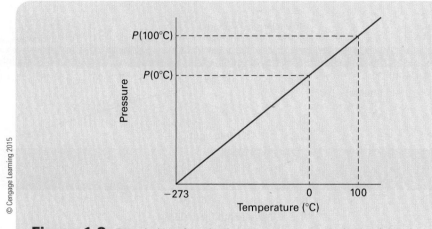

Figure 1.2: Description of an absolute temperature on the basis of the ideal gas law.

on the properties of water, as in the case of the Celsius scales, where the zero point is the freezing point of water. This zero is not absolute zero temperature, and this means that the temperature scale is not an absolute scale. To describe an absolute temperature scale using the size of the degree as defined on the Celsius scale, the ideal gas law can be used. Because a given volume of an ideal gas has a pressure that is linearly related to the absolute temperature by equation (1.27), when the temperature goes to absolute zero, then the pressure goes to zero. Figure 1.2 shows a plot of pressure as a function of temperature measured in Celsius for an ideal gas. Extrapolating the temperature until the pressure goes to zero gives the value of absolute zero on the Celsius scale. This is the basis for defining the temperature scale in Kelvin, as shown in the figure where $K = °C + 273$.

1.4b The First Law of Thermodynamics

Energy conservation, as it applies to thermodynamic systems, may be viewed in terms of Figure 1.3. Gas is contained in a cylinder that is sealed with a piston. If energy is supplied in the form of heat to the system by, for example, a flame, then two possible situations may be considered;

1. If the piston is held fixed, then the internal energy of the gas is increased by the addition of energy, and this is manifested by an increase in temperature.
2. If the piston is allowed to move, then some of the energy may be used to lift the piston, thereby doing work against the force of gravity.

In general,

$$Q = \Delta U + W, \tag{1.28}$$

where Q is the energy input in the form of heat, ΔU is the change in the internal energy of the gas as indicated by the change in temperature, and W is the mechanical work that is done. Equation (1.28) forms the basis for understanding energy conversion processes involving heat in closed systems.

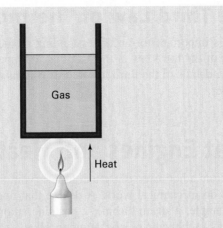

Figure 1.3: Example of the conservation of energy and an illustration of how thermal energy can be used to do mechanical work.

1.4c The Second Law of Thermodynamics

There are many ways of stating the second law of thermodynamics. In addition to the statement given at the beginning of this section (a closed system will move toward equilibrium), the second law may also be stated as follows:

> Heat naturally flows from a hot place to a cold place.

or

> The entropy of the universe always increases.

To see that these three statements are just alternate ways of looking at the same phenomenon, consider the behavior of a hot piece of metal and a cold piece of metal that are brought into thermal contact. This system will attain thermal equilibrium by transferring heat from the hot metal to the cold metal until the two pieces of metal are at the same temperature. When the two pieces of metal are at the same temperature, they are in thermal equilibrium, as defined by the zeroth law. Entropy is a measure of disorder. Two pieces of metal at different temperature represent a state of higher order (i.e., the hot metal atoms are separated from the cold metal atoms) than if they were in thermal equilibrium (i.e., at the same temperature). Thus, attaining equilibrium increases the overall entropy of the universe.

Thermal energy can be used to do work only if heat flows from hot to cold. An analogy is the conversion of potential energy to kinetic energy. Consider, for example, water in a reservoir at some height above a hydroelectric generating station. As long as the water remains in the reservoir, no work is done, and no electricity is generated. When the water is allowed to run downhill and through the station, electricity is generated. In this way, gravitational potential energy is converted into kinetic energy and subsequently into electrical energy. In the same way, the thermal energy contained in a piece of hot material can be converted into other forms of energy if it flows toward something that is at a lower temperature. This is the principle of operation in a heat engine, as described in the next section.

1.4d The Third Law of Thermodynamics

The third law is important for an understanding of the efficiency of thermodynamic processes. Part of the third law is the fact that a temperature of absolute zero cannot be attained. The details of the third law and its origins are not relevant and will not be discussed further.

1.5 Heat Engines and Heat Pumps

If heat moves from a hot reservoir to a cold reservoir, some of the thermal energy can be extracted to do mechanical work. A device that does this is called a *heat engine*, such as, for example, a steam turbine, a gasoline automobile engine, or a jet engine. Figure 1.4 shows a schematic of a heat engine. Heat is removed from a hot reservoir (at temperature T_h). Some of this heat is deposited in a cold reservoir (at temperature T_c), and some is used to do useful mechanical work, W. If the heat removed from the hot reservoir is Q_h, and the heat deposited in the cold reservoir is Q_c, then

$$Q_h = Q_c + W. \tag{1.29}$$

This expression follows along the lines of equation (1.28) and is a direct result of the conservation of energy as stated by the first law of thermodynamics. From a practical standpoint, Q_h can be the energy produced by burning gasoline as in an automobile engine. In this case, Q_c would be the excess heat transferred into the atmosphere, and W is the mechanical work that is done. The efficiency of this process, η, is the ratio of the useful work done, W, to the total input energy, Q_h. In percent, this is written as

$$\eta = 100 \frac{W}{Q_h}. \tag{1.30}$$

From equation (1.29), it is seen that

$$W = Q_h - Q_c \tag{1.31}$$

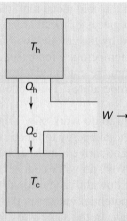

Figure 1.4: Operation of a heat engine to convert thermal energy into mechanical energy.

and the efficiency becomes

$$\eta = 100 \left(1 - \frac{Q_c}{Q_h} \right).$$

(1.32)

A consequence of the work of the French engineer Sadi Carnot is that the ratio of Q_c to Q_h can be expressed as the ratio of the reservoir temperatures:

$$\frac{Q_c}{Q_h} = \frac{T_c}{T_h}$$

(1.33)

This expression allows equation (1.32) to be written as

$$\eta = 100 \left(1 - \frac{T_c}{T_h} \right).$$

(1.34)

This form is convenient because T_c and T_h are more easily measured quantities than Q_c and Q_h. It is essential in equation (1.33) that the temperatures be measured on an absolute temperature scale (typically Kelvin). Measuring temperatures in Celsius in this equation will yield incorrect results. The efficiency as stated by equation (1.34) is known as the *ideal Carnot efficiency* and is the maximum efficiency attainable by a heat engine utilizing hot and cold reservoirs with temperatures T_h and T_c, respectively. It is seen that 100% efficiency is achieved only if T_c is zero degrees (on an absolute scale) which cannot be achieved. It is also seen in Figure 1.4 that Q_c cannot be larger than Q_h. From Carnot's relationship, $Q_c > Q_h$ would imply $T_c > T_h$ and would be inconsistent with the definition of hot and cold. Although the Carnot efficiency of a heat engine is the maximum theoretical efficiency, real heat engines typically operate at efficiencies that can be much less than the Carnot efficiency.

A *heat pump* is basically just the opposite of a heat engine; it uses mechanical energy to move heat from a cold reservoir to a hot reservoir. A schematic of a heat pump is shown in Figure 1.5. Conservation of energy requires that

$$W + Q_c = Q_h.$$

(1.35)

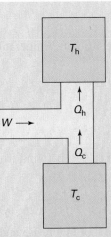

Figure 1.5: Operation of a heat pump that uses mechanical energy to transfer heat.

The ratio of heat deposited into the hot reservoir to the amount of work done can be defined as

$$COP = \frac{Q_h}{W}. \tag{1.36}$$

This quantity is referred to as the *coefficient of performance (COP)* of the heat pump, and it is obvious from Figure 1.5 that this quantity will be greater than 1 (or, in percent, >100%). Using equation (1.31), equation (1.36) can be rewritten to give the ideal Carnot coefficient of performance as

$$COP = \frac{1}{1 - (Q_c/Q_h)}. \tag{1.37}$$

Using equation (1.33), this equation becomes

$$COP = \frac{1}{1 - (T_c/T_h)}. \tag{1.38}$$

Heat pumps have practical applications for the transfer of heat from a cold reservoir to a hot reservoir. Air conditioners and refrigerators are examples of such devices. A heat pump can also heat a house on a cold day by moving heat from the outside to the inside. In this case, the relatively cold air outside is the cold reservoir, and the relatively warm air inside is the hot reservoir. Heat pumps specially designed for this purpose are in fairly common use in many parts of North America where winter temperatures are not extreme. Although heat pumps can be economically attractive for heating purposes, a careful consideration of capital costs, maintenance costs, local climate, and other factors is necessary to assess their viability. Applications of heat pumps are discussed further in Section 17.6.

Example 1.7

Consider an ideal heat pump operating with a cold reservoir temperature of $-20°C$ (e.g., outside on a cold winter day) and a hot reservoir at $+20°C$ (e.g., inside a home). Calculate the ideal (Carnot) coefficient of performance for this heat pump.

Solution

Noting that it is essential to use absolute temperatures in these expressions, we convert from °C to K as

$$T_c = -20°C + 273 = 253 \text{ K}$$

and

$$T_h = 20°C + 273 = 293 \text{ K}$$

The coefficient of performance (*COP*) is found from equation (1.38) to be

$$COP = \frac{1}{1 - (T_c/T_h)} = \frac{1}{1 - (253 \text{ K}/293 \text{ K})} = 7.325$$

This may be viewed in the context of Figure 1.5 and shows that the heat deposited in the hot reservoir (the room) is about seven times the mechanical work done.

1.6 Electricity Generation

Much of the energy used by society is in the form of electricity. The current section provides a brief overview of how electricity is presently generated and distributed. Figure 1.6 shows the distribution of energy sources used to generate electricity worldwide. It can be seen that the majority of electricity comes from fossil fuels, nuclear energy, and hydroelectricity. A very small component, about 1%, comes from other sources (solar, wind, tidal, biofuels, etc). Fossil fuels and nuclear are similar in that they produce electricity from heat by first converting it into mechanical energy (using, for example, a turbine) and then into electricity (using a generator). Hydroelectricity is produced by converting the gravitational potential energy associated with water at higher elevations into mechanical energy and then into electricity. The details of other electricity production methods are quite variable and are discussed in detail throughout the remainder of the text. The present section concentrates on methods (specifically fossil fuels and nuclear) that convert heat into mechanical energy and then into electricity.

The conversion of heat into mechanical energy has been discussed in some detail in the previous section. Any device that works in this manner is a heat engine, and its ultimate efficiency is limited by the temperatures involved, as described by Carnot. The goal in achieving a high efficiency is to maintain the hot reservoir at as high a temperature as possible and to maintain the low temperature reservoir at as low a temperature as possible.

Two approaches are commonly used for the generation of electricity from fossil fuels: thermal generation and combustion turbines. In *thermal generation*, the fuel is burned to produce heat, which is used to produce steam, which, in turn drives a turbine. These facilities most commonly use coal but can, in principle, run on any combustible fuel. *Combustion turbines* use the hot gas produced by the combustion itself to drive the turbine. Combustion turbines commonly use natural gas as a fuel but can use any highly volatile liquid fuel (i.e., gasoline). Nuclear energy is used to produce electricity exclusively by the thermal method and, aside from differences in how the heat is produced, follows closely along the lines of thermal generation from coal. In the present section, the technology of thermal generation and combustion turbines is considered.

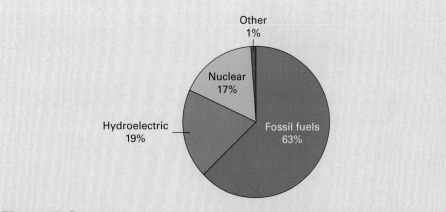

© Cengage Learning 2015

Figure 1.6: Proportions of primary energy sources used for the production of electricity worldwide.

1.6a **Thermal Electricity Generation**

One of the basic requirements of a thermal generation facility is a sustainable cold reservoir. Typically, this is the ocean, a large lake or river, or the atmosphere (by means of cooling towers). The general diagram of a generating station that makes use of a body of water as the cold reservoir is illustrated in Figure 1.7, and a photograph of a medium-sized generating facility is shown in Figure 1.8. The hot reservoir is steam in the boiler, which is produced by burning a fossil fuel or by a nuclear reaction. The steam at high temperature and high pressure does work as it passes through the turbine, causing the rotor assembly to rotate. The energy extracted as the steam passes through the turbine results in a reduction of both the temperature and the pressure of the steam. The typical design of the rotor of a large turbine is shown in Figure 1.9. Excess heat is transferred into the cold reservoir through a heat exchanger. The mechanical energy is converted into electricity by means of a generator. The output of these facilities is measured in watts electrical (W_e), where the subscript e stands for "electrical". The efficiency of converting thermal energy into electricity is constrained by the Carnot efficiency of converting heat into mechanical energy. (The conversion of mechanical energy into electricity is very high, typically ~90% or so). If water is used as the cold reservoir, there is a clear lower temperature limit of 0°C for T_c. Cooling towers, as illustrated in Figure 1.10, use the atmosphere as the

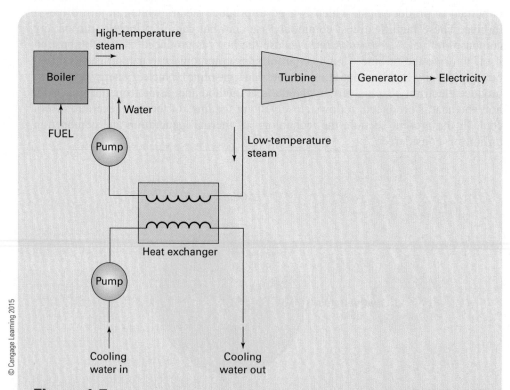

© Cengage Learning 2015

Figure 1.7: Schematic diagram of a typical fossil fuel or nuclear power generating station.

Richard A. Dunlap

Figure 1.8: Tuft's Cove generating station in Dartmouth, Nova Scotia. This is a 350-MW$_e$ facility that uses oil or natural gas as fuel.

cold reservoir. Although the value of T_c for cooling towers is not necessarily limited to 0°C, it is typically somewhat more variable than for water cooling. Typically, net efficiencies around 35–45% are achieved in the production of electricity from fossil fuels and around 30% for nuclear fuel. It is important to realize that the energy content of the fossil or nuclear fuel must be roughly three times the output in watts electrical because of the efficiency of the heat engine.

Figure 1.9: Rotor assembly of a turbine showing multiple stages of turbine blades.

Figure 1.10: Cooling towers typically used for thermal generating stations.

1.6b Combustion Turbines

Combustion turbines are sometimes used for the production of electricity from liquid or gaseous fossil fuels (often natural gas). The hot (burning) gas is fed directly into a turbine (much like a jet engine) and turns the turbine as it decreases in temperature and pressure. These facilities tend to be smaller than thermal (coal-fired) fossil fuel stations, and

© Robert Shantz/Alamy

Figure 1.11: Natural gas-fired combustion turbines at the Pyramid Generating Station operated by Tri-State G&T Inc. near Lordsburg, New Mexico.

the cost of natural gas, gasoline, or oil is typically higher (per joule) than it is for coal. Combustion turbines, as shown in Figure 1.11, are often used to supplement coal-fired or nuclear electricity production during periods of high demand (Chapter 18) because they have relatively short start-up times compared to thermal generating facilities.

1.6c Distribution of Electricity

Electricity is distributed from the generating station to the end user by means of a system of electrically conducting wires. Because all wires (other than superconductors) have resistance, some electrical energy is lost and converted into heat during transmission. From the previous discussion of electrical energy, it is easy to understand the approach to minimizing resistive losses during power transmission. For a given wire resistance, R, the power loss from equation (1.24) is

$$P_{\text{loss}} = \frac{I^2}{R},$$ **(1.39)**

where I is the current in the wire. The resistance of the wire is determined by the material from which the wire is made from (usually aluminum), the diameter of the wire, and its length. If a certain amount of power needs to be transmitted, P_{trans}, over some distance, then from equation (1.23),

$$P_{\text{trans}} = VI,$$ **(1.40)**

where V is the voltage. From this expression, the current is given by

$$I = \frac{P_{\text{trans}}}{V},$$ **(1.41)**

and substituting into equation (1.39) gives

$$P_{\text{loss}} = \frac{P_{\text{trans}}^2 R}{V^2}.$$ **(1.42)**

Richard A. Dunlap

Figure 1.12: Small power distribution transformer on a utility pole in a residential area.

Thus, it is clear that the power loss is inversely proportional to the square of the voltage. Transmitting electric power through a transmission cable is the most efficient when the voltage is as high as practical. This is the approach taken to electric power distribution: The longer the distance that the power needs to be transmitted (and the greater the resistance), then (generally) the higher the voltage. Because the voltage for residential use is typically 220 V, the voltage must be stepped down using a power distribution transformer. Figures 1.12 and 1.13 show power distribution transformers used for changing the voltage (up or down) for distribution over large distances or for residential use.

Richard A. Dunlap

Figure 1.13: Transformers at an electric power distribution facility.

1.7 Summary

This chapter has shown that power is a measure of the energy produced or utilized per unit of time. Energy can exist in a variety of forms. It is convenient, from a practical standpoint, to think of these forms as kinetic energy associated with a moving object, gravitational potential energy associated with a mass that is vertically displaced, thermal energy associated with an object at an elevated temperature, chemical energy associated with the electronic bonds between atoms, nuclear energy resulting from the bonding of neutrons and protons in a nucleus, electrical energy corresponding to a flow of electric current, and electromagnetic energy associated with photons.

This chapter has also provided an introduction to the basic principles of thermodynamics: the definition of temperature, the conservation of energy, the tendency of all systems to achieve thermal equilibrium, and the inability to reach absolute zero temperature. These principles were applied to a thermodynamic system to describe the behavior of a heat engine in terms of its Carnot efficiency and a heat pump in terms of its coefficient of performance.

Some of the ways in which primary energy sources can be transformed into forms that are suitable for applications have been discussed. Examples are the conversion of the chemical energy in fossil fuels into heat energy by combustion or the conversion of mechanical energy in the wind into electricity. These conversion processes rely on the development of appropriate technologies but are also governed by fundamental physical laws. For example, more efficient automobile engines can be developed, but ultimately the limiting factor for efficiency is the laws of thermodynamics.

Finally, the chapter overviewed the basic principles for the production and distribution of electricity. The thermal generation of electricity by fossil fuels or nuclear energy is limited by the Carnot efficiency of a heat engine. On the basis of Ohm's law, the chapter has shown that losses are minimized when electricity is transmitted at high voltage.

Problems

1.1 Compare the energy scales of mechanical, chemical, and mass energy by calculating (a) the energy associated with dropping 1 kg of coal through a vertical distance of 100 m, (b) burning 1 kg of coal, and (c) converting the mass of 1 kg of coal into energy.

1.2 A simple way of looking at the energy associated with the combustion of methane [as shown in equation (1.15)] is to view the oxidation of the carbon by equation (1.13) and the oxidation of hydrogen by equation (1.19). Based on the energies involved in these processes, discuss the validity of this approach.

1.3 One cubic meter of water is poured off a 100-m high bridge. If the change in gravitational potential energy is converted into electricity with an efficiency of 90%, how long can this energy illuminate a standard 60-W light bulb?

1.4 One gram of methane is burned, and the heat is used to raise the temperature of 1 kg of water. If the initial temperature of the water is 25°C, what is the final temperature?

1.5 (a) Estimate the energy required to raise the temperature of a cup of coffee from room temperature to 60°C. (b) Estimate the mass energy contained in a cup of hot coffee.

1.6 What is the energy [in electron volts (eV)] of a green photon?

1.7 An average person can produce considerable power output for short periods of time. However, over an entire 8-hour workday, a person might be able to average 100 W output of physical power. If energy is valued at $0.12/kWh (about the average cost of electricity from a public utility), what is the monetary value of a person's physical work per day?

1.8 Calculate the ideal Carnot efficiency of a steam turbine that utilizes steam at a temperature of 575°C and ejects water to the environment at a temperature of 35°C.

Bibliography

F. S. Aschner. *Planning Fundamentals of Thermal Power Plants.* Wiley, New York (1978).

G. J. Aubrecht II. *Energy: Physical, Environmental, and Social Impact* (3rd ed.). Pearson Prentice Hall, Upper Saddle River, NJ (2006).

R. A. Dunlap. *Experimental Physics—Modern Methods.* Oxford University Press, New York (1988).

E. L. McFarland, J. L. Hunt, and J. L. Campbell *Energy. Physics and the Environment* (3rd ed.). Cengage, Stamford, CT (2007).

K. Wark *Thermodynamics* (4th ed.). McGraw-Hill, New York (1983).

CHAPTER 2

Past, Present, and Future World Energy Use

Learning Objectives: After reading the material in Chapter 2, you should understand:

- The energy needs of humanity throughout history.
- The current energy distribution and relationship to economic, geographical, climate, and industrial factors.
- The principles of exponential growth.
- The Hubbert model of resource utilization.
- Resource limitations to energy production and use.

- Limits of technology on energy production and use.
- Economic factors that limit energy use.
- Social factors affecting energy production.
- Environmental aspects of energy use.
- Political factors affecting energy use.
- The integration of new energy technologies with existing technology.

2.1 Introduction

Before the use of fire, humans relied only on the energy that their bodies produced to do the work needed to live their lives. This energy came from the food that each person consumed. If an average daily food consumption was 8.6 MJ, then this translated into an average continuous power utilization of $(8.6 \times 10^6 \text{ J})/(86,400 \text{ s/d}) \approx 100$ W, about the heat and light output of a large light bulb. As humans implemented energy from other sources to improve their lives—first burning wood and then domesticating animals, followed by burning fossil fuels—the power utilization of each human grew. Total world energy use grew both because of the needs of the individual and because of a growing population. Throughout most of the past, humans used sources of energy that came from nature without any awareness (at least on a global scale) of any limitations to the availability of energy or any adverse environmental consequences of its use. In recent years, an awareness of both these factors arose and prompted the investigation of mechanisms for conserving energy resources, utilizing alternative sources, and mitigating the undesirable consequences of energy production and use. Planning for the future requires an analysis of past energy use, an evaluation of available resources, and a projection of future needs. The present chapter overviews the ways in which past energy use can be understood and the future can be anticipated.

2.2 Past and Present Energy Use

The average rates of energy use per person at different stages of human development are summarized in Table 2.1. This table illustrates the significance of technological advances on humanity's need for energy. It shows that the current per-capita power in the United States, 11.7 kW per person (continuous average), is more than 100 times the average power produced by a person's body. The utilization of energy sources from nature (i.e., the burning of fossil fuels) supplements our own body's energy production by this amount.

A detailed breakdown of energy use in the United States is illustrated in Figure 2.1. The relative importance of the various energy sources shown in the figure is fairly typical of industrialized nations, although some variability exists, as discussed in detail in future chapters dealing with specific sources of energy. A comparison of the situation in the United States and that worldwide is shown in Figures 2.2 and 2.3.

Nations vary substantially in the amount of energy they use. Figure 2.4 shows the relationship between per-capita energy use and per-capita gross domestic product (GDP). Overall, the relationship between per-capita energy use and per-capita GDP is, as expected, a direct consequence of the degree of industrialization. However, the figure shows some minor variations that can be fairly well understood. For example, among industrialized nations, Switzerland and Spain fall below the average line (less energy use relative to the GDP), while Canada and Norway fall above the line (more energy use relative to the GDP). Some relevant factors that have an effect on energy use are:

1. Climate.
2. Population density.
3. Type of industries.

Table 2.1: Estimated average power used per person as a function of the stage of human development (worldwide) or as a function of year (for the United States, 1850–2000). The power values are calculated from the primary energy use.

society	power consumption (W)
hunter	100
use of fire	200
domestication of animals	500
Renaissance	1160
1850	4880
1900	5340
1950	7300
1960	8180
1970	11,000
1980	11,250
1990	11,000
2000	11,730

© Cengage Learning 2015

Source: Data adapted from G. J. Aubrecht II. *Energy: Physical, Environmental, and Social Impact* (3rd ed.). Pearson Prentice Hall, Upper Saddle River, NJ (2006).

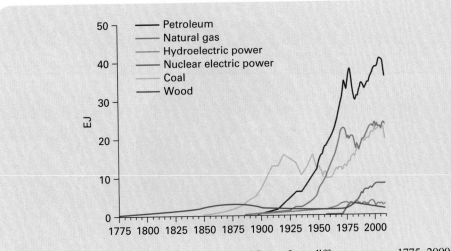

U.S. EIA/DOE

Figure 2.1: Annual energy use in the United States from different sources, 1775–2009.

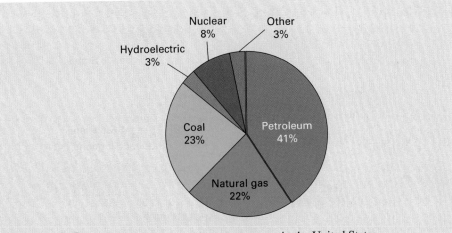

© Cengage Learning 2015

Figure 2.2: Current distribution of energy sources in the United States.

Canada, for example, has a cool climate (resulting in greater heating costs), a low population density (resulting in greater transportation costs), and a fairly large fraction of heavy manufacturing (resulting in greater energy needs relative to product value). Norway, as well, has a severe climate. By comparison, Spain has a more moderate climate, while Switzerland has a higher population density and a greater prevalence of less energy-intensive industries, such as banking and watch making.

Energy use is sometimes given in terms of primary energy use. *Primary energy* refers to the quantity of energy as it is extracted from nature. This might be the energy content of coal that can be extracted by burning the coal or the energy content of solar radiation incident upon the earth. In most cases, the form of energy as it is obtained

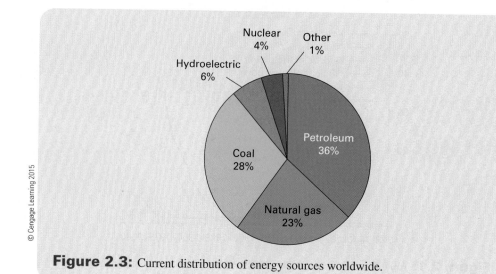

Figure 2.3: Current distribution of energy sources worldwide.

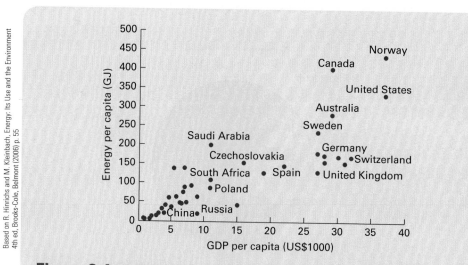

Based on R. Hinrichs and M. Kleinbach, Energy: Its Use and the Environment 4th ed, Brooks-Cole, Belmont (2006) p. 55

Figure 2.4: Relationship of annual per-capita energy use and per-capita gross domestic product (GDP) for different nations.

from the environment is not the same as the form required for an application, which is the energy after it is converted to a useful form and put to use by the end user. Converting energy from one form to another inevitably involves losses because the conversion process can never be 100% efficient. It is not that energy is not conserved; it is that it ends up in a form that is different from what is needed, often as residual unconverted energy or as excess heat. If, for example, oil is converted to heat (by burning) to heat a home, this is a fairly efficient conversion process (typically 85% or better). Similarly, converting the energy content of falling water to electricity in a hydroelectric facility is quite efficient (typically 90% or better). On the other hand, converting the energy

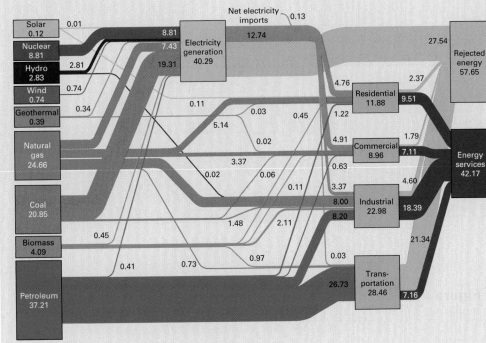

Figure 2.5: Annual energy use (in 2009) in the United States, showing sources, end use, and conversion losses. Units are EJ.

content of coal to electricity by burning the coal and using the heat to boil water to run a generator is an intrinsically inefficient process. Conversion efficiencies will be discussed throughout the book. The relationship of primary to end user energy in the United States is illustrated in Figures 2.5 and 2.6. Figure 2.5 illustrates the significance of losses that occur during the conversion process of changing energy from its primary form into the form needed by the end user.

Clearly, humans use considerable energy, particularly those who live in industrialized nations. On average, a person living in North America might use about 12 kW of power (continuously). Humans obviously consume much more energy than they could produce on their own. To a large extent, this energy use can be accounted for by a simple analysis. Consider, as an example, a person who lives in a temperate region of North America (e.g., Boston) in a single-family, oil-heated home (with one other person) and who commutes to work by automobile (with one passenger). It is easy to account for a significant fraction of this person's energy use in terms of the energy purchased. Three sources are obvious: gasoline for transportation, oil for residential heating, and electricity for residential use. The typical annual use for these two people might be

$$2810 \text{ L gasoline at } 3.48 \times 10^7 \text{ J/L} = 9.8 \times 10^{10} \text{ J},$$
$$2700 \text{ L heating oil at } 3.85 \times 10^7 \text{ J/L} = 1.04 \times 10^{11} \text{ J, and}$$
$$1.2 \times 10^4 \text{ kWh electricity} = 4.32 \times 10^{10} \text{ J.}$$

In a similar climate, the net energy use for heating would be similar for homes heated electrically or with natural gas. The energy content of gasoline or heating oil is close

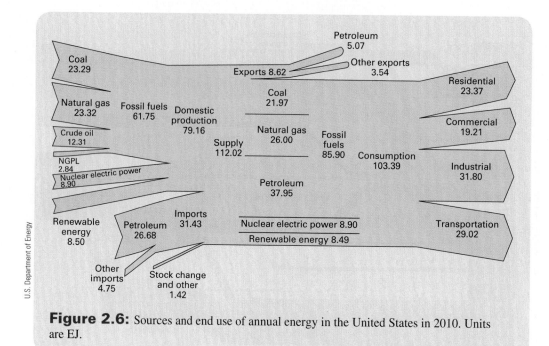

U.S. Department of Energy

Figure 2.6: Sources and end use of annual energy in the United States in 2010. Units are EJ.

to that of the primary energy source (crude oil) because production costs (in terms of energy) are fairly minimal. The energy content of the electricity may (or may not) be close to that of the primary source. If the electricity comes from hydroelectric power, then the efficiency is high. Most electricity in North America comes (at present) from burning coal with an efficiency of around 35–40%. This means that the 4.32×10^{10} J of electrical energy actually represents about 1.1×10^{11} J of primary energy. Thus, for this person, transportation, heating, and electricity represent fairly similar amounts of energy use. This, of course, would not be the same for someone living in, say, Manitoba as it would be for someone living in Florida. Nor would it be the same for a person living in an urban apartment and a person living in a rural single-family home. However, this example gives a rough estimate of this type of analysis and shows that the average (primary) energy consumption per person can be calculated as

$$0.5 \times \frac{0.98 \times 10^{11} \text{ J/y} + 1.04 \times 10^{11} \text{ J/y} + 1.1 \times 10^{11} \text{ J/y}}{3.15 \times 10^7 \text{ s/y}} = 5.0 \text{ kW.} \quad \textbf{(2.1)}$$

This is less than half of the expected value of about 12 kW. A person uses other energy, such as at a place of employment. This might be electricity for lights and computers for office workers or more substantial amounts of energy for equipment for workers in manufacturing industries. Energy is used to manufacture goods that a person uses and to produce food that a person consumes. In general, it is difficult to attribute specific energy use to specific individuals, which is why societal averages are more meaningful.

It may also be obvious from this discussion that everything that humans produce has a cost in energy. For example, if a person buys a loaf of bread at a grocery store, it might be customary to think of the "value" of the bread as, say, $1.49. The $1.49 goes, in part, to pay a farmer for growing the wheat, to pay a trucker for delivering the grain to the bakery, to pay the bakery for making the bread, to pay another trucker for delivering the bread to the store, and finally to pay the store for operating expenses (as well as some profit for everyone). But the bread has a value in energy as well. The farmer used energy (in the form of gasoline or diesel fuel) to run the equipment needed to cultivate the land. The trucker used fuel to transport the grain and bread, the bakery used energy to bake the bread, and the store used energy for lights and heat. If you think about things in depth, you can see that a lot of other hidden energy costs come into play in actually getting the bread on your dinner table. It is also obvious that it becomes difficult to account for energy use on a per-person basis. It would be interesting to ask how the energy value of the bread (i.e., how much energy is needed to produce and distribute it) compares to the amount of energy that the bread provides to a person who consumes it (in terms of its caloric content). In this situation, the overall efficiency of the process (i.e., is a person getting as much energy from eating the bread as is used to produce it?) is less important than the goal of providing food for humankind. (Actually, in North America it requires about eight times as much energy to produce food as the food provides to the consumer.)

This kind of analysis is much more important when dealing with, for example, the energy content of a kilogram of coal. It is customary to think of the cost of energy in terms of dollars per joule, but a far more fundamental consideration is the cost of energy in joules per joule; that is, how much energy is consumed in mining, processing, transporting, and otherwise delivering 1 kg of coal compared to the energy gained by burning it and converting the heat energy into the required form of energy? Although financial considerations generally outweigh energy considerations, the bottom line is that, if it costs more (in energy) to produce the energy that is gained, then there is a net energy loss in the process. This kind of analysis is important in considering the viability of any type of alternative energy source.

This brings up another important consideration: the environmental impact of the things we utilize. We might not think of a loaf of bread (for example) as having any adverse environment impact. However, energy is utilized to produce bread, and if that energy comes from, say, fossil fuels, then there is an environmental impact. To fully analyze the environmental consequences of every object we utilize, it is necessary to consider all stages of its production, use, and ultimate disposal. This kind of approach is referred to as *life cycle analysis* and will be considered in more detail in Chapter 4.

It is also important to understand the future implications of the information in Figure 2.1. The increase in energy use as a function of time may be understood by means of the detailed analysis of growth mechanisms. This topic is considered in the next section.

2.3 Exponential Growth

Energy use is clearly increasing. This is a result of two factors: population increases and an increase in per-capita energy use. In highly industrialized countries, the per-capita energy use is fairly constant (Table 2.1). In this case, the increase in energy use is largely governed by population changes. In the case of developing countries, both population growth and an increase in per-capita energy use are important factors. In fact, the largest component of the increase in world energy use is in rapidly developing countries with large populations (e.g., China and India). In an idealized future, all nations might achieve a similar level of industrialization, and the average per-capita use of energy would be relatively constant worldwide. In this case, population growth

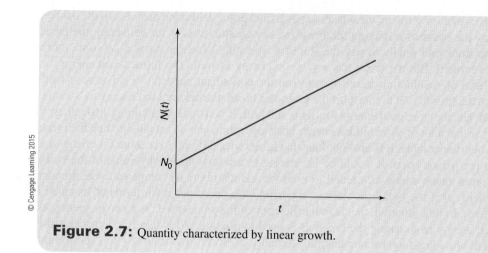

Figure 2.7: Quantity characterized by linear growth.

© Cengage Learning 2015

would be the dominant factor governing energy use. For an assessment of future energy needs, it is important to be able to predict overall changes in energy use in the short term and farther into the future. A general mathematical analysis of growth is a good starting point.

The simplest type of growth is linear growth. For a quantity, N, the time dependence is given by a constant dN/dt as

$$N(t) = N_0 + \frac{dN}{dt}\, t, \tag{2.2}$$

where N_0 is the value at time $t = 0$, is shown in Figure 2.7. On a linear graph, this is described by a straight line with a constant slope (dN/dt).

Another type of growth is exponential growth. The rate of change of the quantity, N, is related to the magnitude of that quantity by the expression

$$\frac{dN}{dt} = aN. \tag{2.3}$$

This may be integrated to give

$$N(t) = N_0 \exp(at). \tag{2.4}$$

A graph of this relationship is shown in Figure 2.8. It is most easy to identify true exponential growth by plotting data on a semilogarithmic scale. Taking the natural logarithm of both sides of equation (2.4) gives

$$\ln[N(t)] = \ln(N_0) + at. \tag{2.5}$$

Converting the natural logarithm to a base 10 logarithm (for ease of data presentation) gives

$$\log[N(t)] = \log(N_0) + at \log(e). \tag{2.6}$$

Thus a semilog plot of $N(t)$ as a function of t is a straight line with a slope of $a\log(e)$, as shown in Figure 2.9. The time required for the quantity N to double (called the doubling time) is found by setting $N(t) = 2N_0$ in equation (2.3) and solving for t_D as

$$t_D = \frac{\ln(2)}{a}. \tag{2.7}$$

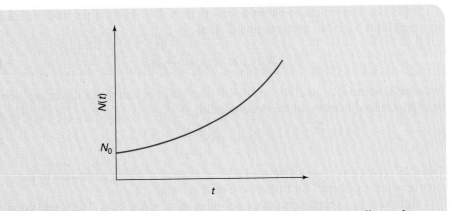

Figure 2.8: Quantity characterized by exponential growth shown on a linear plot.

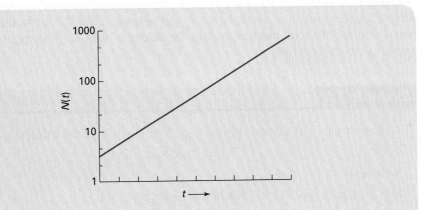

Figure 2.9: Quantity characterized by exponential growth shown on a semilog plot.

Example 2.1

A particular quantity has a doubling time of 20 years. If the quantity has a value of 10^6 at time $t = 0$, what is its value after 5 years?

Solution

Solving equation (2.7) for the constant a gives

$$a = \frac{\ln(2)}{t_D} = \frac{0.693}{20 \text{ y}} = 0.0347 \text{ y}^{-1}.$$

From equation (2.4), the quantity at $t = 5$ y is

$$N(t = 5 \text{ y}) = 10^6 \times \exp(0.0347 \text{ y}^{-1} \times 5 \text{ y}) = 1.189 \times 10^6.$$

It is also convenient to define a growth rate, R (in percent per unit time; e.g., % per year). This is given from equation (2.4) as

$$R = 100 \cdot [\exp(a) - 1]. \tag{2.8}$$

In the case, where R is small (typically less than 10% per unit time), this expression may be approximated in terms of the doubling time as

$$R = 100 \frac{\ln(2)}{t_D}. \tag{2.9}$$

Table 2.2 gives the relationship of doubling times (in years) compared to the growth rate (in percent per year) for true exponential growth.

It is now possible to apply this information to actual data. The world population during the past millennium is shown in Figure 2.10, where the estimated total population as of 2012 is approximately 7.0×10^9. On a linear scale, these data certainly look fairly exponential. In fact, in recent years they are reasonably well described by an exponential with an average annual growth rate of about 1.6%, corresponding to a doubling time of roughly 43 years. If the world population continues to double every 43 years, some predictions about the future can be made. Table 2.3 shows the predicted population and the total mass of humans (assuming a mass of 70 kg per person) for various years in the future.

Example 2.2

A quantity grows at a rate of 1% per year. When will it reach 10 times its current value?

Solution

Solving equation (2.8) for a as a function of R gives

$$a = \ln\left(1 + \frac{R}{100}\right).$$

Substituting in a value of $R = 1$ gives

$$a = \ln(1 + 0.01) = 9.95 \times 10^{-3} \, y^{-1}.$$

From equation (2.4), the quantity as a function of time (relative to the present value) is

$$\frac{N(t)}{N_0} = \exp(at).$$

Solving for t gives

$$t = \frac{1}{a} \ln\left(\frac{N(t)}{N_0}\right).$$

so using $N(t)/N_0 = 10$ and this value for a and solving for t gives

$$t = \frac{\ln 10}{9.95 \times 10^{-3} y^{-1}} = 231 \, y.$$

Table 2.2: Annual growth rates and corresponding doubling times.	
% growth per year	**t_D (y)**
1	69.7
2	35.0
3	23.4
4	17.7
5	14.2
6	11.9
7	10.2
8	9.0
9	8.0
10	7.3
20	3.8
50	1.7
100	1.0

© Cengage Learning 2015

Because the total mass of the earth is about 6×10^{24} kg, these predictions are clearly impossible to make very far into the future; that is, the population of the earth cannot increase exponentially with a growth rate of 1.6% per year for the next 2000 years. Even 1.6% annual growth over the next 650 years would result in a population of approximately 1.5×10^{14}, which corresponds to 1 human per square meter of the earth's land area. Certainly, the exponential growth of the human population will be limited long before this happens. A careful analysis of limits to human population must be based on the ability of the planet to produce enough food to feed its population. Some estimates have suggested a limit to the sustainable human population of about 5 billion (less than the current value). Much depends on assumptions concerning food production technology and the politics of distribution, but uncertainty in the future climate is

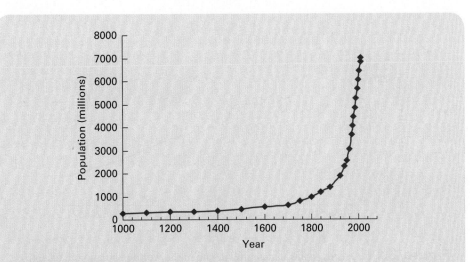

© Cengage Learning 2015

Figure 2.10: Population of the world for the past 1000 years.

Table 2.3: World population and mass of humanity as a function of year for 1.6% annual growth.

year	population	mass (kg)
2012	7.00×10^9	4.90×10^{11}
2200	1.42×10^{11}	9.92×10^{12}
2400	3.48×10^{12}	2.43×10^{14}
2600	8.53×10^{13}	5.97×10^{15}
2800	2.09×10^{15}	1.46×10^{17}
3000	5.13×10^{16}	3.59×10^{18}
3200	1.26×10^{18}	8.82×10^{19}
3400	3.09×10^{19}	2.16×10^{21}
3600	7.58×10^{20}	5.31×10^{22}
3800	1.86×10^{22}	1.30×10^{24}
4000	4.56×10^{23}	3.19×10^{25}

© Cengage Learning 2015

also a significant factor. Most studies indicate that a world population of much more than 10 billion would place an unmanageable strain on food production capabilities. In Figure 2.11, the trends in world population are seen from the plot of annual growth rate as a function of year. Clearly, the trend toward a slower growth rate is evidenced, and this analysis suggests that an equilibrium value will be approached at some point in the not too distant future.

The overall trends in historical and predicted future energy use for some portions of the world's population are shown in Figure 2.12. The significance of increasing population and degree of industrialization in developing countries (e.g China) is clearly evident in the figure. This figure suggests that industrialized nations are in a region of linear growth in their energy but that the rate of growth differs substantially among nations. One prediction for overall world energy use is illustrated in Figure 2.13. Such

Based on the U.S. Census Bureau, http://www.census.gov/population/international/

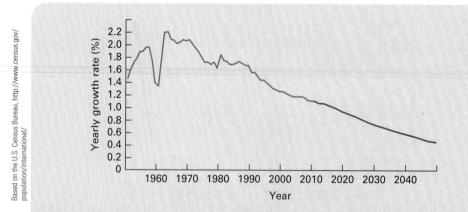

Figure 2.11: Trends in the annual growth rate of the world population in recent years and projected future growth.

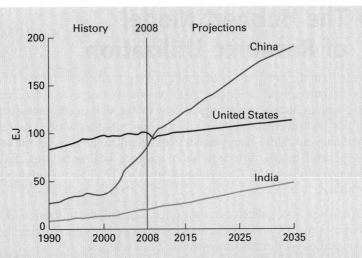

Figure 2.12: Historical and predicted energy use in different components of the world population,1990–2035 (EJ). OECD = Organisation for Economic Co-operation and Development.

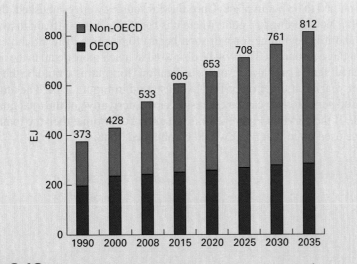

Figure 2.13: Previous and predicted total world energy use up to the year 2035 (EJ).

predictions suggest a global energy requirement of 8 to 10×10^{20} J per year by the year 2050. Ultimately, total energy use will be limited by limitations to human population and to the availability of energy resources. Although this type of prediction serves as a basis for understanding the energy needs of the future, clearly many unknowns and a great deal of uncertainty are involved in the analysis. This will be discussed further in Chapter 21.

2.4 The Hubbert Model of Resource Utilization

It is clear from the previous section that making predictions for the future is not necessarily straightforward or simple. Also obvious is that even minor variations in parameters such as annual percentage of growth can lead to enormously different future predictions. However, the previous discussion provides some general guidance for understanding our future energy needs and the utilization of available resources.

A general model that deals with resource utilization was developed by the American geophysicist, M. King Hubbert, in 1956. The basic assumptions of this model are as follows:

- When it is first realized that a resource is useful, the utilization of that resource begins slowly. This is because efficient procedures for utilizing the resource and an appropriate infrastructure need to be developed.
- Once the appropriate infrastructure has been developed, resource utilization increases.
- When the resource becomes scarce, utilization decreases and eventually stops.

This model can be readily applied to the use of oil. When the usefulness of oil (as a heating fuel and later as a transportation fuel) was realized, the locations of oil deposits had to be determined, oil wells needed to be constructed, and a system for the distribution and use of oil had to be put in place. Once these resources were established, the use of oil increased. It is now getting to a point where it is becoming more difficult to locate new oil resources, and it is beginning (or will soon begin) to be more difficult and expensive to make use of those resources. This will be discussed in much more detail in the next chapter.

Given all that, a graph of resource utilization (for oil this would be the quantity of oil used per year) as a function of time is illustrated in Figure 2.14. The amount of the resource used per unit time can be expressed as the derivative of the total quantity used, Q, as dQ/dt. If the curve in Figure 2.14 is integrated over time, then the total amount of the resource used up to that point would be obtained; that is,

$$Q(t) = \int_0^t \frac{dQ}{dt} dt. \qquad \textbf{(2.10)}$$

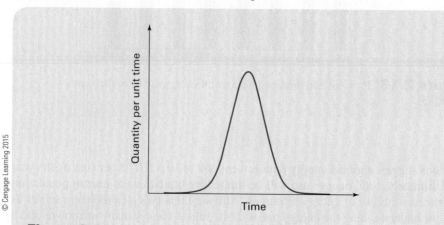

Figure 2.14: Quantity of a resource used per unit time as a function of time.

ENERGY EXTRA 2.1
Limits to growth

Predicting future conditions on earth is, at best, difficult. How humanity interacts with nature will determine the environment of the future. Understanding the consequences of our actions is not straightforward because the ecosystem is very complex. Specifically, with regard to energy utilization, it is important to understand the environmental impact of our actions and the availability and need for resources in the future. Although a carefully considered analysis of our resource utilization can provide substantial insight, there is really no consensus on how best to ensure the best future conditions for humanity.

One of the most influential approaches to modeling the impact of humanity on the environment was published in the 1972 book, *Limits to Growth*, by authors Donella H. Meadows, Dennis L. Meadows, Jørgen Randers, and William W. Behrens III. A revised edition, *Limits to Growth: The 30-Year Update*, by Donnella Meadows, Jørgen Randers, and Dennis Meadows was published in 2004. The basic hypothesis of the authors is that variables such as world population and resource depletion change exponentially in time, whereas technology's ability to increase resources is linear. This work has pointed out the importance of exponential growth in the following way: If the quantity of a resource is R and it is used at a constant rate, C, then the duration of its availability, t_0, is merely

$$t_0 = R/C.$$

If, however, use grows exponentially with a growth rate of, r, then the total resource can be written as the integral over the exponentially increasing use rate (beginning as C at $t = 0$) as

$$R = \int_0^{t_0} Ce^{rt}dt = \frac{C}{r}(e^{rt_0} - 1)$$

and solving for t_0 in terms of R and C gives

$$t_0 = \frac{1}{r}\left[\ln\left(\frac{rR}{C} + 1\right)\right].$$

Consider, as an example, an initial quantity of a resource of 10^{12} kg, which is being used at a constant rate of 10^9 kg per year. Clearly the resource will last for $t_0 = 1000$ y. If however, use increases at an annual rate of 5% or ($r = 0.05$ y^{-1}) then the lifetime of the resource will be

$$t_0 = \frac{1}{0.05\,\text{y}^{-1}}\left[\ln\left(\frac{0.05\,\text{y}^{-1} \times 1000\,\text{y}}{1} + 1\right)\right] = 78 \text{ y},$$

substantially less than the estimate based on constant use. It is clear from this example that making reliable estimates very far into the future is extremely difficult. Minor uncertainties in resource availability or growth rate can yield vastly different results.

Although the approach of the authors of *Limits to Growth* has gained some acceptance, it has also been criticized. These differing opinions only serve to emphasize the complexity of the problem and the difficulty in making accurate predictions. The work, however, has been instrumental in emphasizing the need for the careful scientific analysis of this problem and the interrelationship of the diverse factors affecting long-term ecological changes.

Topic for Discussion

Consider the world population as a function of time, as shown in Figure 2.10. In the context of limited resources, sketch a graph of expectations for the world population over the next 200 to 500 years. What analogies can be drawn with the Hubbert model, as illustrated in Figure 2.15?

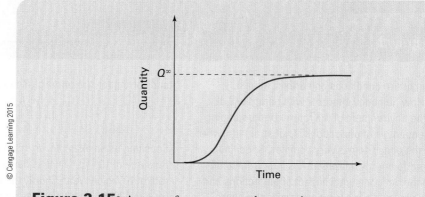

Figure 2.15: Amount of a resource used up to a time, t.

A graph of $Q(t)$ in Figure 2.15 shows the typical sigmoidal behavior expected for resource utilization. As time increases, the curve asymptotically approaches a value Q_∞. This is the total amount of the resource available (or at least economically viable). The application of this model to actual data requires knowledge of the width of the distribution in Figure 2.14 (which can be gained from historical data on use rates) and knowledge of Q_∞ (which can be gained from an analysis of known and speculated unused resources). This approach will be discussed in more detail with regard to the use of fossil fuels in the following chapters.

2.5 Challenges for Sustainable Energy Development

The most reasonable approach to long-term sustainable energy production is not obvious. If a clear solution to the world's future energy needs existed, we would have a well-defined road to achieving our energy goals. Many options need to be considered, and, as discussed in this text, trade-offs must be considered in the development of any new technology. Any viable sustainable energy option must not only make a positive impact on our energy requirements for the present and near future, but it must also have a positive influence on the quality of life for future generations. In general, factors that will influence our choices for new energy technologies to pursue include the:

- Availability of the necessary resources.
- Availability of the necessary technology.
- Consideration of economic factors.
- Consideration of social factors.
- Environmental impact.
- Consideration of political factors.
- Ability to integrate new technology with existing technology.

These factors will now be discussed.

2.5a **Resource Limitations**

The power that is available from alternative energy resources depends on the nature and extent of the resource, as well as the existence of a viable technology to utilize the energy source. Although nonrenewable energy resources, such as fossil fuels or nuclear, are limited in terms of the total energy available, truly renewable resources may be expected to be virtually unlimited in their total availability. However, the power available, even from renewable resources, is limited. Obviously some of the renewable energy technologies discussed in this text, such as solar and wind, are prevalent in most parts of the world, but others, such as geothermal or tidal, are more limited in their distribution. The power available from these resources follows similar patterns. These points are discussed in detail throughout the book and, in particular, in Chapter 21.

It is important to realize, however, that the power available from a particular resource may be extensive, but our ability to harness that energy may be limited by other factors. The utilization of solar energy for the production of electricity is a good example. Solar energy is not infinite because the sun is a finite object, and the earth intercepts only a very small fraction of the energy that the sun produces (see Chapter 8). However, even this small fraction of the sun's energy is very much more than is needed to satisfy all of humanity's needs. To utilize the sun's energy, appropriate devices, such as photovoltaic cells (Chapter 9), are needed to convert electromagnetic energy into electrical energy. One of the common photovoltaic cells for the harvesting of significant solar energy is the CIGS (copper-indium-gallium-selenide) cell. Indium, gallium, and selenium are all relatively rare (and expensive) elements, and they are at least one of the reasons for the economic limitations to solar energy utilization. Table 2.4 shows the quantity of these elements that would be needed to produce photovoltaic cells that could provide all the power necessary for human society. It is clear that, at current production rates, several thousand years would be needed to develop an all–solar energy infrastructure based on CIGS cells. Although production of these elements can be increased, there are ultimately limits to their total availability. Other possible photovoltaic technologies exist, although many of the most promising, from certain standpoints such as efficiency, also suffer from similar resource availability difficulties. This simple example illustrates the materials challenges that often accompany the implementation of alternative energy technologies.

Table 2.4: In, Ga, and Se needed to produce copper-indium-gallium-selenide photovoltaic cells required to generate an average of 1 GW electricity and power sufficient to meet total global needs (about 18 TW). Also given is the 2008 total world production of these elements. (t = tonne = 10^3 kg).

element	to produce 1 GW (t)	to fulfill world energy needs (t)	2008 world production (t)
indium	90	3.8×10^6	140
gallium	30	4.2×10^5	111
selenium	180	7.5×10^6	3000

2.5b **Technological Limitations**

In some instances, for example wind energy, a suitable technology exists that enables us to effectively make use of the resource, although this, in itself, does not ensure that its development will be viable economically and environmentally. Sometimes technological barriers inhibit the development of a resource. In some cases, the lack of a suitable technological infrastructure is the result of a lack of basic scientific knowledge. In some cases, appropriate mechanisms for applying scientific knowledge in a way that is both practical and economical need to be developed. Fusion energy is one example where further basic research is necessary to understand fully the ways in which this energy source can be made viable. Photovoltaics is an example of a field where functioning devices are in common use, but further research is needed to improve efficiency, make them more economically competitive, and make their extensive use feasible from a materials availability standpoint.

The development of renewable energy touches on a very wide variety of diverse fields, ranging from, say, biochemistry (for biofuel synthesis), to semiconductor physics (for photovoltaics), to plasma and nuclear physics (for fusion energy), and to surface science and materials research (for tidal energy). The development of high-temperature superconductors for possible energy-related applications is an example of how scientific and technological advances can be combined to make progress toward improved energy systems. Some of the ways in which superconductors can contribute to energy systems include low-loss power transmission cables, superconducting magnet energy storage, and light-weight, high-output generators for wind energy applications. High-temperature superconductors were first discovered in the lanthanum-barium-copper-oxide system in 1986 by Johannes Georg Bednorz and Karl Alexander Müller. They observed a superconducting transition temperature of 35 K. Less than a year later, researchers discovered superconductivity in the yttrium-barium-copper-oxide system at 92 K. Because this pushed the superconducting transition temperature above the temperature of liquid nitrogen, it opened up the possibility of the simple, inexpensive applications of superconducting materials. Unfortunately, high-temperature superconducting materials are ceramics, and fabricating them in the form of flexible wires from which power lines, magnetic coils, or generator windings can be prepared is not straightforward. In addition, boundaries that typically form between the grains of superconducting material block the flow of current. First-generation high-temperature superconducting "wires" were made from bismuth-strontium-calcium-copper-oxide–based materials. These were expensive and difficult to fabricate. Second-generation high-temperature superconducting "wires" have been developed in recent years and have resolved many of the problems that affected earlier versions. These "wires" are made from yttrium-barium-copper-oxide and are actually thin films of superconducting material deposited onto a metal ribbon substrate, as shown in Figure 2.16. The technology that has been developed for the preparation of these ribbons eliminates difficulties with boundaries between grains and provides a suitable substrate with good mechanical properties for the preparation of long robust conductors. The first commercial test of this technology for power transmission occurred in New York in 2008 (Figure 2.17) and has led the way to possible future developments that will improve the world's energy systems. The general properties of superconductors and their application for energy storage devices will be discussed further in Chapter 18.

In a number of areas, fundamental research (rather than engineering design) is needed to make advances in the development of alternative, renewable energy.

Based on V. Selvamanickam, "Coated Conductors: From R&D to Manufacturing to Commercial Applications," EUCAS | ISEC | ICMC Superconductivity Centennial Conference, September 19-23, 2011, Den Haag, The Netherlands, http://www.superpower-inc.com/system/files/2011_0225-Barcelona+Wind+Seminar_Selva.pdf and http://www.superpower-inc.com/content/2g-hts-wire

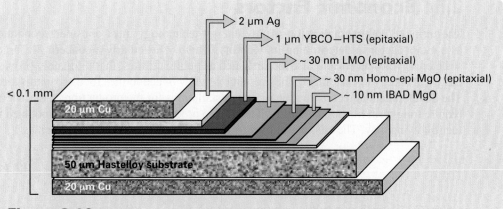

Figure 2.16: Structure of second-generation high-temperature superconducting wire.

Figure 2.17: The world's first superconducting power transmission line in New York in 2008 (connection from distribution station to underground cables).

Some notable examples covered in this text that present significant research challenges are:

- New organic photovoltaic cells that are reasonably efficient and are much more cost-effective than conventional semiconductor based materials (Chapter 9).
- Suitable economical membranes for the exploitation of salinity gradient or osmotic energy (Chapter 14).
- Methods for production of cellulosic ethanol (Chapter 16).
- Efficient non-lithium-based secondary batteries that will provide a cost-effective basis for widespread electric vehicle development (Chapter 19).
- Economical and efficient methods for direct hydrogen production (e.g., solar hydrogen, Chapter 20).

2.5c **Economic Factors**

The ultimate commercialization of any energy technology must consider economics as a major factor in determining its viability. This is particularly a concern for large-scale energy producers and distributors, such as public utilities, that ultimately provide electrical energy to customers. The present overview of economics focuses on this area. The development and construction of new installations for the production of electricity require long-term financial viability. The cost to produce electricity may be modeled in the following way. The cost per kilowatt-hour of electricity generated is given as

$$C = C_{\text{fuel}} + C_{\text{operating}} + \frac{I \cdot CRF}{Rf(8760 \text{ h/y})} \tag{2.11}$$

where C_{fuel} is the cost per kilowatt-hour for fuel, $C_{\text{operating}}$ is the operating and maintenance cost per kilowatt-hour, I is the total capital installation cost, R is the total maximum capacity (in kW), f is the capacity factor, and CRF is the capital recovery factor. The last term on the right-hand side of equation (2.11) gives the contribution to the cost per kilowatt-hour of generated energy that comes from the capital investment costs amortized over the payback period for the facility. The capacity factor is the fraction of the total theoretical capacity that is actually achieved. The capital recovery factor takes into account the accrued interest on the capital investment and is given by

$$CRF = \frac{i(1 + i)^T}{[(1 + i)^T - 1]} \tag{2.12}$$

where i is the annual interest rate expressed as a fraction (i.e., 5.1% would be 0.051) and T is the payback period. The choice of payback period for a particular facility must obviously be shorter than its life expectancy, and 15 years is a common expectation for the payback period of many facilities. Higher-risk (i.e., untested) technologies necessitate shorter payback periods, whereas well established and reliable technologies, such as coal-fired or nuclear facilities, may tolerate longer payback periods.

In general, fuel costs are important for generating technologies such as coal, natural gas or nuclear thermal plants, or combustion turbines. They are typically not of relevance for many renewable technologies, such as solar and wind energy. Contributing factors that may be relevant in equation (2.11) for some systems include both positive and negative factors on the right-hand side of the equation. Positive factors on the right-hand side of equation (2.11) (i.e., those that increase the cost per kilowatt-hour of electricity) include decommissioning costs at the end-of-life cycle. These are probably most notable for nuclear power plants, where radioactive waste disposal must be considered. Negative factors on the right-hand side of equation (2.11) (i.e., those that decrease the cost per kilowatt-hour of electricity) include waste heat recovery sales and end-of-life cycle salvage recoveries. The former may be of relevance, for example, for coal-fired cogeneration plants (Chapter 17), where excess heat from burning coal (i.e., hot steam or water exhausted from the turbines) is sold for heating buildings. In the latter case, for example, rare (and valuable) elements such as indium, gallium, or selenium may be recovered from photovoltaic cells at the end of their life.

The implementation of financial models of energy production depends on a number of factors. A simple approach to the use of equations (2.11) and (2.12) might include an evaluation of fuel costs (for appropriate technologies) based on an assessment of energy content of the fuels and an analysis of conversion efficiencies. Operating

Example 2.3

Calculate the effects of capital recovery costs on the price per kilowatt-hour of electricity produced by a 1.5-MW wind turbine running at a 35% capacity factor. The total installation cost was $2,300,000, and the interest rate is 6.2% over a payback period of 15 years.

Solution

From equations (2.11) and (2.12), the contribution to the cost per kilowatt-hour of generated electricity from the capital recovery term is

$$\frac{I}{Rf(8760 \text{ h/y})} \cdot \frac{i(1+i)^T}{[(1+i)^T - 1]},$$

where $I = \$2,300,000$, $R = 1500 \text{ kW}$, $f = 0.35 \text{ y}^{-1}$, $i = 0.062$ and $T = 15$ years. This equation gives the contribution to the cost of electricity as

$$\frac{\$2,300,000}{(1500 \text{ kW}) \times (0.35 \text{ y}^{-1}) \times (8760 \text{ h/y})} \times \frac{0.062(1+0.062)^{15}}{[(1+0.062)^{15} - 1]} = \$0.052/\text{kWh}.$$

This represents the most significant component to the cost of producing electricity for a wind turbine.

costs depend on a number of factors, including facility design and local labor and materials costs. Infrastructure costs, of course, depend on the energy resource being utilized but can also vary considerably depending on the design of the facility. Some typical examples of (very approximate) capital construction costs per kilowatt of rated capacity for different energy technologies are shown in Table 2.5. The effect of rated capacity on the cost of a wind turbine is illustrated in Figure 2.18. The data show that, for small turbines, there is a clear relationship between the cost per rated capacity and that larger turbines are clearly favorable economically. The net cost per kilowatt-hour generated for large facilities can be estimated on basis of this discussion and is indicated for different energy technologies in Table 2.6.

Table 2.5: Typical capital system costs in the United States in dollars per kilowatt ($/kW) for large electric generating stations using different energy technologies.	
energy resource	infrastructure cost ($/kW capacity)
coal	500
wind	800
natural gas	1000
hydroelectric	1000
geothermal	2500
nuclear (fission)	3000
solar (photovoltaics)	4000

© Cengage Learning 2015

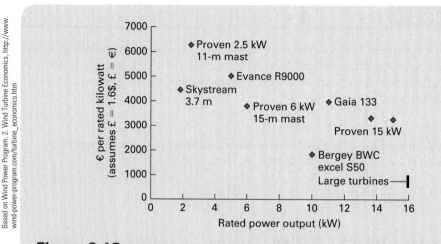

Based on Wind Power Program. 2. Wind Turbine Economics, http://www.wind-power-program.com/turbine_economics.htm

Figure 2.18: Infrastructure cost per unit capacity for a wind turbine installation as a function of total capacity (in Euros per kilowatt-hour capacity).

Variability in the average cost to generate electricity among countries can result from different approaches to electricity generation and differences in national economics. These variations are illustrated in Table 2.7.

Energy producers, such as public utilities, who operate energy production facilities such as coal-fired stations, nuclear plants, or wind farms, depend on equipment suppliers to provide components for these facilities and/or construct them. For alternative energy technologies, much of the development begins as basic scientific research in university or government laboratories funded by government or industry. Commercial viability is the final goal of such a process, and all aspects of the development of new energy resources must work in concert to achieve this goal. Government subsidies and incentives are therefore beneficial during the development stages. Traditionally, the corporate bottom line was always financial profit, and technologies that did not have a foreseeable profit on a suitable timescale were not considered. New developments had to have potential financial benefits that were commensurate with the risk. However, more recently, in many countries a different approach has been implemented; the *triple bottom line* (TLB). The TBL considers three aspects of possible benefits resulting from

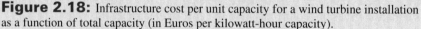

Table 2.6: Typical cost of electricity per kilowatt-hour in the United States in dollars per kilowatt-hour (US$/kWh), as generated by different technologies.	
energy resource	electricity cost (US$/kWh)
coal	0.025
natural gas	0.04
hydroelectric	0.05
wind	0.06
nuclear (fission)	0.065
geothermal	0.08
solar (photovoltaics)	0.30

© Cengage Learning 2015

Table 2.7: Average 2011–2012 residential cost of electricity per kilowatt-hour in equivalent U.S. dollars (US$/kWh) in different countries.

country	average cost (US$/kWh)
United States	0.10[1]
Canada	0.108[2]
France	0.194[3]
United Kingdom	0.22[3]
Australia	0.25[4]
Sweden	0.27[3]
The Netherlands	0.289[3]
Brazil	0.342[5]
Germany	0.365[3]
Denmark	0.404[3]
Japan	0.262[6]
China(Beijing)	0.078[7]
India (Delhi)	0.103[8]

[1] http://www.eia.gov/electricity/monthly/
[2] http://www.electricity.ca/media/Presentations/Electricity%20Pricing%20Presentation_June%202nd_2011.pdf
[3] http://www.energy.eu/#domestic
[4] http://www.abc.net.au/news/2012-03-21/australians-pay-highest-power-prices-says-study/3904024?section=nsw
[5] http://www.aneel.gov.br/area.cfm?idArea=550.
[6] http://thisbluemarble.com/showthread.php?t=36342
[7] http://english.sz.gov.cn/ln/201205/t20120517_1914423.htm
[8] http://www.bsesdelhi.com/docs/pdf/Delhi_Tariff_Economics.pdf

corporate actions, sometimes referred to as *people*, *planet*, and *profit*. These benefits come from the realization that good business practices for long-term sustainability need to consider more than just short-term economic gain.

The *people* aspect of the TBL concerns the fair treatment of workers and maintaining morale by providing fair salaries and benefits and by providing a safe working environment. The people benefits of TBL corporate practices are somewhat difficult to categorize but may include ensuring that suppliers and contractors also have fair worker policies or providing community benefits through educational programs.

The *planet* aspect of TBL is concerned with maintaining environmental quality. A thorough life cycle analysis (Chapter 4) of materials and practices is essential to a complete understanding of the environmental impact of business practices.

The *profit* component of TBL is probably the easiest to quantify. It is based on a traditional corporate accounting approach but may also include an assessment of the economic impact on the community in addition to internal corporate financial benefits.

Profits are essential to the survival of any business. The implementation of policies that promote the people and planet sides of TBL practices are often legal responsibilities of a business, but it should be recognized that being proactive in these areas can be ultimately beneficial to both business and community.

2.5d Social Factors

The public perception of energy is influenced by a number of factors, including economics, comfort, safety, environmental factors, and even aesthetics. There was relatively little public interest in alternative energy sources prior to the energy crisis in the early 1970s, except, perhaps, in the controversy over the risks associated with nuclear energy. Since its beginnings, nuclear energy has been controversial. This debate precedes any significant development of most alternative renewable energy resources (hydroelectric and geothermal are exceptions). Public approval of nuclear energy has generally increased in the

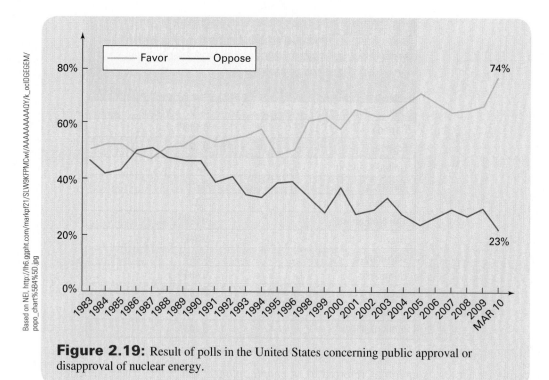

Figure 2.19: Result of polls in the United States concerning public approval or disapproval of nuclear energy.

United States over the past 40 years, as illustrated in Figure 2.19. However, approval typically decreases after a nuclear incident, as evidenced by the anomaly around 1986 when the Chernobyl incident occurred. After a nuclear incident, nuclear energy regulations typically become stricter, and it is likely that the chance of further incidents (at least in the short term) decreases. Only public perception of the risks associated with nuclear energy affects the approval rating. The nuclear incident at the Fukushima reactor facility in Japan illustrates this point. Japan has, in the past, been very pro–nuclear energy. This is not unexpected because it is a highly industrialized, densely populated country with minimal indigenous energy resources. Figure 2.20 shows the results of polls in Japan following the Fukushima incident. Not surprisingly, there was a substantial drop in support for nuclear energy in the ensuing three months. Certainly both public opinion, as well as government policy (in many countries), over the implementation of nuclear energy will continue to be a topic for debate.

In recent years, public concern over energy-related issues has grown. The points of greatest interest to the general public are summarized in the results of a recent poll in the European Union, as shown in Figure 2.21. Certainly the greatest concern is for energy prices and availability. Environmental concerns are lower on the list, as is interest in conservation matters. Another view of the relative importance of economy and environment is illustrated by the results of another survey, as shown in Table 2.8. Residents in the United States, Canada, and the United Kingdom were asked which of the following approaches they favored: one in which the environment was protected at the risk of adversely affecting economic growth and the other in which economic growth was ensured at the risk of environmental damage. In this limited study, at least, environmental

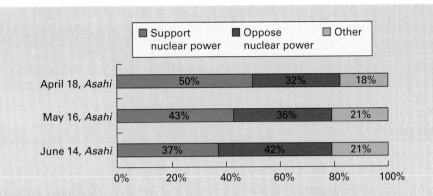

Figure 2.20: Results of polls in Japan concerning approval of nuclear energy following the March 11, 2011 earthquake/tsunami and subsequent nuclear reactor incident.

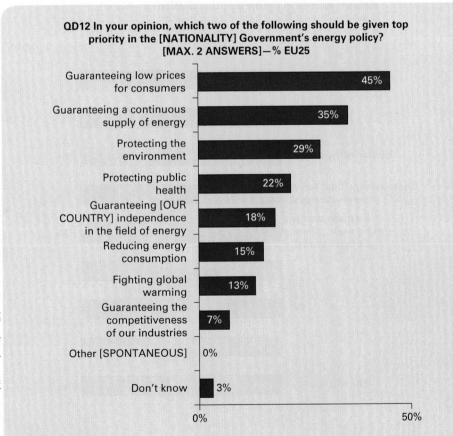

Figure 2.21: Results of a recent European Union poll concerning public priorities on energy-related matters.

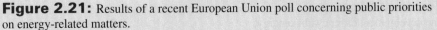

Table 2.8: Results of a recent public opinion poll on relative importance of environmental protection and economic growth.

nation	percent favoring environmental protection	percent favoring economic growth
Canada	55	22
United States	47	26
United Kingdom	40	33

Source: Data from http://www.angus-reid.com/issue/global-warming/.

concerns seem to be important in North America and less so in the United Kingdom, although a significant minority favor economy over environment.

Opinions concerning the need to develop alternative energy technologies (from the same survey as illustrated in Figure 2.19) are shown in Figure 2.22. In the European Union, there is very strong support for the development of solar energy (despite its high

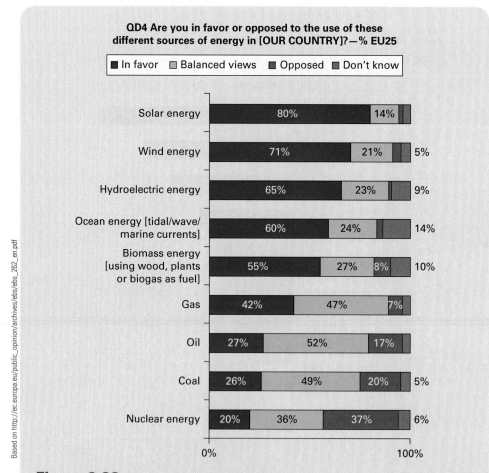

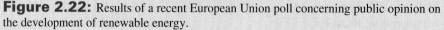

Figure 2.22: Results of a recent European Union poll concerning public opinion on the development of renewable energy.

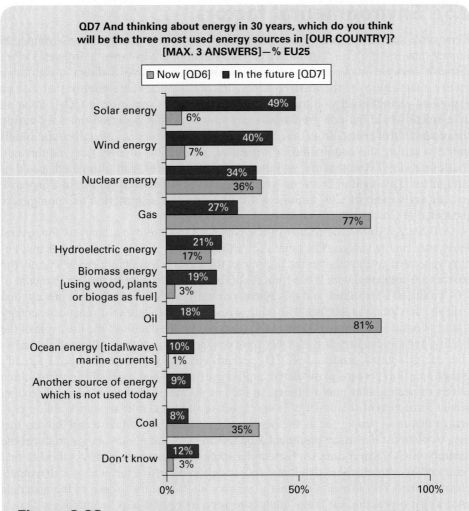

Figure 2.23: Results of a recent European Union poll concerning public opinion on the development of renewable energy.

cost and the general concern over future energy prices). There is also significant opposition to the use of nuclear energy, compared to other energy sources. This is even the case in France, where the vast majority of electricity comes from nuclear energy (21% in favor, 44% neutral, and 33% opposed). The favorable view of solar energy is also accompanied by an optimism that it will provide the majority of energy within the next 30 years and that fossil fuel use will be greatly diminished. This opinion is illustrated by the data shown in Figure 2.23.

While public opinion seems to greatly favor the development of renewable energy sources and there is a positive feeling that this is possibly on an unrealistically short timescale, there is also a reluctance to accept any adverse economic consequences that could result from moving away from inexpensive fossil fuels.

2.5e Environmental factors

Renewable sources of energy are generally considered to have less environmental impact than fossil fuels. A quantitative assessment of environmental impact is often difficult because many environmental factors do not have a direct quantitative metric. One aspect of the environmental impact of renewable energy that can be expressed quantitatively is greenhouse gas emissions (specifically CO_2). Such an assessment would include greenhouse gas emissions that occurred not only during the production of energy from the renewable resource (which is generally quite small) but also during the acquisition of materials, production of necessary equipment, transportation of components, maintenance of the facility, and ultimate disposal of equipment. This kind of cradle-to-grave life cycle is discussed further in Chapter 4, and an assessment of CO_2 emissions for alternative transportation technologies is presented in Chapter 20.

Table 2.9 shows the results of a greenhouse gas life cycle analysis for several renewable energy source. The table shows the quantity of CO_2 produced per unit energy generated, averaged over the lifetime of the facility. Results for renewable energy sources are compared with those from fossil fuels. For fossil fuels, the large amount of CO_2 produced is expected, and this comes primarily from the combustion process that oxidizes the carbon in the fuel. The table gives a range of values for renewable energy sources because the actual amount of CO_2 depends on the details of the life cycle of the facility, and this can vary considerably from one facility to another. Although all renewable energy sources produce less CO_2 per unit electricity generated, none is actually carbon free. It is perhaps particularly surprising that the worst renewable energy source in this respect is solar photovoltaics. In fact, as discussed in the previous subsection, public perception of solar energy is very positive. It is generally the renewable energy source that most people see as the best hope for a sustainable future and the one that they would most like to see promoted. The reason that solar photovolatics produces the quantity of CO_2 that it does (in fact, under some circumstances, nearly a third of that produced by natural gas) is that it is a very materials-intensive technology that extracts energy from a very low-density energy source. In other words, some of the materials used in the production of photovoltaics are very energy intensive to produce. This is typically the case, for example, for relatively rare materials where it is necessary to mine a large quantity of raw material in order to extract a small quantity of the material of interest. Added to this are the facts that photovoltaic materials require high-technology

Based on data from International Energy Agency "Benign energy? The environmental implications of renewables," Paris: OECD (1998)

Table 2.9: Greenhouse gas emissions (CO_2) per unit electrical energy generated for some fossil fuels and renewable resources.

resource	CO_2 (kg/MWh)
coal	955
natural gas	430
solar (photovoltaic)	98–167
wind	7–9
geothermal	7–9
hydroelectric (high head)	3.6–11.6

manufacturing processes and that the low density of solar energy means that a lot of photovoltaic cells need to be manufactured to generate a relatively small amount of electricity. This type of analysis indicates that the environmental impact of utilizing a renewable energy resource, from the greenhouse gas emissions standpoint, is not always obvious.

Other aspects of the environmental impact of renewable energy might be more difficult to quantify but can become apparent by means of careful analysis. As with solar energy, the low energy density of many renewables (compared with fossil fuels) is an important contributing factor in their environmental impact. The extensive land area associated with the generation of electricity by solar photovoltaics, wind, or even hydroelectricity can result in the deforestation or displacement of resources from agricultural development with a resulting impact on the environment and the quality of human life.

Biofuels are a clear example of the complexity of an overall assessment of environmental impact. As discussed in detail in Chapter 16, a careful analysis of all aspects of the impact of biofuel production and use does not necessarily provide a positive incentive for their extensive development.

The low energy density of renewables is also a contributing factor to their relatively high (in some cases) human risk factor, as discussed in detail in Section 6.9. The large quantity of material needed for many renewable technologies means risk associated with the required mining operations, material transportation, manufacturing processes, and related operations, and these typically outweigh the direct risk associated with the energy generation itself. The perception of risk or environmental impact is often an important factor in the formation of public opinion. Although wind energy is one of the least invasive of energy technologies and is generally favored as a viable new energy source, the uncertainty of local effects often results in the resistance of residents to developments in their own neighborhoods. This is sometimes referred to by the acronym *NIMBY* (not in my backyard).

2.5f Political factors

Political decisions on energy-related matters generally follow from a nation's energy policy, as well as the efforts of individual politicians to defend the energy interests of their constituents. Energy policy provides guidelines for energy-related laws and treaties with other countries and directives for government agencies dealing with energy issues. The following items are common components of national energy policies:

- A description of national policies concerning energy generation, transmission and use
- The establishment of energy efficiency and environmental standards related to energy use
- The specification of energy-related fiscal policies, including subsidies, incentives, tax exemptions, and the like, to promote improved energy utilization
- The participation in funding programs for energy-related research and development
- The development of energy-related treaties and agreements with other countries

Specific goals that energy policies typically address the following questions:

- To what extent is energy self-sufficiency expected?
- What future energy production methods are most appropriate?

- How will good energy practices be promoted?
- What environmental impact of energy use is acceptable?
- What is an acceptable energy technology for a transportation system?

The most appropriate means of dealing with these points can vary greatly from one country to another and depend on a number of factors such as:

- Economy.
- Climate.
- Geography.
- Natural resources.
- Population.

Although it may be reasonable for a country with a large land area, low population density, substantial natural resources, and a good economy to strive for energy independence, that goal is not reasonable for a nation that has a small land area, a large population density, and few resources.

While energy policies attempt to establish goals to ensure energy security, the specific approach to achieving these goals needs to be defined in order to implement such policies. The success of an energy policy lies not only in the suitability of its goals but also in the ability of society to overcome the necessary obstacles in order to achieve those goals. For example, the practicality of some U.S. automobile emission standards have been challenged in court by automobile manufactures who claim that they are technologically and economically unrealistic.

General guidelines for the implementation of U.S. energy policy have been outlined in the Department of Energy's 2011 Strategic Plan, which specifies key areas of focus, such as the development of energy efficiency standards for buildings (Chapter 17), increasing renewable energy production, and implementing smart grid technologies. The Obama administration has specified certain quantitative goals for future energy use in the United States, which include:

- Reducing greenhouse gas emissions (compared to the 2005 baseline reductions) 17% by 2020 and 83% by 2050.
- Increasing the number of electric vehicles on the roads to 1 million by 2015.
- Increasing electricity generation from clean energy sources to 80% by 2035.

Although these appear to be laudable and perhaps formidable tasks, the definition of some of the terminology used needs to be clarified. *Electric vehicles* include not only battery electric vehicles [such as the Nissan Leaf (Chapter 19)], but also series or plug-in hybrids [such as the Chevrolet Volt (Chapter 17)]. More surprising, perhaps, is the fact that *clean energy sources* include nuclear, natural gas, and coal.

Many aspects of energy policy are most appropriately considered at the regional, state/provincial, or municipal level (more in Chapter 17). These factors include, for instance, energy conservation, building codes, and other measures that are influenced by local geography and climate.

Overall, establishing government energy policies that are environmentally sound, technologically feasible, and economically viable for a particular nation or region is a challenge that can provide long-term benefits and lead to a sustainable energy future.

ENERGY EXTRA 2.2
Government emission control standards

Concerns over the effects of anthropogenic emissions during energy production are not new. In 1306, King Edward I of England banned coal fires in London to improve air quality. More recently, the health and environmental effects of air pollution were publicized in Henry Obermeyer's 1933 book *Stop That Smoke!* (Harper, New York). In the twentieth century, local or regional regulations concerning emissions became common in order to deal with air quality in many cities. In the 1950s and 1960s, systematic studies were undertaken to identify sources of air pollution. Two important conclusions came out of these investigations: (1) Local sources of pollution had widespread effects outside the region, and (2) a significant portion of air pollution originated from motor vehicles.

The first finding was dealt with by the gradual implementation of regulations at state/provincial and federal government levels. In the United States, the State of California created the California Air Resources Board in 1967, and the federal government established the Environmental Protection Agency in 1970. During the same era, similar government regulatory bodies in Canada, Australia, most of Western Europe, and Japan were also established.

A major focus of such agencies has been the reduction of vehicle emissions by setting standards for new models. Over the years, these standards have become more strict and have required the implementation of new approaches to emission control. Although a number of technological advances have led to a reduction in vehicle emissions over the years, two major developments have played a major role. The first deals with emissions in the engine, and the second deals with emissions in the exhaust gas. In the first case, the *positive crankcase ventilation* (*PVC*) system reduces emissions by directing hydrocarbon-rich fumes from the crankcase into the engine's intake manifold to be burned rather than releasing them to the atmosphere. Implementation of PVC systems

predates the establishment of state and federal regulatory agencies that deal specifically with vehicle emissions. In 1961, pollution concerns prompted the State of California to require that all new passenger vehicles sold in that state must have PVC. By 1964, the implementation of PVC on new passenger vehicles was widespread in the United States.

By the mid-1970s, increasingly strict emission standards (Section 4.3) required a consideration of emissions in the exhaust gas. Exhaust gas emissions are most effectively dealt with using a catalytic converter. This is a device that converts toxic emissions from an internal combustion engine into nontoxic (or at least less toxic) by-products through a stimulated catalytic reaction. Most catalytic converters used on gasoline internal combustion engines are three-way converters, which deal with three types of pollutants in the following ways:

- Reduction of nitrogen oxides to nitrogen and oxygen
- Oxidation of carbon monoxide to carbon dioxide
- Oxidation of hydrocarbons to carbon dioxide and water

Government emission control standards have been effective nationally and even globally and have made substantial reductions in vehicle emissions. Reductions in emissions are discussed in detail in Chapter 4 and are an example of one case where government regulations were effective in dealing with environmental concerns.

Topic for discussion

Consider the details of the chemical reactions for the three processes that occur in a three-way catalytic converter. What are the health and/or environmental concerns of the chemical compounds that are produced?

2.5g Integrating New Technology with Existing Infrastructure

Whatever the mix of energy sources that will be adopted for the future, it is clear that changes must be made in how we produce and utilize energy. The energy technologies that have been developed over many years must give way to the new approaches in order to establish a sustainable infrastructure for the future. As outlined in the Preface, this is a formidable task, and this change must be done in such a way as to maintain the effectiveness of our energy systems.

Transportation is an interesting example of the potential challenges to changes in energy production and use, as well as of the need to consider all relevant aspects of reducing our dependence on fossil fuels. If, for example, future transportation systems are based on battery electric vehicles or fuel cell vehicles, then the elimination of the existing fossil fuel vehicle infrastructure and the development of a new transportation infrastructure must be done in a manner that minimizes or precludes disruption. While the way forward may seem straightforward, certain not so obvious issues need to be considered.

It is clear that the existing operational infrastructure of oil wells, refineries, oil pipelines, and gas stations will need to be replaced by electric charging or hydrogen filling stations. Also, the manufacturing infrastructure for the production of fossil fuel vehicles will need to be converted to the manufacture of battery electric vehicles or fuel cell vehicles. However, the necessary changes to the electricity supply system may constitute an even greater task. A summary of energy sources in an industrialized country (the United States) is illustrated in Figure 2.5. At present, roughly one-third of end user energy is for transportation; the remainder is roughly equally divided between industry and residential plus commercial. The energy for transportation comes almost exclusively from petroleum. If fossil fuel vehicles are replaced by electric vehicles (using either batteries or hydrogen as an electricity storage mechanism), then electric generating capacity must be increased to provide for both our current electricity needs and energy for transportation. Even considering the fact that electric vehicles are more efficient than fossil fuel–powered vehicles (Chapter 20), moving from a fossil fuel transportation economy to an electric transportation economy will require a 60–70% increase in electric generating capacity. A long-term goal of reducing fossil fuel use, from the standpoint of either resource availability or environmental impact, will require a careful consideration of resource availability in order to ensure that our energy needs can be fulfilled. Thus any plan to significantly increase the number of electric vehicles with a corresponding reduction in fossil fuel vehicles will call for suitable increases in electric generating capacity to satisfy the increased demand. The full benefits of fossil fuel vehicle reduction will require that additional generating capacity will come from renewable sources.

An important concern related to the integration of alternative energy technologies with our existing electricity supply grid deals with the intermittent nature of most renewable energy sources. This would include, for example, the daily variations in solar energy, the somewhat less predictable variations in wind energy, and the cyclic periodicity of tidal energy. Hydroelectric and geothermal are alternative resources that offer a more predictable supply of energy; however, geothermal availability is probably not sufficient to make any substantial long-term contribution to global energy use. To make the most efficient use of available resources, it is necessary, as discussed in Chapters 17 and 18, to integrate alternative energy technologies with suitable energy storage techniques and energy from base-load fossil fuel, nuclear, or possibly hydroelectric facilities. A long-term goal of eliminating fossil fuels will require intelligent control of energy production and distribution facilities (i.e., smart grid, as discussed in Chapter 17) to ensure a reliable supply of electrical energy.

2.6 **Summary**

In this chapter, the energy needs of humanity over the years have been reviewed. It was shown that, with the development of technology, came a rapid and significant increase in energy use. The reliance on wood as the major source of energy gave way to coal in the second half of the nineteenth century and then to petroleum in the mid-twentieth century. It has been shown that per-capita energy use in different countries is directly related to economic factors. Climate, population density, and types of industry play an important role in determining energy use. The chapter also summarized the current distribution of energy use, illustrating that fossil fuels account for the vast majority of energy use, followed by nuclear and hydroelectric. Growth mechanisms have been presented, and the analysis of energy resource utilization on the basis of the Hubbert model was presented. This model provides an understanding of the rate at which a resource is utilized in the context of the total quantity of the available resource.

The challenges for the implementation of new energy technologies have been reviewed. These include not only the technological challenges themselves but also those related to a consideration of resource availability, economics, social attitudes, environmental factors, and political policies. Viable energy technologies must be sustainable and economically competitive with traditional alternatives; otherwise, there will be a reluctance to move away from well established technologies. Public opinion of alternative energy is based on a perception of factors such as cost, environmental impact, and risk and is a major factor in determining the acceptance of traditional and new energy sources. A complete environmental analysis of renewable energy technologies is often difficult and indicates the complexity of fully assessing available energy options. Energy policies are designed to promote sound approaches to future energy production and use, and their implementation is most efficient when dealt with at a variety of different government levels.

Problems

2.1 Consider an island with a population density of 50 people/km² (about equal to the present world average). If the annual growth rate is 2%, determine the year in which the population density be equal to 20,000 people/km² (approximately the current population density of Monaco, the world's most densely populated nation).

2.2 A quantity has a doubling time of 110 years. Estimate the annual percent increase in the quantity.

2.3 The population of a particular country has a doubling time of 45 years. When will the population be three times its present value?

2.4   Assume that the historical growth rate of the human population was constant at 1.6% per year. For a population of 7 billion in 2012, determine the time in the past when the human population was 2.

2.5   What is the current average human population density (i.e., people per square kilometer) on earth?

2.6   The total world population in 2012 was about 7 billion, and Figure 2.11 shows that at that time the actual world population growth rate was about 1% per year. The figure also shows an anticipated roughly linear decrease in growth rate that extrapolates to zero growth in about the year 2080. Assuming an average growth rate of 0.5% between 2012 and 2080, what would the world population be in 2080? How does this compare with estimates discussed in the text for limits to human population?

2.7   The population of a state is 25,600 in the year 1800 and 218,900 in the year 1900. Calculate the expected population in the year 2000 if (a) the growth is linear and (b) the growth is exponential.

2.8   The population of a country as a function of time is shown in the following table. Is the growth exponential?

year	population (millions)
1700	0.501
1720	0.677
1740	0.891
1760	1.202
1780	1.622
1800	2.163
1820	2.884
1840	3.890
1860	5.176
1880	6.761
1900	8.702
1920	10.23
1940	11.74
1960	13.18
1980	14.45
2000	15.49

2.9   Consider a solar photovoltaic system with a total rated output of 10 MW_e and a capacity factor of 29%. If the total installation cost is $35,000,000, calculate the decrease in the cost of electricity per kilowatt-hour if the payback period is 25 years instead of 15 years. Assume a constant interest rate of 5.8%.

Bibliography

E. Cassedy and P. Grossman. *An Introduction to Energy: Resources, Technology, and Society* (2nd ed.). Cambridge University Press, Cambridge, MA (1999).

R. Hinrichs and M. Kleinbach, *Energy: Its Use and the Environment* (5th ed.). Brooks-Cole, Belmont, CA (2012).

J. A. Kraushaar and R. A. Ristinen. *Energy and Problems of a Technical Society* (2nd ed.). Wiley, New York (1993).

E. L. McFarland, J. L. Hunt, and J. L. Campbell. *Energy, Physics and the Environment* (3rd ed.). Cengage, Stamford, CT (2007).

R. A. Ristinen and J. J. Kraushaar. *Energy and the Environment.* Wiley, Hoboken, NJ (2006).

V. Smil. *Energy at the Crossroads: Global Perspectives and Uncertainties.* MIT Press, Cambridge, MA (2003).

J. Tester, E. Drake, M. Driscoll, M. W. Golay, and W. A. Peters. *Sustainable Energy: Choosing Among Options.* MIT Press, Cambridge, MA. (2006).

R. Wolfson. *Energy, Environment, and Climate* (2nd ed.). Norton, New York (2011).

World Energy Council. *2010 Survey of Energy Resources,* available online at http://www.worldenergy.org/documents/ser_2010_report_1.pdf.

PART II

Fossil Fuels

For the past 150 years, fossil fuels have formed the largest component of our energy use. During that time, we have used up a substantial fraction of the available fossil fuel resources. These resources are not renewable, and their depletion will ultimately require the implementation of alternative sources of energy. Estimates of the longevity of fossil fuel resources are difficult and depend on a number of factors. The estimates range from a small number of decades for domestic (U.S.) oil to many hundreds of years for coal resources. During the time we have used fossil fuels, we have also made a substantial negative impact on the environment. Some of the damage caused to the earth and its ecosystem may be irreversible.

Clearly, our current use of fossil fuels cannot continue indefinitely. Dwindling supplies will force us to consider other energy options. However, environmental concerns supply the motivation to pursue alternative energy opportunities as a means of reducing the adverse effects of fossil fuel use. The infrastructure of human society is based on the use of substantial quantities of energy. Because most alternative energy sources have relatively low energy densities, developing a sufficient quantity of these resources to meet our needs is an enormous undertaking, as suggested in the Preface. However, to fully appreciate how fossil fuels can be replaced, it is important to understand, in detail, how we currently utilize fossil fuels and the consequences of this use. Chapter 3 provides an overview of our fossil fuel use, and Chapter 4 summarizes its environmental effects.

An offshore oil platform is shown in the photograph. ■

James Jones Jr–Shutterstock.com

Fossil Fuel Resources and Use

Learning Objectives: After reading the material in Chapter 3, you should understand:

- The properties of fossil files and methods for obtaining and processing them.
- The availability of fossil fuels.
- The use of fossil fuels worldwide.
- The application of the Hubbert model to fossil fuel use.

- Enhanced fossil fuel recovery methods.
- The properties and availability of shale oil and tar sands.
- Methods for coal liquefaction and gasification.

3.1 Introduction

Fossil fuels originate from ancient organic matter that has been subject to high temperatures and pressures inside the earth over periods from millions to hundreds of millions of years. Depending on the details of the starting material and the formation conditions, the resulting fossil fuel can be solid (coal), liquid (oil), or gas (natural gas). The age of even the most recent fossil fuel deposits is large compared with the timescale on which we are depleting this resource. Thus it cannot be renewed. Unlike fuels such as wood, which can be used in a manner that is carbon neutral by replacing trees as they are used (Chapter 16), fossil fuel use produces a net release of carbon into the atmosphere. The resulting environmental consequences of fossil fuel use are a serious concern, as discussed in the next chapter. In this chapter, the properties of various fossil fuels are introduced, and the methods by which they are extracted from the earth, processed, and converted into other forms are discussed. Both traditional fossil fuel use, which is currently responsible for about 85% of our energy, as well as possible future methods of enhanced recovery, are presented.

3.2 Oil

As seen in Figures 2.2 and 2.3, oil and its derivatives (e.g., gasoline) are the largest single energy source at present. This is due to the facts that there is an enormous infrastructure for extracting these resources from the earth, processing them, and using them, and that, at present, they remain inexpensive compared with most other sources of energy. One of the major uses of oil and its derivatives is as a fuel for transportation because of their high energy density and the convenience of their liquid (or gaseous) form. Like all fossil fuels, oil was formed as a result of the decomposition of organic plant or animal matter during prehistoric times. Typical oil deposits are about 500 million years old. At that time in the earth's history, life existed primarily in saltwater oceans. and oil deposits are therefore located in regions that were once at the bottom of the seas. Organisms collected on the floors of ancient seas, and these were covered with layers of mud and sand. Over the years, the pressure and temperature resulting from the layers of sediment turned the mud and sand into sedimentary rock and converted the organic material into oil and natural gas. Even the very early stages of the formation of the hydrocarbons associated with fossil fuels take thousands of years. However, the complete process by which petroleum is formed takes many millions of years. During the decomposition process, much of the carbon content of the organic material was lost to the atmosphere in the form of carbon dioxide. Only a very small fraction of the carbon in the original organic matter contributes to the formation of oil. In fact, about 20 tonnes of organic matter is needed to produce one liter of oil. Light hydrocarbon molecules are the constituents of natural gas, and heavier hydrocarbon molecules make up oil. It is common to find these together in the same deposit. As the oil and gas form, they can move through the layers of sedimentary rock and eventually collect together in deposits. A typical deposit might consist of oil and gas in a layer of very porous rock that is trapped between two layers of dense impermeable rock. Figure 3.1 shows a schematic of a typical deposit. As a result of the continental drift that occurs on a timescale of many millions of years, some of the petroleum deposits have remained under the oceans while others have ended up underneath land.

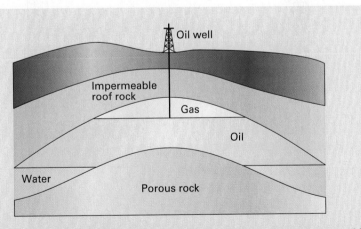

Figure 3.1: Geology of a typical oil deposit where the oil and gas are trapped in porous rock between layers of dense rock. The gas rises to the top, and the oil floats on any water in the deposit.

Example 3.1

Assuming that organic matter has an energy content of about 15 MJ/kg (Chapter 16), estimate the fraction of energy from organic matter that is available in a crude oil deposit.

Solution

From the preceding discussion, 1 L of crude oil is produced from the decomposition of 20 t of organic matter. The original organic matter will have an energy content of

$$(15 \text{ MJ/kg}) \times (1000 \text{ kg/t}) \times (20 \text{ t}) = 3 \times 10^5 \text{ MJ}.$$

The resulting liter of crude oil will have an energy content of 38.5 MJ (from Appendix IV), for a ratio of

$$\frac{38.5 \text{ MJ}}{3 \times 10^5 \text{ MJ}} = 1.3 \times 10^{-4} \text{ (or 0.13\%)}.$$

The use of oil as a fuel became established in the mid-1800s, although it did not become a major component of our energy production until after 1900. This trend is seen for the United States in Figure 2.1, and the rapid increase in oil use during the past century is largely a result of the development of the automobile. Titusville, Pennsylvania, is often cited as the location of the world's first oil well in 1859 (Figure 3.2). However,

Reproduction, copyrighted in 1890, of a retouched photograph showing Edwin L. Drake, to the right, and the Drake Well in the background, in Titusville, Pennsylvania, where the first commercial well was drilled in 1859 to find oil.

Figure 3.2: Oil well drilled in Titusville, Pennsylvania, in 1859 by Edwin Drake.

Courtesy of Martin Dillon

Figure 3.3: Oil well dug in Petrolia, Ontario, in 1858.

Zeljko Radojko/Shutterstock.com

Figure 3.4: Small pump jack oil well.

it has been claimed that oil production began in Petrolia, Ontario, in 1858 (Figure 3.3). Today several types of commercial oil wells are used for different types of oil deposits. Some examples of land-based oil wells are shown in Figures 3.4 and 3.5, and an oil platform for drilling under the oceans is illustrated in Figure 3.6.

Figure 3.5: Oil drilling rig in Wyoming, U.S.

Figure 3.6: Off-shore oil platform.

3.3 Refining

Crude oil, as it is extracted from the earth, is a mixture of a large number of different compounds, mostly hydrocarbons. The hydrocarbons cover a very large range of molecular masses. Table 3.1 lists some common components of crude oil and their properties.

Carbon may bond with hydrogen in many different ways. One important series of hydrocarbons is the alkane series, which has the chemical formula C_nH_{2n+2}. Properties of the alkane series are given in Table 3.2. Some important members of this series are methane (the major component of natural gas), ethane (the basis for ethanol, C_2H_5OH), and octane (an important component of gasoline).

As is obvious from Tables 3.1 and 3.2, the larger the number of carbon atoms per molecule, the higher the boiling point of the hydrocarbon will be. This property forms

Table 3.1: Properties of typical hydrocarbons extracted from oil during the refining process.

name	number of carbon atoms per molecule	state at room temperature	boiling temperature (°C)	uses
natural gas	1–5	gas	−165 to 25	gaseous fuel
petroleum ether	5–7	liquid	25 to 90	industrial solvent
gasoline	5–12	liquid	25 to 200	automobile fuel
kerosene	12–16	liquid	175 to 275	stove and jet fuel
fuel oil	15–18	liquid	< 375	diesel and home heating
lubricating oil	16–20	liquid	> 350	lubrication
grease	>17	semisolid	—	lubrication
paraffin	>19	solid	—	candles
tar	large	solid	—	roofing and paving

© Cengage Learning 2015

Table 3.2: Alkane series of hydrocarbons with the formula C_nH_{2n+2}. The heat of combustion is the HHV (Chapter 1).

n	formula	name	boiling temperature (°C)	molecular mass (g/mol)	heat of combustion MJ/kg	heat of combustion eV per molecule
1	CH_4	methane	−164	16	55.5	9.20
2	C_2H_6	ethane	−89	30	51.9	16.1
3	C_3H_8	propane	−42	44	50.3	23.0
4	C_4H_{10}	butane	0	58	49.5	29.8
5	C_5H_{12}	pentane	36	72	48.7	36.3
6	C_6H_{14}	hexane	69	86	48.1	42.9
7	C_7H_{16}	heptane	98	100	48.1	49.9
8	C_8H_{18}	octane	125	114	46.8	55.3

© Cengage Learning 2015

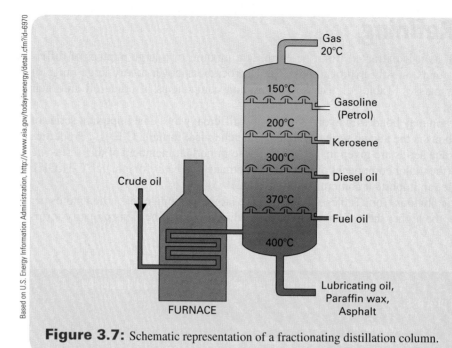

Figure 3.7: Schematic representation of a fractionating distillation column.

the basis of the refining process. Crude oil is heated in a furnace to about 400°C. It then travels up a fractionating column, as seen in Figure 3.7, where its temperature is progressively decreased. At 400°C most of the hydrocarbons are vaporized, and any very heavy hydrocarbons or impurities fall to the bottom of the fractionating column. As the vapor travels up the column, it experiences decreasing temperature, and progressively lighter hydrocarbons condense out and can be extracted. In this way, the crude oil is separated into components of different molecular mass ranges. A photograph of a typical oil refinery showing the fractionating columns appears in Figure 3.8.

The proportions of the components produced in the distillation process are determined by the proportions of the different compounds present in the crude oil. Unfortunately, this ratio of these components is generally not compatible with our relative need for them. For example, gasoline and diesel fuel are in much higher demand than, say, paraffin. To make optimal use of our petroleum resources, it is necessary to convert some of the less needed refinery products to more needed forms. In most cases, this involves breaking down heavy hydrocarbons (like tar and paraffin) to make lighter hydrocarbons (like octane) in a process called cracking. Of the several variations on this process, most involve heating the hydrocarbon to a high temperature to break down the chemical bonds. Some common processes are *steam cracking*, where the heavy hydrocarbon is mixed with steam and heated; *hydrocracking*, where the hydrocarbon is exposed to hydrogen gas; and *catalytic cracking*, where the hydrocarbon is exposed to a catalyst like alumina, silica, or a *zeolite*. The proportions of the various by-products of the cracking process depend on the details of the process used and the temperatures involved.

The reverse process, which involves the sticking together light hydrocarbons into heavier ones, is referred to as polymerization. Ethane, propane, and/or butane may be combined to produce octane or the similar-weight hydrocarbons that comprise gasoline. Light hydrocarbons are heated and react with a catalyst in order to yield the desired reactions.

Figure 3.8: Fractionating columns at Dartmouth Refinery, Dartmouth, Nova Scotia.

Gasoline typically contains about 500 different hydrocarbons that have between 5 and 12 carbon atoms per molecule. Its average density is about 720 kg/m³. A substantial fraction of these hydrocarbon components of gasoline have molecular masses that are similar to octane. When gasoline is sold, it is commonly labeled with an octane rating. This number does not relate directly to the amount of octane in the fuel. It is rather a measure of the efficiency of the combustion of the gasoline in an internal combustion engine relative to the efficiency of combustion of a standard mixture of octane and other hydrocarbons. It is interesting to note that gasoline sold in, for example, New York might have an octane rating of, say, 89, while if the same gasoline is sold in Denver, it might have an octane rating of 87. The variance results from differences in combustion efficiency that are related to differences in air density (i.e., related to altitude).

3.4 Natural Gas

Natural gas is a mixture of light hydrocarbons (typically about 85% methane and 15% ethane) that is gaseous at room temperature (STP). Natural gas may be extracted from deposits in the earth where it is associated with oil, as in Figure 3.1, or from deposits where it occurs on its own. In either case, it is formed in a manner similar to the that of oil from the decomposition of ancient organic matter. The formation of natural gas from organic matter is slightly less efficient than the formation of oil, and only about 0.085% of the carbon in the original organic matter becomes a component of natural gas. Natural gas may also be obtained as a by-product from the distillation of crude oil, as seen in Figure 3.7, where it is extracted as the lightest hydrocarbon component of the oil. The use of natural gas has increased over the years and in the United States is now about on par with the use of coal (Figure 2.1) in terms of net energy production. Natural gas typically contains fewer impurities than heavier hydrocarbons and burns more efficiently, thus creating less pollution. The ideal combustion of methane is given as

$$CH_4 + 2O_2 \rightarrow CO_2 + 2H_2O + 55.5 \text{ MJ/kg}. \tag{3.1}$$

The by-products are nontoxic and have negligible adverse effects other than the role of CO_2 as a greenhouse gas in climate change, as discussed in Chapter 4. Although other possibly undesirable chemicals (e.g., NO_x) can be formed (Chapter 4), the combustion of natural gas is inevitably cleaner than the combustion of gasoline or oil.

Example 3.2

Calculate the mass of CO_2 produced per MJ of energy from the combustion of methane.

Solution

From Table 3.2 and equation (3.1), the combustion of 1 kg of CH_4 produces 55.5 MJ/kg of energy, so 1 MJ is produced by the combustion of 0.018 kg of methane. The molecular mass of methane is 16 g/mol, so that 18 g of methane corresponds to 1.125 moles. Equation (3.1) shows that the combustion of 1 mole of CH_4 produces 1 mole of CO_2. Thus 1.125 moles of CH_4 yields 1.125 moles of CO_2. Because CO_2 has a molecular mass of 44 g/mol, 1.125 moles of CO_2 will have a mass of $(1.125 \text{ mol}) \times (44 \text{ g/mol}) \times (10^{-3} \text{ kg/g}) = 0.049$ kg.

In its gaseous state, natural gas may be transported easily overland through pipes. Transportation across oceans is somewhat more difficult because, at STP, natural gas occupies large volumes. Natural gas can be liquefied to produce liquid (or liquefied natural gas, LNG) by lowering the temperature to below about −165°C. In this state, it occupies only about 1/600th of the volume as in its gaseous state and can be transported across oceans in ships designed for this purpose. Liquid natural gas should not be confused with *natural gas liquids*, which are heavier hydrocarbons (liquid at room temperature) that are mixed with natural gas from some deposits and which are extracted during the distillation process.

3.5 Coal

Coal is formed, much as oil and natural gas are, over extended periods of elevated pressure and temperature. In the case of coal, however, the organic material originates from terrestrial plant matter. The earliest extensive plant growth on land occurred about 350 million years ago, and these early forests are the origin of the oldest coal deposits. Coal formation is, relatively speaking, an efficient process in which typically about 0.8% of the original carbon in the plant matter ends up as coal.

Table 3.3: Types of coal and their properties. These are typical values; the actual values can be quite variable, and the ranges overlap. Most of the noncarbon content of coal is in the form of volatile compounds, with a significant fraction being water.

type	carbon (%)	moisture content (%)	energy content (MJ/kg)
anthracite	90	10	35
bituminous	55	20	31
sub-bituminous	45	30	23
lignite	25	45	14

© Cengage Learning 2015

Coal exists in several varieties depending on age and formation conditions. The oldest and also the hardest coal is anthracite, and this has the highest carbon content. Typically, younger coals are softer and have lower carbon content. Examples of the major categories (or ranks) of coal are given in Table 3.3. These are often further divided into subcategories. Bituminous and sub-bituminous coals are the most common (comprising about 71% of coal in the United States), followed by lignite (28%) and anthracite (1%).

Coal may be found at various depths and in deposits of different geometries. It may, as a result of continental movements, exist under the oceans. In this case, it is not an economically viable resource because recovery methods are quite different from those used for a liquid or gaseous resource, like oil or natural gas, that exist beneath the ocean. About half the coal that occurs on land is also not economically recoverable because it is either too deep or occurs in very thin veins.

In the United States, coal was second to wood as a source of energy up until about 1880, when it became the most important energy resource. In the 1940s, the use of oil (including gasoline) surpassed the use of coal. At present, coal remains about even with natural gas as the second most common source of energy in the United States. In North America, most of the coal mined is used to generate electricity. A rough breakdown of coal use in recent years in the United States is given in Table 3.4.

It can be seen from the table that the second most common use of coal in the United States is the production of coke. Coke is made by heating coal to about 1200°C in the absence of air. Volatile materials are driven off, and more or less pure carbon is left behind. This carbon, or coke, is used primarily in the smelting of steel. Iron ores, which are comprised primarily of iron oxides, are heated to high temperature with the coke. By the reaction

$$Fe_aO_b + bC \rightarrow aFe + bCO, \tag{3.2}$$

the iron oxide is reduced, releasing carbon monoxide and leaving pure iron behind.

Table 3.4: Current use of coal in the United States.

use	% of total
electricity generation	70
coke production	17
export	10
other (residential heating, industrial processes, etc.)	3

© Cengage Learning 2015

Example 3.3

Using the information in Figure 2.5 for 2009 and the information in Table 3.4, estimate the total number of tonnes of coal produced annually in the United States (assume the coal is all bituminous).

Solution

From Figure 2.5, the contribution of coal to the annual electricity generation in the United States is about 19.3 EJ. Note that this is the primary energy content of the coal burned, not the electric output. Using the energy content of bituminous coal given in Appendix IV, 31 MJ/kg = 31×10^9 MJ/t

The coal contribution to electricity generation requires

$$\frac{19.3 \times 10^{18}\,\text{MJ}}{31 \times 10^9\,\text{MJ/t}} = 6.21 \times 10^8\,\text{t}$$

Since this corresponds to 70% of the U.S. coal production (Table 3.4), the total coal production is

$$\frac{6.2 \times 10^8\,\text{t}}{0.7} = 8.9 \times 10^8\,\text{t}.$$

This is in reasonable agreement with the actual value of 1.06×10^9 t, as given in Section 3.6.

3.6 Overview of Fossil Fuel Resources

It is fairly straightforward to determine how much of each of these fossil fuel resources is being used at present. It is also fairly easy to determine how much has been used in the past. It is not so easy to determine how much more is available in the earth for future use. It is even more difficult (assuming we knew how much is remaining) to estimate how long these remaining resources will last. The past and present use for each conventional fossil fuel resource is now discussed, along with estimates of the remaining resources.

3.6a Oil

Oil production and use for different countries and regions of the world are summarized in Table 3.5. North America uses much more oil than it produces. This is a direct result of the fact that the United States imports most of its oil. Canada, by comparison, produces very nearly the same amount of oil as it uses. Other regions, such as Africa and the Middle East, export most of their oil production. Oil is a primary energy source, and the energy available per year from the world oil production of about 3×10^{10} bbl (bbl = barrels = 158.97 L) is 180 EJ. Historical trends of U.S. oil use are illustrated in Figure 3.9. Clearly in recent years, the United States has been less able to meet its oil needs from domestic sources than it has in the past. Some insight into the reasons for this behavior can be gained by comparing the details of U.S. oil production with, for example, oil production in Saudi Arabia. In terms of total production, the United States produces somewhat less oil per year than Saudi Arabia (2.45×10^9 bbl versus 3.96×10^9 bbl). However, there are about 600,000 oil wells in the United States, compared with about 1000 oil wells in Saudi Arabia. Thus U.S. oil wells (on average) produce much less oil than Saudi Arabian oil wells.

Table 3.5: Daily oil production (from 2008). Major producers are shown by country. Amounts include natural gas liquids.

country/region	daily production (10^3 bbl)	country/region	daily production (10^3 bbl)
Algeria	1993	Kazakhstan	1554
Angola	1894	Malaysia	754
Egypt	722	other Asia	1308
Libya	1846	**Asia total**	**10,149**
Nigeria	2170	Norway	2456
Sudan	480	Russia	9886
other Africa	1250	United Kingdom	1526
Africa total	**10,355**	other Europe	880
Canada	3201	**Europe total**	**14,748**
Mexico	3158	Iran	4504
United States	6734	Iraq	2423
other North America	214	Kuwait	2784
North America total	**13,307**	Oman	763
Argentina	723	Qatar	1378
Brazil	1899	Saudi Arabia	10,846
Columbia	618	Syria	351
Ecuador	514	United Arab Emirates	2980
Venezuela	2566	Yemen	317
other South America	195	other Middle East	43
South America total	**6515**	**Middle East total**	**26,389**
Azerbaijan	914	Australia	556
China	3795	other Oceania	101
India	820	**Oceania total**	**657**
Indonesia	1004	**world total**	**82,120**

Based on data from World Energy Council ...

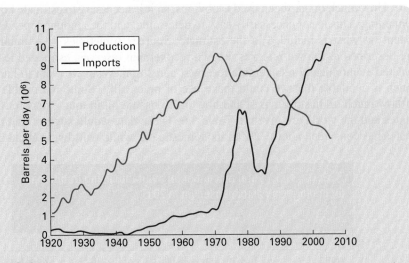

http://wasatchecon.wordpress.com/2010/11/28/
us-oil-production-versus-imports-1920-2005/

Figure 3.9: Percent of oil used in the United States that is imported from other countries.

Example 3.4

Calculate the average rate of oil production (in L/min) from a typical oil well in the United States and a typical oil well in Saudi Arabia.

Solution

In the United States, the average production per well per day is

$$\frac{2.45 \times 10^9 \,\text{bbl/year}}{600{,}000 \text{ wells} \times 365 \text{ day/year}} = 11.2 \text{ bbl/day per well.}$$

Since 1 bbl = 159 L, this production rate corresponds to

$$\frac{11.2 \text{ bbl/day} \times 159 \text{ L/bbl}}{1440 \text{ min/day}} = 1.2 \text{ L/min,}$$

equivalent to a fairly slowly running water faucet.

In Saudi Arabia, the production is

$$\frac{3.96 \times 10^9 \text{ bbl/year}}{1000 \text{ wells} \times 365 \text{ day/year}} = 10{,}850 \text{ bbl/day per well.}$$

This production rate corresponds to

$$\frac{10{,}850 \text{ bbl/day} \times 159 \text{ L/bbl}}{1440 \text{ min/day}} = 1200 \text{ L/min}$$

The differences illustrated in Example 3.4 between U.S. and Saudi Arabian oil production are related to oil well design; a typical continental U.S. oil well is shown in Figure 3.4, and Middle Eastern oil wells are similar to the design in Figure 3.5. These design differences result from the fact that oil has been utilized much longer in the United States than in the Middle East, and the oil wells are very mature. Basically, this means that oil reserves in the United States are getting close to being depleted.

Trends in world oil production can be viewed quantitatively in the context of the Hubbert model. This analysis requires the knowledge of the total amount of oil available. How much oil has been used is obvious, so it is necessary to estimate how much is remaining. This is not necessarily easy to determine, and the amount depends to a large extent on what is included in the estimate. Sources sometimes give amounts which are described as known reserves, known recoverable reserves, estimated reserves, estimated economically recoverable reserves, and so on. What we want to know is how much is available that we could make use of practically. Some reasonable estimates of how much oil has been used and how much usable oil in total exists for the United States and the world are given in Table 3.6. These data would suggest that the United States has consumed about 72% of its domestic oil, while worldwide about 34% of oil

Table 3.6: Total amount of oil available using traditional technology and amount used up to 2005.		
region	total available	used as of 2005
United States	2.1×10^{11} bbl = 1300 EJ	1.6×10^{11} bbl = 980 EJ
world	2.5×10^{12} bbl = 18,000 EJ	1.0×10^{12} bbl = 6200 EJ

© Cengage Learning 2015

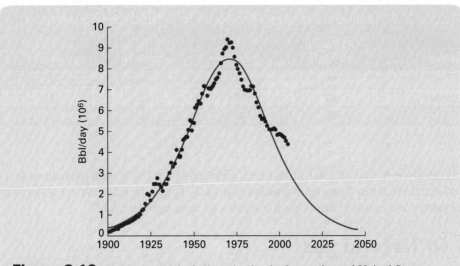

Figure 3.10: Hubbert model of oil production in the continental United States (10^6 bbl/day).

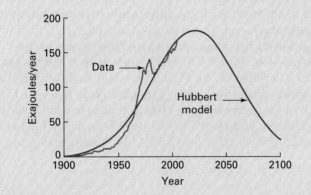

Figure 3.11: Hubbert model of oil use worldwide.

Based on http://www.theoildrum.com/node/2689

Based on E. L. McFarland, J. L. Hunt and J. L. Campbell. Energy, Physics and the Environment. Cengage Learning, 2007.

resources have been utilized. These results, and oil use in general, are seen to be consistent with the Hubbert model by plotting oil production (per year) as a function of year. These data are seen for the United States and the world in Figures 3.10 and 3.11, respectively. Clearly, the data (aside from a few anomalous years) follow a Gaussian-like curve. The area under the curve up to the present represents the total amount of oil used so far, and the area under the total curve represents the total amount of oil available. In both cases, the longevity of oil resources is obvious from an inspection of the time axis.

A more quantitative analysis of the longevity of oil resources requires some knowledge of future growth rates in oil use. Table 3.7 gives the approximate lifetime of oil resources for the world, the United States, and Canada, assuming various annual growth rates in oil use. The lifetime is seen to be a sensitive function of growth rate, and the actual measured growth rates suggest that conservation efforts in the United States are effective. The effects of energy conservation are suggested by Figure 3.12, where

Table 3.7: Lifetime of traditional oil resources for different annual growth rates.

region	0% annual growth (y)	2% annual growth (y)	5% annual growth (y)	actual growth rate 1995–2005 (%)
United States	45	33	24	−2.1
Canada	200	80	48	2.6
world	75	45	31	1.7

it is clearly seen that oil production trends changed in the 1970s with the realization of resource limitations. The world situation will, in the future, be substantially affected by the energy use trends in developing countries, as seen in Figure 2.11.

3.6b Natural gas

Natural gas has become attractive in recent years due to the smaller amount of pollution it produces compared to that from the combustion of oil or coal. The current production and use rates of natural gas by country and region are summarized in Table 3.8. The total world production is approximately 3.05×10^{12} m^3 per year, corresponding to about 100 EJ. As with oil, the world production of natural gas is very nearly the same as world use. There is, however, much less difference between production and use on a regional scale than there is for oil. This is largely due to the greater difficulty and expense in transporting a gaseous fuel (rather than a liquid fuel) across oceans.

An investigation of Figure 2.1 would suggest that natural gas production in the United States is leveling off (or even decreasing) and that production is near the peak on the Hubbert production curve. Estimates place the total U.S. resources at around 30×10^{12} m^3, corresponding to 1100 EJ (compare with Table 3.6 for oil), and the total world resource at around 7700 EJ. Estimates of the lifetime of natural gas for different annual growth rates for Canada, the United States, and the world are

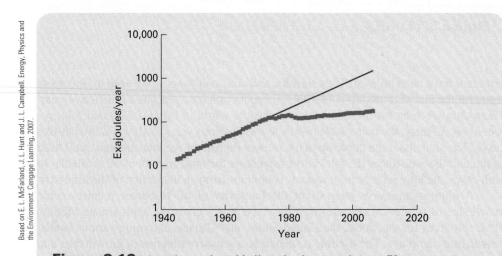

Figure 3.12: Actual annual world oil production over the past 70 years or so.

Table 3.8: Annual natural gas production worldwide (from 2008). Major producers are shown by country. Production amounts are net values that take into account production losses.

country/region	annual production (10^9 m^3)
Algeria	86.5
Egypt	48.3
other Africa	67.0
Africa total	**201.8**
Canada	167.5
Mexico	46.6
United States	574.4
other North America	39.7
North America total	**828.2**
Argentina	45.2
Venezuela	24.1
other South America	43.3
South America total	**112.6**
China	76.1
India	32.2
Indonesia	70.0
Malaysia	57.3
Pakistan	37.5
Thailand	28.8
Turkmenistan	66.1
Uzbekistan	63.4
other Asia	93.8
Asia total	**525.2**
Netherlands	80.0
Norway	99.2
Russia	621.3
United Kingdom	68.2
other Europe	76.5
Europe total	**945.2**
Iran	116.3
Qatar	77.0
Saudi Arabia	80.4
United Arab Emirates	50.2
other Middle East	58.7
Middle East total	**382.6**
Australia	47.5
other Oceania	4.1
Oceania total	**51.6**
world total	**3047.2**

Based on data from World Energy Council. . . .

Table 3.9: Lifetime of traditional natural gas resources for different annual growth rates.

region	0% annual growth (y)	2% annual growth (y)	5% annual growth (y)	actual growth rate 1995–2005
United States	35	27	20	−0.2%
Canada	55	38	27	1.6%
world	75	46	32	2.6%

given in Table 3.9. Except for Canada's somewhat greater oil resources, the estimates for oil and natural gas are similar.

3.6c Coal

Coal production and use for various regions is given in Table 3.10. Total world production is about 6.7×10^9 t per year, corresponding to about a quarter of world energy. The

Table 3.10: Annual coal production worldwide (from 2008). Major producers are shown by country.

country/region	annual production (10^6 t)	country/region	annual production (10^6 t)
South Africa	251.0	Bosnia-Herzegovina	11.2
other Africa	4.4	Bulgaria	28.8
Africa total	**255.4**	Czech Republic	60.1
Canada	68.1	Germany	194.4
Mexico	11.5	Greece	65.7
United States	1061.8	Poland	144.0
other North America	0	Romania	35.2
North America total	**1141.4**	Russia	326.5
Brazil	6.6	Serbia-Montenegro	37.4
Columbia	73.5	Spain	10.2
other South America	7.3	Ukraine	59.7
South America total	**87.4**	United Kingdom	18.1
China	2782.0	other Europe	29.1
India	515.8	**Europe total**	**1020.4**
Indonesia	229.0	**Middle East total**	**2.6**
Kazakhstan	104.9	Australia	397.6
North Korea	33.4	other Oceania	4.9
Thailand	18.0	**Oceania total**	**402.5**
Turkey	78.8	**world total**	**6739.2**
Vietnam	39.8		
other Asia	27.8		
Asia total	**3829.5**		

ENERGY EXTRA 3.1
Fracking

One method of increasing the productivity of a natural gas or an oil well is referred to *fracking*, which is short for "hydraulic fracturing." By this method, introduced in the oil and gas industry in the late 1940s, rock around a well is fractured to allow natural gas or oil to flow into the well. The rock is fractured by injecting a fluid at high pressure into the well. Wells often extend vertically downward from the surface, and, once they enter a layer where natural gas or oil resources are present, they turn horizontally to access as much of the deposit as possible. In this horizontal region of the well, fracking is an effective way to increase well productivity. Once the fracture has been created, a *proppant* is injected into the well. The proppant is a material, such as sand, that fills the crack but is porous. This material allows oil or gas to flow into the well but supports the fracture to

prevent it from closing under the pressure of the rocks above. The fracking process is illustrated in the diagram in this box. The fracking technique is in common use in the natural gas industry, and it has been estimated that as many as 90% of the natural gas wells in the United States employ this method.

The liquid most commonly used for fracking is water, but it includes a number of additives to alter its properties. About 750 different chemical additives have been used in the industry, many of which are toxic or carcinogenic. A partial list of additive classes is shown in the table. The proppant itself is commonly sand, which is chosen for particular properties such as grain size and composition to optimize its effectiveness. It may also include a radioactive tracer to allow for observation of its distribution.

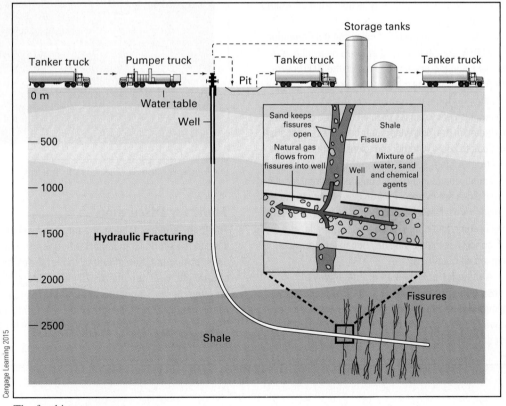

The fracking process

Continued on page 82

Energy Extra 3.1 continued

adapted from information at
http://en.wikipedia.org/wiki/Hydrolic_fracturing

A partial list of fracking fluid additives		
additive	purpose	example material
acid	improves entry of fluid into rock	hydrocloric acid
surfactant	reduces fluid surface temperature	propanol
corrosion inhibitor	reduces corrosion of well equipment	methanol
biocide	reduces bacterial growth	glutaraldehyde
scale inhibitor	inhibits precipitation of minerals from the fluid	ethylene glycol

Because of these materials, there are health and safety concerns over fracking. The fracking fluid may be returned to surface, or it may be left underground. If it is returned to the surface, it must be dealt with properly because of potential health effects. If it is left underground, it could spread and enter the drinking water supply. Also, methane (natural gas) can propagate through fractures and enter the water supply. Buildings have exploded as a result of water that contained dissolved methane (possibly from a fracked well).

A 2004 study by the U.S. Environmental Protection Agency (EPA) concluded that there was no evidence of health risks due to fracking. As a result, the 2005 Energy Policy Act states that oil companies do not have to make known their (proprietary) formulas for fracking fluids. However, this policy causes problems if toxic chemicals are detected in drinking water near drilling sites because it is difficult to know whether the toxins come from the well or from another source. Although proposals have been recently made in the U.S. Congress, no real regulatory guidelines are in place for fracking. However, due to health concerns, New York has banned fracking. Fracking is used in some (but not all) provinces in Canada, and the public concern is that its use may become more widespread in that country.

Although fracking is an effective way of increasing the productivity of wells, it remains very controversial because the health risks are unclear. Improved regulation and environmental studies are needed to alleviate public concerns over this practice.

Richard A. Dunlap

Public expression of objections to fracking.

Topic for Discussion

The long-term availability of adequate freshwater supplies in the world is an important concern. Fracking uses large quantities of water; in fact, fracking a single natural gas well can utilizes 20 million liters of water (or even more). To put this in perspective, make a rough estimate of the length of time that this amount of water could provide the freshwater supply for a typical single-family residence.

Based on data from E. L. McFarland, J. L. Hunt and J. L. Campbell. Energy, Physics and the Environment. Cengage Learning, 2007.

Table 3.11: Lifetime of traditional coal resources for different annual growth rates.				
region	0% annual growth (y)	2% annual growth (y)	5% annual growth (y)	actual growth rate 1995–2005
United States	1500	170	88	0.3%
Canada	1800	180	92	−1.5%
World	1200	160	84	2.5%

extrapolation of coal use into the future is much less certain than it is for oil or natural gas. This is because the expected lifetime for coal is much longer than for these other resources, and minor differences in growth rate will have a much greater effect. As well, our ability to foresee the needs of society further into the future becomes less certain. Table 3.11 gives estimates for the lifetime of coal resources for different growth rates for the world, the United States, and Canada. Clearly, if the growth rate is low, then coal has the potential to last for a considerable time compared with oil and natural gas. The application of the Hubbert model to the use of coal, as illustrated in Figure 3.13, shows that the use of coal has not yet reached its peak and that, in comparison to the analysis for oil (Figures 3.10 and 3.11) the peak for coal is very much broader.

3.7 Enhanced Oil Recovery

It is clear from the discussion in the previous chapter that fossil fuel resources are limited and that oil and natural gas are more limited than coal. In this section, several topics related to alternative approaches to fossil fuel use are discussed: (1) methods for extracting a larger fraction of oil from a reservoir, (2) other sources of oil, and (3) methods

Based on A. Valero and A. Valero "Physical geonomics: Combining the exergy and Hubbert peak analysis for predicting mineral resources depletion," Resources, Conservation and Recycling 54 (2010) 1074–1083.

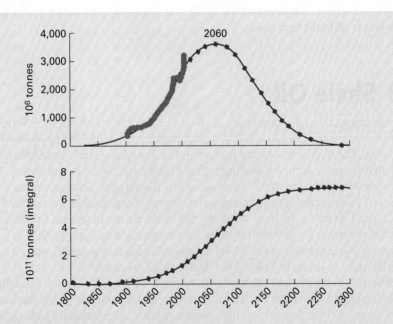

Figure 3.13: Application of the Hubbert model to the world utilization of coal. Data are plotted as the mass of oil with equivalent energy content.

of converting coal into liquid or gaseous hydrocarbons so that they may be used more conveniently for transportation needs.

Simple, or primary, oil extraction involves allowing the oil to flow from the reservoir under its own natural pressure. Typically only 15–20% of the oil in a deposit can be obtained by this method because there is insufficient pressure to force the remaining oil out of the ground.

Secondary recovery makes use of a second well drilled near the primary well into which water or gas is injected to pressurize the deposit and to force the oil out of the primary well. This method can typically recover an additional 25% or so of the oil in the reservoir and is in common use for wells in the United States. The fraction of the oil in the reservoir that can be extracted depends on the rate at which the oil is extracted. Optimizing the fraction of oil that can be recovered is not generally compatible with an extraction rate that best meets supply needs. This fraction also depends on the nature of the deposit and the nature of the oil. Although all crude oil is fairly viscous, deposits containing lighter varieties of oil allow for more efficient removal. The main difficulty in extracting all of the oil from a deposit is its viscosity and the surface tension between the oil and the porous rock where it resides.

Tertiary recovery involves methods that can effectively deal with these problems. Surfactants, which are chemicals that allow oil and water to mix (like detergents), can be injected into the well along with water, and this enables the oil to get "unstuck" from the pores in the rock. Another technique is to inject steam into a secondary well. This pressurizes the oil, forces it toward the primary well, and heats it, thereby reducing its viscosity and allowing it to flow more easily. Fireflooding is a technique where some of the oil in the deposit is actually burned *in situ*. This heats the remaining oil and forces it out of the well. This approach will be discussed further in the next section. These tertiary techniques tend to be expensive, either financially (as in the case of surfactants) or in terms of energy (as in the case of steam injection). In the latter case, the energy cost is about one-third of the energy value of the oil extracted. The future viability of these techniques is difficult to assess.

3.8 Shale Oil

Shale oil comes from a sedimentary rock (*oil shale*) and results from relatively recent (50 million years ago) hydrocarbon deposits produced from decaying aquatic life in ancient lakes. The hydrocarbons are in the form of *kerogens*. These have chain-like molecules that are more complex and intertwined than those found in petroleum. They also tend to contain more impurities, such as sulfur. Heating the kerogens breaks down the chains and yields petroleum-like molecules. It was recognized in the nineteeth century that oil could be extracted from oil shale. Oil shale occurs in several locations, and the estimated resources for different regions are summarized in Table 3.12. It is interesting to compare these numbers to the total estimated recoverable traditional oil resources, 2.5×10^{12} bbl, as given in Table 3.6. Clearly shale oil can be significant.

The actual use of shale oil has been minimal. Figure 3.14 shows shale oil production for countries that have utilized this resource. The amount of oil that can be extracted from oil shale depends on the specific deposit. A good-quality oil shale deposit would produce upward of 0.6 bbl (100 L) per tonne of shale. The processing of several tens of millions of tonnes of shale per annum would amount to less than 0.1% of traditional oil production.

Table 3.12: Estimated in-ground shale oil resources.	
location	**resources (10^9 bbl)**
Congo	100
Morocco	53
United States	3707
Brazil	82
Italy	73
Russia	248
Jordan	34
Australia	32
rest of world	457
total	**4786**

© Cengage Learning 2015

Example 3.5

Compare the energy content of 1 t of good-quality oil shale with the energy content of 1 t of bituminous coal.

Solution

From Appendix IV, the energy content of 1 L of crude oil is 38.5 MJ. Therefore, 1 t of good-quality oil shale which contains 100 L/t will yield

$$(38.5 \text{ MJ/L}) \times (100 \text{ L}) = 3.85 \times 10^3 \text{ MJ}.$$

One tonne of bituminous coal has energy content as given in Appendix IV of 3.1×10^4 MJ. Thus, good-quality oil shale has only about 12% of the energy content of coal.

Table 3.12 shows that the United States has the vast majority of oil shale resources worldwide, yet there has been virtually no commercial exploitation of these resources. As Figure 3.14 illustrates, about 80% of the global shale oil production is from Estonia. In fact, Estonia obtains about 90% of its energy from shale oil and is the only country that derives a major component of its energy from this source. The experiences in this country serve as a good case study for the possible implications of potential larger scale shale oil production elsewhere.

To assess the possibilities for shale oil use, it is important to consider the methods by which oil is extracted from oil shale. The shale is heated to about 500°C to extract the kerogens and break down the hydrocarbon chains. The resulting mix of hydrocarbons can be refined in much the same way that crude oil is refined. This process requires considerable quantities of water. In fact, using current technology, the production of 1 L of shale oil requires about 3 L of water. As well, the production of any useful quantities of oil requires the processing of considerable amounts of shale. It is also interesting to note that during the process of crushing the shale and extracting the oil, the volume of the shale actually increases by about 35%. The requirement for large quantities of water

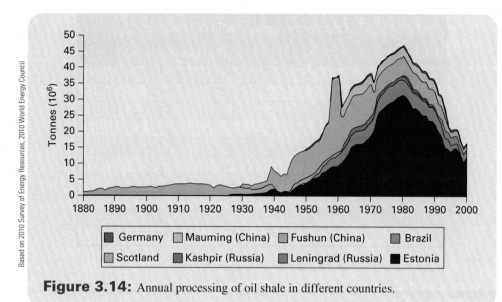

Based on 2010 Survey of Energy Resources, 2010 World Energy Council

Figure 3.14: Annual processing of oil shale in different countries.

and the need to dispose of significant waste material are factors that affect the viability of shale oil utilization. An evaluation of the desirability of shale oil production must include (at least) the following points:

- The availability of an adequate water supply
- The ability to effectively dispose of spent shale
- The possibility of adverse environmental consequences of mining the shale
- The cost, both financial and in terms of the energy used to produce shale oil

In the United States, most of the shale oil deposits occur near the boundaries between Utah, Wyoming, and Colorado. These locations are illustrated on the map in Figure 3.15. This region has relatively little rainfall, although there is an important farming industry. Water is provided for agricultural activities by the White River and the Green River. The quantity of water needed for any significant shale oil production could have severe implications for agriculture, and it is neither practical nor economical to transport the quantities of water that are needed from other locations.

The problem of spent shale disposal and restoration of the land after mining must also be considered. Because oil shale deposits are near the surface, the environmental consequences can be more severe than for, say, coal mining. The impact on the environment includes exposure of previously buried materials that may contaminate the water supply, increased erosion, and the distribution of particulates in the air during processing. Mining and the extraction of oil from shale produces potentially harmful emissions, including carbon dioxide. In this respect, it is perhaps significant that in 2002 it was estimated that 97% of the air pollution in Estonia originated from the power industry. By comparison, in 1998, electricity generation in Canada resulted in 20% of the country's SO_2 emissions and 10% of the NO_x emissions. Overall total greenhouse gas emissions from shale oil mining and processing are greater than they are for an equivalent amount of traditional fossil fuel. Because current U.S. government policies prohibit government agencies from purchasing oil that is produced by methods that

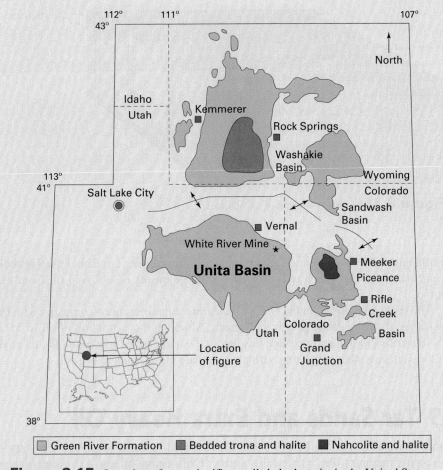

Figure 3.15: Location of most significant oil shale deposits in the United States.

release more greenhouse gases than conventional petroleum, shale oil initiatives must deal with this difficulty.

From a purely financial standpoint, the viability of shale oil production may be questionable. A study by the RAND Corporation concluded that, as of 2005, the cost of producing shale oil was in the range of $70–95 per bbl. At that time, the world market price for crude oil was about $40 per bbl. Fluctuations in oil prices (often increases), along with technological improvements and long-term amortization of infrastructure costs, may at some point make shale oil economically competitive.

An alternative method, *in situ retorting,* of extracting oil from oil shale has been considered. This process is similar to fireflooding. In this case, as illustrated in Figure 3.16, a well is drilled into the oil shale deposit, and a second well is drilled nearby. Air and natural gas are injected into the second well and burned. The heat propagates outward, heating the oil shale and driving the hydrocarbons out through the first well. Although this is a less efficient method of extracting the oil from the oil shale, it avoids

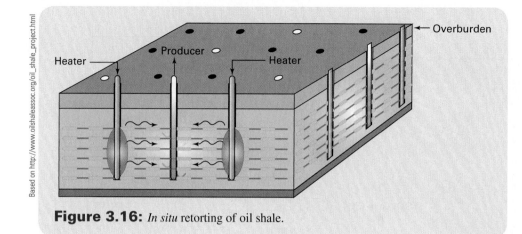

Based on http://www.oilshaleassoc.org/oil_shale_project.html

Figure 3.16: *In situ* retorting of oil shale.

the problems and costs associated with mining and transporting large quantities of rock. It is unclear if this approach will provide a suitable means of utilizing oil from this resource.

The shale oil industry in the United States has sometimes been compared with the early stages of the tar sands project in Canada. Although the extraction of oil from tar sands (described in the next section) has met with some degree of economic and environmental success, the long-term usefulness of shale oil remains to be seen.

3.9 Tar Sands and Extra Heavy Oil

Oil that is extracted from oil wells using conventional recovery methods may vary in its viscosity, ranging from light to fairly heavy. Extra-heavy oil also exists and is more difficult to extract from the earth. This extra-heavy oil is sometimes categorized in two ways: extra-heavy oil and natural bitumen (or tar sands), the latter deposits being a mixture of extra-heavy oil and sand. In both cases, the possibility of using these resources lies in our ability to extract the oil in an economical and environmentally conscious way.

The estimated world resources in extra-heavy oil and natural bitumen are summarized in Tables 3.13 and 3.14. These are each close to the estimated total oil available from traditional sources, although it may be uncertain how much of this oil is actually recoverable. It is clear from the tables that virtually all extra-heavy oil exists in Venezuela, whereas about two-thirds of the oil available from tar sands exists in Canada.

© Cengage Learning 2015

Table 3.13: Estimated extra-heavy oil resources.	
location	10^9 bbl
Venezuela	2012
rest of world	38
total	**2050**

Table 3.14: Estimated oil resources from natural bitumen (tar sands).	
location	**10^9 bbl**
Nigeria	425
Canada	1625
Kazakhstan	250
Russia	200
rest of world	56
total	**2256**

© Cengage Learning 2015

Extra-heavy oil deposits are located in the Orinoco Oil Belt in North Eastern Venezuela, as illustrated in Figure 3.17. Commercial production of extra-heavy oil from this region began in 2001. Due to the political situation in Venezuela, details of current production are somewhat unclear, although approximately 40 million L/day are being produced. Future production in excess of 100 million L/day is anticipated.

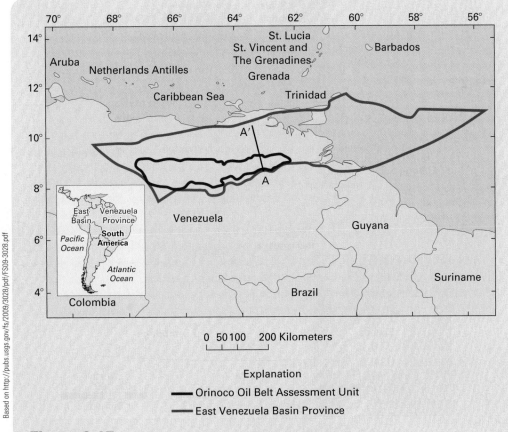

Based on http://pubs.usgs.gov/fs/2009/3028/pdf/FS09-3028.pdf

Figure 3.17: Location of the Orinoco Oil Belt in Venezuela

The existence of tar sands in Canada has been known since the eighteenth century. However, it was not until the middle of the twentieth century that interest in utilizing this resource became serious, and the first commercial production of oil from tar sands began in Alberta in 1967. The Canadian tar sands deposits occur almost exclusively in three locations in Alberta (the Athabasca Oil Sands, the Peace River Oil Sands, and the Cold Lake Oil Sands), as indicated on the map in Figure 3.18.

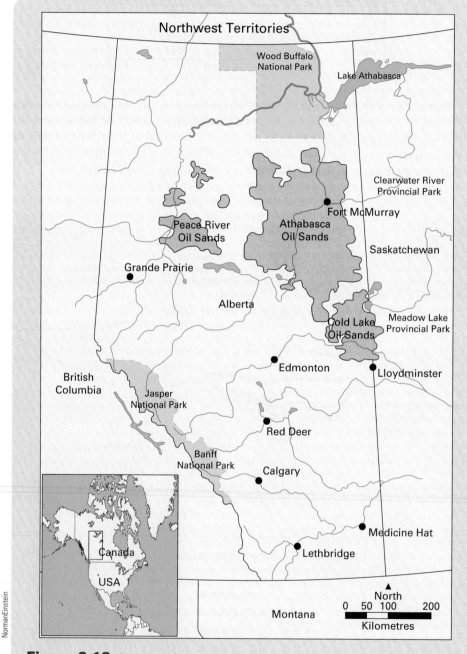

Figure 3.18: Location of tar sands oil deposits in Alberta, Canada.

Current Canadian production of oil from tar sands accounts for more than a quarter of all Canadian oil production and is expected to reach close to 500 million L/day within the next five years.

3.10 Coal Liquefaction and Gasification

Coal resources are much more plentiful than oil or natural gas resources. Although the use of liquid or gaseous fuels is convenient for transportation applications, the use of a solid fuel is somewhat more problematic. Subject to considerations discussed in the next chapter, coal is a possible means of generating electricity for quite some time. The development of suitable electric vehicle technology could allow coal to be used as a source of energy for transportation. However, the inevitable losses due to intrinsically low conversion efficiencies may be a consideration. Another alternative for the use of coal for transportation purposes involves the conversions of the carbon atoms in the coal into a gaseous or liquid hydrocarbon. These processes are referred to as coal gasification and coal liquefaction, respectively.

The basic process of coal gasification was discovered around 1780. It was commercialized in the early twentieth century and was used to produce gas for distribution to homes and industry for heating and lighting purposes. A similar process by which the carbon in wood could be converted into a combustible gas for use as an automobile fuel was used to deal with gasoline shortages during World War II (Figure 3.19). Because *coal gas* is of lower energy content and contains more impurities (and hence produces more pollution) than natural gas (which is fairly pure methane), its use was mostly discontinued after natural gas came into common use in the 1940s. In more recent years, coal gasification has not been used extensively because

Figure 3.19: Automobile manufactured by Adler with wood gas generator attached to the back (Adler Diplomat equipped with Imbert Holzgas Generator, 1941).

of the potentially low quality of the fuel produced and the comparatively low cost of traditional oil and natural gas that can be used for similar purposes. However, some commercial utilization of coal gas has taken place in South Africa since the 1960s. Current research in the development of coal gasification facilities is aimed at alleviating the problem of dwindling oil and natural gas supplies. A basic description of the coal gasification process follows.

The basic process of coal gasification involves heating the coal in a *gasifier*, which induces the following sequence of processes:

Pyrolysis is the driving off of volatile compounds from the coal, leaving a more carbon-rich material.

Combustion involves the incomplete oxidation of the carbon to produce carbon monoxide by the reaction

$$2C + O_2 \rightarrow 2CO. \tag{3.3}$$

Some carbon dioxide can also be produced, but it is advantageous to minimize this.

Gasification is the process by which the remaining carbon and the carbon monoxide are reacted with steam by the reactions

$$C + H_2O \rightarrow H_2 + CO \tag{3.4}$$

and

$$CO + H_2O \rightarrow CO_2 + H_2. \tag{3.5}$$

The resulting gas is typically about 50% CO and 50% H_2, which can be burned to produce energy by the reactions

$$2CO + O_2 \rightarrow 2CO_2 \tag{3.6}$$

and

$$2H_2 + O_2 \rightarrow 2H_2O. \tag{3.7}$$

The remaining components of the coal gas, mostly CO_2, add relatively little to the energy production. More complex processes involving catalytic reactions can be used to convert the coal gas into a fuel that is approximately 90% methane.

Coal liquefaction is an alternative approach for producing a transportation fuel from coal. Liquefaction can be by means of either a direct or an indirect process. In the direct processes, coal is ground into a fine powder and then mixed with a solvent to form a slurry containing about one-third to one-half coal. The slurry is then heated in a hydrogen atmosphere under a pressure of about 15 MPa to about 400°C for about an hour. Chemical reactions between the carbon in the coal and the hydrogen atmosphere produce a variety of liquid hydrocarbons. Catalysts can be used to improve the reaction rates. Low-quality coals, such as lignite and sub-bituminous, work best in this process, whereas very hard coals such as anthracite are mostly unreactive. Liquids produced by this process tend to have fairly low molecular masses but can be polymerized to produce fuels similar to gasoline and diesel fuel. Typically, about half the weight of the coal is converted into a liquid fuel.

Indirect liquefaction first produces a gas from the coal by a process involving the reaction of the coal with steam and oxygen at high temperature (as discussed). In the so-called *Fischer-Tropsch process*, coal gas is reacted with a catalyst to produce liquid hydrocarbons that have a wide range of molecular masses.

The direct route is generally more efficient and has been the subject of research efforts in recent years. All coal liquefaction processes release substantial quantities of CO_2, and even if they can be made financially and energetically viable, their desirability must be considered in the context of environmental factors, as discussed in the next chapter.

3.11 Summary

This chapter has reviewed the properties of fossil fuels. Oil has been utilized as a source of energy for more than 150 years and can be refined into various gaseous, liquid, and solid hydrocarbons. Fuel oil (or diesel fuel) and gasoline are the most commonly used oil products.

An analysis based on the Hubbert model suggests that annual oil production worldwide is near its maximum. It seems obvious that oil production rates will begin to decrease in the foreseeable future. This decrease has already been seen in the United States where oil reserves are much more mature than they are in the Middle East. Models suggest that within the next 50 years, traditional oil resources will be nearing the end of their availability.

Natural gas consists primarily of methane. Its use has increased in recent years due to its availability and the fact that it produces less greenhouse gas and pollution per unit energy than do other fossil fuels.

Coal resources are more extensive than oil or natural gas, and an evaluation of their future availability on the basis of the Hubbert model indicates that the longevity of coal is substantially greater than that of other fossil fuels.

Other, less traditional, fossil fuel resources include shale oil and tar sands. Although shale oil resources are extensive, particularly in the United States, the viability of this energy resource is uncertain from both an economic and environmental standpoint. Thus, commercial utilization has been minimal in the past, and the future development of this resource is unclear. Tar sands and extra-heavy oil resources are also very extensive, and their use has been successfully commercialized, particularly in Venezuela (for extra-heavy oil) and in Canada (for tar sands). These resources together may more than double the future availability of oil.

The availability of extensive coal resources, as well as the desirability of liquid and gaseous fuels for transportation use, makes the prospect of coal liquefaction or gasification attractive. The technology for these processes has been known for many years but has seen limited utilization. As long as crude oil prices remain relatively low, the economic viability of coal liquefaction or gasification is questionable.

Problems

3.1   Write down the chemical formulae for the combustion of the alkanes in Table 3.2 with $n = 2$ to $n = 8$. Determine the mass of CO_2 produced per megajoule of energy.

3.2   Locate the current world price of oil (US$/bbl) and the current retail price of regular gasoline in your area (per liter or per gallon, as appropriate). What is the markup in the price per unit energy between crude oil and retail gasoline?

3.3   Locate the current world price of oil (US$/bbl), coal (bituminous coal, per tonne), and natural gas (per 1000 m^3). For each fossil fuel, calculate the cost of energy per gigajoule assuming that the fuel can be converted to usable energy with an efficiency of 100%.

3.4   Assume that all of the United States' annual electricity requirement of 3×10^{12} kWh is produced by coal-fired generating stations operating at a net overall efficiency of 40%.
   (a)   How many tonnes of coal are burned per second? (Assume the coal is all bituminous)
   (b)   Assuming that coal is 100% carbon, how many tonnes of CO_2 will be produced each year?

3.5   On land, coal is transported primarily by train. A typical large coal train may be about a mile long and may consist of 120 cars, each holding 110 tonnes of coal. As each ton of coal has an equivalent energy content (in terms of the stored chemical energy it contains), a moving coal train represents a flow of (chemical) energy or a power. For a coal train traveling at 100 km/h, calculate the equivalent power in watts.

3.6   How many tonnes of oil shale which produces 120 L/t would be needed to produce the same energy as 1 tonne of bituminous coal?

3.7   Assume that the total energy needs of a person in the United States (Chapter 2) is satisfied by burning coal. If each person is responsible for the mining, transportation, processing, and burning of their own coal, how much coal must each person process, on average, per day?

3.8   If shale oil replaced coal in the United States, how many years would the U.S. resources last at the current rate of use?

Bibliography

G. J. Aubrecht II. *Energy: Physical, Environmental, and Social Impact* (3rd ed.). Pearson Prentice Hall, Upper Saddle River, NJ (2006).

G. Boyle, B. Everett, and J. Ramage (Eds.). *Energy Systems and Sustainability*. Oxford University Press, Oxford (2003).

K. S. Deffeyes. *Beyond Oil: The View from Hubbert's Peak*. Hill and Wang, New York (2005).

K. S. Deffeyes. *Hubbert's Peak: The Impending World Oil Shortage*. Princeton University, Princeton, NJ (2001).

M. E. Eberhart. *Feeding the Fire: The Lost History and Uncertain Future of Mankind's Energy Addiction*. Harmony Books, New York (2007).

J. A. Fay and D. S. Golomb. *Energy and the Environment* (2nd ed.). Oxford University Press, New York (2012).

R. Heinberg. *Power Down: Options and Actions for a Post-Carbon World*. New Society, Gabriola Island, Canada (2004).

R. Hinrichs and M. Kleinbach. *Energy: Its Use and the Environment* (5th ed.). Brooks-Cole, Belmont (2012).

M. King Hubbert. "Energy Resources of the Earth." *Scientific American* **225** (1971). pp. 60–70.

J. A. Kraushaar and R. A. Ristinen. *Energy and Problems of a Technical Society* (2nd ed.). Wiley, New York (1993).

M. Simmons. *Twilight in the Desert: The Coming Saudi Oil Shock and the World Economy.* Wiley, New York: (2005).

V. Smil. *Energy at the Crossroads Global Perspectives and Uncertainties.* MIT Press, Cambridge (2003).

Environmental Consequences of Fossil Fuel Use

Learning Objectives: After reading the material in Chapter 4, you should understand:

- The causes and effects of thermal pollution.
- The causes and types of chemical pollution.
- Principles of the greenhouse effect.
- The reasons for global climate change.
- Methods of carbon sequestration.

4.1 Introduction

Fossil fuel use has numerous possible environmental consequences. A specific example described in the last chapter was the geographic consequences of oil shale mining and its effects on water resources. Similar effects can be related to coal mining and oil drilling. However, this chapter concentrates primarily on three effects of fossil fuel use that are directly related to the production of energy from these fuels: thermal pollution, chemical/particulate pollution, and greenhouse gas emission.

4.2 Thermal Pollution

The production of useful energy from fossil fuels almost always involves combustion. The heat produced by burning fossil fuels is sometimes used directly for residential or commercial heating or for industrial processes. In these cases, the heat eventually finds its way into the environment. The effect of this thermal energy is most noticeable in relation to chemical pollution in urban areas as will be discussed in this chapter. In other cases, the heat produced by the combustion of fossil fuels is, at least partially, converted into mechanical energy in a heat engine. This mechanical energy is most commonly used for transportation (automobiles, trucks, etc.) or for the generation of electricity. In either case, the thermodynamic efficiency of the conversion of heat into mechanical energy is limited by the Carnot efficiency. For an automobile engine, the overall efficiency is fairly low, typically around 17%. For an electric generating station, the efficiency can be as high as 40%. This latter efficiency is similar to that obtained by nuclear power plants, as discussed in the next few chapters, because these also use heat engines to convert the thermal energy released by nuclear reactions into mechanical energy. Thus, the basic principles for heat disposal discussed in this section are applicable to nuclear power plants as well. The more efficient the facility is, the greater the fraction of energy that is converted into mechanical energy and the less the fraction of energy that is transferred into the environment as waste heat. Therefore, efficiency factors, such as those discussed in Chapter 1, are important to consider.

The traditional method of disposing of waste heat from a fossil fuel or nuclear generating station is to use a body of water as the cold reservoir (Figure 1.7). In the simple case, *once-through cooling* is used; cold water is pumped from the body of water, heated in a heat exchanger, and then transferred back into the cooling reservoir. The use of water as the cold reservoir for a heat engine has several attractive advantages:

- Water has a high specific heat, so it can contain a substantial quantity of thermal energy with a minimal increase in temperature.
- Water has a high thermal conductivity, so the transfer of heat to the water is efficient.
- Large reservoirs (such as the oceans) or somewhat smaller ones (such as lakes or rivers) are readily available in many parts of the world.
- The temperature of the water remains relatively stable (in comparison to the temperature of the air) as a function of time of the year.
- The infrastructure required for the use of water as the cold reservoir for a heat engine is simple, straightforward, and relatively inexpensive.

However, the release of heat into a body of water can have significant undesirable environmental effects. This heat release can change the ecology in rivers and more particularly in lakes or enclosed ocean bays. This excess heat is important not only in terms of the possible increase in the average temperature of the body of water but also in terms of the vertical distribution of heat (i.e., the temperature profile) in the water. In a lake, the surface layer consists of warmer water with a layer of cooler, denser water below. The oxygen content of the cooler water is greater. The details of the effects of introducing waste heat into a body of water are complex, but the ecology of the lake can be affected in three clear ways:

- Changes in oxygen content resulting from temperature changes can affect biological processes in organisms in the lake.
- Changes in temperature can affect chemical reaction rates.
- Changes in the temperature profile can affect the natural seasonal mechanisms that mix the water in the lake.

These factors can have profound effects on the ecology of the region and have been a major factor in promoting the use of the atmosphere as a cold reservoir for electricity generating stations. In fact, in some countries, regulations have required the implementation of air cooling for new power plant construction. Air cooling uses cooling towers (Figure 1.10) to transfer waste heat to the atmosphere. There are two major designs of cooling towers: wet cooling towers and dry cooling towers. *Wet cooling towers* use excess heat to evaporate water. The latent heat of vaporization of the water is responsible for carrying away the excess waste heat. In *dry cooling towers*, hot water (carrying the waste heat) is circulated through pipes that are designed to effectively transfer heat to the atmosphere. This is basically like the design of a radiator for an automobile engine. For both methods, air may be circulated through the tower either by natural convection or by fans.

The use of cooling towers for disposing of waste heat is not without environmental effects. The changes in temperature and humidity around cooling towers (particularly wet cooling towers) can cause a localized region of increased fog and precipitation. Dry cooling towers are the preferred method of waste heat disposal from an environmental (if not economic) perspective. In all cases, there is a trade-off involving construction costs, operational costs, efficiency, and environmental effects.

4.3 Chemical and Particulate Pollution

Chemical and particulate pollution from burning fossil fuels can fall into several categories (depending on the chemical involved):

- Carbon monoxide (CO)
- Nitrogen-oxygen compounds (NO_x)
- Hydrocarbons
- Sulfur dioxide (SO_2)
- Particulates

The formation and significance of each are now discussed.

4.3a Carbon Monoxide

Carbon monoxide is produced by the incomplete oxidation of carbon during combustion. This corresponds to the chemical reaction

$$2C + O_2 \rightarrow 2CO. \tag{4.1}$$

This reaction typically results when insufficient oxygen is present to produce CO_2. In the United States, about 10^{11} kg of CO are released to the atmosphere every year as a

---

Example 4.1

If all the CO produced in the United States from the burning of fossil fuels was converted into CO_2 before its release, what would be the mass of this additional CO_2? Compare this with the approximately 3×10^{12} kg of CO_2 that is released directly in the United States per year.

Solution

It is stated in the text that fossil fuel burning in the United States releases 10^{11} kg CO annually.

If this is converted into CO_2, the mass of the CO_2 is given in terms of the molecular masses of CO (28 g/mol) and CO_2 (44 g/mol) as

$$\frac{10^{11} \text{kg} \times 44 \text{ g/mol}}{28 \text{ g/mol}} = 1.57 \times 10^{11} \text{ kg}.$$

This represents about 5% of the total CO_2 emissions. Compare this with CO_2 produced directly from coal burning in the United States as presented in Example 4.4.

Table 4.1: Sources of CO released to the atmosphere as a result of fossil fuel use.

source	% CO released
vehicles	60
industrial	10
waste disposal	8
agricultural burning	8
forest fires	8
other	6

© Cengage Learning 2015

Table 4.2: U.S. emission standards for CO, NO_x and hydrocarbons in grams per kilometer.

year	CO	NO_x	HC
1960 (precontrol estimate)	52.2	2.5	6.58
1970	21.1	3.1	2.5
1975	9.3	1.9	0.93
1980	4.3	1.2	0.25
1981	2.1	0.62	0.25
1983	2.1	0.62	0.25
1994	2.1	0.25	0.16
2001	2.1	0.12	0.078
2004	1.1	0.04	0.056

© Cengage Learning 2015

result of burning fossil fuels. A rough breakdown of the sources of CO in the United States is shown in Table 4.1.

Carbon monoxide is an important factor in pollution because of its toxic properties, which result from its particular reactivity with hemoglobin. CO is about 200 times as reactive with hemoglobin as O_2. This means that inhaled CO tends to displace O_2 that is dissolved in the blood, leading to adverse health effects. Typically, exposure for an hour or so to different concentrations of CO in air results in the following symptoms:

- 100 ppm: headache/confusion
- 300 ppm: nausea/unconsciousness
- 600 ppm: death

The long-term effects of low-level CO exposure are not clearly known. Because the major source of CO is from vehicles, much progress has been made in recent years in reducing the CO emissions from gasoline engines. CO and some other emissions, as specified by U.S. standards for automobiles, are shown in Table 4.2. The implementation of emission control standards has clearly made a significant reduction in CO emission, but this is to some extent offset by the greater number of vehicles on the road and the greater number of kilometers driven per vehicle per year.

Example 4.2

There about 250 million personal motor vehicles in North America. If each vehicle is driven an average of 10,000 km per year, what is the total reduction in CO emissions per year between precontrol vehicles and vehicles that meet the 2004 U.S. emission standard?

Solution

The difference between precontrol and 2004 standard vehicles is

$$52.2 \text{ kg/km} - 1.1 \text{ kg/km} = 51.1 \text{ g (CO) per km}.$$

For 250 million vehicles, each driving 25,000 km, the total annual CO reduction is

$$(0.0511 \text{ kg/km}) \times (2.5 \times 10^8 \text{ vehicles}) \times (10,000 \text{ km/vehicle}) = 1.3 \times 10^{11} \text{ kg}.$$

CO levels in urban areas have decreased during the years since emission control standards were in place. There are clear fluctuations during the day that depend on transportation patterns. Figure 4.1 shows average CO concentrations as a function of the time of day in Denver, Colorado. A clear correlation between vehicle activity in the morning and afternoon is seen, and the dispersion of CO into the upper atmosphere is seen on a timescale of a few hours. The graph also illustrates the progress that has been made in reducing vehicular CO emissions.

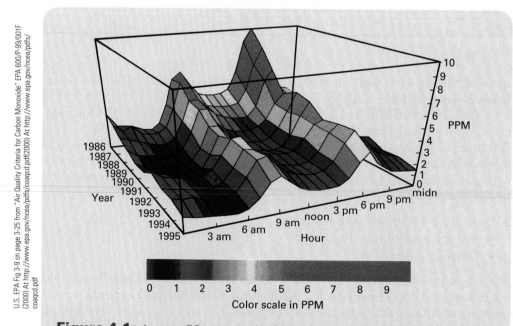

U.S. EPA Fig 3-9 on page 3-25 from "Air Quality Criteria for Carbon Monoxide" EPA 600/P-99/001F (2000) At http://www.epa.gov/ncea/pdfs/coaqcd.pdf(2000) At http://www.epa.gov/ncea/pdfs/coaqcd.pdf

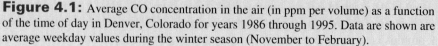

Figure 4.1: Average CO concentration in the air (in ppm per volume) as a function of the time of day in Denver, Colorado for years 1986 through 1995. Data are shown are average weekday values during the winter season (November to February).

Table 4.3: Typical NO$_x$ emission sources in the United States..	
source	% NO$_x$
vehicles	35
coal burning	20
natural gas burning	25
other	20

© Cengage Learning 2015

4.3b Nitrogen Oxides

Nitrogen oxides are formed during the combustion of fossil fuels (or any other materials) when the temperature is sufficiently high (above about 1100°C). This is because nitrogen in the atmosphere becomes oxidized by the reaction

$$N_2 + O_2 \rightarrow 2NO. \tag{4.2}$$

The nitric oxide (NO) formed by this reaction is a colorless gas that is mildly toxic to humans. The NO itself is not the most significant aspect of pollution from nitrogen oxygen compounds. On the timescale of a few hours, the NO reacts with ozone (O_3, which is reasonably plentiful in the atmosphere) to form nitrogen dioxide by the reaction

$$NO + O_3 \rightarrow NO_2 + O_2 \tag{4.3}$$

NO_2 is a highly toxic brown gas. It is responsible for the brownish color of smog that accumulates in urban areas. At low concentrations, it is an irritant causing respiratory and eye problems. At higher concentrations, it causes lung, liver, and heart damage. Because NO readily converts to NO_2 in the atmosphere, these species are generally lumped together as NO_x. A rough breakdown of the sources of NO_x is shown in Table 4.3.

The reduction of vehicle emissions of NO_x as a result of U.S. emission standards is indicated in Table 4.2. Daily fluctuations of NO_x levels in an urban area as a result of daily variation in human activity is clearly indicated in Figure 4.2. The conversion of

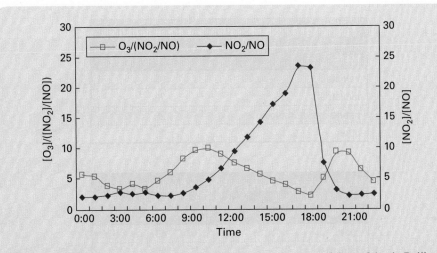

Based on Wen-xing Wang, Fa-he Chai, Kai Zhang, Shu-lan Wang, Yi-zhen Chen, Xue-zhong Wang and Ya-qin Yang "Study on ambient air quality in Beijing for the summer 2008 Olympic Games" Air Qual Atmos Health 1 (2008) 31–36

Figure 4.2: $O_3/(NO_2/NO)$ and NO_2/NO ratios as a function of time of day in Beijing.

NO to NO_2 by equation (4.3) is seen by the increase in NO_2 levels at the expense of the decrease in O_3 levels during the day.

4.3c Hydrocarbons

Hydrocarbons (HC) are released during the burning of fossil fuels as a result of incomplete combustion. This is a respiratory and eye irritant and, in higher concentrations, can cause lung disease. A rough breakdown of the sources of hydrocarbon emissions is given in Table 4.4. The reduction in U.S. hydrocarbon emissions from vehicles that has resulted from emission control standards is shown in Table 4.2.

4.3d Sulfur Dioxide

Sulfur dioxide (SO_2) results primarily from the combustion of sulfur-containing fuels and to a lesser extent from industrial processes. A breakdown of SO_2 sources is shown in Table 4.5. Coal, as shown, is the major source of SO_2 because all coal contains some concentration of sulfur, and this is released as SO_2 during combustion. Other fossil fuels, such as oil and natural gas, contain smaller concentrations of sulfur impurities, and these impurities can largely be eliminated before the fuel is used.

SO_2 is a respiratory and eye irritant but also has adverse environmental effects for another reason. It reacts with oxygen in the atmosphere by the following reaction:

$$2SO_2 + O_2 \rightarrow 2SO_3. \tag{4.4}$$

This is followed by a reaction with water vapor to produce sulfuric acid:

$$SO_3 + H_2O \rightarrow H_2SO_4 + O_2. \tag{4.5}$$

The formation of sulfuric acid in the atmosphere results in so-called *acid rain*. This has particular adverse effects on buildings and painted surfaces (e.g., automobiles). The emission of SO_2 from coal generating stations can be reduced by reacting the exhaust gas with CaO or $CaCO_3$, in a device known as a *scrubber*, prior to releasing it to the atmosphere.

Table 4.4: Sources of hydrocarbon pollution.

source	% HC
vehicles	35
industrial	20
natural gas burning	25
other	20

© Cengage Learning 2015

Table 4.5: Sources of SO_2 pollution.

source	% SO_2
coal	65
industrial	25
other	10

© Cengage Learning 2015

4.3e **Particulates**

Particulates are basically dust particles that appear as smoke during the burning of some fossil fuels (or wood). They also result from some industrial processes and natural processes such as forest fires and volcanoes. Particulates are relatively unimportant for the combustion of oil, gasoline, and natural gas, but they are much more significant during the combustion of coal. Particulates act primarily as a respiratory irritant, although, because their chemical composition is uncertain, particulates that contain toxic compounds can have more serious health implications. Because most coal that is used for energy production is burned in coal-fired electric generating stations, it is relatively easy to deal with this type of pollution, and most coal-fired stations utilize systems for reducing particulate pollution. These systems fall into two major categories; *mechanical filters (baghouses)* and *electrostatic precipitators*. The former are basically filters that remove particles in the exhaust gas. The filters are shaken to collect the particles in a container (Figure 4.3). An electrostatic precipitator (Figure 4.4) uses electrostatically charged plates to collect the particles from the exhaust.

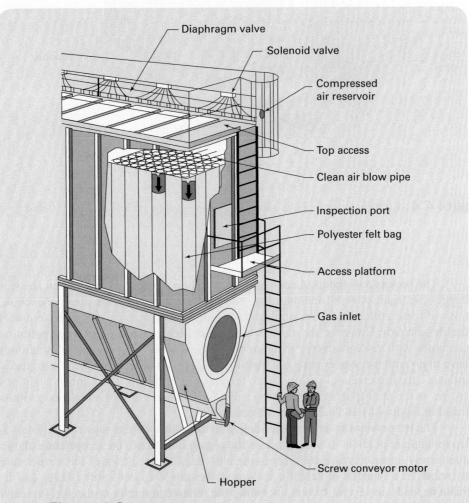

Based on http://www.neundorfer.com/knowledge_base/baghouse_fabric_filters.aspx

Diaphragm valve
Solenoid valve
Compressed air reservoir
Top access
Clean air blow pipe
Inspection port
Polyester felt bag
Access platform
Gas inlet
Screw conveyor motor
Hopper

Figure 4.3: Baghouse with filters for particulate removal.

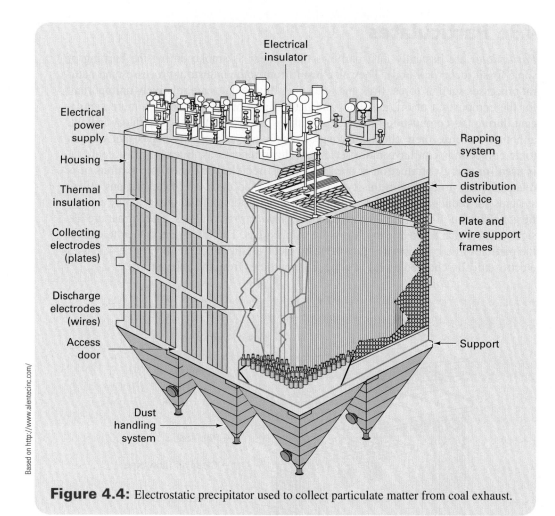

Based on http://www.alentecinc.com/

Figure 4.4: Electrostatic precipitator used to collect particulate matter from coal exhaust.

The presence of chemical and particulate pollution, particularly in urban areas, is largely the result of fossil fuel burning. The health implications are clear; concentrations of these pollutants in the atmosphere are often near or even exceed government health standards. However, the most serious health problems can occur when pollution combines with certain adverse meteorological conditions. These weather conditions have to do with the temperature as a function of altitude in the atmosphere and greatly influence the dispersion of pollution into the upper atmosphere. The normal conditions present in the atmosphere that are responsible for the dispersion of pollutants, as illustrated in Figures 4.1 and 4.2, will be discussed first.

The temperature as a function of height in the atmosphere is shown in Figure 4.5. Up to a height of about 10 km, which is the region of interest, the temperature shows a relatively linear decrease with increasing altitude. The rate at which the temperature decreases as a function of height is known as the *adiabatic lapse rate* (ALR), and its value depends on several factors, including the relative humidity of the air. Typically the ALR is of the order of about 10°C/km or 0.01°C/m.

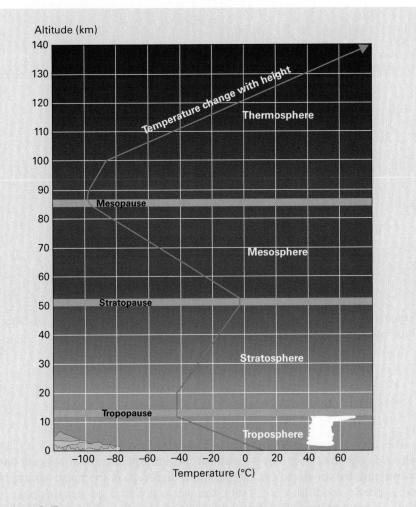

Figure 4.5: Height variations of the temperature of the atmosphere.

NOAA, http://www.srh.noaa.gov/jetstream/atmos/atmprofile.htm

To describe the effects of the ALR on air movement, let us consider a parcel of air with a volume, V, at the earth's surface. The pressure, P, and temperature, T, are determined by its equilibrium condition with the surrounding air. If we lift this parcel of air, then, as it rises, it will cool at the ALR in order to remain in equilibrium. It will also expand due to the decreasing atmospheric pressure. Under certain meteorological conditions, the actual temperature of the atmosphere may decrease either more quickly or more slowly than the ALR (Figure 4.6). If the actual temperature decreases faster than the ALR, then a parcel of air that is raised from the surface and cools at the ALR will be warmer than the surrounding air. This warmer air will tend to rise and will be replaced by cooler air from above. This promotes mixing and is an unstable condition. If, on the other hand, the actual atmospheric temperature decreases more slowly than the ALR, then a parcel of air that is raised above the surface of the earth will be cooler than its surroundings and will tend to sink back to the surface. This inhibits mixing and is a stable condition.

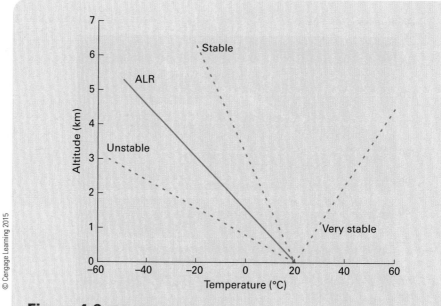

Figure 4.6: Examples of temperature variations in the atmosphere that are faster (unstable) or slower (stable) than the ALR.

This analysis applies to the warm air released from the exhaust of, say, an automobile or a coal-fired generating station. If the atmospheric conditions are unstable, the warm polluted air rises, and the pollution disperses in the upper atmosphere. If the atmospheric conditions are stable, the warm polluted air remains near ground level. In an extreme case, the temperature near the earth may actually increase as a function of height (up to some altitude). This is a very stable condition and is referred to as a *temperature inversion*. This situation is not uncommon because, during the normal daily warming and cooling cycle, the ground cools faster than the air in the evening, giving rise to a cooler region near the earth's surface. This temperature inversion normally lasts for a few hours and is a contributing factor in trapping pollution near the earth's surface in urban areas. This situation is the reason that exhaust from factories and power plants (which is warm, polluted air) is released from chimneys (smokestacks) above ground level so that it is above any possible near-ground temperature inversions. (Smelter chimneys are often close to the height of the Empire State Building.)

In rare cases, meteorological conditions cause a temperature inversion that may last for several days, trapping pollutants near ground level. This can have serious health effects. A well-known situation of this type occurred in London in 1952, where a temperature inversion lasted for an extended period of time. The correlation between the death rate (above a normal rate of about 250 individuals per day) and the level of pollution is clearly shown in Figure 4.7. The most recent analysis of this occurrence indicates that the excess number of deaths that resulted from the presence of pollution was around 12,000.

ENERGY EXTRA 4.1
Oil spills

An oil spill is the release of oil (or other liquid petroleum product) into the environment. The terminology is generally used to indicate a spill that occurs as a result of human activity and does not include natural seepage of oil from underground. We generally think of oil spills as those that result from seagoing oil tankers or offshore oil drilling activities. Oil spills, however, also include spills from terrestrial oil wells. Spills are generally accidental, but several notable intentional spills have occurred during wartime. Oil spills have a significant negative impact on the environment. Fish, birds, marine mammals, invertebrates, turtles, and plant life are all at considerable risk directly from the toxic properties of petroleum and indirectly as a result of overall ecological changes.

Economic consequences of oil spills occur at a variety of levels.

The largest accidental oil spills are shown in the table, which does not include intentional spills during the 1991 Persian Gulf War. These intentional spills included damaged oil wells that either gushed oil onto the ground or were set ablaze, as well as oil intentionally discharged from oil terminals or tankers into the ocean. The total oil introduced into the environment from these events has been estimated to be as much as 2 billion barrels (more than 200 times the largest accidental spill). The well-known *Exxon Valdez* disaster in 1989 released between 260,000 and 750,000 bbl of oil and is the largest marine spill in the United States except for the *Deepwater Horizon* accident.

The largest accidental oil spills (>2,000,000 bbl).

name	type	location/country	year(s)	bbls (approx)
Lakeview Gusher	terrestrial well	California, USA	1910–1911	9,000,000
Deepwater Horizon	drilling platform	Gulf of Mexico, USA	2010	4,500,000
Ixtox I	drilling rig	Gulf of Mexico, Mexico	1979–1980	3,400,000
Atlantic Express/Agean Captain	tankers	Trinidad and Tobago	1979	2,105,000
Fergana Valley	terrestrial well	Uzbekistan	1992	2,090,000

It is sometimes difficult to determine the volume of oil released, particularly for marine accidents. The volume of oil from a tanker is, of course, limited by the volume of the tanker. The *Atlantic Express/Agean Captain* disaster is the most significant tanker disaster and actually involved the collision of two oil tankers, both of which contributed to the spill. The oil released from an offshore well is more difficult to determine but can be estimated in several ways. One approach is to consider the appearance of the oil on the surface of the ocean. Different thicknesses of oil have different optical properties, and, of course, different thicknesses represent different volumes of oil per unit surface area. The next table summarizes how the appearance of the oil can be used to estimate volume per surface area. It is possible to estimate the surface area covered with oil from aerial or satellite imagery (see the photograph of the *Deepwater Horizon* spill as observed from NASA's Terra satellite) and therefore to obtain the total volume of oil. This can result in an underestimate of the amount of oil released because it measures only the oil that is floating on the water's surface.

Thickness and appearance of oil on the surface of water.

thickness (nm)	quantity (bbl/km^2)	appearance
38	0.233	barely visible
76	0.459	silvery, without color
150	0.944	faint color
300	1.82	bright bands of color
1000	6.10	dull color
2000	12.3	much darker color

Continued on page 108

Energy Extra 4.1 continued

The *Deepwater Horizon* oil spill as viewed from space on May 24, 2010. The extent of the spill is shown in the insert map.

Topic for Discussion

If an oil tanker spilled 1 million bbl of oil in the ocean and the oil formed a uniform layer 500 nm thick, how large an area would be covered?

4.4 The Greenhouse Effect

Although pollution resulting from the use of fossil fuels is a serious problem, the release of greenhouse gases causes more global and long lasting adverse effects. A consideration of the effects of greenhouse gases begins with an analysis of the equilibrium temperature of the earth. Virtually all the energy that establishes the equilibrium temperature of the earth comes as radiation from the sun. Geothermal energy coming from the interior of the earth accounts for only about 0.1% of the total heat and can be ignored in the present analysis.

The simplest analysis considers a planet without atmosphere and an incident radiation from the sun of S [in watts per square meter (W/m^2)]. Radiative energy from the sun that is incident on the planet is either absorbed by the planet or is reflected back into space. The fraction of the incident radiation that is reflected by a planet is called the *albedo*, a, so the fraction that is absorbed is $(1 - a)$. The absorbed energy heats the surface, and this warm surface radiates energy out into space. When the absorbed power

Based on J. Fenger. "Air pollution in the last 50 years–from local to global." Atmospheric Environment 43 (2009) pp. 13–22

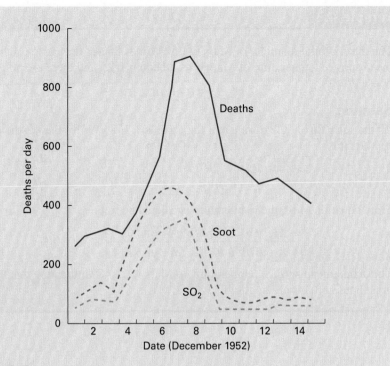

Figure 4.7: Correlation between measured pollutants and death rate in London in 1952.

(i.e., the incident energy per unit time) and the radiated power are equal, then the temperature of the planet achieves equilibrium. From the viewpoint of the sun, the planet appears as a disk with an area of πR^2, where R is the radius of the planet. Thus the solar power absorbed by the planet is

$$P_{\text{absorbed}} = (1 - a)S\,\pi R^2. \tag{4.6}$$

The power radiated into space by the warm surface is given as a function of the surface temperature, T, by the Stefan-Boltzmann law as

$$P_{\text{radiated}} = 4\pi R^2 \sigma T^4. \tag{4.7}$$

Here σ is the Stefan-Boltzmann constant (5.67051×10^{-8} W·m^{-2}·K^{-4}), and $4\pi R^2$ is the total surface area of the planet. (The physics of this process will be discussed further in Chapter 8.) Equating (4.6) and (4.7) and solving for temperature gives

$$T = \left[\frac{(1 - \alpha)S}{4\sigma} \right]^{1/4}. \tag{4.8}$$

It is important in this analysis to use temperatures that are measured on an absolute temperature scale (e.g., Kelvin). The solar flux at the outside of the earth's atmosphere is, 1.367 kW/m^2, and the earth's albedo is about 0.3, so equation (4.8) gives the equilibrium temperature of the surface of the earth (without atmosphere) of 254 K (or about $-19°$C). Clearly this value is too low in terms of our current climate and in terms of the possibility of the evolution of water-based life on earth. What is missing from this approach is the fact that the earth has an atmosphere.

Example 4.3

The mean orbital diameter of Mars is 1.523 times that of the earth. Calculate the mean surface temperature of Mars if its albedo is 0.25 and there are no greenhouse effects.

Solution

As the solar flux at the distance of the earth is 1.367 kW/m^2, then at the distance of Mars it is

$$\frac{1.367 \text{ kW/m}^2}{(1.523)^2} = 0.589 \text{ kW/m}^2.$$

From equation (4.8), the temperature is calculated to be

$$\left[\frac{(1 - 0.25) \times (589 \text{ W} \cdot \text{m}^{-2})}{4 \times 5.67051 \times 10^{-8} \text{ W} \cdot \text{m}^2 \cdot \text{K}^{-4}}\right]^{1/4} = 210 \text{ K}$$

The actual mean surface temperature on Mars is 210 K, indicating that the greenhouse effects are negligible.

When an atmosphere is present, the problem becomes somewhat more complex. Incident radiation from the sun may be reflected, absorbed, or transmitted by the atmosphere and/or reflected or absorbed by the surface of the earth. More importantly, radiation emitted by the planet may be reflected, absorbed, or transmitted by the atmosphere. To properly analyze the effects of the atmosphere, it is necessary to consider the wavelength of the radiation involved. The wavelength of the radiation emitted by a body is related to its temperature (see Chapter 8). The radiation that is incident on the earth is produced by the surface of the sun, which is at about 6000 K. This radiation has a relatively short wavelength, and much of it is in the visible part of the spectrum (see Chapters 1 and 8). The radiation emitted from the surface of the earth comes from a body that is at about 300 K. This radiation has a relatively long wavelength and is in the infrared part of the spectrum. The transparency of the atmosphere is a function of the wavelength of the radiation involved. Certain molecular species that exist in the atmosphere do not interact significantly with short-wavelength radiation but readily absorb long-wavelength radiation. Thus, these molecules allow the short-wavelength sunlight to pass through the atmosphere easily and arrive at the surface of the earth but prevent some portion of the long-wavelength radiation emitted by the surface of the earth from escaping back into space. To maintain the equilibrium condition, requiring equations (4.6) and (4.7) to be equal, the temperature, T, in equation (4.7) must be greater. This means that the temperature of the surface of the earth will be higher than that predicted by the simple model that excludes the atmosphere. Thus, heat is trapped by the atmosphere, giving rise to the so-called *greenhouse effect* and resulting in an average temperature for the earth that is increased by more than 30 K.

Certain molecular species in the atmosphere are most effective at absorbing infrared radiation than others, and these are the major contributors to the greenhouse effect (Table 4.6).

Table 4.6: Greenhouse gases in the earth's atmosphere. The relative absorption is normalized to the absorption per molecule for CO_2. Radiative forcing is a measure of the overall effectiveness of a particular gas at altering the energy balance in the atmosphere, that is, its ability to contribute to global warming. Data are shown for the most abundant chlorofluorocarbon (CFC), CCl_2F_2.

molecular species	approximate current concentration in atmosphere	relative infrared absorption per molecule	radiative forcing (W/m^2)
carbon dioxide (CO_2)	390 ppm	1	1.85
methane (CH_4)	1.8 ppm	25	0.51
nitrous oxide (N_2O)	320 ppb	298	0.18
CFC (CCl_2F_2)	530 ppt	10,900	0.17

© Cengage Learning 2015

Carbon dioxide and methane are both natural components of our atmosphere that are produced by natural biological processes. Also, carbon dioxide results from the complete combustion of any carbon containing fuel by the reaction

$$C + O_2 \rightarrow CO_2. \tag{4.9}$$

Nitrous oxide (sometimes referred to as laughing gas because of its medical use) is largely the result of the agricultural use of certain fertilizers. Chlorofluorocarbons come exclusively from human activity (industry, etc.) but are not directly related to the production of energy from fossil fuels. The concentrations of the various greenhouse gases determine the equilibrium temperature of the surface of the earth. Because the production of energy by burning fossil fuels produces CO_2, it is important to consider the relationship (if any) between human activities and the earth's temperature.

Example 4.4

According to Figure 2.6, 21.97 EJ of energy from coal combustion are used per year in the United States. What is the mass of CO_2 released?

Solution
From equation (2.10), the energy release from the combustion of carbon is

$$C + O_2 \rightarrow CO_2 + 32.8 \text{ MJ/kg}.$$

Thus 21.97 EJ of energy represent the combustion of

$$\frac{2.197 \times 10^{13} \text{ MJ}}{32.8 \text{ MJ/kg}} = 6.70 \times 10^{11} \text{ kg of carbon.}$$

Since the molecular mass of carbon is 12 and the molecular mass of CO_2 is 44 g/mol, the combustion of this amount of carbon releases

$$\frac{(6.70 \times 10^{11} \text{ kg}) \times (44 \text{ g/mol})}{12 \text{ g/mol}} = 2.46 \times 10^{12} \text{ kg of } CO_2.$$

4.5 Climate Change

The earth's temperature and the concentration of CO_2 in the earth's atmosphere in the past can be determined from an analysis of air bubbles trapped in the ice in Greenland and in the Antarctic. The results of some of these studies are shown in Figure 4.8. These data indicate a clear correlation between atmospheric CO_2 concentration and the earth's temperature and show the natural cyclic trends over the past half million years or so of the earth's history.

To look at more recent trends in atmospheric CO_2, we can include the results from direct measurements. Figure 4.9 shows CO_2 levels for the past 1000 years as determined by ice core and direct atmospheric measurements. More detailed direct measurements over the past 50 years or so are shown in Figure 4.10. These figures indicate that, in the past century or so, the CO_2 concentration has risen substantially above even the highest values during the past half million years, and the general continuing upward

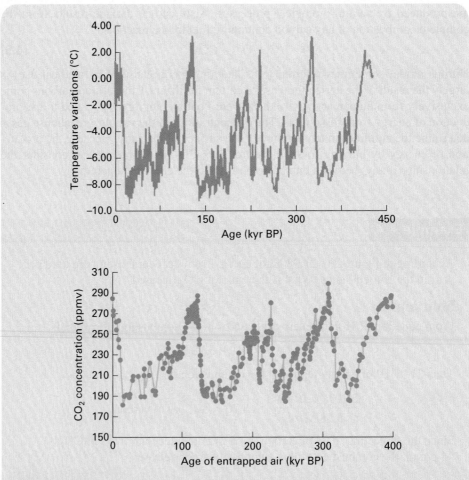

Based on http://cdiac.oml.gov/trends/temp/vostok/graphics/graphics/tempplot5.gif and http://cdiac.oml.gov/trends/co2/graphics/vostok.co2.gif

Figure 4.8: Measured temperature fluctuations and CO_2 concentrations as measured from ice core data at Vostok, Antarctica in 1000s of years before present (kyr BP).

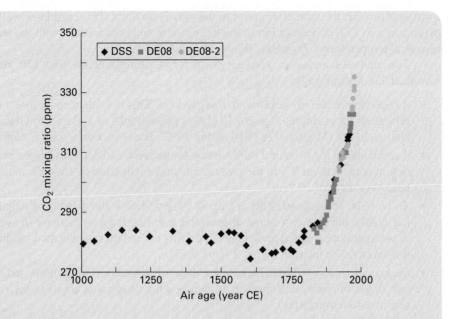

Based on http://cdiac.ornl.gov/trends/co2/graphics/lawdome.gif

Figure 4.9: Atmospheric CO_2 concentration for the past 1000 years, taken at Law Dome, Antarctica.

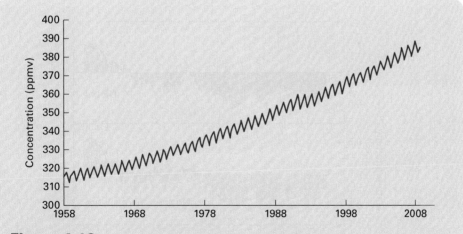

Based on http://cdiac.ornl.gov/trends/co2/graphics/ SIOMLOINSTITUTHRU2008.JPG

Figure 4.10: Measured atmospheric CO_2 concentration at Mauna Loa Observatory in Hawaii. The annual oscillations are due to fluctuations resulting from the seasonal growth of vegetation (which absorbs CO_2).

trend is clear. Compare the recent value of about 390 ppm in Figure 4.10 with the peak values in Figure 4.8 of about 290 ppm.

It is important to try to correlate these changes in CO_2 concentration over the past century or so with changes in the earth's temperature. Figure 4.11 shows the measured mean temperature variations for the past 150 years. Although there is much scatter from year to year, the increasing trend in the data is obvious. This correlates well with the increase in CO_2 concentration in the atmosphere and is consistent with the general historical temperature–CO_2 trends shown in Figure 4.8.

The gradual increase in average world temperature over the past 150 years is manifested in several ways:

- *A reduction in the size and number of glaciers:* This is evidenced by direct observation; for example, Figure 4.12 shows photographs of a glacier in Glacier National Park (Montana) in 1910 and in 1997. The reduction in size is clear.

- *A reduction in the area and thickness of Arctic sea ice:* There has been a reduction in area of about 9% in the past decade and a reduction of 15–40% in thickness over the past 30 years.

- *An increase in sea level:* This is a result of two factors: the melting of Antarctic ice and the increase in volume of seawater resulting from an increase in average sea temperature (related to changes in density). The latter is indicated by direct sea temperature measurements (Figure 4.13).

- *Biological changes:* These include the dates of annual bird migrations and certain stages of plant development, as well as color changes in corals (which are sensitive to temperature).

- *Increased geographical ranges of certain plants and animals.*

- *Thawing of the permafrost in the Arctic.*

- *Weather changes, such as more frequent El Niño events.*

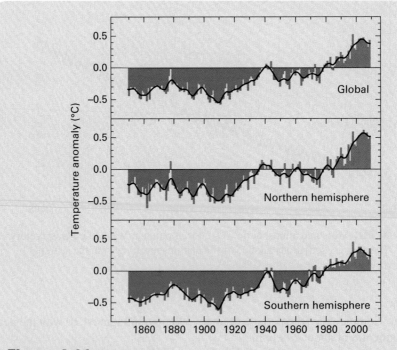

Figure 4.11: Mean global temperature measured relative to the average for the period 1850–2009.

Figure 4.12: Reduction in the size of the Grinnell Glacier in Glacier National Park between 1900 (top) and 2008 (bottom).

The relationship between average world temperature and CO_2 concentration (and that of other greenhouse gases as well) seems clear. The general scientific consensus is that these changes are linked to human activity, specifically to the release of CO_2 into the atmosphere as a result of fossil fuel burning. Other anthropogenic factors such as deforestation have also been cited as contributing to global warming.

The consequences of global warming can be profound. If a continuation of these trends can be expected, increases in CO_2 levels will yield further increases in average global temperature. Predictions of the severity of these effects depend on a number of

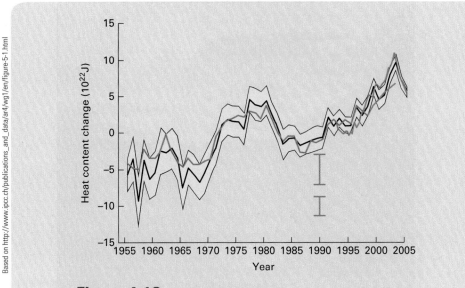

Based on http://www.ipcc.ch/publications_and_data/ar4/wg1/en/figure-5-1.html

Figure 4.13: Global annual ocean heat content since 1955.

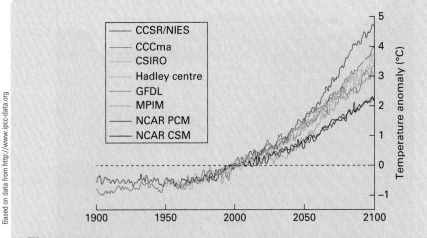

Based on data from http://www.ipcc-data.org

Figure 4.14: Predicted average world temperature increase as a result of CO_2 emissions based on various models.

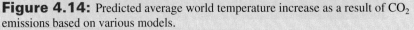

parameters but are typically in the range of a 2–5°C increase in average global temperature over the next century (Figure 4.14).

4.6 Carbon Sequestration

Because the combustion of all carbon based fuels releases CO_2, a reduction in fossil fuel use will result in a reduction in CO_2 emissions. The consequences of this action for changes that have resulted from fossil fuel use are not entirely clear because the

ENERGY EXTRA 4.2
Life cycle assessment

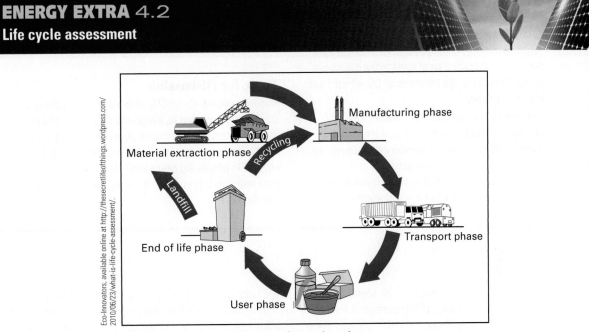

Eco-Innovators, available online at http://thesecretlifeofthings.wordpress.com/2010/06/23/what-is-life-cycle-assessment/.

Phases in the life cycle of a manufactured product.

In Chapter 3, it was seen that an energy value is associated with all products that humans use. A more thorough analysis of the environmental impact of a product throughout its life is referred to as *life cycle assessment, life cycle analysis,* or sometimes *cradle-to-grave analysis.* This type of analysis considers all the ways in which the production, use, and disposal of an item influences our environment. As shown in the figure, the life of a product may be divided into several phases. The process begins with the extraction of raw materials from nature. These natural resources are then processed and manufactured into a product that is transported to the user. At the end of the useful life of the product, it is either discarded as waste or recycled (or in a few cases refurbished, reused, etc.).

An assessment of each stage of the life cycle must consider the relevant inputs and outputs. For example, the manufacturing process may be described in terms of material input (e.g., steel and plastic), resource input (e.g., electricity), waste output (e.g., CO_2, waste heat, scrap plastic), and material output (e.g., an automobile), whereas the use phase may be described by material input (automobile), resource input (gasoline), waste output (CO_2 emissions), and material output (used automobile).

It may seem reasonable to minimize the environmental impact at each stage of the life cycle, but this may not always be the best approach. For example, the production of a more fuel efficient automobile (i.e., lower CO_2 emissions during the use phase) may involve more extensive production processes (i.e., more CO_2 emission in the production phase), and/or less easily recyclable materials at the end of life phase. Thus, for a detailed product assessment, it is important to consider the environmental impact at each phase and how these impacts are interrelated.

The actual life cycle assessment follows these steps:

- *Scope and goals*: Where do we start, and where do we end? For an automobile, do we start with iron ore, or do we start with processed steel? We might also just consider the use phase (i.e., the resources needed to run the car during its lifetime, gas, oil, etc.) and its waste products (the CO_2, NO_x, etc. produced).

- *Inventory analysis:* Can we assess all materials that go into automobile manufacture, or should we concentrate on the major ones (steel, plastic, etc. that make up 98% of the mass)? If the latter, are we ignoring minor components (e.g., electronics) that are energy intensive (per unit mass) to produce?

Continued on page 118

Energy Extra 4.2 continued

- *Impact assessment:* What is the environmental impact? For the materials phase of an automobile, we might consider how many kilograms of CO_2 are produced in processing 1 kg of steel. In the use phase, we might consider the kilograms of CO_2 emitted per kilometer driven.

- *Interpretation:* What do these results mean, and can they be used to modify the production and/or use of an item in order to optimize its effectiveness and minimize its adverse impacts?

  Overall, this kind of analysis is not simple or even well–defined, but it is important in understanding how our actions influence the environment around us.

Topic for Discussion

Find information about CO_2 emissions during the production of steel, that is, kilograms of CO_2 emitted per kilogram steel manufactured. Assuming that a 1.5-t family car is made primarily of steel, how much CO_2 is emitted in producing that steel? How does this compare with the CO_2 emissions during one year of average driving?

reversibility of some changes is not certain. Another approach is to minimize the effects of the carbon released to the environment. Reforestation may have some positive effects, but the sequestration of carbon emissions from fossil fuel burning is essential for CO_2 reduction if we continue the widespread use of these fuels. This is most easily accomplished when dealing with emissions from a relatively small number of major power plants rather than much more numerous portable sources of carbon emissions, such as automobiles.

The magnitude of this problem is emphasized by a simple calculation of the amount of CO_2 produced by fossil fuel combustion. The combustion of coal provides a good example. One kilogram of bituminous coal has a volume of about 7.7×10^{-4} m^3 and contains about 550 g of carbon. If this is completely oxidized, it will produce $(550 \text{ g}) \times (44 \text{ g/mol})/(12 \text{ g/mol}) = 2000$ g of CO_2. At standard temperature and pressure (STP), this will occupy a volume of more than 1 m^3. The total world coal usage of about 5×10^{12} kg per year amounts to an annual emission of about 10^{13} kg of CO_2. This quantity of CO_2 at STP will occupy a cube about 18 km on a side. Three of the major approaches to sequestering or confining this gas are storage:

- Underground.
- In the oceans.
- In solid form.

Of course, the storage of CO_2 is not ultimately beneficial if the CO_2 escapes, so it is important to consider this possibility for any proposed storage method. The specifics of these methods are now discussed.

Underground Storage CO_2 may be pumped into cavities underground for storage. Possible underground cavities are depleted oil wells and depleted natural gas wells. In fact, at present in the United States about 60×10^6 m^3 of CO_2 per day are pumped underground. Much of this is for the purpose of enhancing oil recovery rather than carbon sequestration as such. Although CO_2 pumped into a depleted natural gas well can be expected to remain there indefinitely (as the well trapped natural gas underground for millions of years), it is uncertain whether CO_2 pumped into a depleted oil well will remain in place. It is possible that over time the CO_2 could diffuse through porous rock and escape. Studies to answer this question are needed.

Ocean Storage CO_2 may be disposed of in the oceans in several ways. It may be pumped into the ocean as gaseous CO_2, it may be liquefied and pumped into the ocean, or it may be released in the form of solid compounds. CO_2 released in the ocean near the surface tends to escape from the surface and return to the atmosphere. CO_2 released at a sufficient depth may be trapped in the lower colder layers of water that mix very little with the warmer surface layers. Two factors need to be considered: the longevity of ocean sequestration and its effects on the ocean environment. Some work has suggested that up to about 20% of CO_2 pumped deep into the oceans may diffuse out to the atmosphere on a timescale of about 300 years. Considerable uncertainty in this behavior still exists. Also, effects on ocean life are uncertain. Some studies suggest that marine organisms tend to avoid regions of increased CO_2 concentration, and other studies have indicated increased organism mortality due to ocean CO_2 sequestration.

Solid Storage CO_2 may be reacted with calcium- or magnesium-containing chemicals to produce fairly stable solids that contain the carbon atoms. The two most likely candidates for this type of storage are calcium oxide and magnesium oxide. These are both naturally occurring minerals that are quite common components of the earth's crust. Table 4.7 gives some of the relevant properties. Carbon storage occurs as a result of the reaction of these oxides with CO_2 by the processes

$$CaO + CO_2 \rightarrow CaCO_3 \tag{4.10}$$

and

$$MgO + CO_2 \rightarrow MgCO_3. \tag{4.11}$$

Both processes are exothermic, as indicated in Table 4.7. Therefore, the compounds on the right are stable, and the CO_2 is permanently captured. The major problem with the reactions in equations (4.10) and (4.11) is that they proceed very slowly at room temperature and atmospheric pressure (STP). To capture carbon at a useful rate, it is necessary to increase the rate of the reaction. The successful use of this method for carbon capture requires the development of an industrial-scale process that is efficient, environmentally acceptable, and economical (both from a financial and an energy standpoint). Certainly the viability of carbon sequestration must be studied in more detail, and its use as a method of ameliorating the effects of continued fossil fuel use has yet to be established.

Table 4.7: Properties of naturally occurring calcium and magnesium oxides as carbon sequestration materials. The *enthalpy of formation* is the heat (or energy) associated with the reaction where the negative sign indicates that the reaction is exothermic.

oxide	% earth's crust	carbonate	enthalpy of formation (kJ/mol)
CaO	4.9	$CaCO_3$	−179
MgO	4.36	$MgCO_3$	−117

© Cengage Learning 2015

Table 4.8: Carbon emissions for the top 20 CO_2-producing countries in 2007.

country	carbon/year (10^6 t/y)	population (10^6)	carbon/year per-capita (kg/y)	GDP (US$$10^9$)	carbon/year per $GDP (kg/(y·$))
China	1782	1321.8	1348	2527	0.71
United States	1592	301.1	5287	13,160	0.12
India	440	1129.9	389	805	0.55
Russia	419	141.4	2963	733	0.57
Japan	342	127.5	2682	4883	0.07
Germany	215	82.4	2609	2875	0.07
Canada	152	33.4	4551	1089	0.14
United Kingdom	147	60.8	2418	2346	0.06
South Korea	137	49	2796	897	0.15
Iran	135	65.4	2064	193	0.70
Mexico	128	108.7	1178	743	0.17
Italy	124	58.1	2134	1785	0.07
South Africa	118	44	2682	201	0.59
Saudi Arabia	110	27.6	3986	282	0.39
Indonesia	108	234.7	460	265	0.41
Australia	102	21.1	4834	645	0.16
France	101	61.1	1653	2151	0.05
Brazil	100	190	526	967	0.10
Spain	98	40.4	2426	1084	0.09
Ukraine	87	46.3	1879	82	1.06

A final point concerning carbon emissions that will factor into our assessment of energy policy in different countries is illustrated in Table 4.8. We would expect that population would be a major factor in determining the carbon emissions of a country. We might also expect that other factors, such as those reflected in the per-capita energy consumption discussed in Chapter 2, would influence carbon emissions. These would include the degree of industrialization and types of industry [both of which correlate with the gross domestic product (GDP)]. Carbon emissions per dollar of GDP (Table 4.8) may be an indicator of a country's commitment to an effective energy policy. Of course, climate is also a factor. However, taking all these points into consideration, we see some interesting features in this table, particularly when viewed together with the information presented in Figure 2.4. France has the lowest carbon emissions per dollar GDP of any country in the table and scores slightly better than similar neighboring European countries, such as Spain and Italy. Brazil scores quite well compared with other countries with similar climates and similar per-capita GDPs, such as Mexico. Canada, which in many ways is similar to Australia, except for a much colder climate, actually has the lower carbon emissions of these two countries. It may be possible to understand some of these discrepancies in terms of each country's approach to satisfying their energy needs and the natural energy resources available to each. Factors that contribute to these features certainly include the facts that France produces nearly all its electricity by nuclear reactors (Chapter 6), Brazil has a well developed national commitment to biofuels (Chapter 16), and Canada obtains the largest fraction of its electricity from hydroelectric power (Chapter 11).

4.7 Summary

This chapter reviewed the reasons why energy production and use produce pollution. Any energy generation method that utilizes a heat engine to convert thermal energy to mechanical energy (often with the final objective of producing electricity) generates excess heat that is released into the environment. These methods include nuclear- and fossil fuel–fired generating stations, as well as vehicles operating on internal combustion engines. Although the effects of thermal pollution are typically localized, thermal generating stations that release excess heat into rivers, lakes, or the ocean can have substantial effects on the local ecology. Chemical pollution results from the combustion of fossil fuels, either in the process of generating electricity or for transportation applications. Carbon monoxide results from the incomplete combustion of carbon compounds, and nitrogen oxides are produced when nitrogen in the air is subject to elevated temperatures. Studies have shown that a significant fraction of these pollutants come from vehicle emissions. Most governments have implemented emission control standards to mitigate this problem. Sulfur compounds and particulate matter result from impurities and unburned components of fossil fuels and are most significant during the combustion of coal for electricity generation. Chemical or mechanical methods are in common use to remove these components of exhaust gas from coal-fired stations before it is released to the atmosphere.

The greenhouse effect results from the presence of molecules in the atmosphere that selectively absorb radiation at particular wavelengths. Sunlight, which penetrates the atmosphere, is absorbed by the earth and reirradiated at longer wavelengths that are more readily absorbed in the atmosphere, leading to an increase in temperature at the surface of the earth. Carbon dioxide is produced by the combustion of any hydrocarbon and is an effective greenhouse gas. The continued use of fossil fuels will lead to an increase in greenhouse gases in the atmosphere, and this is the major contributor to global warming. Carbon sequestration is a possible means of reducing greenhouse gas emissions. Carbon dioxide from fossil fuel combustion can be stored underground in pressurized caverns or dispersed in the oceans. Perhaps the most attractive method of sequestering carbon is the formation of carbonates by solid state reaction. More work on this technology needs to be done to develop an efficient and economical process.

Problems

4.1 Pluto has no atmosphere and has an average orbital diameter that is 39.2 times that of the earth. The mean surface temperature is 37 K. Estimate the albedo of Pluto.

4.2 Assume that coal has a sulfur content of 3% by weight. If all the sulfur is converted into SO_2 during the combustion process, how much SO_2 is produced per tonne of coal? How much is produced per megajoule of energy produced?

4.3 A typical gasoline-powered vehicle requires about 3.7 MJ of primary energy (i.e., energy content of the gasoline) to travel 1 km. How many kilograms of CO_2 are produced annually by a vehicle that is driven 40,000 km per year?

4.4   Denver has a land area of about 400 km². Using Figure 4.1, make the assumption that all CO is uniformly distributed in the atmosphere up to a height of 300 m. Estimate the average total mass of CO in Denver's atmosphere. *Note:* The density of air is 1.204 kg/m³.

4.5   There are about 250 million personal vehicles in the United States. Assume that each drives 25,000 km per year on average. What is the average reduction in NO_x emissions per square kilometer per year resulting from the implementation of the 2004 U.S. emission standards compared with the situation before any standards were introduced (1960)?

4.6   (a) Calculate the mass of CO_2 produced per kilogram of methane burned.
(b) At most natural gas wells, excess natural gas is generally burned rather than released to the atmosphere. Methane released during the decomposition of waste at landfill sites is also sometimes burned. Discuss the positive or negative environmental effects of these practices.

4.7   If approximately 0.4% of the carbon emitted from a gasoline internal combustion engine is CO rather than CO_2, how many liters of gasoline would need to be burned to reach a lethal level of CO in a cubic volume 100 m on a side? *Note:* The density of air is 1.204 kg/m³.

4.8   Suppose 10^{13} kg per year of CO_2 emitted by coal-fired generating stations was to be sequestered by calcium oxide (CaO, density 3350 kg/m³) according to the reaction

$$CaO + CO_2 \rightarrow CaCO_3.$$

What volume of CaO would be needed?

Bibliography

R. Alley. *The 2-Mile Time Machine: Ice Cores, Abrupt Climate Change and Our Future.* Princeton University Press, Princeton, NJ (2002).

S. B. Alpert. "Clean coal technology and advanced coal-based power plants." *Annual Review of Energy and the Environment* **16** (1991): 1.

G. J. Aubrecht II. *Energy: Physical, Environmental, and Social Impact* (3rd ed.). Pearson Prentice Hall, Upper Saddle River, NJ (2006).

R. E. Balzhiser and K. E. Yeager. "Coal fired power plants for the future." *Scientific American* **257** (1987): 100.

W. Burroughs. *Climate Change: A Multidisciplinary Approach.* Cambridge University Press, Cambridge, MA (2001).

C. J. Campbell, *The Coming Oil Crisis,* Multi-Science Publishers, London (1988).

W. Collins, R. Colman, J. Haywood, M. R. Manning, and P. Mote. "The physical science behind climate change." *Scientific American* **297** (2007): 64.

B. P. Eliasson, W. F. Riemer, and A. Wokaun (Eds.). *Greenhouse Gas Control Technologies.* Pergamon Press, Amsterdam (1999).

J.A. Fay and D. S. Golomb. *Energy and the Environment.* Oxford University Press, New York (2002).

D. Goodstein. *Out of Gas: The End of the Age of Oil.* W.W. Norton, New York (2004).

J. Hansen. "Defusing the global warming time bomb." *Scientific American* **290** (2004): 68.

R. J. Heinsohn and R. L. Kabel. *Sources and Control of Air Pollution.* Prentice-Hall, Upper Saddle River, NJ (1999).

H. J. Herzog and E. M. Drake. "Carbon dioxide recovery and disposal from large energy systems." *Ann. Rev. Energy. Environ.* **21** (1996): 145.

H. Herzog, B. Eliasson, and O. Kaarstad. "Capturing greenhouse gases." *Scientific American* **282** (2000): 72.

R. Hinrichs and M. Kleinbach. *Energy: Its Use and the Environment* (5th ed.). Brooks-Cole, Belmont, CA (2012).

M. Hoffert, K. Caldeira, G. Benford, et al. „Advanced technology paths to global climate stability: Energy for a greenhouse planet." *Science* **298** (2002): 981.

S. Holloway. "An overview of the underground disposal of carbon dioxide." *Energy Conversion and Management* **38** (1997): 193.

J. Houghton. *Global Warming: The Complete Briefing.* Cambridge University Press, Cambridge, MA (2004).

J. A. Kraushaar and R. A. Ristinen. *Energy and Problems of a Technical Society* (2nd ed.). Wiley, New York (1993).

V.A. Mohnen. "The challenge of acid rain." *Scientific American* **259** (1988): 30.

E. Parsons and D. Keith. "Fossil fuels without CO_2 emissions." *Science* **282** (1998): 1053.

K. Schnell and C. Brown. *Air Pollution Control Technology Handbook.* CRC Press, Boca Raton, FL (2002).

V. Smil. *Energy at the Crossroads Global Perspectives and Uncertainties.* MIT Press, Cambridge, MA (2003).

B. Sorensen. *Renewable Energy: Its Physics, Engineering, Environmental Impacts, Economics and Planning* (2nd ed.). Academic Press, London (2002).

K. Wark, C. F. Warner, and W. T. Davis. *Air Pollution, Its Origin and Control.* Addison-Wesley, Reading, MA (1998).

T. Wigley, R. Richels, and J. Edmonds. "Economic and environmental choices in stabilization of atmospheric CO_2 concentrations." *Nature* **379** (1996): 240.

E. J. Wilson and D. Gerard (eds.). *Carbon Capture and Sequestration.* Blackwell, Oxford (2007).

PART III

Nuclear Energy

After fossil fuels, nuclear energy is the largest component of the world's current energy use. In the early days of commercial nuclear fission reactor development, the hope was that nuclear power would alleviate any energy concerns for the future. Part of this conception arose as a result of a 1954 speech to the National Association of Science Writers by the then chairman of the United States Atomic Energy Commission, Lewis L. Strauss. Strauss stated, "Our children will enjoy in their homes electrical energy too cheap to meter," implying that electricity from fission reactors could be distributed without charge or for a flat fee. In an earlier speech, he had said, "[I]ndustry would have electrical power from atomic furnaces in five to fifteen years." This prediction certainly came true, although some have suggested that the term *atomic furnaces* referred to fusion energy, not fission energy. Although nuclear energy has never become "too cheap to meter," it has become economically competitive with many other established energy technologies. The initial growth of the nuclear power industry, however, did not continue much past the mid-1980s. This was particularly true in the United States, where concerns over security, safety, and waste disposal were significant factors forming both public opinion and government policy about nuclear energy. Major nuclear reactor accidents—Three Mile Island, then Chernobyl, and most recently Fukushima—have demonstrated the validity of safety concerns. Also, it became apparent that the simple and easy approach to nuclear reactors would not provide a long-term solution for our energy needs.

An appreciation of the reasons for changing opinions and policies over nuclear energy from fission reactors requires a detailed understanding of their operation, the environmental consequences, the safety risks of their use, and the availability of nuclear resources. This part of the book reviews these concepts. It also provides an overview of possible future approaches to nuclear fission energy that address concerns over safety and the limited longevity of the resource. Finally, the possibilities of fusion energy are presented. We begin with an overview of the basic principles of nuclear physics.

The Monju Fast Breeder Reactor in Japan in the photograph is a sodium-cooled fast breeder reactor with a rated capacity of 280 MW$_e$. Construction of this reactor began in 1983, and it became operational in 1994. However, a sodium leak shortly thereafter caused a fire, and the reactor was shut down. It was restarted briefly in 2010, but, due to a series of incidents (none of which released radioactive material), public opposition, and legal battles, the reactor remained shutdown for most of its life. In fact, it has been estimated that the reactor has generated electricity for only about 1 hour since it was constructed, and its future remains uncertain. ■

AFP/Getty Images

Some Basic Nuclear Physics

Learning Objectives: After reading the material in Chapter 5, you should understand:

- The composition and stability of the nucleus.
- The relationship of the binding energy of the nucleus to its properties.
- The statistics of nuclear decay processes.
- Alpha, beta, and gamma decay processes.
- The reactions between neutrons and nuclei.

5.1 Introduction

The chemical potential energy associated with chemical bonding can be converted into kinetic energy during a chemical reaction such as combustion. The energy release in a chemical reaction corresponds to a few electron volts (eV) per atom or typically a few tens of megajoules per kilogram (MJ/kg) of fuel (see Chapter 1). The nuclear potential energy associated with nuclear bonding can be converted into kinetic energy during nuclear reactions. The energy release in this reaction can be up to about 200 megaelectron volts (MeV) per nucleus, corresponding to about 10^8 MJ/kg of fuel. These numbers emphasize the potential commercial importance of nuclear energy and have motivated the development of both fission reactors (Chapter 6) and the potential development of fusion reactors (Chapter 7).

An understanding of the differences between the interactions that bond electrons to the nucleus of an atom and the interactions that bond the neutrons and protons together inside the nucleus is necessary in order to appreciate the differences in the energy scales between these two types of processes. The basic physics of nuclear interactions leads to an understanding of the properties of the nucleus and to how nuclear processes can be utilized to produce energy in a controlled manner. The present chapter provides an overview of some basic nuclear physics with an emphasis on the properties that are important for the development of commercial power reactors.

5.2 The Structure of the Nucleus

An atom consists of a nucleus and the electrons bonded to it by the Coulombic interaction (or electrostatic interaction for the purposes of the present discussion). The nucleus itself consists of neutral *neutrons* and positively charged *protons* (except the nucleus of the lightest hydrogen isotope, which is just a proton). A nuclear species (or *nuclide*) is a nucleus with a specific number of neutrons, N (the *neutron number*), and a specific number of protons, Z (the *atomic number*). The atomic number determines the identity of the element. The total number of *nucleons, A*, which is a collective name for both neutrons and protons, is

$$A = N + Z \tag{5.1}$$

and is referred to as the *mass number*. A neutral atom also has Z electrons bound through the Coulombic interaction to the nucleus to cancel the Z positive charges associated with the protons. A nuclide is represented by the terminology $^A_Z E_N$, where A, Z, and N are as previously defined, and E is the name of the element. For example, $^{13}_6 C_7$ would represent the nucleus of a carbon atom with 7 neutrons ($N = 7$) and 6 protons ($Z = 6$), for a total of 13 nucleons ($A = 13$). This terminology is somewhat redundant because the element's name (carbon) specifies the value of Z, and N is related to A by the expression $N = A - Z$. Thus, the terminology ^{13}C provides all the necessary information and is commonly used, although the more complete form is sometimes useful for bookkeeping when dealing with nuclear reactions and decays, as will be explained.

Continuing with the example of carbon, all carbon atoms have nuclei that contain 6 protons. However, not all carbon atoms have nuclei containing 7 neutrons. Some carbon nuclei may have more or fewer neutrons. In fact, carbon nuclei are known that contain between 3 and 10 neutrons (Table 5.1). The family of nuclides with the same number of protons (i.e., the nuclides corresponding to the same element) are referred to as *isotopes* of that element. Thus, the table lists the eight known isotopes of carbon. Some of these isotopes are stable (i.e., they last indefinitely) and occur in nature in the relative proportions given by their natural abundances. The table shows that nearly all carbon atoms occurring in nature have nuclei with 6 neutrons, whereas the remaining have 7 neutrons. Other isotopes of carbon are unstable and spontaneously decay to another nuclide (more about decay processes later). The half-life is the time required for half of a collection of a specific type of nuclei to decay to something else. The shorter

Table 5.1: Summary of the properties of known carbon nuclei, $Z = 6$.

N	A	mass (u)	natural abundance (%)	half-life
3	9	9.031040087	0	127 ms
4	10	10.01685311	0	19.3 s
5	11	11.01143382	0	20.4 m
6	12	12.00000000	98.9	∞
7	13	13.00335484	1.1	∞
8	14	14.00324199	~0	5730 y
9	15	15.01059926	0	2.45 s
10	16	16.01470124	0	747 ms

the half-life, the more unstable the nucleus is. Having too many or too few neutrons produces an unstable nucleus. Beyond certain limits (i.e., fewer than 3 or more than 10 neutrons for carbon), a nucleus cannot form, even for a short period of time. The masses listed in the table are the masses of neutral atoms of carbon of the specified isotopes. These include the masses of the 6 protons, the N neutrons, and the 6 atomic electrons. They also include the mass equivalent of the binding energies (discussed in the next section). Atomic masses are typically specified for different nuclides because they are easier to measure than the masses of the nuclei by themselves. Atomic masses are measured in *atomic mass units* (abbreviated u) where the u is defined by setting the mass of the neutral ^{12}C atom to be exactly 12.0 u. In traditional metric units, the atomic mass unit corresponds to $1.6605402 \times 10^{-27}$ kg.

The same general features as in Table 5.1 are also seen for nuclei with other numbers of protons (i.e., different elements). The ability of various combinations of neutrons and protons to form stable nuclei or unstable nuclei or not to form nuclei at all can be summarized by plotting N as a function of Z (Figure 5.1). This graph is referred to as a *Segrè plot*. It is important to note that, for light nuclei, the stable combinations of neutrons and protons occur for the cases where N and Z are more or less equal. This has been seen to be the case for carbon. For heavier nuclei, the Segrè plot shows that stable nuclei occur when the neutron number is greater than the proton number. This is evidenced by the departure of the stability region from the $N = Z$ line in the figure. This feature is an essential factor in the production of nuclear energy.

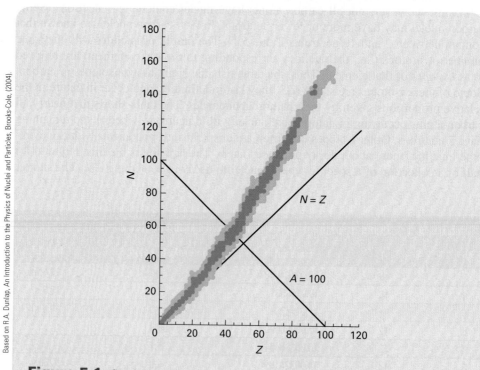

Figure 5.1: Segrè plot of stable nuclei (dark blue area) and unstable nuclei (light blue area).

5.3 Binding Energy

It is important to understand why the particles in a nucleus bond together. There are some parallels between the bonding of nucleons in the nucleus and the bonding of electrons in an atom, where the negatively charged atomic electrons are bonded to the positively charged nucleus by the Coulombic interaction. For the particles in the nucleus, (the neutral neutrons and the positively charged protons), the Coulombic interaction is repulsive (between the protons) and does not influence the neutrons.

Other interactions (besides the electromagnetic interaction) also exist in nature. The gravitational interaction is the one we experience most directly. There are, in fact, four known interactions in nature: (1) the strong (or hadronic) interaction, (2) the electromagnetic interaction, (3) the weak interaction, and (4) gravity (Table 5.2).

Each interaction can be characterized in terms of its relative strength, its range, and the types of objects that it acts on. Gravity is much weaker than any of the other interactions and is of no importance in a description of nuclear physics. It is generally of significance only for large (massive) objects (relative to the nucleus) and at large distances (compared to nuclear dimensions). Interestingly, various interactions act on different types of objects, and we have direct experience of this behavior. For example, gravity acts on masses (independent of whether charges are present). The electromagnetic interaction acts on charges and magnetic moments (which are manifestations of charge) independent of their mass. The gravitational interaction between an object and the earth is manifested as its weight, and the electromagnetic interaction between permanent magnets is manifested as a detectable force between them. This example also demonstrates another property related to interactions: the attractive or repulsive nature of the interaction. Gravity is always an attractive force, whereas the electromagnetic interaction can be either attractive or repulsive.

The positively charged protons that comprise a nucleus experience a repulsive Coulombic interaction, but they (along with the neutrons) also experience a much stronger attractive strong interaction. The strong interaction, which acts only on a certain type of subatomic particle (hadrons), holds the nucleons together in the nucleus. The hadrons are a class of particles that includes neutrons and protons (but not electrons, which are leptons). Examples of subatomic particles are given in Table 5.3. As Table 5.2 shows, the strong interaction is very short ranged; that is, it causes a very strong attractive force between the neutrons and protons within the nucleus but drops off to zero very quickly outside the nucleus. The total force holding the nucleus together is the net result of the large, attractive strong component acting between all the nucleons and a much smaller repulsive Coulombic interaction acting between the charged protons. This total interaction represents the binding energy holding the nucleus together. From Einstein's relation,

$$E = mc^2, \tag{5.2}$$

Table 5.2: The four known interactions in nature and some of their properties.			
interaction	relative strength	range (m)	acts on
strong	1	10^{-15}	hadrons
electromagnetic	10^{-2}	long	charges
weak	10^{-5}	10^{-18}	leptons and hadrons
gravitational	10^{-39}	long	masses

Table 5.3: Some examples of subatomic particles and their classification and properties. [Interactions (gravity excluded): S = strong, E = electromagnetic, W = weak.]

particle	class	mass (u)	charge	interactions
neutron	hadron	1.008664904	0	S, E, W
proton	hadron	1.007276470	+e	S, E, W
electron	lepton	0.0005485799	−e	E, W
positron	lepton	0.0005485799	+e	E, W
electron neutrino	lepton	~0	0	W
electron antineutrino	lepton	~0	0	W

© Cengage Learning 2015

there is an equivalence between mass and energy, so a binding energy, B, is equivalent to a mass m,

$$m = \frac{B}{c^2} \tag{5.3}$$

B is defined to be explicitly positive, and, the larger the value of B, the more strongly the nucleus is bound together. The mass associated with the binding energy decreases the mass of a system of particles. This means that a nucleus comprised of N neutrons and Z protons will have a mass of

$$m_{\text{nucleus}} = Nm_n + Zm_p - \frac{B}{c^2}, \tag{5.4}$$

where m_n and m_p are the masses of the neutron and proton, respectively. The mass of the electrons, Zm_e, and the mass equivalent of the Coulombic binding energy of the electrons, $-b/c^2$, can be added to give the mass of the neutral atom:

$$m_{\text{atom}} = Nm_n + Z(m_p + m_e) - \frac{B}{c^2} - \frac{b}{c^2}. \tag{5.5}$$

A practical application of equation (5.5) is illustrated in Table 5.4. Here the masses of atoms with nuclei containing 49 nucleons are given. These 49 nucleons can be any of the combinations of neutrons and protons, as shown in the table. The table shows that ^{49}Ti has the smallest mass (and hence the greatest binding energy) and is the most stable, that is, it is the only stable nuclide with $A = 49$. As the mass increases for fewer or more neutrons, the mass of nuclei with $A = 49$ increases, and thus the binding energy decreases. The table also shows that this results in increasingly unstable nuclei (as seen by the decreasing half-life). Details of decay processes will be discussed in the next section.

Table 5.4: Atomic masses of nuclides with $A = 49$.

nuclide	N	Z	mass (u)	half-life	decay
^{49}Ca	29	20	48.9556733	8.8 m	β^-
^{49}Sc	28	21	48.95002407	57.5 m	β^-
^{49}Ti	27	22	48.94787079	∞	stable
^{49}V	26	23	48.94851691	330 d	ec
^{49}Cr	25	24	48.95134114	41.9 m	β^+, ec

© Cengage Learning 2015

An important relationship between units used in nuclear physics can be derived by considering the mass of the neutron, 1.008664904 u, as given in Table 5.3. In conventional metric units, this is

$$m_n = 1.008664904 \text{ u} \times 1.6605402 \times 10^{-27} \text{ kg/u} = 1.6749286 \times 10^{-27} \text{ kg.} \quad \textbf{(5.6)}$$

The units on the right-hand side can be multiplied by $(m^2/s^2)/(m^2/s^2)$ (i.e., unity) to get

$$1.6749286 \times 10^{-27} \text{ kg} \cdot (m^2/s^2)/(m^2/s^2) = 1.6749286 \times 10^{-27} \text{ J}/(m^2/s^2), \quad \textbf{(5.7)}$$

where the definition of the joule has been used. Multiplying by the speed of light $(c = 2.997925 \times 10^8 \text{ m/s})$ squared gives the energy equivalent of the neutron mass:

$$E = 1.6749286 \times 10^{-27} \text{ J}/(m^2/s^2) \times (2.997925 \times 10^8 \text{ m/s})^2 = 1.50535 \times 10^{-10} \text{ J.} \quad \textbf{(5.8)}$$

Converting this to millions of electron volts (MeV) gives

$$E = 1.50535 \times 10^{-10} \text{ J} \times (1.602177 \times 10^{-13} \text{ MeV/J}) = 939.565 \text{ MeV.} \quad \textbf{(5.9)}$$

This is the energy equivalent of the mass of the neutron. From equations (5.2) and (5.3), it is seen that

$$1.008664904 \text{ u} = 939.565 \text{ MeV}/c^2 \quad \textbf{(5.10)}$$

or

$$\frac{939.565 \text{ MeV}/c^2}{1.008664904 \text{ u}} = 931.494 \text{ MeV/u.} \quad \textbf{(5.11)}$$

This is a convenient form of units for the speed of light squared (rather than m^2/s^2), which is commonly used in the field of nuclear physics. It will be used in the remainder of the present chapter.

Example 5.1

Calculate the total binding energy associated with a ^{49}Ti atom.

Solution

From equation (5.5), the binding energy may be written as

$$B = (Nm_n + Z(m_p + m_e) - m_{atom})c^2.$$

Using $N = 27$, $Z = 22$, and m_{atom}, as given in Table 5.4, and the masses of the neutron, proton, and electron, as given in Table 5.3, the binding energy is

$$B = (27 \times 1.008664904 \text{ u} + 22 \times (1.007276470 \text{ u} + 0.0005485799 \text{ u})$$
$$- 48.94787079 \text{ u}) \times 931.494 \text{ MeV/u}$$

$$= 426.8 \text{ MeV.}$$

5.4 Nuclear Decays

A collection of unstable nuclei eventually decays into something more stable. If $N(t)$ nuclei of the initial type are present at time t, then the change in the number of nuclei per time interval Δt is defined as

$$\Delta N = -\lambda N(t)\Delta t, \tag{5.12}$$

where the change is negative because the number of nuclei of the initial type is decreasing and λ (the *decay constant*) is a measure of the probability that a given nucleus will decay during the time interval Δt. The greater the value of λ, the more likely the nucleus is to decay, and the faster the decay process will be. The differential form of equation (5.12) can be written as

$$dN = -\lambda N(t)dt. \tag{5.13}$$

This can be integrated to give

$$N(t) = N_0 e^{-\lambda t}, \tag{5.14}$$

where N_0 is an integration constant representing the number of nuclei present at time $t = 0$. The half-life, $t_{1/2}$, is found from equation (5.14) by setting $t = t_{1/2}$ and setting $N(t_{1/2}) = N_0/2$. This gives

$$t_{1/2} = (\ln 2)/\lambda. \tag{5.15}$$

Example 5.2

A nuclide decays with a half-life of 2 days. If 10^8 nuclei of this nuclide are present at the beginning of an experiment, how many remain after 5 days?

Solution

From equation (5.15), the decay constant is given in terms of the half-life as

$$\lambda = \frac{\ln 2}{t_{1/2}},$$

so

$$\lambda = \frac{0.693}{2 \text{ d}} = 0.347 \text{ d}^{-1}$$

Using $N_0 = 10^8$ in equation (5.14) with $t = 5$ d gives

$$N(t = 5 \text{ d}) = 10^8 \times \exp(-0.347 \text{ d}^{-1} \times 5 \text{ d})$$
$$= 10^8 \times \exp(-1.733) = 1.77 \times 10^7 \text{ nuclei.}$$

The rate at which the nuclei decay as a function of time can be found by differentiating equation (5.14) to get

$$\left|\frac{dN(t)}{dt}\right| = \lambda N(0)e^{-\lambda t}. \tag{5.16}$$

Measuring the number of decays per unit time (Figure 5.2) allows for a determination of the half-life, the decay constant, and the number of nuclei present at $t = 0$. The definition of the half-life as the time at which the count rate drops to half of its initial value gives $t_{1/2} = 4.1$ minutes for the example shown in Figure 5.2. From equation (5.15), λ may then be obtained from the measured half-life. Equation (5.16) may be written for $t = 0$ as

$$\left|\frac{dN(0)}{dt}\right| = \lambda N(0). \tag{5.17}$$

This equation allows the value of $N(0)$ to be obtained from a measure of the number of decays per unit time at $t = 0$.

The details of specific decay processes provide insight into the information given in Table 5.4. It is found that ^{49}Sc decays with a half-life of 57.5 minutes to ^{49}Ti. The ^{49}Sc nucleus has 28 neutrons and 21 protons, and the ^{49}Ti nucleus has 27 neutrons and 22 protons. Thus, it would seem that this decay process corresponds to the conversion of one of the ^{49}Sc neutrons to a proton. This is the *beta decay* (β^- *decay*) process and results in the emission of an electron during the conversion of a neutron to a proton as

$$n \rightarrow p + e^- + \bar{\nu}_e, \tag{5.18}$$

where the $\bar{\nu}_e$ is an electron antineutrino (Table 5.3). An early name for an electron was a beta particle, and the observation of electrons as by-products of this decay is the reason for the name of the decay process. This decay is the result of the weak interaction (Table 5.3) because the electron is a lepton and is not subject to the strong interaction. In the case of the β^- decay of ^{49}Sc the process is

$$^{49}\text{Sc} \rightarrow {}^{49}\text{Ti} + e^- + \bar{\nu}_e. \tag{5.19}$$

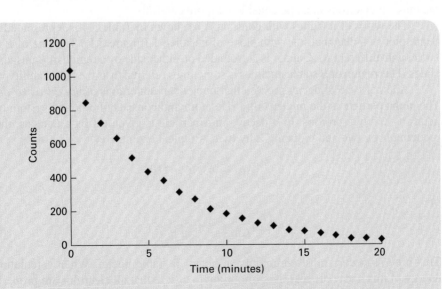

Figure 5.2: Measured decays per unit time for a radioactive source.

In a decay process, the nuclide on the left-hand side is referred to as the *parent*, and the nucleus on the right-hand side is referred to as the *progeny* (or daughter). Analogously, the conversion of a proton to a neutron, referred to as β^+ *decay*, results in the emission of a positron (or antielectron) and an electron neutrino:

$$p \rightarrow n + e^+ + \nu_e. \tag{5.20}$$

This process can be seen in the β^+ decay of ^{49}V:

$$^{49}\text{V} \rightarrow {}^{49}\text{Ti} + e^+ + \nu_e. \tag{5.21}$$

Electron capture (ec, Table 5.4) is equivalent to β^+ decay. The distinction between these processes is not of great relevance to the present discussion and will not be discussed in detail (both processes will be subsequently referred to as β^+ decay). Nuclei that minimize the mass (i.e., maximize the binding energy) with respect to Z for a constant value of A are referred to as *beta stable* (or β stable) because it is energetically unfavorable for them to decay by either β^- decay or β^+ decay.

Beta decay can be viewed more quantitatively, and the energy associated with the decay process can be calculated. The difference in the binding energy of the parent and the progeny is converted into kinetic energy according to Einstein's relation

$$E = \Delta m c^2. \tag{5.22}$$

From Table 5.4, the change in mass that occurs during the β^- decay of ^{49}Sc can be converted to energy as

$$E = (48.95002407 \text{ u} - 48.94787079 \text{ u}) \times 931.494 \text{ MeV/u} = 2.01 \text{ MeV}. \tag{5.23}$$

This decay energy is given up as kinetic energy, partly to the progeny nucleus but mostly to the electron and the antineutrino. It is important in such reactions to properly account for all electrons in the problem. It is convenient in the example of β^- decay that no extra electron masses must be included. In other decay processes, such as β^+ decay, the mass of electrons must be considered more carefully.

The series of nuclides with constant A as shown in Table 5.4 is represented in the Segrè plot as a diagonal line with slope of negative 1 (Figure 5.1). The line of β stable nuclei with different N, Z, and A is surrounded on either side by a region of β unstable nuclides. This region of β stable nuclides is sometimes referred to as the beta stability valley.

Another type of decay process that nuclei can undergo is *alpha decay* (α *decay*). The alpha particle is the nucleus of a ^{4}He atom and consists of a bound system of two neutrons and two protons. In α decay, an α particle is given off by a nucleus, thereby reducing N by two and reducing Z by two. A typical example is

$$^{241}\text{Am} \rightarrow {}^{237}\text{Np} + {}^4\text{He}$$

or

$$^{241}\text{Am} \rightarrow {}^{237}\text{Np} + \alpha. \tag{5.24}$$

This process occurs most commonly in heavy β stable nuclei. When calculating the energy of this decay process, it is important to ensure that all electrons are properly accounted for. This is most conveniently done if the atomic masses of the parent and the progeny are used along with the mass of the ^{4}He atom (Example 5.3).

Example 5.3

Calculate the total energy associated with the reaction shown in equation (5.24) if the mass of the ^{4}He atom is 4.00260325 u and the masses of the ^{241}Am and ^{237}Np atoms are 241.0568229 u and 237.0481673 u, respectively.

Solution

Subtracting the masses on the right-hand side of the equation from the mass on the left-hand side of the equation and converting into energy units:

$$E = (241.0568229\ \text{u} - 237.0481673\ \text{u} - 4.00260325\ \text{u})$$
$$\times\ 931.494\ \text{MeV/u} = 5.64\ \text{MeV}.$$

It is important to note that, if equation (5.24) is written with a ^{4}He atom on the right-hand side, the number of electrons on the two sides of the equation cancel out, and their masses do not have to be included explicitly in the calculation if atomic masses are used.

The relationship of β decay and α decay with the data on the Segrè plot is seen by expanding a region of the plot as shown in Figure 5.3. This region is above the α-stability limit, so even nuclei that are β stable are not stable against α decay. Alpha decays are represented by a diagonal line with a length of 2 units and a slope of 1 pointing downward. The example of the decay process given by equation (5.24) is illustrated in the figure. Beta decays are represented by diagonal lines of 1 unit length with a slope of ± 1. For a β^- decay, for example,

$$^{235}\text{Pa} \rightarrow\ ^{235}\text{U} + \text{e}^- + \bar{\nu}_\text{e}, \tag{5.25}$$

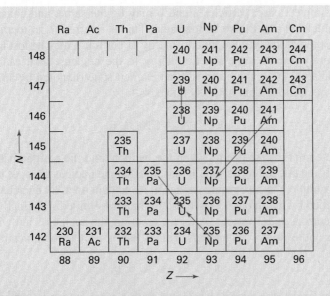

Figure 5.3: Region of the Segrè plot between N = 142 to 148 and Z = 88 to 96, corresponding to A = 230 to 244. An example of α-decay as given in equation (5.24) and examples of β^+ and β^- decays as given in equations (5.25) and (5.26), respectively are shown. Neutron capture by ^{238}U (discussed in the next section) is also illustrated.

as shown in the figure, the decay proceeds downward along the line. For a β^+ decay, as an example,

$$^{235}\text{Np} \rightarrow {}^{235}\text{U} + \text{e}^+ + \nu_\text{e}, \tag{5.26}$$

as shown in the figure, the decay proceeds upward along the line.

Nuclei, like atoms, can exist in their ground state or in an excited state. *Gamma decay* (γ-decay) is the emission of a high-energy photon (or gamma ray) from a nucleus when it undergoes a transition from an excited state to a lower energy state. This process often follows β-decay or α-decay because these decay processes often leave the progeny nucleus in an excited state. Normally, excited states of a nucleus undergo gamma decay processes until the nucleus is in its ground state.

5.5 Nuclear Reactions

Decay processes occur spontaneously in unstable nuclei. Generally, one unstable nucleus appears on the left-hand side of equations such as (5.18), and two or more by products appear on the right-hand side. In a nuclear reaction, a nucleus reacts with a particle (such as an alpha particle or neutron), and two or more by-products result. A simple, generic reaction is

$$a + A \rightarrow B + b. \tag{5.27}$$

This represents a particle (a) incident on a nucleus (A) resulting in the emission of a particle (b) and a resulting nucleus (B). This is often written in shorthand notation as A(a, b)B.

In the simplest case, particles a and b may be the same, and conservation laws require that nuclei A and B also be the same. This is referred to as *scattering*, and the result is that some of the original kinetic energy of particle a may be lost to the nucleus. This may merely impart some kinetic energy to the nucleus, or it can (if sufficient energy is available) leave the nucleus in an excited state (denoted by an asterisk). An example is a neutron incident on a ^{14}N nucleus:

$$n + {}^{14}\text{N} \rightarrow {}^{14}\text{N}^* + n. \tag{5.28}$$

A more complex case occurs when the incident and the emitted particles are not the same. This means, for conservation reasons, that the nucleus changes identity. Consider, as an example, the case where a neutron is incident on a ^{14}N nucleus. The neutron may be absorbed by the nucleus, and a proton may be emitted. Originally, the ^{14}N nucleus had 7 neutrons and 7 protons. After the reaction, the nucleus is left with 8 neutrons and 6 protons and is now a ^{14}C nucleus. This is the (n, p) reaction given by

$$n + {}^{14}\text{N} \rightarrow {}^{14}\text{C} + p. \tag{5.29}$$

As with decays, the energy associated with a reaction can be calculated by subtracting the total of the masses on the right-hand side from the total of the masses on the left-hand side and converting the mass difference into energy units. For reactions that involve charged particles—protons, α particles, and the like—it is important, when using atomic masses, to properly account for all electrons. A calculation of the energy for the reaction shown in equation (5.29) gives 0.63 MeV. This means that 0.63 MeV of energy is given up when this reaction occurs and that this excess energy appears as kinetic energy of the proton and ^{14}C nucleus. A reaction that gives up energy is called *exothermic*.

We might also consider the reaction of a proton incident on a ^{13}C nucleus where a neutron is emitted:

$$p + {}^{13}C \rightarrow {}^{13}C + n. \tag{5.30}$$

A calculation of the energy for this reaction gives -3.00 MeV. It is important to note that this energy is negative. This kind of reaction is called an *endothermic* reaction, and energy must be supplied (at least 3.00 MeV of it) to allow the reaction to take place. This energy can come from the kinetic energy of the incident proton if (for example) the proton is supplied by a particle accelerator.

A reaction which will be important for the discussion of nuclear reactors is the (n,γ) reaction, sometimes referred to as *neutron capture*. Here a neutron is absorbed by a nucleus. The nucleus is left in an excited state, and it subsequently decays back to the ground state by γ decay. An important reaction of this type (discussed further in the next chapter) is

$$n + {}^{238}U \rightarrow {}^{239}U + \gamma. \tag{5.31}$$

Example 5.4

Calculate the energy given up in the reaction shown in equation (5.31). Use the atomic masses $m(^{238}U) = 238.0507826$ u and $m(^{239}U) = 239.0542878$ u.

Solution
Following equation (5.22), the energy may be written in terms of the difference between the total mass of the left-hand side of the equation and the total mass on the right-hand side. The neutron mass is obtained from Table 5.3:

$E = [m(^{238}U) + m_n - m(^{239}U)]c^2$

$\quad = (238.0507826 \text{ u} + 1.008664904 \text{ u} - 239.0542878 \text{ u}) \times 931.494 \text{ MeV/u}$

$\quad = 4.78 \text{ MeV}$

In the case of neutron reactions, such as the (n, γ) reaction shown in equation (5.31), the identity of the element does not change, so the number of electrons on the right- and left-hand sides of the equations is properly accounted for by using atomic masses.

The energy release in the reaction given in equation (5.31) is primarily given up to the γ ray. If the neutron has kinetic energy when it is absorbed by the ^{238}U nucleus, then this adds to the possible γ ray energy.

5.6 Summary

The present chapter has shown that the nucleus consists of neutral neutrons and positively charged protons. The nucleus is held together by its binding energy, which is the net result of the repulsive Coulombic interaction between the protons and the attractive strong interaction between all nucleons. While light nuclei tend to have approximately equal numbers of neutrons and protons, heavy nuclei require a greater proportion of neutrons to be stable. Nuclei with a ratio of neutrons to protons that is outside of the range of stability, as represented on a Segrè plot, are unstable and decay toward a more stable configuration of nucleons.

This chapter also presented the statistics of decay processes and showed that these are described by exponential behavior with a characteristic lifetime. The various decay processes—alpha decay, beta decay, and gamma decay—by which unstable nuclei decay toward the most stable nucleon configuration have been overviewed in this chapter. Alpha decay corresponds to the release of an alpha particle, or ^{4}He nucleus, from an unstable nucleus. Beta decay corresponds to the conversion of a proton to a neutron (or vice versa), along with the release of an electron (or positron) and an antineutrino (or neutrino). Finally, gamma decay corresponds to the de-excitation of the neutrons and/or protons in a nucleus to a lower energy level with the emission of a photon (gamma ray). The methods by which the energy associated with these reactions can be calculated on the basis of nuclear masses have been described.

Nuclear binding energy can be released through nuclear reactions, and these processes provide a mechanism for harnessing the enormous energy associated with nuclear interactions. This chapter described one of the most important types of nuclear reactions related to nuclear energy production, that is, the reactions between nuclei and neutrons.

Problems

5.1 Tabulate the number of neutrons, protons, and electrons for neutral atoms of the following nuclides: ^{29}Si, ^{44}Sc, ^{49}V, ^{210}Bi, and ^{241}Pu.

5.2 Calculate the total binding energy for a ^{49}Sc atom and a ^{49}V atom. Compare with the results from Example 5.1 for ^{49}Ti, and show that the most stable nucleus has the greatest binding energy.

5.3 A sample of ^{137}Cs (half-life = 30.1 years) decays at a rate of 10^{10} decays per second. How long will it take the decay rate to decrease to 10^7 decays per second?

5.4 Plot the atomic mass for nuclides with $A = 49$ as a function of Z. This is generally referred to as the mass parabola. Show for the ^{49}Ca $\rightarrow$ ^{49}Sc $\rightarrow$ ^{49}Ti decay series that the half-life becomes longer as the mass difference becomes smaller.

5.5 What are the decay products of the β^- decay of the following nuclides: ^{3}H, ^{14}C, ^{64}Cu, and ^{125}Sn?

5.6   Calculate the energy released during the α decay of ^{230}Th. The following atomic masses may be of use: $m(^{230}\text{Th}) = 230.0331266$ u, $m(^{228}\text{Th}) = 228.0287313$ u, $m(^{228}\text{Ra}) = 228.0310641$ u, and $m(^{226}\text{Ra}) = 226.0254026$ u.

5.7   Nuclide A has a half-life of 4 days, and nuclide B has a half-life of 16 days. At the beginning of an experiment, a sample contains 2^{18} nuclei of nuclide A and 2^{12} nuclei of nuclide B. How long will it take before the numbers of nuclei of the two nuclides are equal?

5.8 ^{64}Cu can decay by both β^- decay and β^+ decay. Draw an appropriate portion of a Segrè plot along the lines of Figure 5.3 showing these two decay processes.

Bibliography

D. Bodansky. *Nuclear Energy: Principles, Practices, Prospects.* American Institute of Physics, New York (1996).

M. G. Bowler. *Nuclear Physics.* Pergamon Press, Oxford (1974).

W. E. Burcham. *Elements of Nuclear Physics.* Longman, London (1979).

W. E. Burcham. *Nuclear Physics: An Introduction* (2nd ed.). Longman, London (1973).

B. L, Cohen. *Concepts of Nuclear Physics.* McGraw-Hill, New York (1971).

W. N. Cottingham and D. A. Greenwood. *An Introduction to Nuclear Physics.* Cambridge University Press, Cambridge (1986).

A. Das and T. Ferbel. *Introduction to Nuclear and Particle Physics.* Wiley, New York (1994).

R.A. Dunlap. *An Introduction to the Physics of Nuclei and Particles.* Brooks-Cole, Belmont, CA (2004).

H. A. Enge. *Introduction to Nuclear Physics.* Addison-Wesley, Reading, MA (1966).

H. Frauenfelder and E. M. Henley. *Sub-Atomic Physics.* Prentice-Hall, Englewood Cliffs (1974).

K. S. Krane. *Introductory Nuclear Physics.* Wiley, New York (1988).

W. E. Meyerhof. *Elements of Nuclear Physics.* McGraw-Hill, New York (1967).

R. L. Murray. *Nuclear Energy: An Introduction to the Concepts, Systems, and Applications of Nuclear Processes.* Pergamon Press, New York (1993).

W. S. C. Williams. *Nuclear and Particle Physics.* Clarendon Press, Oxford (1991).

Energy from Nuclear Fission

Learning Objectives: After reading the material in Chapter 6, you should understand:

- The relationship between nuclear binding energy and the mechanism for extracting nuclear energy by fission.
- Differences between spontaneous and induced fission and the importance of the Coulomb barrier.
- Fission processes in the isotopes of uranium.
- Critical reactions and thermal reactor control.
- The types of thermal fission reactors and their properties.
- The world use of fission energy.

- The availability and production of uranium worldwide.
- Nuclear reactor safety and the reasons and consequences of nuclear accidents.
- Methods for nuclear waste disposal.
- New designs of thermal reactors with improved safety features.
- The principles of operation and advantages of a fast breeder reactor.

6.1 Introduction

One of the first important steps in the development of nuclear energy came in 1932 when James Chadwick discovered the neutron. In 1938, Lise Meitner, Otto Hahn, Fritz Strassman, and Otto Robert Frisch discovered that nuclear reactions involving neutrons could induce fission and liberate substantial amounts of energy. The development of the first nuclear reactor at the University of Chicago by Enrico Fermi in 1942 demonstrated that a controlled fission reaction could be sustained. The development of this reactor, however, was not motivated by the need for sustainable energy but by the desire to produce nuclear weapons. The development of nonmilitary fission reactors for the production of electricity was pursued after World War II. The first fission reactor to generate electricity, the Obninsk Nuclear Power Plant in the USSR, became operational in 1954. It produced about 5 MW$_e$. The first commercial nuclear power station was opened in 1956 in Sellafield, U.K., and in 1962 the Shippingport Reactor in Pennsylvania became the first operational commercial nuclear reactor in the United States. The idea that nuclear energy could provide inexpensive and virtually unlimited electric power prompted the rapid increase in the number of nuclear power reactor facilities over the next 30 years. Over most of the last

25 years or so, the growth of the nuclear power industry has been much slower and has even been negative in some areas. This chapter describes the physics and engineering of nuclear fission reactors and provides some insight into the reasons for the history of the nuclear power industry, as well as its future.

6.2 **The Fission of Uranium**

When a decay or an exothermic nuclear reaction occurs, the decrease in the total mass of the constituents involved is converted into energy. This energy is manifested as kinetic energy of the progeny nucleus and emitted particles, and it can, in principle, be converted into useful energy. This decrease in mass is associated with an increase in binding energy [equation (5.3)]. Because these processes conserve the total number of nucleons (i.e., number of neutrons plus number of protons), the average binding energy per nucleon is a good measure of the kinetic energy that can be liberated. In Figure 6.1, the average binding energy per nucleon is plotted as a function of the number of nucleons in the nucleus.

Except for a few anomalous peaks, the figure shows that, as the number of nucleons increases, the average binding energy per nucleon increases sharply for light nuclei, followed by a broad maximum and then a slow decrease for heavy nuclei. This behavior allows substantial amounts of energy to be obtained from certain types of nuclear reactions. Consider, for example, a ^{238}U nucleus. This nucleus has a binding energy of 7.57 MeV per nucleon or a total binding energy for the 238 nucleons of about 1800 MeV. If, for some reason, this nucleus were to split in half, forming two nuclei containing 119 nucleons each, then, according to the figure, each of those nuclei would

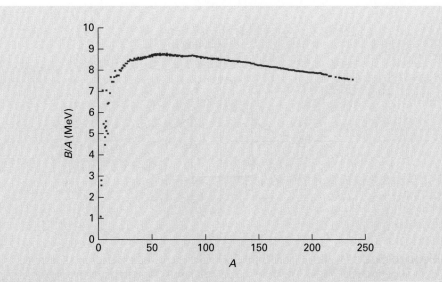

Based on R.A. Dunlap An, Introduction to the Physics of Nuclei and Particles, Brooks-Cole, Belmont (2004)

Figure 6.1: Average binding energy per nucleon as a function of the number of nucleons.

have an average binding energy per nucleon of 8.50 MeV for a total binding energy of the two smaller nuclei of about 2020 MeV. This difference in binding energies, about 220 MeV, would be given up to the smaller nuclei as kinetic energy. The splitting of a heavy nucleus into two smaller fragments is referred to as *fission*. The situation just described is symmetric fission because the two fission fragments have the same number of nucleons. In this simple example, the energy measured in MeV per uranium fission (220 MeV per fission) can be converted to joules per kilogram of uranium to obtain 8.9×10^7 MJ/kg. This can be compared to the burning of carbon, which produces 32.8 MJ/kg. Thus, this simple model shows that nuclear fission produces about 2.7×10^6 times as much energy per kilogram of fuel as combustion. Putting this in different terms, a gram of uranium produces as much energy as almost 3 t of coal. This clearly justifies an interest in nuclear power. Another nuclear reaction—fusion, or the sticking together of light nuclei—deals with the sharp increase on the left-hand part of Figure 6.1 and is discussed in detail in Chapter 7.

It is important to consider why a uranium nucleus might undergo fission. In principle, any process that releases energy (i.e., is exothermic) can occur on its own spontaneously. This can certainly happen for a uranium nucleus because this process is clearly exothermic. However, in a collection of uranium nuclei, it occurs only very rarely. The reason for this can be seen by examining the forces holding the nucleus together. When the nucleons are close together in the nucleus, the attractive strong interaction overcomes the repulsive Coulombic interaction between the charged protons to bind the particles together. This can be represented by viewing the nucleons as sitting in a deep potential well (Figure 6.2). However, to combine separate individual particles (at least protons) to form a nucleus, it is necessary to overcome the Coulombic repulsion so that the particles are close enough together for the strong interaction to take over. The so-called Coulomb barrier responsible for this behavior is shown in the figure. For the particles in a nucleus to split apart (during fission) to form two smaller nuclei, they must get through (or over) the Coulomb barrier. This is fairly difficult for the nucleons in a nucleus, but it does happen occasionally and is referred to as *spontaneous fission*.

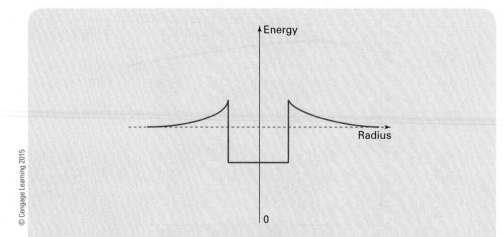

© Cengage Learning 2015

Figure 6.2: Simplified model of the potential energy of a nucleus as a function of distance from the origin. The deep square well results from the attractive strong interaction, and the portion outside the square well results from the repulsive Coulombic interaction for the protons.

One way of increasing the likelihood of fission is to provide some additional energy to the nucleons. This can be done by bombarding the nucleus with neutrons.

To appreciate how this relates to the operation of a nuclear reactor, it is necessary to look in more detail at the properties of uranium. Natural uranium consists of two isotopes; ^{235}U and ^{238}U. These occur with natural abundances of 0.72% and 99.28%, respectively. The greater abundance of ^{238}U compared with ^{235}U is directly a result of its longer half-life, 4.5×10^9 years compared with 7.0×10^8 years. The reaction of a neutron with a ^{238}U nucleus was shown in the last chapter:

$$n + {}^{238}U \rightarrow {}^{239}U + 4.78 \text{ MeV.} \tag{6.1}$$

This provides an additional 4.78 MeV of energy (if the incident neutron has minimal kinetic energy). A similar reaction for a neutron incident on a ^{235}U nucleus is

$$n + {}^{235}U \rightarrow {}^{236}U + 6.54 \text{ MeV;} \tag{6.2}$$

that is, substantially more energy becomes available. An actual calculation shows that the most energetic neutrons and protons in the nucleus are at energies about 6 MeV, below the maximum height of the Coulomb barrier at the edge of the nucleus. The energy provided by an incident neutron (even with very small kinetic energy) to a ^{235}U nucleus is sufficient for the newly formed ^{236}U nucleus to get over the Coulomb barrier. This means that the ^{236}U nucleus is unstable and can undergo fission. This is not the case for ^{238}U unless the incident neutron has sufficient kinetic energy (about 1.4 MeV) to push the nucleus over the Coulomb barrier. Nuclides like ^{235}U that undergo fission when struck by a low-energy neutron are referred to as *fissile*; those that do not are referred to as *nonfissile*. Fission that results from the interaction of the nucleus with an incident neutron is called *induced fission*. Compare the results shown in Example 6.1 for a fissile nuclide with that shown in Example 5.4 for a nonfissile nuclide.

Example 6.1

Use the following known atomic masses of ^{235}U and ^{236}U to obtain the energy release shown in equation (6.2):

$$m({}^{235}U) = 235.0439231 \text{ u}$$

and

$$m({}^{236}U) = 236.0455619 \text{ u.}$$

Solution

The mass of the neutron is given in Appendix II. The energy release in the reaction when the initial kinetic energy on the left-hand side is negligible is given by the energy equivalent of the mass difference between the left-hand side and right-hand side of the equation:

$E = [m_n + m({}^{235}U) - m({}^{236}U)]c^2$

$\quad = [1.008664904 \text{ u} + 235.0439231 \text{ u} - 236.0455619 \text{ u}] \times 931.494 \text{ MeV/u}$

$\quad = 6.54 \text{ MeV.}$

6.3 Nuclear Reactor Design

The fission of uranium just described, particularly that involving ^{235}U, might explain how uranium can be used to obtain energy from nuclear reactions, but a number of important points need to be considered in more detail. Two of these are (1) where do the neutrons come from, and (2) what is the purpose of the ^{238}U? The origin of the neutrons is considered first.

In a piece of ^{235}U, a few nuclei will undergo spontaneous fission; although this is unlikely. However, a 1-kg piece of ^{235}U contains about 2.6×10^{24} uranium nuclei, and a number of these will undergo spontaneous fission, even in a very short time. The simple case of symmetric fission is not very likely. More commonly, one of the fragments is larger than the other. Typically, one fragment with about 137 nucleons and one fragment with about 96 nucleons are produced. All fission processes do not produce the same fission fragments. The distribution of fragment sizes, known as the *fission yield*, is shown in Figure 6.3.

A typical example of an induced fission process in ^{235}U is

$$n + {}^{235}U \rightarrow {}^{236}U \rightarrow {}^{137}I + {}^{96}Y + 3n, \tag{6.3}$$

where the ^{236}U nucleus exists only momentarily before breaking up into the two fragments.

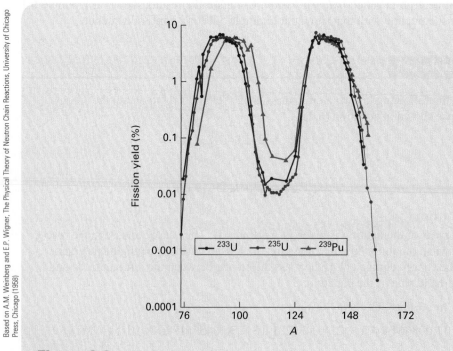

Based on A.M. Weinberg and E.P. Wigner, The Physical Theory of Neutron Chain Reactions, University of Chicago Press, Chicago (1958)

Figure 6.3: Fission yield for some fissile nuclei.

Example 6.2

Calculate the actual fission energy release from the induced fission process shown in equation (6.3), using the following atomic masses:

$$m(^{235}U) = 235.0439231 \text{ u.}$$

$$m(^{137}I) = 136.91787084 \text{ u.}$$

$$m(^{96}Y) = 95.91589779 \text{ u.}$$

Solution

The excess energy is given by the energy equivalence of the difference between masses on the left-hand side of the equation and those on the right-hand side:

$$E = \{[m_n + m(^{235}U)] - [m(^{137}I) + m(^{96}Y) + 3m_n]\}c^2$$

$$= [(1.008664904 \text{ u} + 235.0439231 \text{ u}) - (136.91787084 \text{ u} + 95.91589779 \text{ u}$$

$$+ 3 \times 1.008664904 \text{ u})] \times 931.494 \text{ MeV/u}$$

$$= 179.6 \text{ MeV.}$$

The energy calculated in Example 6.2 is slightly less than the simple calculation for symmetric fission. Two features of equation (6.3) are important. First, 3 leftover neutrons are not connected to either fission fragment; second, the fission fragments are not β stable. The reasons for these properties can be readily seen from the Segrè plot in Figure 5.1. Light β stable nuclei have approximately equal numbers of neutrons and protons, whereas heavy β stable nuclei have more neutrons than protons. Thus, when a heavy β stable nucleus breaks into two smaller nuclei, there are more neutrons than are necessary to form two lighter β stable nuclei. Some of these neutrons, typically 2 or 3 (on the average about 2.5 per fission), are released and are referred to as *prompt neutrons* because they appear at the time of the fission (more about this in the next section). The rest of the neutrons are attached to the fission fragments, but there are still too many neutrons compared with the number of protons to produce nuclei that are β stable. These β unstable nuclei undergo β decay (along with some γ decay) until they reach a point where enough neutrons have been converted to protons to make them β stable.

In principle, the extra neutrons produced in equation (6.3) could be incident on other ^{235}U nuclei and induce further fissions, possibly resulting in a chain reaction that could produce substantial quantities of energy. It is important to consider how these neutrons interact with other uranium nuclei and how the chain reaction can be made to continue in a controlled manner. When the neutrons are released by a fission reaction such as equation (6.3), they have kinetic energies of about 2.0 MeV. If these neutrons are released within a piece of uranium, several things can happen to them. Each neutron may undergo one of the following processes:

1. It can be absorbed by a uranium nucleus and induce fission, thereby producing more neutrons.

2. It can be absorbed by a uranium nucleus and undergo the (n,γ) reaction, thereby lowering the excess energy available to the nucleus to a value below the energy of the Coulomb barrier. In this case, the neutron is captured as described in the previous chapter and cannot contribute to further fission.

3. It can exit from the piece of uranium without undergoing reaction 1 or 2, in which case it cannot contribute to further fission.

4. It can interact with other particles in the material and merely lose some of its kinetic energy. In this case, it still has the possibility of interacting as in 1, 2, or 3.

Because, on average, 2.5 fission neutrons are produced for one induced fission event, if one of these neutrons goes on to induce another fission and 1.5 are lost as in processes 2 and 3, then the chain reaction will continue in a controlled way. The reactor is then said to be *critical*. If the piece of uranium is very small, many neutrons will likely escape (process 3). If the uranium is above a certain size (the *critical mass*), then a chain reaction is possible. The critical mass can be calculated by knowing the probability for reaction 1. This is related to the fission cross section, shown as a function of energy for ^{235}U and ^{238}U in Figure 6.4. A spherical piece of pure ^{235}U has a critical mass of about 52 kg, corresponding to a radius of about 8.7 cm. For a mass less than 52 kg, the chain reaction dies out because too many neutrons escape before inducing further fissions. For a mass of ^{235}U greater than 52 kg, the chain reaction gets out of control because more than one neutron from each fission goes on to induce other fissions. In this case, the amount of energy released becomes very large in a very short period of time. It is not possible to maintain the mass of a piece of ^{235}U at just the critical mass so that the reaction is maintained without getting out of control because these processes occur too quickly. The approach to reactor control depends on a detailed understanding of the behavior of ^{238}U and the properties of the fissions fragments as are now discussed.

Figure 6.4 shows that the fission cross section for ^{238}U drops off very rapidly below about 2 MeV. This is because ^{238}U is nonfissile, and additional kinetic energy from the neutron is necessary to induce fission. If the neutron loses very much of its original energy, it will not be able to induce fission in ^{238}U. The figure also shows that

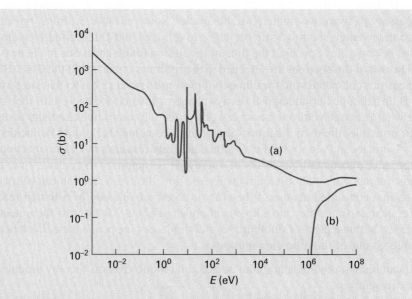

Figure 6.4: Fission cross sections in barns for (a) ^{235}U and (b) ^{238}U nuclei. The cross section is the apparent area of a uranium nucleus as seen by an approaching neutron and is a measure of the probability that a reaction will occur.

the greatest chance of inducing fission in uranium occurs when a very low-energy neutron is incident on a ^{235}U nucleus. The probability here is more than 1000 times greater than it is for ^{235}U at high energy or for ^{238}U at any energy. Thus, although in natural uranium, ^{235}U accounts for less than 1% of all uranium nuclei, it is much more likely that induced fission will occur for a low-energy neutron incident on ^{235}U than for a neutron of any energy incident on ^{238}U. So the best approach to producing fission in a predictable way is to slow the neutrons down while preventing them from interacting with any uranium nuclei. Once their energy has been reduced to a very low value, they are allowed to be incident on a collection of uranium nuclei, where they are almost certain to induce fission in a ^{235}U nucleus. This procedure ensures that the neutron will interact with a ^{235}U nucleus and induce fission rather than being lost by undergoing a reaction (as in reaction 2) with either ^{235}U or ^{238}U.

This approach can be implemented by making the pieces of uranium (the fuel elements) in the reactor fairly small (i.e., much less than the critical mass). This means that when a neutron is emitted from a fission event, it quickly exits the fuel element process 3) and travels through a substance (not containing uranium) known as a *moderator* that reduces the energy of the neutron. The moderator lowers the energy of the neutron until the level is comparable to the thermal energy of the atoms in the moderator, around 0.03 eV. The neutrons are then referred to as *thermal neutrons*, and the reactor that operates on this principle is referred to as a *thermal reactor*, or *thermal neutron reactor*. When the neutrons enter another fuel element, they quickly induce fission in ^{235}U nuclei. The flow of neutrons between fuel elements is controlled by control rods made of a material (often cadmium or boron) that easily absorbs neutrons (without undergoing any undesirable nuclear reactions). The general design of a nuclear reactor core is shown in Figure 6.5. Each control rod can be moved so that a larger or a smaller portion

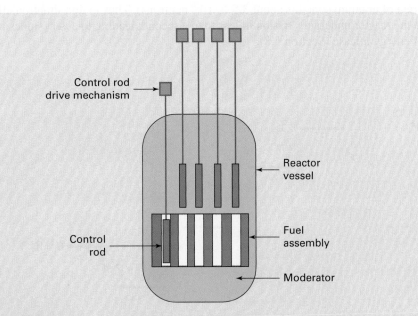

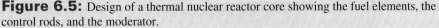

Figure 6.5: Design of a thermal nuclear reactor core showing the fuel elements, the control rods, and the moderator.

of each fuel element is exposed to its neighboring fuel elements, thus allowing more or less fission to occur. The neutron density and temperature in the reactor can be monitored, and the control rods can be moved in such a way as to either increase or decrease the neutron flow between fuel elements in order to maintain the chain reaction. This condition is met when exactly 1 fission neutron (on average) goes on to induce an additional fission. How a controlled fission reaction is maintained is discussed in the next section.

6.4 Fission Reactor Control

The basic principles of reactor control require a careful consideration of the timescale of the processes that take place in the reactor. Once a neutron has been absorbed by a nucleus, the actual fission process itself is very fast, on the order of 10^{-8} seconds, and this is the timescale on which a prompt neutron is emitted. The time that a neutron takes to travel through the moderator is much longer, perhaps on the order of 10^{-4} seconds. This is still a very short time compared to how long it takes to physically move a control rod, which is typically done by some mechanical or hydraulic method in response to variations in the operating conditions. If it takes 1 second to move a control rod in response to an increase in neutron flux, the reactor could quickly become unstable. As an example, consider the case when the fission time constant (i.e., the time between neutron emission and subsequent induced fission) is 10^{-4} seconds and the number of fission neutrons inducing fission suddenly changes from 1.00 to 1.01. In the first 10^{-4}-second interval, the number of fission neutrons in the reactor increases by a factor of 0.01 (or 1%); in the next 10^{-4}-second interval, the number of fission neutrons increases to 1.01^2 of its initial value, and so on. After 1 second, the number of neutrons has increased to $1.01^{10,000}$, or about 10^{43} times the initial value. This is much larger than the total number of neutrons associated with all the uranium nuclei in a fuel rod (a kilogram of uranium contains about 3.5×10^{26} neutrons). Thus in this situation all the fuel in the reactor undergoes fission in less than a second, leading to a very undesirable rate of energy release (Section 6.8).

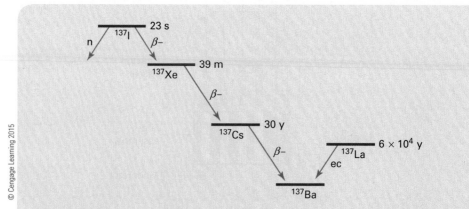

© Cengage Learning 2015

Figure 6.6: β decay of a fission fragment with 137 nucleons (^{137}I) to β stable ^{137}Ba showing delayed neutron emission from ^{137}I.

To resolve this difficulty, it is important to look more carefully at the properties of the fission fragments. The fission fragments are inevitably β unstable and have too many neutrons. Normally, these nuclei β decay by converting neutrons to protons until they become stable. Sometimes during this process, one of the neutrons that is part of the nucleus of one of the fragments (and that is not very tightly bound) escapes prior to the occurrence of the β decay process. This is referred to as *delayed neutron emission* and is illustrated for a fission fragment of mass 137 in Figure 6.6. It can be seen from the figure that the timescale for delayed neutron emission (in this case) is about 23 seconds. On the average, it is about 10 seconds or so for typical fission fragments and is substantially longer for the timescale for processes involving prompt neutrons. For ^{235}U fission, about 2% of the neutrons produced are delayed neutrons. Reactors are typically designed to be about 99% prompt critical. The additional 1% of neutrons needed to reach the critical condition are delayed neutrons, and these are easily controlled on a timescale compatible with methods for adjusting the control rods.

6.5 Types of Thermal Neutron Reactors

Several types of thermal reactors are in current use worldwide. They differ primarily in the material used as the moderator and the way in which heat is extracted from the reactor. The former point also has some effect on the details of the fuel used. The requirements for a moderator material are as follows:

1. The material must be fairly dense (i.e., liquids and solids are alright, but gases are not).
2. The nuclei in the material must be fairly light. This will optimize the amount of energy lost by a neutron when it interacts with a nucleus in the moderator. (*Note:* This can be understood in terms of simple mechanics. If a neutron collides with a similarly sized object (e.g., a proton), the other object recoils, carrying away some of the neutron's energy. If a neutron collides with a very massive nucleus, it scatters more or less elastically and loses very little of its energy.)
3. The material should merely slow down the neutrons rather than absorb them.
4. The interaction of the neutrons with the moderator should not produce any undesirable or hazardous materials.
5. The material should be as nontoxic as possible, chemically stable, and relatively inexpensive.

Unfortunately, no moderator materials are perfect in all these respects. However, three materials do reasonably well at satisfying these criteria and are in common use in nuclear reactors. These materials are water (H_2O), heavy water (D_2O), and graphite (carbon). A number of different reactor designs are in use worldwide, but most fall into the basic categories of water-moderated reactors (developed in the United States), heavy water–moderated reactors (developed in Canada), and graphite-moderated reactors (developed in the United Kingdom, France, and Russia).

Water-moderated reactors may be boiling water reactors or pressurized water reactors. The former are used as power reactors. The latter may also be used as power reactors or (in smaller versions) for nuclear-powered ships and submarines. The

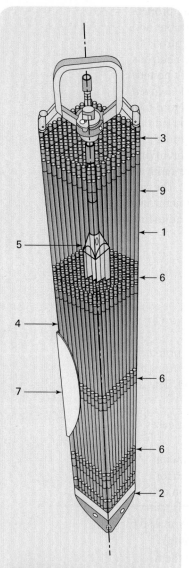

Figure 6.7: BWR fuel assembly. (1) fuel assembly, (2) base plate, (3) head, (4) fuel rod bundle, (5) water channel, (6) spacer grids, (7) fuel channel and (9) full-length fuel rods.

boiling water reactor (BWR) is perhaps the simplest nuclear reactor. The uranium (in the form of UO_2) is formed into cylindrical fuel pellets about a centimeter in diameter and a centimeter long. These are packed into metal tubes about 3–4 m long. The tubes are combined into fuel bundles (the fuel elements just referred to) of typically about 100 tubes (Figure 6.7). The fuel bundles are placed in the core, surrounded by the moderator, and separated by control rods. A power reactor contains a few hundred (up to about 800) fuel bundles, and the total mass of uranium in the core is of the order of 100,000 kg. In a BWR, the water (the moderator) is also used to produce steam for running turbines. Because water is not ideal in all respects as a moderator (specifically, it does not ideally satisfy requirement 3), the uranium used in these reactors must have a higher than natural concentration of ^{235}U. Usually, the fuel is enriched to about 3% ^{235}U. A typical design is illustrated in Figure 6.8. Reactors of this type typically produce about 1000 MW_e of power. It is important to remember that a nuclear reactor, like a fossil fuel generating station, is a heat engine and that its overall efficiency is limited by the Carnot efficiency, typically around 35–40%, so 1000 MW_e corresponds to the production of about 2.5–3 GW of heat energy by the reactor.

Pressurized water reactors (PWRs) are similar to BWRs. The design of the fuel elements is similar except that typically about 200 rods make up the bundle and the core contains about 150–200 bundles (Figure 6.9). In a PWR, the water used as the moderator is kept under pressure to prevent it from boiling. The superheated water transfers heat through a heat exchanger to water that is not kept at high pressure and that is allowed to boil and produce steam; this steam in turn operates the turbines (Figure 6.10). This increases the complexity of the reactor design (compared to BWRs) but reduces the possibility of radioactive material from the reactor core making its way into the environment. A photograph of a PWR is shown in Figure 6.11.

The *heavy water–moderated reactor* was developed in Canada and is known as the CANDU (Canadian Deuterium Uranium) reactor. The design is similar to a PWR; the moderator is held under pressure and does not boil. Rather, it transfers heat through a heat exchanger to water that boils to produce steam to drive the turbines. The fuel bundles are smaller than those in BWRs and PWRs, are about 1.5 m long, and typically contain 37 fuel rods (Figure 6.12). This reactor uses the naturally abundant ^{235}U. Traditionally, CANDU reactors have produced about 700 MW_e, although a newer design produces about 1200 MW_e.

Graphite-moderated reactors, developed in the United Kingdom and France, use a gaseous coolant to transfer the heat from the reactor to water through a heat exchanger. Both helium and CO_2 have been used as a coolant. A schematic of the design is shown in Figure 6.13. Various designs use either natural or enriched (about 2% ^{235}U) fuel.

The water-cooled graphite-moderated reactor has been used extensively in Russia. It is referred to as a RBMK reactor after the Russian name for its design (*Reactor Bolshoi Moschnosti Kanalynyi*). The fuel is ^{235}U enriched to 1.8%, and the fuel elements are very long, typically about 7 m. An illustration of the fuel bundle is shown in Figure 6.14, and a schematic of the reactor design is shown in Figure 6.15. The reasons

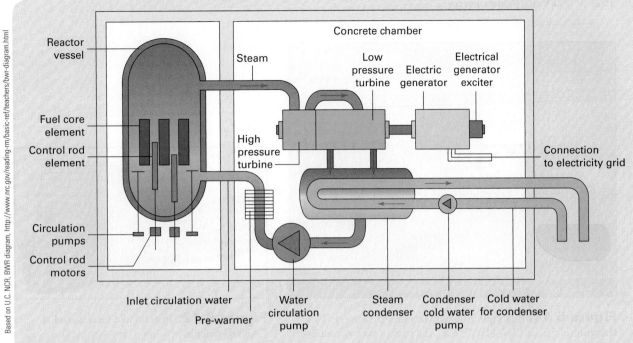

Concrete chamber

Reactor vessel

Steam

Low pressure turbine

Electric generator

Electrical generator exciter

Fuel core element

Control rod element

High pressure turbine

Connection to electricity grid

Circulation pumps

Control rod motors

Inlet circulation water

Pre-warmer

Water circulation pump

Steam condenser

Condenser cold water pump

Cold water for condenser

Figure 6.8: Schematic of the design of a BWR.

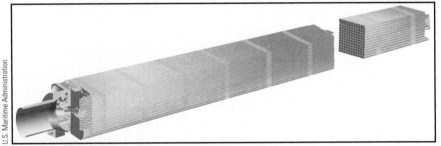

Figure 6.9: Fuel bundle for a PWR.

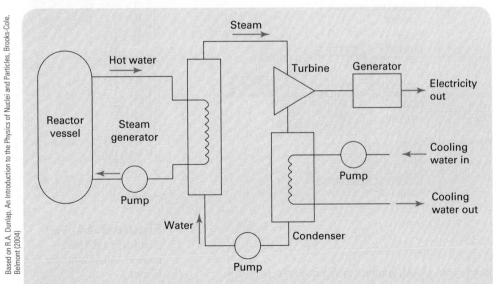

Steam

Hot water

Turbine

Generator

Electricity out

Reactor vessel

Steam generator

Cooling water in

Pump

Pump

Cooling water out

Water

Pump

Condenser

Figure 6.10: Schematic of a PWR.

Figure 6.11: Seabrook Nuclear Generating Station in New Hampshire U.S., a 1244-MW$_e$ PWR. The reactor core is contained inside the dome-shaped containment building at the right.

Figure 6.12: Fuel bundle for a Canadian CANDU reactor.

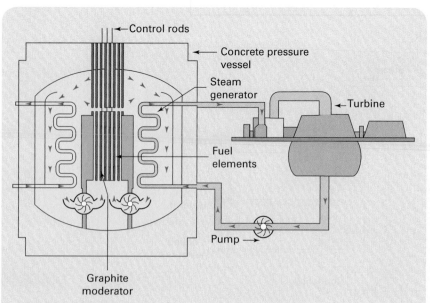

Figure 6.13: Schematic of a gas-cooled, graphite-moderated nuclear reactor.

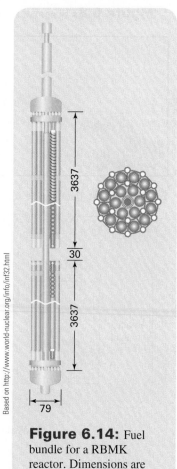

Figure 6.14: Fuel bundle for a RBMK reactor. Dimensions are in mm.

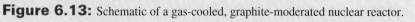

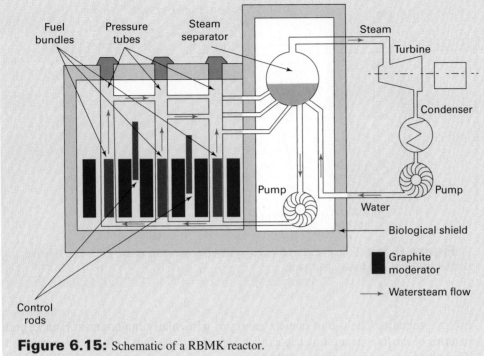

Based on http://www.world-nuclear.org/info/inf31.html

Figure 6.15: Schematic of a RBMK reactor.

for this design and the details of its operation are discussed later in this chapter with regard to the accident involving a reactor of this type at Chernobyl.

6.6 Current Use of Fission Energy

Beginning with the first commercial nuclear power reactors in the late 1950s, there was a substantial and consistent growth of the nuclear power industry that spanned the next 30 years. Figure 6.16 shows the number of reactor facilities in the United States as a function of year since the beginning of the nuclear power industry. The growth illustrated in the figure during the 1970s was due to the perception that nuclear energy was an almost limitless and reasonably priced source of electricity. This situation changed in the mid-1980s, and little new development of nuclear energy has occurred in North America since that time. Figure 6.17 shows the number of new nuclear generating facility starts world wide as a function of year. There is a substantial decline in the 1980s to nearly zero in the 1990s, followed by a slight increase in growth over the past few years. The reasons for the decline in the growth in the nuclear power industry, at least to a substantial degree, are safety and environmental concerns (Sections 6.9 and 6.10).

The current use of nuclear energy is summarized in Table 6.1. Obviously, nuclear energy is a much more important contribution to the generation of electricity in Europe than in most other parts of the world. Japan and South Korea also have notable nuclear

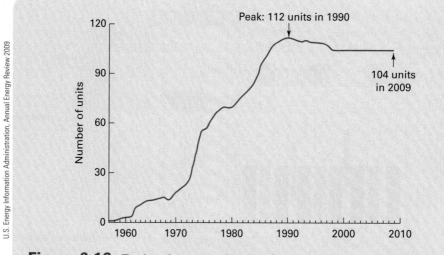

U.S. Energy Information Administration, Annual Energy Review 2009

Figure 6.16: Total nuclear generating capacity (operable units, 1957–2009) in the United States as a function of year.

energy programs. The use of nuclear energy is particularly important in France, both in terms of the total generating capacity and as a fraction of total electricity generation. This ambitious approach to non-carbon-producing electricity generation in France may be reflected in the country's carbon emissions (shown in Table 4.8 and discussed in the previous chapter).

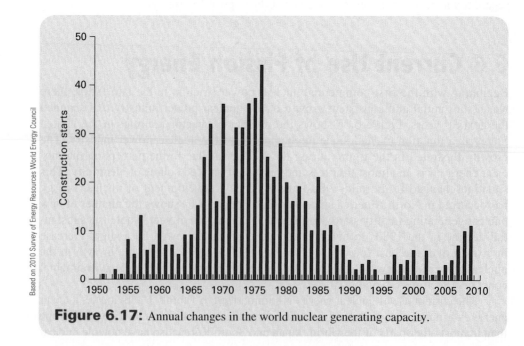

Based on 2010 Survey of Energy Resources World Energy Council

Figure 6.17: Annual changes in the world nuclear generating capacity.

Table 6.1: Summary of operational nuclear power plants worldwide as of 2010.			
country	number of reactors	total capacity (MW$_e$)	% total electric generation (2009)
Argentina	2	935	6.9
Armenia	1	376	45.0
Belgium	7	5863	51.7
Brazil	2	1766	2.9
Bulgaria	2	1906	35.9
Canada	18	12,577	14.8
China	11	8438	1.9
Czech Republic	6	3678	33.8
Finland	4	2696	32.9
France	59	63,260	75.2
Germany	17	20,470	26.1
Hungary	4	1859	43.0
India	18	2708	2.2
Japan	54	46,823	28.9
Korea (South)	20	17,647	34.8
Mexico	2	1300	4.8
Netherlands	1	482	3.7
Pakistan	2	425	2.7
Romania	2	1300	20.6
Russia	31	21,743	17.8
Slovakia	4	1711	53.5
Slovenia	1	666	37.8
South Africa	2	1800	4.8
Spain	8	7450	17.5
Sweden	10	8958	37.4
Switzerland	5	3238	39.5
Taiwan	6	4949	20.7
Ukraine	15	13,107	48.6
United Kingdom	19	10,097	17.9
United States	104	100,683	20.2
world total	437	370,187	~17

Based on data from 2010 Survey of Energy Resources World Energy Council.

6.7 Uranium Resources

Like fossil fuels, uranium is a limited resource. At present, as seen in Figure 1.6 and Table 6.1, nuclear energy accounts for about 17% of the world's production of electricity. That corresponds to about 6% of all primary energy consumption. The rate at which uranium is presently being consumed is well-known. An estimate of the longevity of nuclear energy can be made on the basis of this information, predictions for future uranium use, and an estimate of mineral resources.

Table 6.2: Proved resources of uranium as of 2009 recoverable at less than US$260 production cost.	
country	resources (10^3 t)
Australia	1677
Kazakhstan	832
Russia	568
Canada	542
United States	473
South Africa	296
Namibia	284
Brazil	277
Niger	277
Ukraine	221
China	170
Uzbekistan	115
Jordan	114
India	82
Mongolia	50
other	328
world total	6306

Based on data from 2010 Survey of Energy Resources World Energy Council.

As with most resources, estimating the amount of available uranium is not necessarily straightforward. Certainly, the more one is willing to pay, the more resource will be available for use. As a basis for determining uranium resources, a cutoff at US$260 per kilogram extracted using current technology is sometimes used. The available uranium resources for some countries are shown in Table 6.2. About two-thirds of the world resources of uranium exist in Australia, Kazakhstan, Russia, Canada, and the United States. Canada leads in uranium production, producing about one-third of all the uranium mined and processed. Table 6.3 gives information about the major uranium-producing nations.

Unlike the situation with fossil fuels, where production and use are very nearly the same and differ only by fluctuations in the amount of stored resources, uranium production and use are substantially different (Figure 6.18). In recent years, reactors have consumed nearly twice as much uranium annually as has been produced. This discrepancy is made up with stored uranium, reprocessed fuel from reactors, and military uranium from decommissioned weapons. Reprocessing is a method by which the fissionable component of spent fuel is extracted for reuse in a reactor. This fuel is largely plutonium, and the reluctance of many governments to reprocess spent fuel but rather to merely store it (Section 6.10) is based on security concerns because plutonium extracted from reprocessed fuel can be used for weapons purposes. This is discussed further in Section 6.12 on fast breeder reactors.

If nuclear reactors continue to supply the same quantity of power as they currently do and production has to meet the reactor requirements, then a simple calculation based on available resources (about 2,500,000 t from Table 6.2) and the current demand (about 70,000 t per year from Figure 6.18) indicates a lifetime for these resources of

Table 6.3: Production of uranium in 2008 for major producing countries.

country	production (t/y)
Canada	8995
Kazakhstan	8513
Australia	8425
Namibia	4388
Russia	3510
Niger	3028
Uzbekistan	2326
United States	1492
Ukraine	834
China	790
South Africa	570
other	1009
world total	**43,880**

Based on data from 2010 Survey of Energy Resources World Energy Council.

about 35 years. The use of military uranium, reprocessed fuel, and the development of new technologies for power reactors that would allow for the use of lower-grade fuel could extend this estimate by a substantial amount. The use of lower-grade ore is discussed further later in this chapter. It should be noted, however, that nuclear energy accounts for only about 6% of world energy use, so even with a greatly increased lifetime, nuclear energy (in its present form) could not satisfy all of our energy requirements for more than a small number of decades.

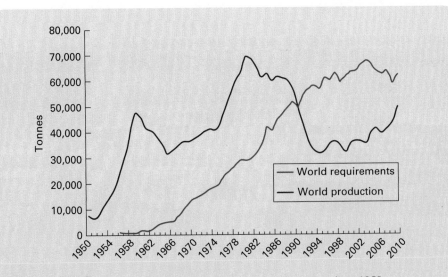

Based on 2010 Survey of Energy Resources World Energy Council.

Figure 6.18. Uranium production and reactor requirements since 1950.

6.8 Nuclear Safety

Although nuclear energy production does not release carbon into the environment, as does the burning of fossil fuels, it is not without its environmental concerns. Because electricity is produced from nuclear energy by a heat engine, heat pollution is a local concern around generating stations. The greatest concerns are, however, for the safety of reactors and the disposal of spent fuel, the first of which is discussed in this section. The security of radioactive materials that could have military uses is also a concern, and this is dealt with briefly in ensuing sections.

The safety of nuclear power plants, or at least the public perception of their safety, was a major factor in the subsiding growth of nuclear energy that began more than 20 years ago. The possibility and consequences of a nuclear accident need to be considered very carefully, and the three most significant accidents to date—Three Mile Island, Chernobyl, and Fukushima Dai-ichi—serve as useful case studies for this analysis.

6.8a Three Mile Island

Three Mile Island Nuclear Generating Station is in Eastern Pennsylvania and consists of two PWRs (Figure 6.19). Four large cooling towers are readily seen in the figure, and the two reactor containment buildings appear near the center. Construction on the Three Mile Island plant began in 1968. The first reactor, TMI-1, which is the left-most reactor in the photograph, began commercial operation in 1974. The second reactor, TMI-2, which was involved in the accident, began operation in 1978 and is the farthest to the right in the figure.

A typical PWR has three water or steam loops (Figure 6.10). The primary loop contains water that is pressurized to prevent it from boiling. It has three functions: to moderate the neutrons, to keep the reactor core from overheating by transporting heat out of the core, and to transfer that heat to the secondary loop. The secondary

Centers for Disease Control and Prevention's Public Health Image Library

Figure 6.19: Three Mile Island Nuclear Generating Station prior to the accident of 28 March 1979.

(water/steam) loop contains water that is allowed to boil and produce steam, which is used to run the turbines. The third, or cooling, loop contains water and is used to transfer excess heat to the environment. In the case of the Three Mile Island reactor, this cooling loop dissipates heat to the atmosphere by means of cooling towers.

The accident at TMI-2 began at 4:00 a.m. on March 28, 1979 (exactly one year, to the minute, from its initial start-up). The accident began as a failure of the main feed water pump in the secondary (non-nuclear) loop. See the schematic of the reactor design in Figure 6.20. This prevented heat from being removed from the primary loop, and the temperature of the water and hence its pressure increased. The pressurized relief valve in the primary loop opened to lower the pressure by releasing steam. Unfortunately, a failure in the system operating this valve caused the valve not to close when the pressure returned to normal, and excess water in the form of steam was released from the valve. This resulted in a significant decrease in cooling water from the primary loop (known as a LOCA, *loss of coolant accident*), and temperature of the core of the reactor increased. Before the reactor was brought under control, part of the core melted, and radioactive material from the fuel elements contaminated the water in the primary loop. Although the containment vessel around the core of the reactor was not breached, some radioactive material (6.3×10^{11} Bq, Bq = becquerel = 1 decay per second) was released to the environment as contaminated steam through the pressure release valve. Although TMI-1 was not involved in this accident and continues to operate as a commercial power reactor, TMI-2 has been shut down since the accident.

Studies by a number of U.S. government agencies (Nuclear Regulatory Commission; Environmental Protection Agency; Department of Health, Education and Welfare; Department of Energy; and Commonwealth of Pennsylvania) have considered the environmental and health consequences of the Three Mile Island accident. The general

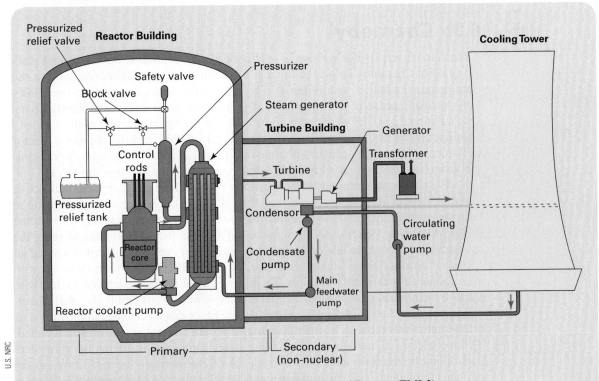

Figure 6.20: Simplified schematic of the Three Mile Island Reactor (TMI-2).

conclusion of these studies has been that there were no immediate adverse health effects caused by the accident and that any long-term effects would be negligible. Specifically, it was found that the average radiation exposure to the population of about 2 million in eastern Pennsylvania was about 10^{-5} Sv (Sv = Sievert). By comparison, the radiation exposure from a chest X-ray is about 6×10^{-5} Sv. The maximum exposure for residents in the immediate area of the reactor was 10^{-3} Sv. This value is about the same as the average annual exposure from background radiation in the area, and the additional exposure is about equivalent to spending a year living in a high-altitude city such as Denver. A risk analysis indicates that no additional cancer deaths are expected as a result of the accident.

Three Mile Island is the worst nuclear power reactor accident in the United States and was the result of a series of unfortunate equipment failures that operators did not deal with appropriately. No new nuclear power reactor construction was started in the United States after the Three Mile Accident (although construction of one reactor, Watts Bar 2 in Tennessee, which was started in 1973, is expected to be completed in 2015). A decrease in reactor construction had started in the United States in the early and mid-1970s, and construction of a number of planned reactors had already been cancelled prior to the Three Mile Island accident. However, the Three Mile Island accident was only one of a number of factors that influenced nuclear policy. There were growing concerns over nuclear safety as a result of a number of less serious accidents involving military and commercial reactors, as well as concerns over nuclear security and waste disposal. Other factors were a reassessment of fossil fuel use after the oil crisis in the early 1970s, a shift in the United States toward importation of energy resources rather than domestic production, and changes in federal pollution standards that promoted the use of inexpensive coal resources. Perhaps one of the most significant impacts of the Three Mile Island accident has been a shift in the public perception of nuclear power.

6.8b Chernobyl

A much more serious nuclear power reactor accident occurred at the Chernobyl reactor in 1986. Chernobyl is located in what was then the Soviet Union (and is now the Ukraine near its border with Belarus). The generating station consisted of four 1000-MW_e nuclear reactors of the water–cooled, graphite-moderated (Russian RBMK) type. Two additional reactors were under construction at the site at the time of the accident. The first reactor at Chernobyl became operational in 1977, and the number 4 reactor (which experienced the accident) became operational in 1983. Two major factors contributed to the accident; an intrinsically unsafe reactor design and operator error.

The reactor at Chernobyl was designed to produce both electricity and radioactive ^{239}Pu for nuclear weapons use. To extract the ^{239}Pu from the reactor while continuing to keep the reactor running to produce electricity, certain compromises in the safety of the design were made. Specifically, to remove some of the fuel rods without shutting the reactor down, a large amount of space above the reactor core was needed. To avoid an extensive structure around the reactor core, early RBMK reactors were designed with compromised containment structures. The philosophy of this approach was to deal with safety by minimizing the possibility of an accident without considering the need to contain an accident if one did occur. Another design factor that contributed to the problems at Chernobyl was the way in which electricity needed to run the reactor was supplied. This electricity was used to run the cooling water pumps and operating the computers controlling the reactor operation. Although the reactor normally operated from power provided through the grid, backup generators were in place to take over in the case of a grid power outage. There were concerns that the delay between a grid power outage and full generator capacity was unacceptable. It was, in fact, the awareness of this situation that prompted a test on the evening of April 25, 1986 to investigate the possibility of using electricity produced onsite by different mechanisms in the event of a power outage. During a routine maintenance shutdown, several safety systems were intentionally disabled in order to conduct the test. The RBMK is inherently unstable at low power levels, and during the test several indications that the reactor was out of control were ignored.

When, in the early morning of April 26, the operators did decide to shut down the reactor for safety reasons, the power output fluctuated, causing pressure to build up in the cooling system and bursting some cooling water pipes. The water and steam that was released reacted with the hot reactor core that contained the graphite moderator and the zirconium tubes containing the fuel rods. This reaction decomposed the water, releasing hydrogen. The hydrogen ignited, causing an explosion that blew off the top of the reactor building. This explosion did not result from the reaction of nuclei, as is the case for a nuclear weapon, but rather from the chemical reaction of hydrogen that had been produced in the reactor. However, a substantial quantity of radioactive materials from the core was released to the environment. The results of this explosion are shown in Figure 6.21. Because graphite (pure carbon) is flammable,

© RIA Novosti/Alamy

Figure 6.21: Chernobyl reactor 4 after the accident of April 26, 1986.

Time & Life Pictures/Getty Images

Figure 6.22: Concrete sarcophagus around Chernobyl reactor number 4.

the resulting fire needed to be dealt with. To contain the remains of the reactor core, a concrete enclosure (called the Sarcophagus) was subsequently constructed around reactor 4 (Figure 6.22).

In the immediate aftermath of the Chernobyl accident, 237 people suffered from acute radiation sickness. These were workers and firefighters at the reactor, and 29 of these died from radiation exposure. Two additional deaths resulted directly from the explosion and fire. A study in 2005 identified an additional 25 subsequent cancer deaths that could be directly attributed to the accident.

The longer-term effects of the Chernobyl accident are unclear. A significant amount of radioactive material was released to the environment, about 1.8×10^{18} Bq in total (compare with 6.3×10^{11} Bq for Three Mile Island). A sizable area around the reactor site was evacuated and will remain uninhabitable for many years (Section 6.10 for information about the longevity of radioactive waste). At present, the so-called Exclusion Zone, which was once home to about 120,000 people, encompasses an area about 30 km in radius around the reactor site. Due to weather conditions at the time, the largest portion of radioactive material was transported into what is now Belarus, with smaller amounts into Ukraine and Russia. The distribution of residual radiation as of 1996 is shown on the map in Figure 6.23. The figure shows that contamination has spread over an area several hundred kilometers across and that many areas, even up to more than 100 km from the site, remain uninhabitable. Measurable radioactive material was detected over most of Europe and the eastern part of the Soviet Union. In fact, a large number of people over an extended area have been exposed to an increased level of radiation (Table 6.4). The Chernobyl reactor facility is located on the Pripyat River, which provides water to area reservoir facilities, and elevated radiation levels were recorded in drinking water supplies for up to a few

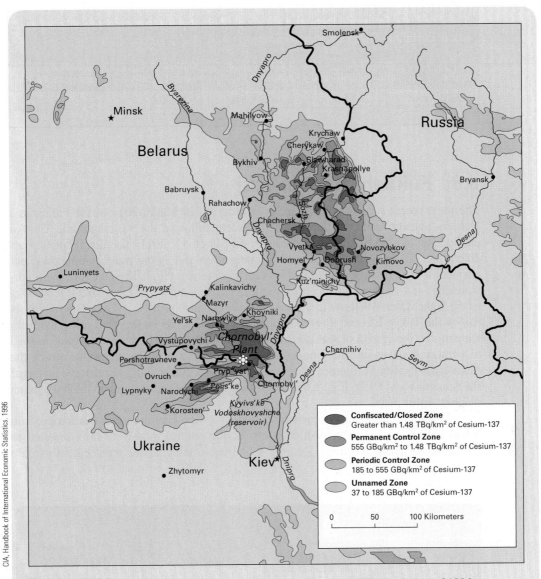

Figure 6.23: Radiation levels in the contaminated area around Chernobyl as of 1996.

months after the accident. The ultimate effects of this exposure on the population of the area are difficult to determine as the relationship between low levels of radiation and increased health risk are uncertain. However, an increased incidence of cancer is unquestionably one of the results of excess radiation exposure. (Risk factors associated with various aspects of energy production are discussed in the next section.) An estimate of the effects of the Chernobyl accident on cancer deaths in Europe and the Soviet Union are summarized in Table 6.4. Although the percentage increase in cancer deaths is quite small (1–2% of all cancer deaths), the actual number of additional deaths is substantial (many thousands).

Table 6.4: Estimated long-term effects of Chernobyl in terms of increased cancer risk in the former Soviet Union and Europe.				
region	population affected (millions)	natural cancer deaths	chernobyl cancer deaths	increase in cancer deaths (%)
Soviet Union	279	35,000,000	6500	1.9
Europe	490	88,000,000	10,400	1.2

© Cengage Learning 2015

6.8c Fukushima Dai-ichi

The most recent serious nuclear accident occurred in March 2011 at the Fukushima Dai-ichi nuclear facility in Japan. This accident followed the 9.0 magnitude Tōhoku earthquake and the tsunami that it caused on March 11, 2011. The details of this accident are still being evaluated, and in many ways this is the most complex nuclear accident to date because it involved six reactors at the facility. The layout of the Fukushima Dai-ichi nuclear reactors prior to the accident is shown in Figure 6.24. It consists of six light water BWRs with a total generating capacity of 4.7 GW_e. The facility is run by the Tokyo Electric Power Company (TEPCO). The facility was protected from tsunamis by the system of seawalls, as shown in the figure. The seawalls were designed to withstand waves up to 5.7 m in height. An analysis of the tsunami that struck the Fukushima facility 51 minutes after the earthquake occurred indicated that the height of the tsunami was 13.1 m. Figure 6.25 shows the impact of the tsunami on the Fukushima nuclear power plant.

When the earthquake occurred, reactors 1–3 were operational, and reactors 5 and 6 had been shut down for maintenance (i.e., the control rods were fully inserted to eliminate the chain reaction). Reactor 4 had been defueled for replacement of core

National Land Image Information (Color Aerial Photographs), Ministry of Land, Infrastructure, Transport and Tourism

Figure 6.24: Fukushima Dai-ichi nuclear facility in Japan prior to the nuclear accident of 2011. The reactors are the square buildings, with number 6 still under construction. A portion of the seawall structure is seen to the right.

Figure 6.25: Tsunami approaching Fukushima Dai-ichi nuclear facility near the number 5 reactor on March 11, 2011.

components, and the fuel rods were placed in storage in a separate pool in the reactor building. Reactors 1–3 shut down automatically at the time of the earthquake, and emergency electrical generators started up to maintain the water pumps and control systems needed to cool the reactors. Fuel in a reactor that has been shut down or even used fuel stored outside the reactor must be cooled because of heat resulting from the radioactive decay of the fission by-products (Figure 6.6). When the tsunami breached the seawall system (Figure 6.25), the power lines connecting the reactor facility to the grid were destroyed, and he generators and control systems were flooded and rendered inoperative.

Following the loss of cooling water, the cores of reactors 1–3 experienced complete meltdowns on a timescale of a few days. Overheating of the zirconium-containing fuel rod tubes released hydrogen gas, as in the Chernobyl incident, according to the reaction

$$Zr + 2H_2O \rightarrow ZrO_2 + 2H_2. \qquad \textbf{(6.4)}$$

The hydrogen release led to explosions and fires in these three reactors, causing substantial damage to the reactors and buildings.

Hydrogen buildup in reactor 4 also led to explosions and fires in that building. Current evidence suggests that there is minimal damage to the fuel rods stored in the spent fuel pool in reactor 4. Because no fuel was in the core at the time of the accident, there was no core meltdown, and cooling of the core is not necessary. The damage to reactors 3 and 4 about ten days after the tsunami is shown in Figure 6.26.

Reactors 5 and 6 overheated but did not experience a core meltdown. Because building panels were removed to vent the hydrogen, there was no explosion in either of these reactors. Cooling water was restored to these two reactors, and they are in safe shutdown mode with no structural damage.

Figure 6.26: Damage to Fukushima Dai-ichi nuclear facility reactors 3 (left) and 4 (right) on March 20, 2011.

Overall structural damage, radiation leaks, and a generally damaged electrical grid hindered efforts to control the nuclear accident at the Fukushima facility. Complete containment and cleanup of the facility are likely to take many years or even decades. Some estimates have suggested that it might be a century before the fuel can be removed from the facility and properly dealt with.

The environmental impact of Fukushima is certainly substantial, and the release of radioactive material has been significant. This radioactive contamination has had the most substantial effect in Japan, but elevated levels of radioactive material have been observed throughout the world. This material consists primarily of ^{131}I and ^{137}Cs, with half-lives of 8 days and 30 years, respectively. Because of the long half-life of ^{137}Cs, its long-term effects are not known, although some estimates have suggested about a 1% long-term increase in the cancer rate of the population in the accident area. Studies have indicated that the total release of radioactive material to the environment may be in the range of 10–20% of that released in the Chernobyl accident. Two nuclear plant workers were hospitalized for severe (although not life-threatening) radiation exposure, and several dozen workers have been injured in non-radiation-related incidents.

The *International Nuclear and Radiological Event Scale (INES)*, established by the International Atomic Energy Agency (IAEA), is used for the categorization of nuclear events that may have an impact on people or the environment. Chernobyl and Fukushima have both been categorized as Level 7 events (the highest level) and are the only two nuclear accidents to warrant this classification. Although both incidents are categorized as Level 7, the details in terms of their radiation release and long-term effects are quite different. It is generally acknowledged that Fukushima is the most complex nuclear accident because of the number of reactors involved. Three Mile Island has been classified as a Level 5 event. There have been no Level 6 events at commercial power reactors, although there has been at least one other Level 5

ENERGY EXTRA 6.1
Nuclear accidents and the INES classifications

In 1990 the *International Atomic Energy Agency* established the *International Nuclear and Radiological Event Scale (INES)* for the categorization of nuclear events that may have an impact on people or the environment. The scale has eight levels (numbered 0–7), with 7 being the most serious. Levels 4–7 are designated *accidents*, and Levels 1–3 are designated *incidents*. Level 0 is an operating deviation without safety significance. The scale may be viewed as a pyramid, as shown in the figure, where less significant events occur quite frequently, and more significant events occur progressively less often. The scale is

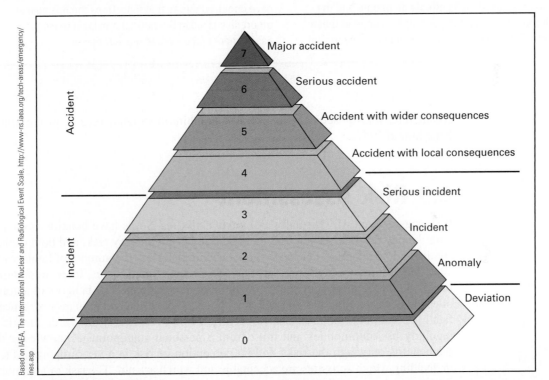

Based on IAEA, The International Nuclear and Radiological Event Scale, http://www-ns.iaea.org/tech-areas/emergency/ines.asp

The INES event triangle.

Level	Designation	Typical characteristics
7	major accident	major release of radioactive material with widespread health and environmental effects
6	serious accident	significant release of radioactive material with impact on people and the environment
5	accident with wider consequences	limited release of radioactive material with some impact on people and the environment
4	accident with local consequences	minor release of radioactive material with possible impact on local food sources
3	serious incident	exposure in excess of ten times the legal annual limits for radiation workers
2	incident	exposure in excess of the legal annual limits for radiation workers
1	anomaly	minor problems with reactor safety components with possible exposure in excess of the legal annual limits for the general public
0	deviation	deviation of reactor operation with no safety significance

© Cengage Learning 2015

Continued on page 168

Energy Extra 6.1 continued

intended to be logarithmic, similar to the Richter scale used to classify earthquakes. However, unlike an earthquake whose magnitude may be determined by a single quantitative measure of the amplitude of shaking, the evaluation of a nuclear event involves a more subjective analysis based on the study of a number of factors. As a result, a definitive classification of nuclear accidents is possible only well after the incident.

The table lists the INES levels and their characteristics. The evaluation of the severity of an event is based on its effects on people and the environment and on the geographic scope of these effects. The three nuclear accidents at commercial power reactors (Three Mile Island, Chernobyl, and Fukushima) have been the most widely publicized in the media. However, a number of serious (up to Level 6) accidents have occurred at military and research reactors around the world.

Topic for Discussion

Commercial nuclear power has been utilized for more than 50 years. Review the major nuclear accidents that have occurred during that time. Does the Rasmussen report discussed in Section 6.9 provide a reasonable assessment of the risks of nuclear power?

commercial reactor accident (a fire at a graphite-moderated reactor in the United Kingdom in 1957).

6.9 **Risk Assessment**

Risk is associated with virtually all human activities. Some have benefits associated with them as well. Assessing risks and benefits is not an easy task, and both of these are difficult to quantify. One way of quantifying risk is by the number of fatalities per year. The use of automobiles is an example. In the United States, there are approximately 42,000 fatalities per year from automobile accidents. The benefits of automobiles include such things as convenience, financial benefits, and perhaps avoidance of fatalities that might otherwise have occurred. However, nearly all North Americans regularly use automobiles and thus, from a personal standpoint, acknowledge that the benefits outweigh the risks. Public perception of risk is a very important factor in deciding which activities are acceptable and which are not. The risk of automobile fatalities is a very well-defined quantity; the year-to-year fluctuations in the number of fatalities are comparatively small. The risk of driving an automobile is (more or less) the same from one year to the next. The public also perceives the risk on a per-event basis. A fatal automobile accident incurs perhaps one or two deaths, maybe a few. There are no catastrophic automobile accidents involving thousands of fatalities. An activity that might result in an accident involving one or two fatalities seems to be acceptable even if it is certain that this will happen tens of thousands of times every year. A single catastrophic event that claims 42,000 lives seems unacceptable even if it occurs only once a year.

The production of energy always involves risk. These risks might be divided into two categories: (1) the risk to workers directly related to the production of energy and (2) the risk to the general public. The occupational risks involved in the production of electricity from a natural resource can be roughly divided into several categories that include (1) the extraction of the resource from the earth, (2) the processing of the resource, (3) its transportation, and (4) the operation of the generating station.

Table 6.5: Anticipated number of fatalities per GWy_e of electricity generated from coal and uranium.

resource	mining	processing	transportation	generating station	total
coal (deep mines)	1.7	0.02	2.3	0.01	4.0
uranium	0.2	0.001	0.01	0.01	0.22

Cengage Learning

A comparison of the generation of electricity from coal and from uranium would show some similarities and some differences in the occupational risks. In both cases, there would be a risk to miners from accidents or from exposure to hazardous materials (i.e., coal dust in one case and radioactive materials in the other). In both cases, there would be a risk to workers involved in processing the resource, both from industrial accidents and from exposure to coal dust or radioactive materials. Similarly, in both cases, transportation workers could be involved in accidents. For generating station workers, one might imagine, accidents, fire, explosions, and other risks as factors for the coal-fired station and nuclear accidents for the production of electricity from uranium. It is important to realize, however, in comparing these two cases that one gram of uranium produces as much electricity as 3 t of coal and that this would certainly influence our perspective of, for example, the significance of transportation accidents. Table 6.5 gives the results of one study of the anticipated number of fatalities among energy workers for the production of 1 GWy_e of electricity from coal and uranium. Clearly these estimates suggest that the occupational hazards of producing energy from coal are much greater than those of producing electricity from uranium.

The risk to the general public from different energy production methods is important. The most obvious risk associated with the production of electricity from coal would be air pollution. The long-term effects of possible climate change are difficult to assess and are not considered in this section. Because the effects of air pollution are fairly localized, the influence of pollution from a coal-fired power plant depends considerably on the local population density. The amount of pollution is also dependent on the implementation of processes for cleaning the exhaust. Some estimates of the anticipated number of fatalities per gigawatt-year of electricity produced from coal are shown in Table 6.6. Methods to reduce pollution in the exhaust gas are effective at reducing fatalities, as is the choice of station location.

Table 6.6: Anticipated fatalities per gigawatt-year of electricity produced by coal under different plant conditions.

location	scrubbers	fatalities
urban	no	74
rural	no	11
urban	yes	28
rural	yes	7

Based on data from Kraushaar and Ristinen ...

Table 6.7: Decrease in life expectancy due to some risks.	
factor	decreased life expectancy
smoking 1 pack per day	7 y
urban rather than rural living	5 y
overweight by 25%	3.6 y
nuclear power plants as of 1970	less than 1 minute

Based on data from H. A. Bethe

For nuclear power, three factors should be considered: (1) health effects related to the disposal of radioactive waste, (2) radioactive emissions during normal plant operation, and (3) nuclear accidents. Nuclear waste is discussed in some detail in the next section. Under normal operating conditions, the radioactive material in a nuclear power plant is very well contained, and strict guidelines exist for the amount of radiation that the general public can be exposed to. In fact, studies have shown that radiation exposure from emissions from a coal-fired power plant may be similar or even greater than that from a nuclear power plant. This is a result of the fact that coal contains uranium and thorium impurities, and some of these are released to the environment with the exhaust.

One way of assessing the risk of different activities is to estimate the resulting average decrease in life expectancy. Table 6.7 shows the results for this type of analysis, suggesting that lifestyle changes can have a much greater effect on the health of the general public than the proliferation of nuclear power.

For most people, the greatest concern related to nuclear power is the possibility of a nuclear accident. Certainly this fear was exacerbated first by the Three Mile Island accident and then by Chernobyl. The Rasmussen report (produced for the U.S. Nuclear Regulatory Commission) considered the safety of the different types of reactors in use in the United States (BWRs and PWRs). This study was made in 1975, before either the Three Mile Island or the Chernobyl accident occurred. It considered the possibility of a nuclear accident in comparison to the possibility of other anthropogenic and natural disasters. Disasters may be categorized by the frequency with which they are anticipated to occur and by the number of resulting fatalities. For example, fires that produce one fatality are frequent, fires that produce 100 fatalities happen every few years, and fires that produce 1000 fatalities are uncommon. Rasmussen's analyses for various anthropogenic disasters and for various natural disasters are summarized in Figures 6.27 and 6.28, respectively.

For the more hazardous events, such as airplane crashes and tornadoes, there are substantial data to compare with the curves in these figures. However, little information is available for comparison with the predictions for nuclear energy safety. Basically, the only clear accident fatalities are from Chernobyl. The plots in Figure 6.28 would suggest that this event was somewhat anomalous and that further disasters of this magnitude are highly unlikely. There is some controversy over some of the details of the Rasmussen report, but the general consensus is, even nearly 40 years later, that nuclear power is not as unsafe as public perception might make it seem.

The overall risk, to both occupational workers and the general public, of various energy sources has been studied in detail by Inhaber (1979). These results have been adapted to give an overall relative risk factor for the various energy production methods shown in Table 6.8. The analysis presented here is consistent with the comparison of coal and nuclear energy already presented. Perhaps surprisingly, energy sources like solar (photovoltaics) are fairly high on the list. (Solar energy is discussed in detail in

Reactor Safety Study: An assessment of Accident Risks in U.S. Commercial Nuclear Power Plants U.S. Nuclear Regulatory Commission (1975) WASH-1400 (NUREG 75/014) p. 2

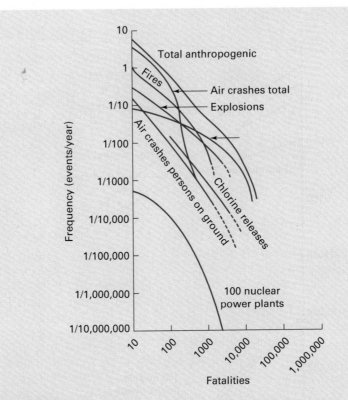

Figure 6.27: Relationship for frequency as a function of fatalities for anthropogenic disasters as predicted by the Rasmussen report.

Chapters 8 and 9.) The risk is relatively small to the general public but fairly substantial for occupational workers. The basic reason for this feature of solar energy is the very low energy density of solar radiation at the surface of the earth and the fairly low efficiency of photovoltaic cells. Generally speaking, it would require a solar collector with an area of about 40 km^2 to provide the power produced by an average-sized nuclear power plant. This requires an enormous amount of material to be processed and a large

Based on data from Kraushaar and Ristinen . . .

Table 6.8: Normalized relative total risk (to occupational workers and the general public) of producing electricity by different methods.

electricity source	relative risk
coal	100
oil	67
wind	33
solar (photovoltaic)	23
methanol (biofuel)	10
hydroelectric	1.5
nuclear	0.3
natural gas	0.2

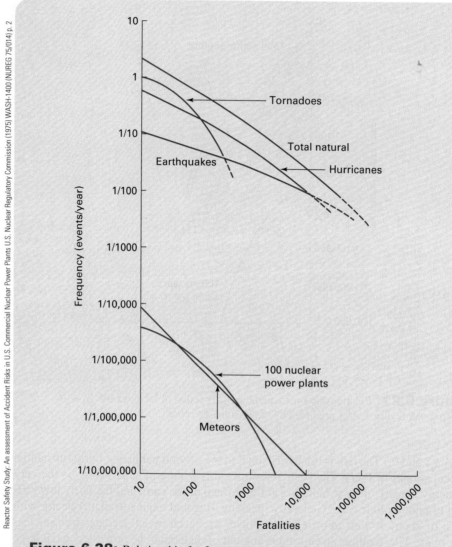

Reactor Safety Study: An assessment of Accident Risks in U.S. Commercial Nuclear Power Plants U.S. Nuclear Regulatory Commission (1975) WASH-1400 (NUREG 75/014) p. 2

Figure 6.28: Relationship for frequency as a function of fatalities for natural disasters as predicted by the Rasmussen report.

time commitment for construction. The amount of material required and the necessary construction times are compared for different energy sources in Figure 6.29. The figure clearly shows that some of the alternative energy sources to be discussed in later chapters need to be analyzed very carefully.

6.10 Waste Disposal

Waste disposal is another serious concern for the use of nuclear energy. Typically, reactors need to be refueled after 6–12 months of operation because the ^{235}U in the core becomes sufficiently depleted and other nuclides are formed that interfere with

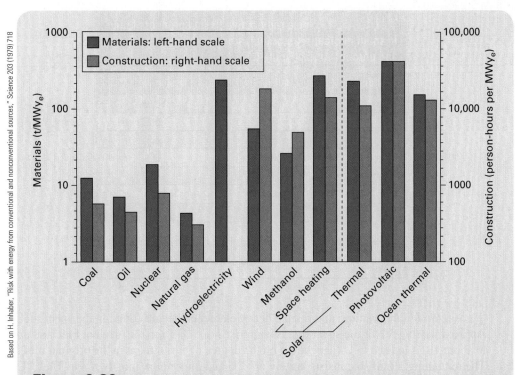

Based on H. Inhaber, "Risk with energy from conventional and nonconventional sources," Science 203 (1979) 718

Figure 6.29: Material and construction time resources necessary to produce 1 MWy$_e$ of electricity by different methods. Note that the vertical scale is logarithmic.

the normal operation of the reactor. The spent fuel is highly radioactive and must be handled appropriately. As explained previously, some spent fuel is reprocessed and made into new fuel for reactors. At present, the majority is stored in safe locations. Spent fuel contains a mixture of various nuclides that result from the nuclear processes in the reactor. The significance of the different nuclides in the spent fuel depends on their concentration, their decay process, and their half-life. Nuclides with fairly short half-lives are not a concern because the radioactivity decays away rapidly. Very long-lived nuclides are not the biggest concern because they decay very slowly and therefore produce comparatively weak radiation. It is the nuclides with half-lives in the tens to hundreds of years that cause the most concern. These decay rapidly enough that they produce considerable radiation but slowly enough that storing them for many half-lives is a long-term commitment (e.g., hundreds of years).

The spent fuel from a nuclear reactor is a mixture of nuclides of two different types: fission products and actinides. *Fission products* account for about 3% of the mass of the spent fuel (discussed in some detail in the previous chapter) and are exemplified by equation (6.3). The fission products undergo β decay until a stable nuclide is reached. The first few β decays in the chain occur fairly quickly, and it is the last decay or two before β stability is achieved that are the most problematic. Figure 6.6 shows such a sequence of β decays for $A = 137$. As seen there, ^{137}I decays to ^{137}Cs within a few minutes, but the ^{137}Cs β decay to ^{137}Ba has a half-life of 30 years. Table 6.9 gives the properties of the principal, fairly long-lived β unstable nuclides that are found in spent reactor fuel.

Table 6.9: Some radioactive fission products found in spent nuclear reactor fuel, along with their half-lives and percent yields (Figure 6.3).

nuclide	half-life (y)	fission yield (%)
^{155}Eu	4.76	0.08
^{85}Kr	10.76	0.22
^{90}Sr	28.9	4.5
^{137}Cs	30	6.3
^{151}Sm	90	0.53
^{99}Tc	211×10^3	6.1
^{126}Sn	230×10^3	0.11
^{93}Zr	1.5×10^6	5.5
^{135}Cs	2.3×10^6	6.9
^{107}Pd	6.5×10^6	1.2

© Cengage Learning 2015

Actinides are uranium and related heavy elements that are found in the spent fuel. Approximately 95% of the spent fuel is comprised of ^{238}U from the original fuel, and up to about 1% is unused ^{235}U (depending on the degree of enrichment in the original fuel). The remaining 1% or so consists largely of ^{236}U and plutonium, mostly ^{239}Pu. The ^{236}U is produced from ^{235}U by neutron capture, followed by γ decay:

$$n + {}^{235}U \rightarrow {}^{236}U + \gamma. \tag{6.5}$$

The ^{236}U has an α decay half-life of 24 million years and is therefore stable on the timescale of the reactor operation. The ^{239}Pu is produced by neutron capture from ^{238}U by the following reaction:

$$n + {}^{238}U \rightarrow {}^{239}U + \gamma. \tag{6.6}$$

The ^{239}U β decays to ^{239}Np and then to ^{239}Pu on a timescale of a couple of days:

$$^{239}U \rightarrow {}^{239}Np + e^- + \bar{v}_e \rightarrow {}^{239}Pu + e^- + \bar{v}_e. \tag{6.7}$$

^{239}Pu is β stable and has an α-decay half-life of 24,000 years. The existence of ^{239}Pu in spent reactor fuel causes security concerns for fuel reprocessing because the ^{239}Pu can be used to make weapons. The existence of ^{239}Pu is, however, the basis on which energy can be extracted from ^{238}U, and the method by which the longevity of uranium resources for thermal neutron reactors may be extended. It also forms the basis for the operation of the fast breeder reactor.

Generally speaking, the average life of the activity from fission products is shorter than that of the actinides. So, although the activity comes primarily from fission products at the time the fuel is removed from the reactor, after a sufficiently long time, the actinides are responsible for the majority of the activity (Figure 6.30).

The public perception of the problem of radioactive nuclear waste may not be entirely realistic. Because the production of energy from nuclear reactions requires small amounts of fuel (compared to processes that make use of chemical energy), the actual amount of nuclear reactor waste generated is relatively small in terms of mass

Example 6.3

Describe the by-products of the α decay of ^{239}Pu.

Solution

The ^{239}Pu nucleus contains 94 protons and 145 neutrons. Because the α particle (which is the nucleus of the ^{4}He atom) contains 2 protons and 2 neutrons, the conservation of proton and neutron number during α decay requires that the progeny nucleus must have $(94 - 2) = 92$ protons and $(145 - 2) = 143$ neutrons. This is a ^{235}U nucleus; thus the α decay process would be

$$^{239}\text{Pu} \rightarrow {}^{235}\text{U} + \alpha$$

or

$$^{239}\text{Pu} \rightarrow {}^{235}\text{U} + {}^4\text{He}.$$

or volume. Low-level waste is generated not only from reactors and fuel processing (e.g., uranium refining) but also from medical activities, research, industry and other sources. It is the high-level radioactive waste from reactors that is most problematic. Figure 6.31 shows the total amount of nuclear waste generated by reactors to date far and extrapolated into the future (assuming the continued use of nuclear energy at the current rate and a continuation of current reprocessing policies). The total amount of high-level reactor waste generated to date by the commercial reactor program world-wide (about 300,000 t) amounts to the volume of a cube 40 m on a side; about the volume of coal burned by one typical coal-fired generating station in 4 days. Thus we see that the volume of high-level waste is not very large. It is merely a question of how to safely and securely contain it until it is no longer hazardous. A similar problem exists for hazardous industrial chemicals that are produced in quantities much larger than those of radioactive waste and that, in many cases, are distributed in the environment

Based on J.L. Zhu and C.Y. Chen, "Radioactive waste management: World overview," IAEA Bulletin 4/1989 p. 5

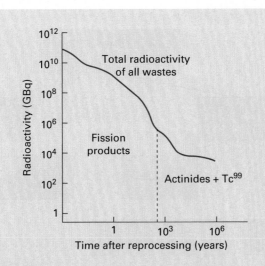

Figure 6.30: Activity as a function of time in spent reactor fuel.

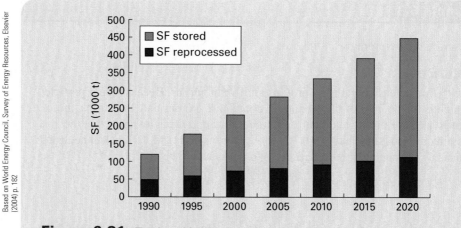

Based on World Energy Council, Survey of Energy Resources, Elsevier (2004) p. 182

Figure 6.31: Total worldwide mass of nuclear waste produced by commercial power reactors. (SF = spent fuel).

with relatively little concern for the long-term consequences. A major difference is that an arsenic atom (for example) always remains an arsenic atom and is always be toxic; a radioactive nucleus ultimately decays into something that is not radioactive.

Thus far, there has been little consensus on the best solution for nuclear waste disposal. At present, most radioactive waste produced by commercial power reactors is stored in facilities at the reactor site, although this is not considered to be a final solution to the storage problem. Several approaches can be taken to nuclear waste disposal. The waste can be stored until it has sufficiently low activity to be of relatively little concern or it can be eliminated from our environment. One proposal to eliminate nuclear waste from our environment is to shoot it into space. Another is to use neutron irradiation or some other appropriate means to transmute the radioactive nuclides into stable ones. Both of these are likely to be economically inviable. Figure 6.32 summarizes the possibilities of dealing with nuclear waste. Most likely, continental underground

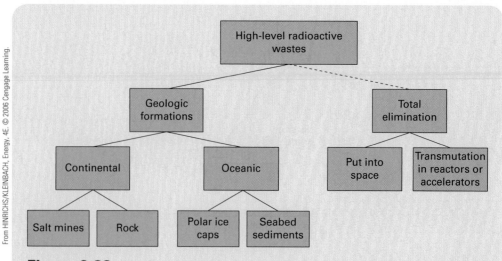

From HINRICHS/KLEINBACH, Energy, 4E. © 2006 Cengage Learning.

Figure 6.32: Possible options for radioactive waste disposal.

storage will, in the near future at least, be the most practical. One approach that can help to alleviate the storage problem is to separate the relatively short-lived fission products from the generally longer-lived actinides. The fission products will need to be stored for about 300 years before they are nonradioactive enough to be disposed of by simpler means. The actinides will need to be stored (unless they are reprocessed into new fuel) more or less indefinitely. It should be pointed out that, at present in the United States, the quantity of high-level radioactive waste produced by military activities far exceeds that produced by commercial power reactors.

6.11 Advanced Reactor Design

Although there has been little expansion of the nuclear power industry in the past 30–40 years, there has been renewed interest in constructing new reactors in recent years. This interest stems from changes in the public view of viable energy sources for the future and developments that will alleviate much of the concern for public safety. An analysis of the principal nuclear accidents to date indicates that operator error is as major a concern (as at Three Mile Island and Chernobyl) as is faulty design (Chernobyl). However, more recent events have shown that natural disasters (e.g., Fukushima) must also be considered as a possible cause of a nuclear accident. New reactor designs are much more conscious of safety systems to avoid accidents as well as improved systems for containing radioactive materials in the event of an accident.

Also, several more innovative designs for reactors are being developed. Perhaps the most promising of these is the *pebble bed reactor*. This is similar in some ways to the British reactor design because it is a graphite-moderated, gas- (He-) cooled reactor. Rather than being in the form of rods, the uranium fuel elements are in the form of spheres about 0.5 mm in diameter. Each is coated with a temperature-resistant silicon carbide layer (Figure 6.33). About 15,000 of these individually

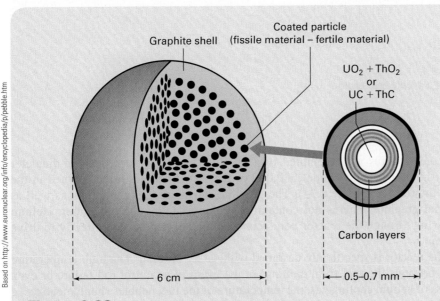

Based on http://www.euronuclear.org/info/encyclopedia/p/pebble.htm

Figure 6.33: Fuel element for a pebble bed reactor.

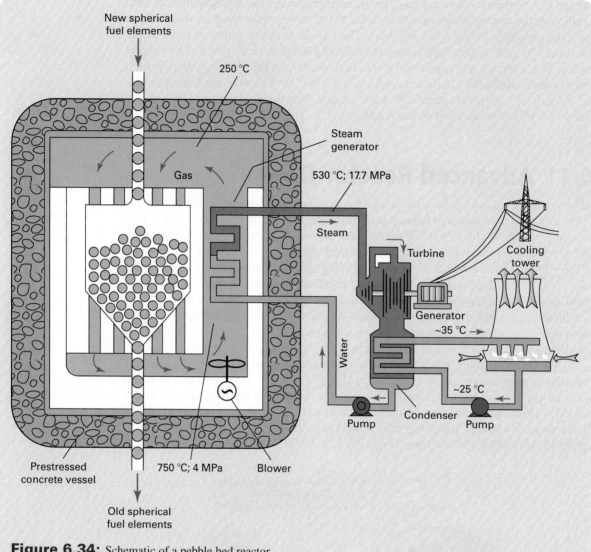

Figure 6.34: Schematic of a pebble bed reactor.

coated fuel spheres are assembled in a graphite matrix to form a 6-cm-diameter sphere (the fuel pebble). About 300,000 of these fuel pebbles are contained in the core of the reactor (Figure 6.34). The graphite surrounding the uranium, as well as additional graphite spheres not containing uranium, act as the moderator. Helium gas flowing through the reactor core acts as a coolant and is used directly to drive gas turbines.

The reactor is specifically designed with safety in mind. The high-temperature coating prevents the fuel from melting. In the event of a control rod failure or even total loss of helium cooling gas, the temperature of the fuel pebbles should not exceed the melting temperature of the carbide coating. These reactors are typically small

(150 MW$_e$ or so) and more compact than traditional designs. They may be the basis for thermal reactor designs in the future and a revival of nuclear power.

6.12 **Fast Breeder Reactors**

In the preceding, reactor designs, most of the uranium in a reactor will clearly go unused. This is the ^{238}U. A *fast breeder reactor* (FBR) is designed to make use of much of the potential energy associated with the ^{238}U by converting it into fissile reactor fuel, as in the reactions in equation (6.6). The fast breeder reactor consists of a core containing fissile nuclei, either ^{235}U or ^{239}Pu. The latter material is more practical in the long term because it is produced (bred) by the reactor. The core is surrounded by a breeder blanket of primarily ^{238}U. The core is enriched to about 25% in fissionable nuclei, enabling a controlled fission reaction to be maintained using fast neutrons. The fast neutrons have a higher relative cross section for capture by ^{238}U than do thermal neutrons. Thus some of the excess neutrons can be used to produce new fissionable material in the breeder blanket. To ensure that neutrons in the core remain at high energy, no moderator is used, precluding the use of water (light or heavy) as a coolant because these would both moderate the neutrons. The critical parameter for an FBR is the *breeding ratio*. This is the ratio of the number of fissile nuclei produced from nonfissile ^{238}U in the breeder jacket to the number of fissile nuclei that undergo fission in the core. Ideally, the breeding ratio should be greater than one so that the reactor will produce more fissile fuel than it consumes (in addition to producing heat).

Fast breeder reactors are primarily categorized by the type of coolant they use: gas cooled, liquid metal cooled, and molten salt cooled. Liquid sodium, lead, and a lead-bismuth mixture have been investigated as possible coolants for the liquid metal–cooled variety. All three designs have been considered, and laboratory prototype liquid metal and molten salt reactors have been constructed. However, only the liquid metal fast breeder reactor (LMFBR) using sodium as a coolant has been implemented in a full-scale version. A typical design is shown in Figure 6.35. Since the 1960s, approximately 20 functional LMFBRs have been built, and several of these are still operational. Most of these have been fairly small (~200 MW$_e$) reactors. An LMFBR was constructed in Michigan and became operational in 1963. It ceased operation in 1972, and the U.S. program on fast breeder reactors was subsequently discontinued. Other countries (France, Russia, India, and Japan) have active FBR development programs and, in some cases, operational reactors. Further technological development may be necessary to ensure the reliability and safety of FBRs.

Clearly, the advantage of FBRs is the ability to extract energy from a substantial fraction of the available ^{238}U. Because ^{238}U is 140 times as abundant as ^{235}U, we might expect a similar increase in the longevity of uranium resources if FBRs replace thermal neutron reactors. This means that the world uranium resources would provide energy at the current rate of nuclear energy production for perhaps 5000 years rather than 35, as previously indicated for thermal reactors. The situation is actually much better than this. The requirements for the breeder blanket are less demanding than those for thermal reactor fuel, meaning that lower-quality uranium ore would become viable. Estimates have suggested that at current usage rates our uranium resources could fuel fast breeder reactors for the next 50,000 years. Even with a more active FBR program, resources could be sustained for many thousands of years.

ENERGY EXTRA 6.2
Advanced fission reactor designs

Since nuclear generated electricity first in appeared in the 1950s, the design of fission reactors has continued to develop. The very earliest reactors are referred to as first-generation, or generation I, reactors, and, aside from a small number of units still operational in the United Kingdom, all of these reactors have been decommissioned. Second-generation (generation II) reactors are presently in the most common use. Generation III reactors are sometimes referred to as advanced reactors, and some have been operational worldwide since the late 1990s. Generation III reactors are evolutionary designs that follow from generation II reactors but incorporate additional design features that include:

- Lowered production costs.
- Increased ease of operation.
- Improved safety and security.
- More rugged design and increased impact resistance.
- Greater utilization of fuel.
- Longer life expectancy.

A number of generation III reactors are currently operational worldwide, and these designs will play the most important role in the development of nuclear power over the next decade or two.

Generation III+ reactors, such as the pebble bed reactor, tend to be more radical in their design and strive to further the improved cost and safety aspects of generation III reactors. A number of generation III+ designs (like the pebble bed reactor) are in the latter stages of development, and some of them are expected to be online in a few years. A feature that many generation III+ reactors have in common is a modular design. Rather than aiming for increased-capacity reactors, many of which are now well in excess of 1-GW$_e$ output, generation III+ reactors tend to be in the few-hundred-megawatt range. These individual reactor modules can be combined into larger facilities.

Another interesting concept is the so-called personal nuclear reactor. These designs developed

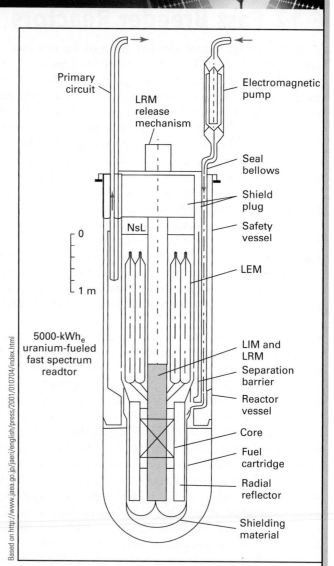

Based on http://www.jaea.go.jp/jaeri/english/press/2001/010704/index.html

Internal design of the Rapid-L reactor.

to some extent out of ideas on how a future lunar base might be powered. The requirements for such a unit are that it would be easy to operate, self-contained, safe, small, and near zero maintenance and that it would not require refueling for extended periods of time. These same criteria would lead to a small reactor that would be appropriate for powering an apartment or office building or a remote area that could not

Continued on page 181

Energy Extra 6.2 continued

be connected to the grid. The Tokai Research Establishment, Japan Atomic Energy Research Institute, has promoted the design of the 200-kW$_e$ Rapid-L reactor, shown in the figure. The dimensions on the diagram indicate the small size of the reactor, and its total mass is expected to be around 8000 kg. It is designed to operate continuously for ten years without refueling. Whether such an approach to nuclear power is economical and practical and whether it gains public support and is consistent with government policies remain to be seen.

Topic for Discussion

A Rapid-L nuclear reactor is expected to be economically viable. It operates with minimal maintenance costs and has a capacity factor of 50%. If the charge for electricity is \$0.15 per kWh$_e$ (slightly more than the usual rate for fossil fuel–generated electricity), what is the maximum infrastructure cost if the payback period has to be ten years or less? How does this installation cost, in terms of dollars per W$_e$, compare with typical installation costs for large-scale (i.e., ~1-GW$_e$) nuclear generating facilities.

The ability to breed fissile material from nonfissile material opens up another interesting possibility, the use of naturally occurring ^{232}Th as a fuel. ^{232}Th, ^{238}U, and ^{235}U are the only three naturally occurring actinide nuclides. Because of its longer half-life, 1.4×10^{10} years, compared with 4.5×10^9 years for ^{238}U, ^{232}Th is more plentiful

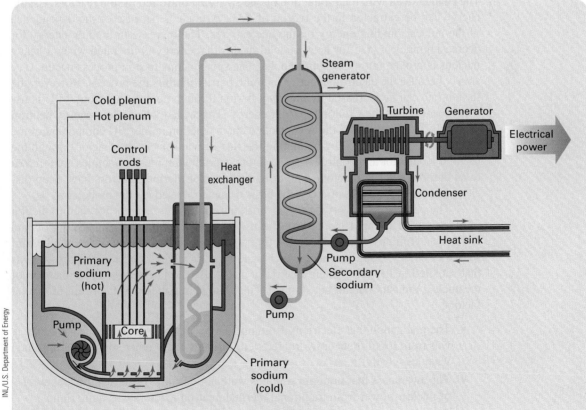

Figure 6.35: Schematic of a sodium-cooled LMFBR.

INL/U.S. Department of Energy

on earth than either of the uranium isotopes. It is nonfissile and naturally decays by α decay according to the process

$$^{232}\text{Th} \rightarrow\ ^{228}\text{Ra} + \alpha \tag{6.8}$$

with an energy release of 4.081 MeV. ^{232}Th may be converted into a fissile nucleus by the following process: Firstly, neutron capture produces ^{233}Th:

$$\text{n} +\ ^{232}\text{Th} \rightarrow\ ^{233}\text{Th} + \gamma. \tag{6.9}$$

The ^{233}Th then undergoes two β decays:

$$^{233}\text{Th} \rightarrow\ ^{233}\text{Pa} + \text{e}^- + \bar{v}_e \rightarrow\ ^{233}\text{U} + \text{e}^- + \bar{v}_e, \tag{6.10}$$

(Pa = protactinium), with half-lives of 22 minutes and 27 days, respectively. The ^{233}U is β stable and decays naturally with a half-life of 1.6×10^5 years. It is therefore stable long enough to serve as a nuclear reactor fuel, and, because it is fissile, can be used to produce energy by induced fission. Thorium is a resource that has attracted interest in recent years and may be a factor in future nuclear reactor development.

6.13 Summary

The chapter dealt with the production of energy from controlled fission reactors. Usable energy can be extracted in the fission of a heavy nucleus because of the dependence of the nuclear binding energy on the nuclear size. The increase in binding energy for decreasing nuclear size for heavy nuclei means that energy is liberated when a heavy nucleus is broken into two lighter nuclei. The spontaneous fission process occurs very slowly because the electromagnetic Coulomb barrier inhibits the process. However, the fission rate can be increased by inducing fission through neutron reactions. ^{235}U is fissile, meaning that the Coulomb barrier energy is exceeded when a low-energy neutron is absorbed. ^{238}U is not fissile. A controlled chain reaction in the ^{235}U component of the reactor fuel is maintained by regulating the flux of neutrons in the reactor through the use of a moderator and control rods. Reactors operating on these principles are referred to as thermal neutron reactors. The different types of thermal reactors have been discussed, and these are classified on the basis of material used for the moderator.

The chapter also overviewed the world use of fission reactors for the production of electricity. The rapid increase in the use of nuclear energy during the 1960s and 1970s was prompted by the hope that this would provide an inexpensive and plentiful source of energy. Some countries, such as France, developed nuclear power programs that satisfied nearly all of their electricity needs. By the late 1980s, the expansion of the nuclear industry had all but stopped. This change in policy was the result of several factors:

- The realization that electricity from nuclear energy was no less expensive than it was from fossil fuels or hydroelectricity (In fact, these other sources were somewhat less costly.)
- The awareness that uranium supplies were not unlimited and that the longevity of nuclear power from traditional thermal neutron reactors was quite finite
- The concern for the security of nuclear materials, a major factor for policies concerning fast breeder reactors and fuel reprocessing in many countries

- The growing concern for the safety of nuclear reactors (The potential for wide reaching health and environmental consequences of nuclear accidents became apparent after the Three Mile Island and Chernobyl incidents.)
- Concerns over the environmental consequences of nuclear waste disposal

In more recent years, the attitude toward nuclear power has warmed somewhat, even to the point of expansion and/or future planned development of nuclear power capabilities in many parts of the world. This direction is certainly influenced by dwindling fossil fuel supplies and an increased awareness of their environmental effects, the development of new fission reactors that are inherently safer, and a more positive attitude toward fuel reprocessing.

This chapter also discussed the potential risks associated with nuclear reactors. This discussion presented the results of risk assessment studies but also focused on an evaluation of the actual nuclear accidents that occurred at Three Mile Island (1979), Chernobyl (1986), and Fukushima (2011). The recent Fukushima accident in Japan, however, has already had an effect on government policy in some countries. Switzerland and Germany are both phasing out nuclear power activities, and voters in Italy have rejected plans to revive nuclear activities.

The limitations of the availability of uranium resources have been discussed. Thermal reactors make use of only the energy associated with ^{235}U, which comprises less than 1% of naturally occurring uranium. Fast breeder reactors utilize high-energy neutrons to convert nonfissile ^{238}U into fissile ^{239}Pu. By this method, the available energy from natural uranium can be increased by a factor of 100 or more. These reactors are in limited use worldwide, and the possibilities of increasing fast breeder reactor use and of fuel reprocessing have been presented. The possible use of naturally occurring ^{232}Th as a fuel in nuclear reactors was discussed.

Problems

6.1   What is the mass of ^{235}U that, when undergoing induced fission, would produce the same amount of thermal energy as 1 tonne of coal?

6.2   **a.** Calculate the mass-energy for one kg of ^{235}U.
   b. Calculate the total nuclear binding energy for 1 kg of ^{235}U.
   c. Calculate the energy from nuclear fission of 1 kg of ^{235}U.

6.3   Using the following masses, show that ^{239}Pu is fissile:

$$m(^{239}\text{Pu}) = 239.0521565 \text{ u.}$$

$$m(^{240}\text{Pu}) = 240.0538075 \text{ u.}$$

Note: This property of ^{239}Pu allows it to be used in a thermal neutron reactor in the same way that naturally occurring ^{235}U is used.

6.4 ^{137}Cs is a common and particularly troublesome fission fragment produced in thermal neutron reactors. The difficulty with ^{137}Cs arises because of its long half-life (30 years) in the β decay chain (Figure 6.6). Calculate the time required for ^{137}Cs in nuclear reactor waste to decay to 10^{-6} of its original activity.

6.5   Calculate the α decay energy of ^{239}Pu [$m(^{239}$Pu$) = 239.0521565$ u].

6.6   Low-grade uranium ore contains a fraction of a percent (by weight) of uranium; very high-grade ore (as is found in some deposits in Canada) can contain up to 20% uranium. If energy is extracted from uranium by fission of ^{235}U (0.72% of naturally occurring uranium), then how large a percentage of uranium in ore is necessary to make a tonne of uranium ore equivalent (in terms of energy) to a tonne of coal?

6.7   A possible process for induced fission of ^{235}U is

$$n + {}^{235}U \rightarrow {}^{236}U \rightarrow {}^{138}Xe + {}^{96}Sr + 2n.$$

Using the following atomic masses, calculate the energy released in this process:

$$m(^{138}Xe) = 137.9139885 \text{ u.}$$

$$m(^{96}Sr) = 95.92168047 \text{ u.}$$

6.8   Use the information in Figure 6.27 to argue that it is more likely to die in a nuclear accident with 10 fatalities than in one with 1000 fatalities.

Bibliography

D. Bodansky. *Nuclear Energy: Principles, Practices, Prospects.* American Institute of Physics, New York (1996).

B. L. Cohen. *The Nuclear Energy Option—An Alternative for the 90s.* Plenum Press, New York (1990).

B. L. Cohen. "The nuclear reactor accident at Chernobyl, USSR." *American Journal of Physics* **55** (1987): 1076.

B. L. Cohen. "Breeder reactors: A renewable energy source." *American Journal of Physics* **51** (1983): 75.

B. L. Cohen. "The disposal of radioactive wastes from fission reactors." *Scientific American* **236** (1977): 21.

R. A. Dunlap. *An Introduction to the Physics of Nuclei and Particles.* Brooks-Cole, Belmont, CA (2004).

M. M. El-Wakil. *Powerplant Technology.* McGraw-Hill, New York (1984).

J. A. Fay and D. S. Golomb. *Energy and the Environment.* Oxford University Press, New York (2002).

S. Glasstone and A. Sesonske. *Nuclear Reactor Engineering: Reactor Design Basics, Volume 1.* Springer, New York (1994).

S. Glasstone and A. Sesonske. *Nuclear Reactor Engineering: Reactor Systems Engineering, Volume 2.* Springer, New York (1994).

D. R. Inglis. *Nuclear Energy—Its Physics and Its Social Challenge.* Addison-Wesley, Reading, MA (1973).

H. Inhaber. "Risk Evaluation." *Science* **203** (1979): 718.

R. A. Knief. *Nuclear Engineering.* Taylor and Francis, Washington, DC (1992).

J. A. Kraushaar and R. A. Ristinen. *Energy and Problems of a Technical Society* (2nd ed.). Wiley, New York (1993).

J. Lamarsh. *Introduction to Nuclear Engineering* (2nd ed.). Addison-Wesley, Reading, MA (1983).

H. W. Lewis. "The safety of fission reactors." *Scientific American* **242** (1980): 53.

J. N. Lillington. *The Future of Nuclear Power*. Elsevier, Amsterdam (2004).

R. L. Murray. *Nuclear Energy: An Introduction to the Concepts, Systems, and Applications of Nuclear Processes*. Pergamon, New York (1993).

R. L. Murray. *Understanding Radioactive Waste*. Battelle Press, Columbus (1989).

R. L. Murray. *Introduction to Nuclear Engineering*. Prentice Hall, Englewood Cliffs, NJ (1961).

A. V. Nero Jr. *A Guidebook to Nuclear Reactors,* University of California Press, Berkeley (1979).

R. G. Steed. *Nuclear Power: In Canada and Beyond*. General Store, Renfrew (2007).

J. Tester, E. Drake, M. Driscoll, M. W. Golay, and W. A. Peters. *Sustainable Energy: Choosing Among Options*. MIT Press, Cambridge, MA (2006).

D. F. Torgerson, K. R. Hedges, and R. B. Duffy. "The evolutionary CANDU reactor—Past, present and future." *Physics in Canada* **60** (2004): 341.

J. Weil. "Pebble bed design returns." *IEEE Spectrum* **38** (2001): 37.

World Energy Council. *2010 Survey of Energy Resources*, available at http://www.worldenergy.org/documents/ser_2010_report_1.pdf

Energy from Nuclear Fusion

Learning Objectives: After reading the material in Chapter 7, you should understand:

- The properties of fusion reactions and the production of fusion energy.
- The design and operation of magnetic confinement reactors.
- The design and operation of inertial confinement reactors.
- The importance of the Lawson criterion.
- The design of a fusion power reactor.
- Progress toward a viable fusion reactor.

7.1 Introduction

In the previous chapter, it was seen that the use of either fast breeder reactors or thermal reactors with effective fuel reprocessing, would provide much of our energy needs for a substantial period of time. The decision to utilize nuclear fission energy must deal with concerns over reactor safety, nuclear waste disposal, and the security of nuclear materials. An alternative approach that makes use of the enormous energy associated with the nuclear force is fusion energy. Figure 6.1 shows that the binding energy per nucleon of a nucleus increases with nuclear size for very light nuclei. Thus, binding together two light nuclei to produce a heavier nucleus (up to about $A = 55$) is an exothermic process and can produce usable energy. Fusion has several significant advantages over fission, such as

- A potentially inexpensive and plentiful supply of fuel.
- Reactions that are inherently easier to control and are therefore much safer.
- Substantially reduced environmental hazards from reactor by-products.

However, at present, fusion power is not technologically feasible because of the fundamental differences between the fission and fusion processes. This chapter reviews the physics of nuclear fusion and overviews the efforts to produce a viable fusion reactor.

7.2 Fusion Energy

The bombardment of a fissile nucleus with a low-energy neutron provides enough excess energy to put the nucleus in an energy level that is above the Coulomb barrier. This causes the nucleus to undergo induced fission, and the fission fragments are repelled from each other by the repulsive Coulombic interaction between the protons. To fuse two light nuclei, they must be pushed together against the Coulomb force. The nuclei must have enough energy and be in proximity of one another for a long enough time to get through or over the Coulomb barrier. Once the nuclei are close enough together long enough, there is a probability that the strong interaction will take over and fuse the nuclei, thus releasing energy. It is certainly straightforward to accelerate nuclei to these energies, even in very modest particle accelerators, and to collide them with other nuclei to produce fusion reactions. Unfortunately, in such a situation, the energy expenditure for the accelerator is substantially greater than the energy gain from the fusion. Thus, although this is a useful way of learning about fusion reactions in the laboratory, it is not a practical means of obtaining energy. So, to make use of fusion energy, it is necessary to create conditions in the laboratory that are compatible with fusion but that do not require an excessive expenditure of energy. This generally means making a collection of nuclei that is dense enough and hot enough to undergo fusion.

From a practical standpoint, it is desirable to reduce the Coulombic repulsion by using nuclei with as few positively charged protons as possible. Thus, although it might be possible, in principle, to fuse two aluminum nuclei (13 protons each) to form an iron nucleus (26 protons), it is not a productive way to approach this problem.

The simplest fusion process might appear to be the fusion of two protons, or the so-called *p-p process*; that is, the fusion of two hydrogen (^{1}H) nuclei. However, two protons cannot form a bound state, and p-p fusion requires a simultaneous β^+ decay process in which one of the protons is converted to a neutron to give

$$p + p \rightarrow d + e^+ + \nu_e. \tag{7.1}$$

Here d is the deuteron, or the nucleus of a ^{2}H atom, that is, a bound pair consisting of a neutron and a proton. Because β decay is involved (equation 5.18), the weak interaction is, at least partly, responsible for p-p fusion. As a result, it is very difficult to cause this reaction to occur.

Example 7.1

Calculate the energy associated with the fusion process shown in equation (7.1).

Solution

The mass difference between the left- and right-hand sides of the equation is converted into energy by the relationship.

$$E = (2m_p - m_d - m_e)c^2.$$

Using values for the masses from Appendix II gives (recall that the positron is the antiparticle of the electron and that their masses are identical)

$$E = (2 \times 1.007276470 \text{ u} - 2.013553214 \text{ u} - 0.000548579903 \text{ u})$$
$$\times 931.494 \text{ MeV/u} = 0.42 \text{ MeV}.$$

Continued on page 188

Example 7.1 continued

Normally the fusion process is followed by the annihilation of the positron with an electron. This annihilation process is

$$e^- + e^+ \rightarrow 2\gamma,$$

where the mass of the photons on the right-hand side of the equation is zero. Thus, the energy released is

$$E = (m_e + m_e)c^2 = (2 \times 0.000548579903 \text{ u}) \times 931.494 \text{ MeV/u} = 1.02 \text{ MeV}.$$

The total energy related to the p-p fusion process is therefore

$$0.42 \text{ MeV} + 1.02 \text{ MeV} = 1.44 \text{ MeV}.$$

Fusion processes involving deuterons (i.e., nuclei of ^{2}H atoms) are of importance, and the simplest of these is *d-p fusion*:

$$\text{d} + \text{p} \rightarrow {}^3\text{He} + \gamma \qquad (Q = 5.49 \text{ MeV}), \qquad \textbf{(7.2)}$$

where, for the purpose of energy calculations, the ^{3}He on the right-hand side refers to the nucleus of a ^{3}He atom. Alternatively, atomic masses may be used for the process $^2\text{H} + {}^1\text{H} \rightarrow {}^3\text{He} + \gamma$. In either case, the number of electrons must be conserved in the process because weak interactions are not involved. The energy release, Q, is shown in the equation. The most obvious process involving the fusion of two deuterons is the formation of ^{4}He:

$$\text{d} + \text{d} \rightarrow {}^4\text{He} + \gamma \qquad (Q = 23.8 \text{ MeV}). \qquad \textbf{(7.3)}$$

However, this process is unlikely because the large amount of energy released makes the ^{4}He nucleus unstable and generally results in the release of one of the neutrons or one of the protons. Consequently, there are two more likely modes of *d-d fusion*:

$$\text{d} + \text{d} \rightarrow {}^3\text{He} + \text{n} \qquad (Q = 3.3 \text{ MeV}) \qquad \textbf{(7.4)}$$

and

$$\text{d} + \text{d} \rightarrow {}^3\text{H} + \text{p} \qquad (Q = 4.0 \text{ MeV}). \qquad \textbf{(7.5)}$$

A final process of importance is the fusion of a deuteron with a triton (i.e., a ^{3}H nucleus). This is *d-t fusion*:

$$\text{d} + \text{t} \rightarrow {}^4\text{He} + \text{n} \qquad (Q = 17.6 \text{ MeV}). \qquad \textbf{(7.6)}$$

This process releases a substantial amount of energy and is of particular importance, as will be seen, for the operation of a controlled fusion reactor.

An understanding of the fusion processes that produce energy in the sun is helpful. The sun, like most stars, produces energy primarily by fusing four hydrogen nuclei (^{1}H) together into one helium nucleus (^{4}He):

$$4\text{p} \rightarrow {}^4\text{He} + 2\text{e}^+ + 2\nu_e, \qquad \textbf{(7.7)}$$

where two of the fusing protons must be converted into neutrons by weak β^+ decay processes. The four hydrogen nuclei do not fuse simultaneously to form helium. Instead, the helium forms in a series of steps. The first step of this fusion process is the fusion of two protons, as given by equation (7.1). In principle, two deuterons could then fuse according to equation (7.3) to form a ^{4}He nucleus, although, as explained, this is not likely. A more likely process is *p-d fusion*, as given by equation (7.2), to form ^{3}He. Two ^{3}He nuclei will then most likely fuse according to the reaction

$$^3\text{He} + {}^3\text{He} \rightarrow {}^4\text{He} + 2{}^1\text{H} + \gamma \qquad (Q = 12.86 \text{ MeV}). \qquad \textbf{(7.8)}$$

This overall process is the most common method of energy production in the sun and is referred to as the *proton-proton cycle* (or *p-p cycle*). Although other processes are going on as well, the p-p cycle produces about 85% of the energy from the sun. The total energy associated with the p-p cycle is $Q = 26.7$ MeV, most of which is eventually converted into solar radiation. All of the steps in this process occur slowly. Equation (7.1) involves the weak interaction; equations (7.2) and (7.8) are limited by the amount of ^{2}H and ^{3}He in the sun and, in the latter case, an increased Coulomb barrier. At the present stage of the sun's evolution, most of the nuclei are still unreacted ^{1}H. However, equation (7.1) is the most limiting factor in the energy production in the sun. This is actually a good situation because it results in an energy output from the sun that is compatible with the requirements for life on earth and ensures that the sun's hydrogen supply is not depleted too quickly.

One might suspect that if the conditions present in the sun could be reproduced in the laboratory, then a functioning fusion reactor would be possible. This is not true. The sun consists of about 10^{57} protons (more or less). Its total energy output is about 3.8×10^{26} W (or J/s). Since one p-p cycle produces about 27 MeV (or 4×10^{-12} J), about $(3.8 \times 10^{26}\text{W})/(4 \times 10^{-12}\text{J}) \approx 10^{38}$ p-p cycles occur every second, corresponding to the fusing of 4×10^{38} protons. As a fraction of the total number of protons in the sun (about 10^{57}), this corresponds to 1 in $(10^{57}/4 \times 10^{38})$, or 1 in 2.5×10^{18}. If the sun produced energy at a constant rate and was able to fuse all of its protons, then it would exist for 2.5×10^{18} seconds, or about 80 billion years. These assumptions are not exactly true, and the sun's life expectancy is somewhat less than this. However, if a fusion reactor filled with ^{1}H nuclei were constructed that approximated the conditions in the sun, then it would take several tens of billions of years to extract all of the available fusion energy from the fuel. This is obviously not practical and is the reason that p-p fusion is not a consideration for fusion energy production on earth.

To properly assess the usefulness of other fusion reactions for the production of energy, it is necessary to examine a bit more of the physics of these processes. It is clearly advantageous to use isotopes of hydrogen rather than helium because of the reduced Coulomb barrier. In general, the greater the number of neutrons, then the greater the strength of the strong interaction that will fuse the nucleons together once the nuclei get past the Coulomb barrier. This is seen for d-d and d-t fusion in Figure 7.1. Here the relative probability of fusion (the fusion cross section) is plotted as a function of energy. These reactions correspond to equations (7.4) and (7.5) for d-d fusion and to equation (7.6) for d-t fusion. For d-d fusion, the data in the figure are the sum of both reactions.

The differences between d-d and d-t fusion shown in the figure clearly indicate that achieving d-t fusion in the laboratory should be much easier than achieving d-d fusion. For this reason, laboratory experiments have concentrated largely on d-t fusion, although, as discussed in Section 7.4, the development of laboratory d-d fusion may be considered the ultimate goal of research in this area.

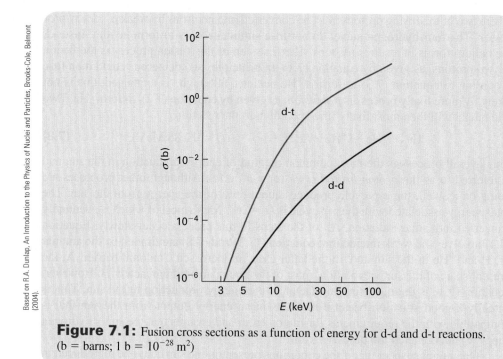

Figure 7.1: Fusion cross sections as a function of energy for d-d and d-t reactions. (b = barns; 1 b = 10^{-28} m^2)

7.3 Magnetic Confinement Reactors

The basic aim of fusion reactor development is to achieve a situation where the energy input into the reactor that is necessary to maintain fusion conditions is less than the energy that is extracted from the reactor. Thus, there is a net gain of energy from the fusion. The traditional approach to fusion reactors is to achieve the necessary conditions by using a very high temperature. This approach seems to be justified by Figure 7.1, where increasing temperature gives rise to an increase in the kinetic energy of the nuclei and a subsequent increase in the probability of fusion. For the thermal energy of a particle to reach, say, 10 keV, as is typical in the figure, the temperature must be about 100 million degrees (K). The major problem with achieving a useful fusion reaction is to obtain this temperature in order to to get the nuclei close enough together and to keep them in that state until a sustained fusion reaction occurs. At these temperatures, all matter is fully ionized and becomes a plasma; that is, the electrons are stripped from the nuclei, and the negatively charged electrons and positively charged nuclei can move about independently. Although the positive and negative charges in the plasma are not bound together, the plasma as a whole contains the same number of positive and negative charges as the initial neutral atoms and remains electrically neutral.

The obvious difficulty in traditional fusion research is the means by which a plasma at such a high temperature can be contained. All solid materials will melt and vaporize long before this temperature is reached. There are two traditional approaches to containing the plasma and obtaining the necessary fusion conditions: magnetic confinement and inertial confinement.

In a *magnetic confinement reactor*, the charged nature of the electrons and ionized nuclei is utilized. Because the ions and electrons in a plasma are free to move independently, their motion can be controlled by the application of a suitable magnetic field. Magnetic confinement reactors utilize magnetic fields to direct the particles in a plasma in order to prevent them from colliding with the walls of the containment vessel. Two basic geometries are used for these devices: a linear geometry and a toroidal geometry.

The *linear* (or *mirror*) *geometry* uses a plasma column that is pinched or closed off at the ends. The plasma is contained in a cylindrical chamber, and an axial magnetic field is provided by coils around the outside of the chamber. Basically the particles travel in a region of comparatively low field along the length of the cylinder and are reflected from the ends of the cylinder by regions of greater field. In general, progress toward the conditions necessary for a sustained fusion reaction in a mirror confinement device has fallen short of that achieved in other reactor designs. As a result, most current fusion research is directed toward the toroidal reactors and inertial confinement reactors.

In a *toroidal reactor*, the plasma column may be closed in the form of a toroid, in which case, the particles travel along toroidal field lines produced by currents in coils around the toroid [Figure 7.2(a)]. The currents produced by the charged particles in this direction are referred to as poloidal currents. In this geometry, the windings of the coils are closer together on the inside of the torus than on the outside, resulting in a stronger magnetic field near the inside. The consequence of this is that the particles slowly spiral outward, toward the region of weaker field and eventually strike the outer wall of the chamber. To compensate for this effect, an additional (poloidal) field is applied [Figure 7.2(b)]. The net field lines are helical in shape, and the particles avoid interaction with the chamber walls as they follow these curved field lines. Probably the most successful reactor based on this design is the tokamak (an acronym for the Russian name of the device: *to*roidal'naya *ka*mera s *ma*gnitnymi *ka*tushkami, or toroidal chamber with magnetic coils (Figure 7.3). In this device, the toroidal current is actually the current associated with the flow of charged plasma particles around the torus. The photograph in Figure 7.4 shows the interior design of a tokamak. One should not necessarily think of a toroid with the general (donut-like) shape illustrated in Figure 7.3 because many designs have a poloidal diameter that

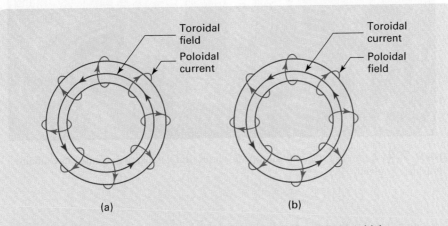

Based on R.A. Dunlap, An Introduction to the Physics of Nuclei and Particles, Brooks-Cole, Belmont (2004).

Figure 7.2: Geometry of currents and magnetic field lines in a toroidal reactor: (a) toroidal field produced by poloidal currents; (b) poloidal field produced by a toroidal current.

Based on R.A. Dunlap, An Introduction to the Physics of Nuclei and Particles, Brooks-Cole, Belmont (2004).

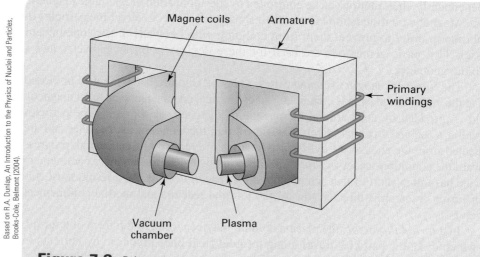

Magnet coils   Armature

Primary windings

Vacuum chamber   Plasma

Figure 7.3: Schematic diagram of a tokamak.

Figure 7.4: Interior of the Joint European Torus (JET), a tokamak located in Culham, Oxfordshire, England.

is not much smaller than the toroidal diameter. In fact, many designs do not have a poloidal cross section that is circular. One of the more successful designs has been the spherical tokamak, which resembles a sphere with a circular hole through it (i.e., the so-called cored-apple geometry).

ITER Organization

Figure 7.5: Cutaway diagram of the ITER reactor.

The most recent and significant development in magnetic fusion experiments has been the International Thermonuclear Reactor (ITER, Figure 7.5). This is joint project of the European Union, China, India, Japan, Russia, South Korea, and the United States. Canada was originally involved in the project but withdrew. This is a tokamak-type magnetic confinement reactor located in southern France. The project was initiated in 2006, and construction is expected to take until 2019.

These reactors use a combination of deuterium and tritium as fuel, and it is hoped that they will achieve conditions that will initiate a sustained fusion reaction. In this condition, some of the energy produced by the fusion can be used to maintain the conditions of the plasma, and the remainder can be extracted as heat. The design goals of the ITER are to produce 500 MW of heat output for 50 MW of energy input. In an operational power reactor, excess heat is used to generate steam to drive turbine/generators and generate electricity.

7.4 Inertial Confinement Reactors

Inertial confinement refers to the situation where the fusion fuel is confined by inertial forces in the plasma itself. Most experiments that fall into this category are referred to as laser fusion experiments. A pellet of fuel (in most cases, a mixture of deuterium and

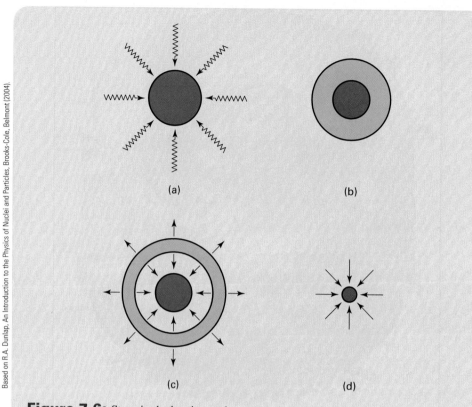

(a)

(b)

(c)

(d)

Figure 7.6: Steps in the heating and compression of a fuel pellet in an inertial confinement fusion reactor. The various stages of fusion are described in the text.

tritium contained in a capsule about a millimeter in diameter) is bombarded by pulses from several directions at once by high-energy laser beams. The fuel pellet heats rapidly to a temperature that is hopefully suitable for fusion to take place. The actual processes that take place in the pellet as it heats are quite complex. Figure 7.6 shows a simplified depiction. In Figure 7.6(a), the laser radiation is absorbed by the fuel pellet, heating it from the outside. In Figure 7.6(b), the heat propagates through the pellet, transforming the outer portions into a plasma. In Figure 7.6(c), this outer plasma atmosphere is driven off as it heats and expands. This process is referred to as *ablation*. In Figure 7.6(d), the remaining core of the pellet is compressed. This results as the outer portion of the pellet expands in an equal and opposite reaction (according to Newton's laws), pushing the inner portion of the pellet inward. These are the inertial forces referred to in the name of the process. Because of the large amount of energy absorbed from the laser beam, the density of the pellet core can be compressed to densities of several thousand times the density of water. A photograph of a laser fusion experiment is shown in Figure 7.7. It is interesting to note that the laser currently being used for this experiment at Lawrence Livermore National Laboratory in California produces an output of 750 TW. That is 50 times the average power consumption worldwide. However, the laser produces energy in pulses that are only 2.4 ns long. One pulse corresponds to 750 TW $\times$ 2.4 ns = 1.8 MJ, or about the energy used by a typical automobile to travel a few hundred meters.

Lawrence Livermore National Laboratory

Figure 7.7: Photograph of the laser fusion (inertial confinement fusion) system at Lawrence Livermore National Laboratory (California). Note workers in lower part of photograph for scale.

Thus, although this is not really a lot of energy, it is concentrated into a very small volume during a very short period of time.

7.5 Progress Toward a Fusion Reactor

As explained, it is necessary to achieve conditions for the plasma where the density of nuclei is great enough, the temperature is high enough, and the conditions are maintained long enough to cause sufficient fusion to occur in order to sustain a reaction. These conditions can be quantified in terms of the *Lawson parameter*, $n\tau$, which is the product of the number density of nuclei, n, and the containment time, τ. To sustain a fusion reaction, the temperature must reach at least 200 MK, and the Lawson parameter must satisfy the following relationship (the Lawson criterion):

$$n\tau > 10^{20} \qquad \mathrm{s \cdot m^{-3}}. \tag{7.9}$$

The definition of containment time for inertial confinement fusion is fairly clear but is less obvious for magnetic confinement. For magnetic confinement reactors, the magnetic fields that confine the plasma are pulsed, and the confinement time is the duration of the field pulse. Figure 7.8 shows the relationship of the relevant quantities for different fusion reactor designs.

The development of a fission reactor occurred very rapidly after an understanding of the fundamental physics was achieved. Progress in the development of a viable fusion

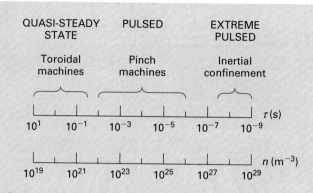

Based on R.A. Dunlap, An Introduction to the Physics of Nuclei and Particles, Brooks-Cole, Belmont (2004).

Figure 7.8: Relationship of the quantities in the Lawson parameter for different types of fusion reactors that are necessary to meet the Lawson criterion.

Example 7.2

For a plasma consisting of deuterons and electrons with a number density of 10^{24} deuterons per m³, what is the actual mass density compared to air (at STP)?

Solution

For an electrically neutral plasma, there is 1 electron per deuteron, so the total mass per m³ will be

$$10^{24}\,\mathrm{m^{-3}} \times [2.013553214\,\mathrm{u} + 0.000548579903\,\mathrm{u}] = 2.014 \times 10^{24}\,\mathrm{u/m^3},$$

where the deuteron and electron masses are given in Appendix II. Converting this to kilograms,

$$2.104 \times 10^{24}\,\mathrm{u/m^3} \times 1.6605 \times 10^{-27}\,\mathrm{kg/u} = 3.34 \times 10^{-3}\,\mathrm{kg/m^3}.$$

The density of air (from the Web) is 1.204 kg/m³. So the plasma is about 0.3% of the density of air.

reactor has been much slower. Work thus far has dealt almost exclusively with the use of deuterium-tritium mixtures for fuel (further comments to follow and in the next section). Experimental laboratory reactors have fulfilled the Lawson criterion, but a situation where there is a net gain in energy from a self-sustaining reaction (sometimes called *ignition*) has not yet been reliably achieved. Progress toward this goal for magnetic confinement reactors and for inertial confinement reactors is illustrated in Figures 7.9 and 7.10, respectively. The light blue area is the region where ignition occurs and net energy is produced. Generally speaking, the progression of data points toward the light blue region represents progress in fusion reactor development as a function of time. For example, in Figure 7.10 the general trend of the points from the lower left to the upper right of the figure represents the development of fusion reactors from the late 1960s using milliwatt lasers to the present decade using terawatt lasers. In the case of inertial confinement, the arrow represents a new and hopeful approach to laser fusion. When the fuel pellet (having been irradiated with the laser radiation) is at its maximum density and temperature,

Based on R.A. Dunlap, An Introduction to the Physics of Nuclei and Particles, Brooks-Cole, Belmont (2004).

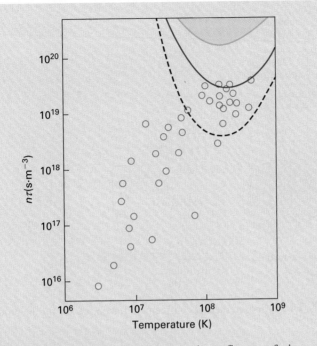

Figure 7.9: Progress toward ignition for magnetic confinement fusion reactors. The broken blue line is the breakeven point if additional energetic particles are injected into the plasma and the red line is breakeven without particle injection.

Based on R.A. Dunlap, An Introduction to the Physics of Nuclei and Particles, Brooks-Cole, Belmont (2004).

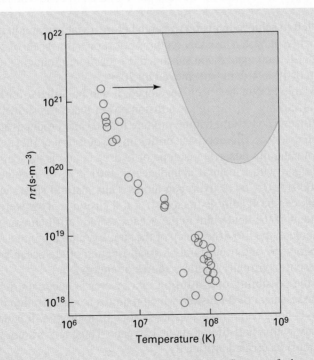

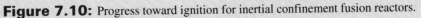

Figure 7.10: Progress toward ignition for inertial confinement fusion reactors.

Based on R.A. Dunlap, An Introduction to the Physics of Nuclei and Particles, Brooks-Cole, Belmont (2004).

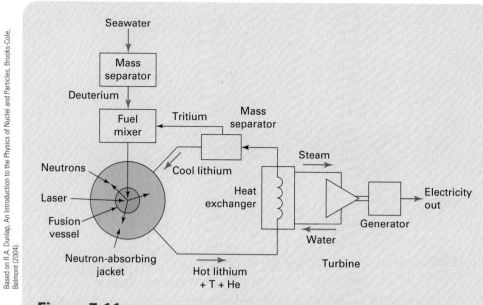

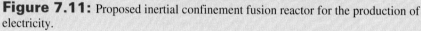

Figure 7.11: Proposed inertial confinement fusion reactor for the production of electricity.

a second laser pulse is aimed at the fuel. This drives the center of the pellet to a higher density and temperature, much as the first pulse did, and pushes the conditions closer to ignition.

Much is left to achieve in fusion research before a viable reaction can be obtained. This is presumably, at best, a task that requires effort on a timescale of several decades. Plans for the ITER anticipate that a sustained d-t reaction, producing a ratio of power out to power in of 10 to 1, will be achieved by 2026.

Once a viable reactor has been developed, it can be utilized for the commercial production of electricity. The design of a fusion power reactor is shown in Figure 7.11. This reactor utilizes a deuterium-tritium mixture for fuel and produces energy by means of the d-t fusion reaction. This approach brings up the question of fuel availability. Deuterium (^{2}H) is a naturally occurring isotope of hydrogen and is present in all naturally occurring hydrogen with a natural abundance of 0.0156%, or 1 deuterium atom for every $100/(0.000156) = 6410$ hydrogen atoms Thus, natural seawater contains deuterium, and this can be readily extracted. If a d-d fusion reactor were feasible, then the energy content of the deuterium in the oceans would fulfill humanity's needs for a period comparable to the life expectancy of the sun. Even taking into account expected increases in energy use and even if only a small fraction of the ocean's deuterium was used, this represents a virtually limitless source of energy.

Because the possibility of d-d fusion in a reactor is uncertain, it is necessary to begin with an assessment of d-t fusion, which requires a consideration of tritium availability. Tritium is not a naturally occurring isotope of hydrogen. It is unstable and decays by β^- decay with a half-life of about 12 years. Thus, any tritium to be used in a fusion reactor must be created artificially.

The reactor design shown in Figure 7.11 illustrates that the region where the fusion occurs is surrounded by a jacket containing lithium. This design, along the lines of

Example 7.3

The volume of water in the oceans is about 1.3×10^9 km³. Calculate the time that this could serve humanity as an exclusive primary energy source by means of d-d fusion at the current rate of energy use.

Solution

The mass of the water in the oceans (in grams) is

$$m = (1.3 \times 10^9 \text{ km}^3) \times (10^9 \text{ m}^3/\text{km}^3) \times (10^6 \text{ g/m}^3) = 1.3 \times 10^{24} \text{ g(H}_2\text{O)}.$$

The total mass of hydrogen is given in terms of the atomic masses of oxygen and hydrogen as

$$m = \frac{2 \times 1 \text{ g/mol}}{2 \times 1 \text{ g/mol} + 16 \text{ g/mol}} \times 1.3 \times 10^{24} = 1.44 \times 10^{23} \text{ g hydrogen}.$$

The atomic percentage of deuterium is 0.0156%, or a fraction of 0.000312 by weight (the atomic mass of deuterium is twice that of normal hydrogen). This means that the oceans contain

$$1.44 \times 10^{23} \times 0.000312 = 4.49 \times 10^{19} \text{ g deuterium}.$$

This corresponds to

$$(4.49 \times 10^{19} \text{ g}) \times \frac{6.02 \times 10^{23} \text{ mol}^{-1}}{2 \text{ g/mol}} = 1.35 \times 10^{43} \text{ atoms of deuterium}.$$

Because one fusion process requires two deuterium atoms, the number of possible fusions is

$$\frac{1.35 \times 10^{43}}{2} = 6.76 \times 10^{42} \text{ fusions}.$$

Each fusion produces between 3.3 MeV and 4.0 MeV, as given in equations (7.4) and (7.5). Using an approximate value of 3.6 MeV/fusion, the total available energy is

$$(6.76 \times 10^{42} \text{ fusions}) \times (3.6 \text{ MeV/fusion}) \times (1.6 \times 10^{-13} \text{ J/MeV}) = 3.9 \times 10^{30} \text{ J}.$$

Because the current rate of use of primary energy (see the Preface) is 5.7×10^{20} J per year, the energy content of deuterium in the oceans will last (at the current usage rate) for

$$(3.9 \times 10^{30} \text{ J})/(5.7 \times 10^{20} \text{ J/y}) = 6.8 \times 10^9 \text{ y}.$$

a liquid metal–cooled fast breeder reactor, transfers the heat produced to water to produce steam, but it also breeds fuel, tritium in this case. Natural lithium consists of about 7% ^{6}Li and 93% ^{7}Li and is a useful material in this respect. Equation (7.6) shows that the d-t fusion reaction produces neutrons. When these neutrons are incident on natural lithium, one of the following reactions can occur:

$$^6\text{Li} + \text{n} \rightarrow {}^3\text{H} + {}^4\text{He} \qquad (Q = 4.78 \text{ MeV}) \qquad \textbf{(7.10)}$$

and

$$^{7}\text{Li} + \text{n} \rightarrow {}^{3}\text{H} + {}^{4}\text{He} + \text{n} \qquad (Q = -2.47 \text{ MeV}). \qquad \textbf{(7.11)}$$

The first reaction is exothermic and has a large cross section, while the second reaction has a smaller cross section and is endothermic. The first reaction is therefore useful for producing tritium from lithium. The availability of fusion power from the d-t reaction is limited by the availability of ^{6}Li nuclei to breed tritium and is also dependent on the design and efficiency of the lithium blanket.

The lifetime of d-t fusion as a means of supplying our energy needs is unclear. There is considerable uncertainty in the amount of lithium available on earth. At present, the demand for lithium is relatively small, and the price is relatively low. This means that there is very little incentive to explore new lithium resources or to develop more efficient methods for its extraction. The best estimate of about 1.7×10^{7} t of useful terrestrial lithium suggests that d-t fusion is an energy resource that could supply our total energy needs for between 500 and 1000 years. Improved lithium blanket technology would extend this somewhat, but, as discussed in Chapter 19, other, more immediate uses of lithium may be a competing factor.

A timeline for the development of an operational commercial fusion power reactor is uncertain. Although many estimates put the development of such a facility somewhere around 2050, much basic scientific development is still required before the technological aspects can be predicted with any degree of accuracy.

ENERGY EXTRA 7.1
Alternative fusion technologies

Most fusion research deals with either magnetic confinement reactors or inertial confinement (laser) fusion. Both approaches strive to attain the best combination of high temperature and density for a plasma of hydrogen nuclei in order to reach the Lawson criterion. Over the years, a variety of other approaches that do not involve high temperatures have also been taken to achieve fusion, mostly without a great deal of success.

Cold fusion by electrolytic methods was reported in the late 1980s by Martin Fleischmann and Stanley Pons. They claimed to have observed fusion during the electrolysis of heavy water (D_2O) using a palladium (Pd) electrode. These results have never been convincingly confirmed, and their validity is viewed with suspicion by the scientific community. In fact, as a matter of policy, the U.S. Patent Office will not consider applications for any inventions claiming a methodology for cold fusion.

In 2002, sonofusion, sometimes called bubble fusion, was first reported by Rusi Taleyarkhan. In this experiment, high-intensity acoustic waves are transmitted through a deuterated (deuterium-containing) organic fluid. Tiny bubbles in the liquid, caused by a beam of neutrons introduced by the researchers, collapsed under the pressure of the sound waves and caused local heating and high pressure of the deuterium containing fluid. Some estimates have suggested that temperatures of 10 million K can be achieved, and this can result in the claimed d-d fusion reaction. Although other reports of similar results have appeared, reproducibility seems to be, at best, unreliable. Considerable controversy surrounds this work, and, although some work in this area is continuing, much of the scientific community seems to categorize this approach with cold fusion.

A scientifically more interesting approach is muon-catalyzed fusion. The theoretical basis for

Continued on page 201

Energy Extra 7.1 continued

muon catalysis was presented by Andrei Dmitrievich Sakharov (Nobel Peace Prize, 1975) and coworkers in the late 1940s. Experimental confirmation of this phenomenon has been reported by numerous researchers, including Luis W. Alvarez (Nobel Physics Prize, 1968), who observed muon-catalyzed p-d fusion. A thorough theoretical description of this process was presented by the well-known physicist John David Jackson. There is little doubt about the scientific validity of muon-catalyzed fusion; the question remains, however, as to whether it is of commercial significance for energy production. A basic description of the process follows.

A muon is a subatomic particle with properties similar to an electron except that its mass is about 207 times greater and is unstable with a lifetime of about 2.2×10^{-6} seconds. Muons are created in certain particle reactions and can be created in relatively low energy hadron-hadron collisions (see Chapter 6) in particle accelerators. When a muon enters into a region of matter (e.g., hydrogen), it can displace an electron in an atom and become bound to the nucleus, forming a muonic atom. Because the muon is much heavier than the electron, the muonic atom is much smaller than a normal atom. Two muonic hydrogen atoms can bind together to form a muonic hydrogen molecule in which the hydrogen nuclei are very close together. Because the muonic molecule lasts for a long time (relatively speaking), the nuclei have an increased chance of fusing. This is particularly the case if the hydrogen nuclei are deuterons and/or tritons. If fusion occurs, energy is produced, and the muon is liberated and can catalyze further fusions (until it ultimately decays).

Because it requires considerable energy to produce muons in an accelerator, each muon must catalyze about 300 (d-t) fusions in order to produce a net energy gain. Actually because electricity is produced by a heat engine at about 40% efficiency, approximately 700–800 fusions per muon would be required. In principle, this is quite possible before the muon decays except for the phenomenon of alpha-particle trapping. Because d-t fusion produces alpha particles (^{4}He nuclei), a muon sometimes gets bound to a ^{4}He nucleus. If this happens, the muon cannot escape and catalyzes no further fusions. Experimental investigations of muon catalysis have achieved about 150 fusions per muon. Although these results fall short of the requirements for net energy production, they do show that alternative approaches to difficult problems can sometimes yield interesting and potentially useful results. Further progress in this field may lead to net energy production, but it is unknown whether this approach can be made economically viable.

Topic for Discussion

The first (and most important) goal of fusion reactor research is to construct a device that produces more energy than it consumes (i.e., a net energy gain). Once this is accomplished, a commercial fusion power reaction might be possible. However, it is important to consider the economics of such an approach. Discuss the necessary considerations for a successful fusion reactor program. The development and economics of fission power reactors is a good basis for comparison.

7.6 Summary

Fusion energy is an attractive alternative to fission energy because safety, waste disposal, and radioactive material security are not major concerns. The scientific and engineering challenges to the development of fusion power are formidable. This chapter described how the dependence of binding energy on nucleon number for light nuclei allowed for the extraction of energy during a fusion process. From a practical standpoint, the difficulty in fusing two nuclei is the need to overcome the barrier that results from the repulsive Coulombic interaction between the charged protons in the two nuclei. Nuclei must be kept together with a sufficiently high density and energy for a sufficiently long period of time to make the probability of fusion high enough to produce useable energy. Two approaches have been taken to meeting these requirements: magnetic confinement fusion and inertial confinement (or laser) fusion. In a magnetic confinement reactor, ionized gas atoms and their electrons form a plasma, which is confined in a magnetic field. In an inertial confinement reactor, pellets of fusion fuel are heated and compressed to high density by bombardment with intense lasers. The conditions necessary for achieving a fusion reaction are described by the Lawson criterion in terms of the particle density and confinement time.

This chapter discussed the various possible fusion reactions that can produce useable energy. Deuterium-deuterium fusion is the most desirable because it makes use of deuterium nuclei, which are a natural component of all hydrogen. The supply of deuterium in the water of the world's oceans would provide a virtually endless supply of fusion fuel. Unfortunately, deuterium-deuterium fusion is very difficult to achieve, and current experimental investigations of fusion power are concentrating on deuterium-tritium fusion. Deuterium-tritium fusion requires the production of tritium from lithium in order to fuel the reactor. The world's supplies of lithium are limited, and the longevity of d-t fusion as an energy source is likely less than that for uranium fission if fuel is reprocessed or fissile material is bred or if thorium is used as a fission fuel.

Much progress has been made in recent years toward achieving a sustainable fusion reaction, although much work is still needed to make the process viable from an energetic and economic standpoint. A possible design of a fusion power reactor based on a heat engine and generator was presented.

Problems

7.1   If all of the world's energy needs were provided by d-t fusion, how much lithium per year would be utilized to produce tritium? Consider the total energy associated with fusion as being equivalent to the world's current primary energy use.

7.2   The total power output of the sun is 4×10^{26} W. Most of this energy is produced within a radius of 175,000 km of its center. If we were able to reproduce the sun's conditions for p-p fusion in a fusion reactor with an active volume of $10 \times 10 \times 10$ m^3, what would be the reactor's power output? Is p-p fusion a reasonable approach to take for a power reactor?

7.3   Using appropriate masses from Appendix II, show that the energy released from equation (7.7) is 26.7 MeV.

7.4   What is the number density of tritons in a 50% d/50% t plasma that has the same mass density as air at STP (1.204 kg/m^3)?

7.5   Consider a body of water 1 km^2 in area and 100 m deep. The center of mass of the water is raised to a height of 300 m. Compare the gravitational potential energy of this body of water to the fusion energy content of the deuterium it contains.

7.6   The ITER is designed to have a magnetic field pulse duration of up to 480 seconds. What corresponding plasma particle density is needed to achieve the Lawson criterion?

7.7   Tritium-tritium fusion is given by the reaction

$$t + t = {}^4\text{He} + 2n.$$

Calculate the energy release for this process. Note the following atomic masses

$$m({}^3\text{H}) = 3.016049268 \text{ u}$$

and

$$m({}^4\text{He}) = 4.00260325 \text{ u}.$$

Ensure that electrons are properly accounted for on the two sides of the equation.

7.8   What is the kinetic energy of a particle in a plasma at a temperature of 200 MK?

Bibliography

U. Columbo and U. Farinelli. "Progress in fusion energy." *Annual Review of Energy* **171** (1992).

R. W. Conn, V. A. Chuyanov, N. Inoue, and D. R. Sweetman. "The International Thermonuclear Experimental Reactor." *Scientific American,* **266** (1992): 103.

R. A. Dunlap. *An Introduction to the Physics of Nuclei and Particles* (Brooks-Cole, Belmont, CA (2004).

H. Furth. "Fusion." *Scientific American* **273** (1995): 174.

G. McCracken and P. Stott. *Fusion: The Energy of the Universe.* Academic Press, New York (2005).

J. Tester, E. Drake, M. Driscoll, M. W. Golay, and W. A. Peters. *Sustainable Energy: Choosing Among Options.* MIT Press, Cambridge (2006).

PART IV

Renewable Energy

There is a clear need to look beyond fossil fuels and nuclear energy for the future. A number of possible viable alternative energy sources are available or are being researched. It is essential to carefully consider the pros and cons of each of these energy technologies. It is not always obvious how much they improve on fossil fuels because of the complexity of fully assessing the environmental impact of infrastructure development and operation and the difficulty of evaluating the economic aspects of particular energy generating methods.

We cannot create energy. Energy production methods are sometimes referred to as energy harvesting technologies because they merely convert primary energy sources from nature (e.g., solar energy, wind, geothermal, etc.) into other forms of energy that we can readily use (e.g., electricity, heat, biofuels etc.). Any technology that converts primary energy into usable energy involves a large number of factors that contribute to its environmental and economic desirability. A systematic approach to the assessment of any potential alternative energy technology is to consider how well it fits the CURVE (*c*lean, *u*nlimited, *r*enewable, *v*ersatile, and *e*conomical). An evaluation of each of these factors provides an objective criterion for determining the viability of certain approaches:

- *Clean:* The environmental impact of an energy source is a crucial factor for its evaluation. A detailed life cycle assessment is beneficial in establishing whether the technology is, in fact, cleaner than fossil fuels.

- *Unlimited:* Of course, no energy source is truly unlimited, but the available quantity of some, like solar energy, is much greater than the total energy use of humankind. The greater the availability, the larger the contribution a resource can make to our future energy needs.

- *Renewable:* Some resources are available in large quantities but do not have an infinite lifetime. This is true, for example, of fossil fuels, which fulfill nearly all our energy needs but will not last indefinitely. They are clearly not renewable. Other resources, such as wave and tidal energy, are renewable but limited in their availability.

- *Versatile:* We use different types of energy for different purposes (e.g., thermal energy for heating buildings or electricity for operating appliances). Of course, one type of energy can be converted into another, but there are always conversion losses, and sometimes the losses can be very large. Although no type of energy can satisfy all our needs, those that have high energy density and are portable fulfill the most demanding applications, such as transportation.

- *Economical:* Any new energy technology must compete with existing technologies, particularly those that are well established, reliable, and economical. This means that fossil fuels are the first point of reference in determining economics. Technologies like solar photovoltaics face a serious challenge because they do not compete economically.

This part of the text considers the most important approaches to alternative energy and evaluates how well they fit the CURVE and how likely they are to find a place in a future fossil-fuel-free energy economy.

The solar tree in the photograph is located in Gleisdorf, Austria, and similar trees exist in many locations worldwide. Solar trees produce electricity by means of photovoltaic (PV) panels, and this electricity can be stored in a battery system. Because of the limited PV surface area, the solar tree is not intended for large-scale energy production. It is more akin to a sculpture that merges artistic design with modern renewable energy technology and serves to promote alternative energy to the public.

© WoodyStock/Alamy

Direct Use of Solar Energy

Learning Objectives: After reading the material in Chapter 8, you should understand:

- The energy content and properties of the sunlight incident on the earth.
- The mechanisms for heat transfer.
- Conductive heat transfer through materials and the use of *R*-values.
- Radiative heat transfer from surfaces.
- The design and operation of a thermal solar collector.

- The energy requirements for residential space heating.
- The description of heating needs on the basis of degree days.
- The storage of thermal energy and the heat capacity of solids.
- The use of passive solar heating techniques as a component of residential heating needs.

8.1 Introduction

Solar radiation is a virtually unlimited source of renewable energy. The simplest technology that makes use of solar energy takes advantage of the heat produced when a material absorbs electromagnetic radiation. This heat may be used directly for space heating or hot water production in residential or office buildings, or it may be stored for use during overcast days or at night. Systems that make use of solar energy for this purpose may be active or passive. *Active systems* use electricity to circulate a fluid, such as water, that absorbs energy from solar radiation and is then used to distribute that energy in the form of heat where it is needed in a building. *Passive systems* merely use the thermal properties of the natural components of a building's structure for the absorption, storage, and distribution of heat in the building.

To understand how solar energy can be utilized, it is first necessary to have knowledge of some of the properties of sunlight, and this chapter begins with a review of the characteristics of solar radiation. It is also necessary to understand the thermal properties of materials to appreciate how solar radiation is absorbed by various materials and converted into thermal energy. This chapter therefore also overviews the thermodynamic properties of the construction materials used in active and passive solar heating systems, as well as those that are appropriate for the storage of thermal energy.

8.2 Properties of Sunlight

The total power output from the sun (as given in Chapter 7) is 3.8×10^{26} W and is emitted isotropically. At the distance of the earth's orbit, this power has a density given by the total power divided by the surface area of a sphere with a radius equal to the average distance between the sun and the earth (1.49×10^{11} m, defined to be one astronomical unit). Minor fluctuations occur during the year because the earth's orbit is an ellipse, not a circle, but these variations are negligible for the present discussion. (Actually, the earth is closer to the sun by about 3% when it is winter in the northern hemisphere.) Thus, the tilt of the earth's axis gives us seasons, not variations in the distance between the earth and the sun. The power density per unit area is therefore given as

$$\frac{P}{A} = \frac{3.8 \times 10^{26} \text{W}}{4\pi \times (1.49 \times 10^{11} \text{ m})^2} = 1367 \text{ W/m}^2. \qquad (8.1)$$

This value of the power per unit area is referred to as the *solar constant* and has been measured using sensors on satellites outside the earth's atmosphere. The solar spectrum is very closely approximated by the radiation from a black body at 6000 K. This power spectrum is shown in Figure 8.1. The *irradiance*, as shown in the figure, is the power per unit area per unit wavelength. The total power per unit area at all wavelengths is given by the integrated area under the curve in the figure. The portion of the spectrum in the visible part of the electromagnetic spectrum is indicated. The small anomalies that occur in addition to the smooth curve expected for black body radiation are due to atomic processes that occur in the gas in the outer portions of the sun's atmosphere.

The total solar power incident on the earth is given by the power density times the apparent area of the earth. To the incident solar radiation, the earth appears as a disk with an area πr^2, where r is the earth's radius (6.731×10^6 m), so

$$P_{\text{total}} = (1367 \text{ W/m}^2) \times 3.14 \times (6.371 \times 10^6 \text{m})^2 = 1.73 \times 10^{17} \text{ W}. \qquad (8.2)$$

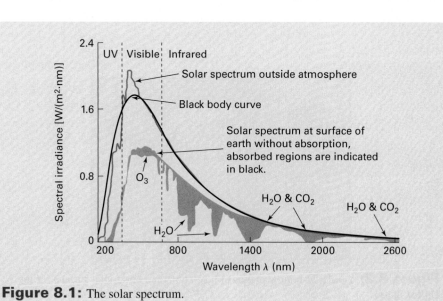

Figure 8.1: The solar spectrum.

Much of the radiation that arrives at the outside of the earth's atmosphere is either reflected back into space or absorbed by the atmosphere. Actually only about half of the total incident sunlight actually gets to the surface of the earth. The power spectrum of the solar insolation at the earth's surface compared with the spectrum of sunlight outside the earth's atmosphere is shown in Figure 8.1. The characteristic features in the spectrum are caused by absorption at certain wavelengths by various atoms and molecules in the atmosphere. The power per unit area that arrives at the surface of the earth, integrated over all wavelengths, is referred to as the *insolation*. The average insolation at the earth's surface is given by the total power in equation (8.2) times the fraction transmitted through the atmosphere and divided by the total surface area of the earth ($4\pi r^2$):

$$\frac{P}{A} = \frac{0.5 \times 1.73 \times 10^{17} \text{W}}{4\pi \times (6.371 \times 10^6 \text{ m})^2} = 168 \text{ W/m}^2. \tag{8.3}$$

This is the solar insolation on a horizontal surface averaged over the entire surface and averaged over all time (day and night). The average integrated daily insolation is a measure of total incident energy per unit area for an entire day and is expressed as $(168 \text{ W/m}^2) \times (86{,}400 \text{ s/d}) = 14.5 \text{ MJ/m}^2$ (Section 8.5).

The insolation at any given location on earth at any given time depends on several factors. The factors that can be readily predicted are

- The time of day.
- The day of the year.
- The latitude.

An additional condition that is difficult to predict exactly is the weather. On average, the insolation at a particular location can be determined on the basis of the average typical weather conditions. For the United States, the insolation at different locations averaged over 24 hours per day for the entire year is shown in Figure 8.2. Although this

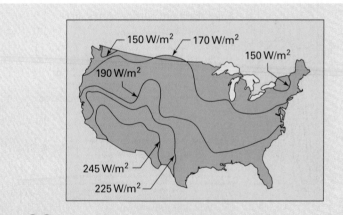

U.S. Department of Energy

Figure 8.2: Yearly 24-hour averages of insolation on a horizontal surface for the United States.

figure shows the general geographical trends for the average insolation, there can be substantial local variations.

The energy of solar radiation can be used in two ways to help satisfy our energy needs: direct use for heating or for the production of electricity. The former use is considered in the present chapter, and the latter is discussed in Chapter 9.

8.3 Heat Transfer

To consider the details of the design of solar collectors, it is necessary to consider how thermal energy is transported. Heat may be transported by conduction, convection, or radiation.

8.3a Conduction

According to the second law of thermodynamics, heat flows from hot to cold, and this basic principle allows us to quantify heat transport by conduction through materials. If a temperature difference exists across a piece of material, then the rate at which energy is transported from the hot side to the cold side (i.e., the power) is:

- Proportional to the magnitude of the temperature difference ($T_h - T_c$).
- Proportional to the cross-sectional area of the material (A).
- Proportional to the thermal conductivity of the material (k).
- Inversely proportional to the thickness of the material (l).

This relation is expressed mathematically as

$$P = \frac{kA(T_h - T_c)}{l}. \tag{8.4}$$

The power is expressed in W, and the thermal conductivity is in $(W \cdot cm)/(m^2 \cdot {}^\circ C)$ when the area is expressed in square meters (m^2), the thickness is expressed in centimeters (cm), and the temperature is in degrees Celsius (°C). For these calculations, it is reasonable to express the temperature in Celsius rather than in degrees measured on an absolute temperature scale (Kelvin) because it is temperature differences and not absolute temperatures that appear in equation (8.4). The thermal conductivities of some common materials (particularly those used in building construction) are given in Table 8.1.

To quantify the ability of a piece of material to conduct heat (or more precisely, its ability to provide insulation), it is convenient to define the *R-value* of a material, which is its resistance to heat flow. The *R*-value is determined from the thermal conductivity, as given in Table 8.1, and the thickness of the material; it is defined as $R = l/k$, so that the heat flow per unit area, as obtained from equation (8.4), may be rewritten as

$$\frac{P}{A} = \frac{T_h - T_c}{R}. \tag{8.5}$$

The *R*-value has units of $(m^2 \cdot {}^\circ C)/W$, sometimes expressed as $(m^2 \cdot K)/W$ or $(m^2 \cdot s \cdot K)/J$. The *R*-value takes into account the thickness of the material, so if a 2-cm. thick piece of

Table 8.1: Thermal conductivities in $(W \cdot cm)/(m^2 \cdot °C)$ for some common materials.

material	$k \, [(W \cdot cm)/(m^2 \cdot °C)]$
aluminum	20,100
iron	4600
concrete	170
brick	71
water	60
glass	59
wood (cross grain)	13
sawdust	5.9
cork	4.2
fiberglass insulation	3.8
polystyrene foam	2.84
air	2.3

© Cengage Learning 2015

Example 8.1

Calculate the power transferred though a piece of aluminum that is 5 cm by 5 cm by 15 cm long when the temperature on the hot end is 100°C and the temperature on the cold end is 20°C. This is the situation when one end of the bar is in contact with boiling water and the other end is at room temperature.

Solution

From equation (8.4)

$$P = \frac{kA(T_h - T_c)}{l},$$

where the following values are used to calculate the power:

$k = 20,100 \, (W \cdot cm)/(m^2 \cdot °C)$ (from Table 8.1).

$A = 5 \text{ cm} \times 5 \text{ cm} = 0.0025 \text{ m}^2.$

$T_h = 100°C.$

$T_c = 20°C.$

$l = 15 \text{ cm}.$

Note that the area of the thermal conductor is given in m^2, whereas the length is given in cm, which, when combined with the value of k, yields power in watts. Why this approach to units is convenient will become apparent when heat transfer through building walls and windows is considered. Thus the power is calculated to be

$$P = 20,100 \frac{W \cdot cm}{m^2 \cdot °C} \times (0.0025 \text{ m}^2) \times \frac{100°C - 20°C}{15 \text{ cm}} = 268 \text{ W}.$$

Example 8.2

Compare the power transferred per unit area through 1 cm thick pieces of iron and glass when the temperature is 20°C on one side and −10°C on the other.

Solution

From equation (8.4)

$$\frac{P}{A} = \frac{k(T_h - T_c)}{l}$$

where the following values are used for both materials:

$$(T_h - T_c) = (20°C - (-10°C)) = 30°C$$
$$l = 1 \text{ cm}$$

For iron $k = 4600 \ [(W \cdot cm)/(m^2 \cdot °C)]$

and for glass $k = 59 \ [(W \cdot cm)/(m^2 \cdot °C)]$

Thus for iron the power transferred will be

$$P/A = (4600 \ [(W \cdot cm)/(m^2 \cdot °C)])(30°C)/(1 \text{ cm}) = 138 \text{ kW/m}^2$$

and for glass

$$P/A = (59 \ [(W \cdot cm)/(m^2 \cdot °C)])(30°C)/(1 \text{ cm}) = 1.77 \text{ kW/m}^2$$

The iron transfers $(138 \text{ kW/m}^2)/(1.77 \text{ kW/m}^2) = 78$ times as much power.

This ratio can easily be seen from just the ratios of the thermal conductivities

$$(4600 \ [(W \cdot cm)/(m^2 \cdot °C)])/(59 \ [(W \cdot cm)/(m^2 \cdot °C)]) = 78.$$

material has an R-value of, say, $R = 2$, then a 4-cm thick piece of the same material will have an R-value of $R = 4$. (This simple approach deals only with conductive heat losses and is not exact if the radiative losses, to be discussed, are included.) As with resistors in an electric circuit, if several materials are placed one after the other (i.e., as resistors in a series circuit), then the R-values are additive:

$$R_{total} = R_1 + R_2 + R_3 + R_4 + \dots \qquad \textbf{(8.6)}$$

Consider a calculation for a piece of window glass. Standard window glass is about 0.3 cm thick, and, using the thermal conductivity for glass given in Table 8.1, one calculates $R = 0.005$. Thus a single pane of window glass provides virtually no insulation on its own. The principal reason that a window reduces conductive heat transfer is that it prevents air flow. A layer of stationary air is created near the surfaces of a pane of glass, and the insulating properties of this nonmoving air are of relevance. When the R-value of this air layer is taken into account, the effective R-value for the pane of glass is about $R = 0.18$. The window also eliminates convective heat transfer from the inside of the building to the outside by preventing air from actually flowing from the inside to the outside. Figure 8.3 illustrates the relationship of different heat loss mechanisms through a window, including conduction (as discussed), convection, and radiation.

Example 8.3

Calculate the effective R-value for a wall comprised of an outer layer of 10 cm of brick, a middle layer of 15 cm of fiberglass insulation, and an inner layer of 2 cm of wood.

Solution

The R-values can be calculated for each of the layers from the values of k in Table 8.1:

$$\text{Brick: } R = \frac{10 \text{ cm}}{71 (\text{W} \cdot \text{cm})/(\text{m}^2 \cdot {}^\circ\text{C})} = 0.14 \text{ m}^2 \cdot {}^\circ\text{C/W}.$$

$$\text{Fiberglass: } R = \frac{15 \text{ cm}}{3.8 (\text{W} \cdot \text{cm})/(\text{m}^2 \cdot {}^\circ\text{C})} = 3.95 \text{ m}^2 \cdot {}^\circ\text{C/W}.$$

$$\text{Wood: } R = \frac{2 \text{ cm}}{13 (\text{W} \cdot \text{cm})/(\text{m}^2 \cdot {}^\circ\text{C})} = 0.15 \text{ m}^2 \cdot {}^\circ\text{C/W}.$$

The total R-value for the wall is $R = (0.14 + 3.95 + 0.15)$ m$^2 \cdot {}^\circ$C/W $= 4.24$ m$^2 \cdot {}^\circ$C/W. This can then be used to calculate the energy transferred per unit time per unit area through the wall. Clearly, in this example, the insulating properties of the wall come almost exclusively from the presence of the fiberglass insulation.

8.3b Convection

Convection is an important component in the transfer of heat and is certainly illustrated by the insulating effects of the dead air layer near a pane of glass, as just explained. Convection causes heat transfer by the movement of a fluid. The physics of this process depends on the properties of the fluid, such as density and viscosity. A quantitative description of convection is very complex and is not considered in this text.

8.3c Radiation

All objects that are at a temperature above absolute zero radiate energy. The total power per unit surface area radiated by a black body at a temperature T is given by the Stefan-Boltzmann law as

$$\frac{P}{A} = \sigma T^4, \tag{8.7}$$

where σ is the Stefan-Boltzmann constant 5.67×10^{-8} W·m^{-2}·K^{-4}. Because this expression deals with the absolute temperature of the object (not a temperature difference), it is essential to represent the temperature in Kelvin. This expression shows that the total amount of radiation produced by a black body is a sensitive function of its temperature. In general, objects are not perfect black bodies and therefore do not radiate as much energy as a black body at the same temperature would. The ratio of the energy radiated by a body to that radiated by a black body at the same temperature is called the *emissivity*, ε. Thus equation (8.7) can be written for a real object as

$$\frac{P}{A} = \varepsilon \sigma T^4 \tag{8.8}$$

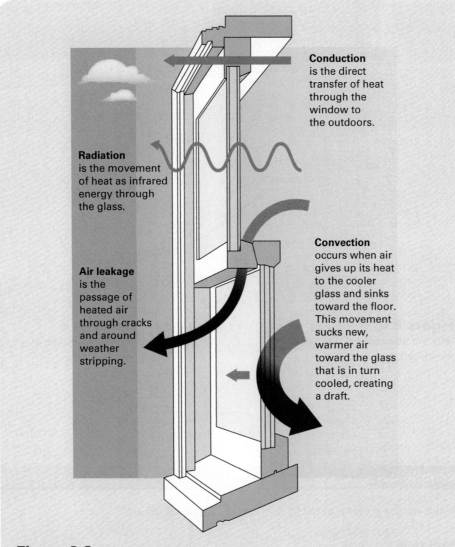

Conduction
is the direct transfer of heat through the window to the outdoors.

Radiation
is the movement of heat as infrared energy through the glass.

Convection
occurs when air gives up its heat to the cooler glass and sinks toward the floor. This movement sucks new, warmer air toward the glass that is in turn cooled, creating a draft.

Air leakage
is the passage of heated air through cracks and around weather stripping.

Figure 8.3: Illustration of heat losses through a window.

where ε can have values from 0 to 1 and is a measure of how effective the surface of a particular material is at radiating energy. The emissivity can vary considerably for different materials, and for a particular material it can vary to some extent as a function of temperature. Most common building materials like glass, wood, steel, most painted surfaces, and so on have emissivities of about 0.9 at around room temperature.

In addition to the temperature dependence of the total energy produced by a surface, the wavelength of the radiation produced is also a function of the temperature. The distribution of wavelengths produced by the sun was shown in Figure 8.1. For surfaces at lower temperatures than the sun, the black body spectrum is shown in Figure 8.4. Clearly, as the temperature decreases, the total radiated power decreases (as evidenced by the area under the curves in the figure), and the wavelength of the maximum in the spectrum increases. For the sun (a black body at about 6000 K), the peak is in the visible

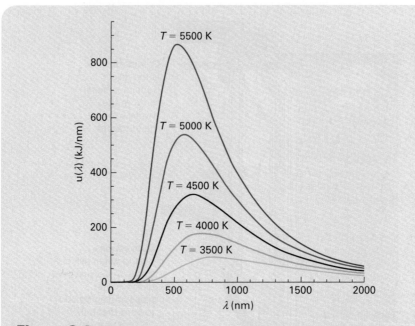

Figure 8.4: Wavelength dependence of black body radiation for surfaces at different temperatures. The vertical axis, u(λ), give the energy density as a function of wavelength.

Based on http://www.britannica.com/EBchecked/topic/643338/Wiens-law

Example 8.4

Calculate the power emitted by 1 m² of the surface of the sun.

Solution

From equation (8.8),

$$\frac{P}{A} = \varepsilon \sigma T^4 \qquad \text{or} \qquad P = A\varepsilon\sigma T^4,$$

where the following values are used:

  $A = 1\ \text{m}^2$.

  $\varepsilon = 1$ (because the sun behaves as a black body).

  $\sigma = 5.67051 \times 10^{-8}\ \text{W·m}^{-2}\text{·K}^{-4}$ (from Appendix II).

  $T = 6000\ \text{K}$ (from the text).

This gives

  $P = (1\ \text{m}^2) \times 1 \times (5.67051 \times 10^{-8}\ \text{W·m}^{-2}\text{·K}^{-4}) \times (6000\ \text{K})^4 = 73.5\ \text{MW}.$

ENERGY EXTRA 8.1
Optical coatings

Passive solar heating relies on the optical properties of windows. It is generally desirable that windows are as transparent as possible in the visible portion of the electromagnetic spectrum. The natural high transmission in the visible and low transmission in the infrared portions of the spectrum exhibited by window glass give rise to passive solar heating inside a building. However, engineering materials with specific optical properties allow for substantially more control of radiative heat flow through windows at the various wavelengths. The transmittance as a function of wavelength is shown in the figure to the right for plain glass and for glass that has been coated with two different materials, a polymer-based coating and a metal-dielectric coating. Over the visible portion of the spectrum (about 350–650 nm), both coated materials are fairly transparent, transmitting about 70% (compared to 90% for plain glass) of the incident radiation. Although glass only begins to become nontransparent well into the infrared (above 2700 nm), the coated glass becomes nontransparent in the very near infrared, thus blocking a much larger portion of long wavelength re-emitted radiation (see Chapter 4 on the greenhouse effect).

The reflectance of the coated and uncoated glass is shown in the figure to the right. Here the behaviors of the two coated materials are seen to be very different. The polymer coated glass is fairly similar to glass, while the metal-dielectric–coated glass becomes very reflective in the infrared. These materials may be utilized to optimize passive solar heating in the winter. As illustrated in the figure, the transmittance

Based on F. Horowitz, M.B. Pereira and G.B. de Azambuja, Appl. Optics 50 (2011) C250.

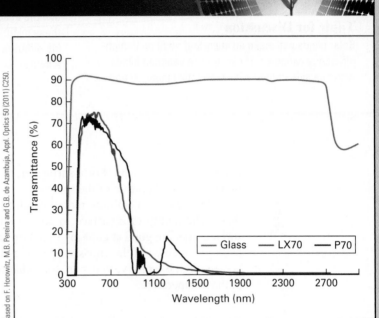

The transmittance of glass and glass coated with a polymer coating (P70) and a metal-dielectric layer (LX70).

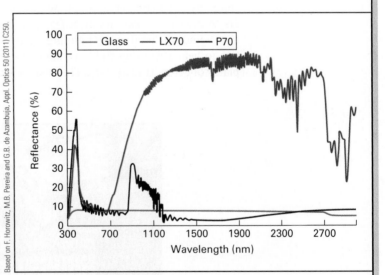

Reflectance of glass and glass coated with a polymer coating (P70) and a metal-dielectric layer (LX70).

Continued on page 216

Energy Extra 8.1 continued

of the coated glass allows light to enter but minimizes long-wavelength radiation heat loss back through it.

Topic for Discussion

Solar blinds can make an important addition to high-efficiency windows. These are like venetian blinds with one side coated with a reflective material and the other side coated with an absorptive material. The blades can be oriented with the reflective side out and the absorptive side in, or vice versa. Discuss how these might be used during the day and night and during different seasons to optimize heating efficiency in a home.

portion of the spectrum. For a surface at, say, 350 K (or about 75°C) the peak is well into the infrared portion of the spectrum.

In addition to emitting radiation, all surfaces absorb radiation that is incident on them. The ability of a surface to absorb energy is its *absorptance, a,* and generally surfaces that are good at emitting radiation are also good at absorbing it. A surface that has a given value of the emissivity, ε, at a certain temperature has an absorptance that is equal to the emissivity, $a = \varepsilon$, for radiation with a wavelength that is characteristic of that temperature.

8.4 Solar Collector Design

Systems for utilizing solar radiation for heating purposes may be either active or passive. An *active system* is one in which heat produced from sunlight is transported and stored by circulating an appropriate fluid (typically water or air) using pumps or blowers. A *passive system* merely relies upon building design to utilize sunlight to heat the air or other material in the building. Active solar heating systems make use of collectors to heat a working fluid. Virtually all residential solar heating systems utilize flat plate collectors (Figure 8.5). The basic design of the flat plate solar collector is shown

Figure 8.5: Flat plate solar collector installation.

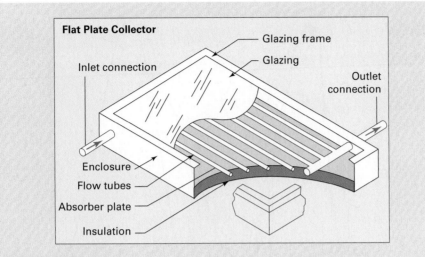

Figure 8.6: Design of a flat plate solar collector.

in Figure 8.6. The collector is basically an insulated box with a window on the front side to allow sunlight to enter. The sunlight is absorbed by a collector plate, and the heat is transferred to water flowing through pipes attached to the plate. The properties of the collector plate are very important for the operation of the collector. Ideally, the plate should absorb as much of the solar radiation as possible and lose as little energy as possible through reirradiation. The relationship between absorptance and emissivity, as previously explained, needs to be considered carefully in the analysis of the flat plate collector. The radiation that is absorbed in the form of sunlight is produced by the sun and has a spectrum that is basically similar to 6000 K black body radiation (Figures 8.1 and 8.4) with a peak wavelength of about 500 nm. The energy that is reirradiated by the collector plate is characteristic of the temperature of the plate (which might be at around 350 K). This reirradiated radiation is in the infrared portion of the spectrum with a peak at about 8000 nm. The ideal design for the collector plate would be to have a large absorptance at 500 nm and a small emissivity at 8000 nm, so that the collector absorbs substantial energy from sunlight but loses very little of it by reirradiation (Figure 8.7). Specially designed collector surface coatings have a high absorptance for visible radiation and low emissivity for infrared radiation. Flat black paint is a simple alternative that works well.

The glass plate on the front of the collector is also an important factor in the collector operation. Glass has a high transmittance for visible radiation but a low transmittance for infrared radiation. Thus, much of the energy that is reirradiated by the warm collector surface is not transmitted through the glass cover and is trapped inside the collector box. This is the same principle as the greenhouse effect discussed in Chapter 4 and is the reason that the interior of an automobile gets hot on a sunny day.

The efficiency of the collector is the ratio of the heat absorbed by the water to the energy incident on the collector. The loss of energy by reirradiation from the hot collector surface is one factor that limits the efficiency of the collector. Heat loss by

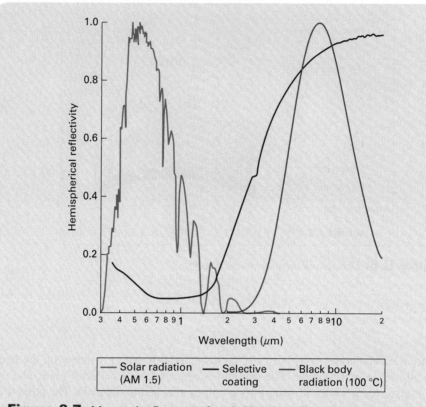

Based on P. Oelhafen and A. Schüller, "Nanostructured materials for solar energy conversion," Solar Energy 79 (2005) 110–121.

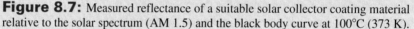

Figure 8.7: Measured reflectance of a suitable solar collector coating material relative to the solar spectrum (AM 1.5) and the black body curve at 100°C (373 K).

conduction through the walls of the box is another contributing factor. Equation (8.8) for radiative heat transfer and equation (8.4) for conductive heat transfer both show that, as the temperature of the collector increases, so does heat loss. At some point, the losses equals the energy incident on the collector, and the efficiency goes to zero. Thus, from an efficiency standpoint, it is advantageous to operate the collector at lower temperatures, and the maximum efficiency occurs when the collector is at the same temperature as its ambient surroundings.

On the other hand, the amount of energy that is actually transported to a home heating system by the water heated in the collector increases with increasing temperature (Section 8.5). From this viewpoint, it would be advantageous to operate the collector at as high a temperature as possible. However, there are practical considerations for the operating temperature based on the properties of the fluid used (e.g., water). If the temperature of the water is high, then the maximum amount of thermal energy is transported through the system. However, when the temperature of the water is high, then the heat losses between the collector and the heating system inside the house are greater. The best compromise between these conflicting criteria is generally to operate the collector at an efficiency of about 50%. This results in a collector plate temperature that is somewhat less than the boiling point of the water.

8.5 Residential Heating Needs

Active solar systems are commonly used for residential space heating and for supplemental water heating. In this section, some requirements for the design of a space heating system are considered. Because solar radiation is present only during the daytime hours, it is necessary to be able to store heat energy for use during the night. Water circulating through the collector is heated and stored in a tank, which also acts as a heat exchanger to provide heat and hot water to the house.

For the design of such a system, it is important to know how much solar energy a particular collector system can produce and how much heat a particular residence needs. The average insolation on a horizontal surface at different locations in the United States shown in Figure 8.2 is not the most useful information for this analysis. First, insolation is less in the winter (because of the angle of the sun), but this is the time of the year when heating needs are the greatest. Second, a horizontal surface is not the ideal orientation for maximizing solar energy collection. Table 8.2 shows the total daily solar energy per square meter for different collector orientations. The perpendicular collector includes a system for moving the collector (during the daylight hours) so that it always remains perpendicular to the incident solar radiation. This provides the maximum amount of energy but adds considerably to the cost, complexity, and presumably reliability of the system. Alternatives are to place the collector facing south (in the northern hemisphere) and at an angle from the vertical so that deviations from perpendicular incidence are minimized. Another option is to position the collector vertically and facing south. The final option is to place the collector horizontally. The table shows the daily insolation for each of these cases.

Table 8.2: Daily integrated insolation at a typical ~40°N latitude location (Allentown, Pennsylvania, latitude 40.65°N). Monthly averages over a 30-year period from 1961 to 1990 are given for flat plate collectors at various angles. *Vertical* means south facing, and *tilted* means south facing and tilted at the optimal angle. Tracking is a two-axis tracking collector that follows the sun during the day.

month	horizontal MJ/m^2	vertical MJ/m^2	tilted MJ/m^2	tracking MJ/m^2
Jan.	6.8	11.2	11.2	13.7
Feb.	9.7	13.0	14.0	16.9
Mar.	13.3	12.2	16.2	20.2
Apr.	16.9	10.8	18.0	23.0
May	19.4	9.4	18.7	25.2
Jun.	21.6	9.0	19.4	27.0
Jul.	21.2	9.4	19.4	27.0
Aug.	18.7	10.4	19.1	25.2
Sep.	15.1	11.9	17.6	21.6
Oct.	11.2	12.2	15.1	18.4
Nov.	7.2	10.1	10.8	13.0
Dec.	5.8	9.4	9.4	11.2

| **Table 8.3:** Relative insolation compared to a perpendicular collector for different orientations on January 21 at 40°N latitude. ||
orientation	% of perpendicular
60° south	89
vertical south	79
horizontal	43

© Cengage Learning 2015

The relative insolation for the different orientations is compared with the ideal perpendicular case for January 21 (typically the coldest time of the year in the northern hemisphere) in Table 8.3. In the interest of economics and simplicity, it is preferable to have a stationary (fixed) collector. The table shows that, at the coldest time of the year, the 60° south-facing collector provides close to 90% of the ideal energy and is the best simple option for system design. Clearly the horizontal orientation is not desirable. Figure 8.8 shows the variations throughout the day for different collector orientations.

This discussion, along with an anticipated collector efficiency in the range of about 50%, provides an estimate for the amount of energy that can be collected for a solar collector of a certain size. The amount of energy that a house needs to fulfill its heating needs depends on several factors: the size of the house, the details of its construction, and the local climate. The climate conditions are expressed as *degree days*. One degree day, sometimes expressed as *heating degree days*, is one day during which the average outside temperature is one degree less than the desired inside temperature. In warmer climates, *cooling degree days* are used as a measure of the

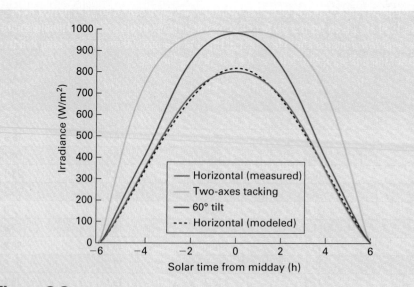

Based on F. Cruz-Peragón, P.J. Casanova-Peláez, F.A. Díaz, R. López-García and J.M. Palomar, "An approach to evaluate the energy advantage of two axes solar tracking systems in Spain," Applied Energy 88 (2011) 5131–5142.

Figure 8.8: Power per unit area incident on a solar collector for different orientations.

energy requirements for air conditioning. In this discussion, only heating degree days are considered.

Degree days, expressed for temperatures in Celsius, that is, *degree days* (°C), are determined on the basis of an ideal inside temperature of 18.3°C. Although this may be less than the target temperature in many buildings, contributions from internal heat sources, as will be discussed, can compensate for this difference. The numbers of degree days for individual days during the year are added together to give the total number of degree days for the year, and this sum can be used as an indication of the annual heating requirements. As an example, consider a cold winter day when the average outside temperature is −5 °C. This day would contribute [18.3°C − (−5°C)] = 23.3 °C degree days towards the annual total.

The average number of degree days per year for various locations in the United States and Canada are shown in Figures 8.9 and 8.10, respectively. More accurate values for some selected cities are given in Table 8.4.

It is now necessary to relate the heat requirements for a building of a particular design to the climate as represented by the number of degree days. We could undertake a detailed calculation of heat loss through the walls, windows, roof, foundation, and so on for the building based on an analysis of the *R*-values of these various building components. However, in 1972, the United States Federal Housing Authority provided an estimate of 67 kJ/m^3 per degree day (°C), for residential heating. Although construction techniques and materials have improved since 1972, this value is useful in acting as a rough guideline for heating needs: The annual heating requirements for a house are given as the volume of the house times the heat requirement per unit volume per degree day times the number of degree days per year for the location of the house.

To design a reasonable solar heating system, it is convenient to look at the daily requirements for the house in Philadelphia in Example 8.5. A winter (February) day, when the average daily temperature is, say, 0°C, represents 18.3 degree days (°C). This will require

$$700 \text{ m}^3 \times 67 \text{ kJ/m}^3 \times 18.3 = 858 \text{ MJ} \qquad \textbf{(8.9)}$$

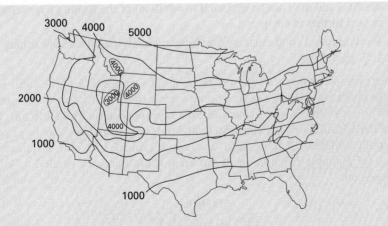

Figure 8.9: Map of the US showing the number of heating degree days (in °C) per year.

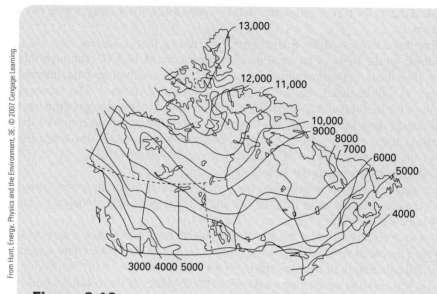

From Hunt, Energy, Physics and the Environment, 3E. © 2007 Cengage Learning.

Figure 8.10: Map of Canada showing the number of heating degree days (in °C) per year.

of heat on that day. This may be an overestimate of the requirements for at least four reasons:

1. Passive solar heating (Section 8.6) has not been considered.

2. There is heat production within a house as a result of other activities: heat generated by people, heat from cooking, heat generated by appliances and electronics, and so on. Some typical numbers are given in Table 8.5 and can account for something of the order of 10% of the heating needs.

3. Solar heating may work in conjunction with another heating method (e.g., oil or electric) and may not be expected to fulfill all of the heating requirements (discussed further in Chapter 17).

4. Well designed contemporary buildings typically require less heat than the 1972 U.S. guideline.

For our example, it might be reasonable to aim for about 60% of the value in equation (8.9), or about 500 MJ, for a solar heating system that would contribute in a substantial way to home heating needs. From Table 8.2 (appropriate for Philadelphia), we see that a total daily insolation (in February) on the order of 14.0 MJ/m^2 can be expected for a 60° south collector. Taking into account a typical collector efficiency of 50% allows for a simple calculation of the area of the collector needed to satisfy these requirements:

$$\frac{500 \text{ MJ}}{14.0 \text{ MJ/m}^2 \times 0.5} = 71 \text{ m}^2. \qquad \textbf{(8.10)}$$

A collector 7m × 10m would fulfill these requirements and would approximately cover the roof area of the house in this example.

Table 8.4: Average number of heating degree days (in °C) per year for selected U.S. and Canadian cities. (averages for 1971–2000).

city	state/province	heating degree days per year (°C)
Honolulu	HI	0
Miami	FL	86
Phoenix	AZ	578
San Diego	CA	602
New Orleans	LA	787
San Francisco	CA	1452
Albuquerque	NM	2378
Portland	OR	2426
Baltimore	MD	2574
New York	NY	2635
St. Louis	MO	2643
Philadelphia	PA	2644
Seattle	WA	2665
Vancouver	BC	2926
Boston	MA	3128
Denver	CO	3404
Detroit	MI	3583
Chicago	IL	3607
Toronto	ON	4066
Portland	ME	4069
Halifax	NS	4367
Minneapolis	MN	4379
Montreal	QC	4575
St. John's	NL	4881
Grand Forks	ND	5281
Regina	SK	5661
Edmonton	AB	5708
Winnipeg	MB	5777
Anchorage	AK	5817
Whitehorse	YT	6811
Mt. Washington	NH	7658
Yellowknife	NT	8256
Barrow	AK	11,052
Resolute	NU	12,526

Example 8.5

(a) Determine the annual heat requirements (in J) for a residence in Philadelphia with 250 m² of heated living space and 2.8-m ceilings.
(b) Estimate the annual heating costs for electric heat (at 100%) if electricity costs $0.11 per kWh.
(c) Estimate the annual heating costs for oil heat (at 85% efficiency) if oil costs $1.10 per liter.

Solution

(a) From Table 8.4, Philadelphia has 2644 heating degree days per year (°C), so the total heat requirement for a house with a volume of 250 m² × 2.8 ft = 700 m³ at 67 kJ/m³ per degree day (°C) is

$$67 \text{ kJ/m}^3 \times 700 \text{ m}^3 \times 2644 = 1.24 \times 10^{11} \text{ J.}$$

(b) The heating requirement can be converted from J to kWh (Appendix III) as

$$(1.24 \times 10^{11} \text{ J})/(3.6 \times 10^6 \text{ J/kWh}) = 34{,}445 \text{ kWh.}$$

At $0.11 per kWh, this will cost $3789.
(c) From Appendix IV, the energy content of 1 L of crude oil (about the same as home heating oil) is 3.85×10^7 J. At 85% efficiency, this will yield about 3.27×10^7 J/L. Meeting the heating requirement of 1.24×10^{11} J will require

$$(1.24 \times 10^{11} \text{ J})/(3.27 \times 10^7 \text{ J/L}) = 3792 \text{ L of oil.}$$

At $1.10 per L, the cost will be about 3792 L × $1.10/L = $4171.

Table 8.5: Typical daily contributions to home heating from internal sources.

source	heat per day MJ
lights	11
cooking	7
appliances	33
electronics	5
people	12
other	5
total	**73**

Example 8.6

(a) Determine the annual heat requirements (in J) for a residence in Winnipeg with 230 m^2 of heated living space and 2.5-m ceilings.
(b) Estimate the annual heating costs for electric heat (at 100%) if electricity costs $0.11 per kWh.
(c) Estimate the annual heating costs for oil heat (at 90% efficiency) if oil costs $0.90 per liter.

Solution

(a) From Table 8.4, Winnipeg has 5777 heating degree days per year (°C), so the total heat requirement for a house with a volume of 230 m^2 × 2.5 m = 575 m^3 at 67 kJ/m^3 is

$$67,000 \text{ J/m}^3 \times 575 \text{ m}^3 \times 5777 = 2.23 \times 10^{11} \text{ J.}$$

(b) The heating requirement can be converted from J to kilowatt-hours (Appendix III):

$$\frac{2.23 \times 10^{11} \text{ J}}{3.6 \times 10^6 \text{ J/kWh}} = 6.18 \times 10^4 \text{ kWh.}$$

At $0.11 per kilowatt-hour, this will cost $6800.

(c) From Appendix IV, the energy content of 1 L of crude oil (about the same as home heating oil) is 3.85 × 10^7 J/L. At 90% efficiency, this will yield about 3.47 × 10^7 J/L. Meeting the heating requirement of 2.23 × 10^{11} J will require

$$\frac{2.23 \times 10^{11} \text{ J}}{3.47 \times 10^7 \text{ J/L}} = 6427 \text{ L of oil.}$$

At $0.90 per liter, this will cost $5784.

8.6 Heat Storage

The preceding discussion provided a basis for a usable design of a residential solar heating system. However, it must be realized that the sun does not shine 24 hours a day and that the insolation value used in the previous section is a daily average. In fact, part of the problem in utilizing solar energy for heating is that in the winter, when the most heat is needed, the duration of sunshine is the least. So the energy that is collected during the day must be stored for use during the night. One may also take the approach that energy collected during sunny days can be stored to supplement heating needs during cloudy days. Heat contained in water that has been heated in a solar collector (Figure 8.6) may be stored by keeping the water in an insulated tank or transferring that heat to another material for storage. (The design of such a system is discussed further in Chapter 17; see Figure 17.16). The amount of heat that can be stored in a material is a function of the material's temperature, its mass, m, and its thermal properties. If a material is cooled from an initial temperature T_i to a final temperature T_f (assuming no phase transitions are involved), then the heat given up is expressed as

$$Q = Cm(T_i - T_f), \tag{8.11}$$

where C is the *specific heat* of the material. It is clear from equation (8.11) that the specific heat is a measure of the thermal energy content of a material per unit mass per unit temperature change. The *volumetric heat capacity* is defined as the heat contained in a material per unit volume per unit temperature and is obtained from the specific heat by multiplying C times the density. Table 8.6 gives the specific heat and volumetric heat capacity of some common materials. Equation (8.11) may also be written as

$$\Delta T = \frac{Q}{Cm} \tag{8.12}$$

and gives the temperature increase of a material when a quantity of heat is applied to it. From equation (8.12), it is clear that a material with a large specific heat experiences a relatively small increase in temperature when heat is applied and that a material with a small specific heat experiences a much larger increase in temperature. Conversely, materials with large specific heats can release significant amounts of heat without undergoing large temperature changes.

Among the common materials listed in Table 8.6, water clearly has the greatest capacity for storing thermal energy, both per unit weight and per unit volume. In addition, it is convenient because it is nontoxic, relatively nonreactive, and inexpensive, and as a liquid, it provides a suitable means of transporting heat. For a space heating system that uses solar energy, it would be appropriate to store an amount of heat that is comparable to what is used daily during the winter. The following analysis provides an idea of the requirements for a useful heat storage system.

To design a system to store the 1.09 GJ calculated in Example 8.7 in a tank of water, it is necessary to know the maximum temperature of the water, T_i, and the

Table 8.6: Specific heats and volumetric heat capacities for some common materials.

material	specific heat J/(kg·°C)	density kg/m³	volumetric heat capacity kJ/(m³·°C)
water	4186	1000	4186
aluminum	895	2700	2416
iron	460	7855	3613
wood (pine)	2800	500	1400
stone (solid)	879	2560	2250
stone (loose)	879	~1500	~1300
brick	920	1800	1656
concrete	653	2300	1502
glass	837	2720	2277
sand	816	1600	1306

Example 8.7

Calculate the amount of energy needed to heat the home in Winnipeg from Example 8.6 on a winter day when the average daily temperature is −10°C.

Solution

An average temperature of −10°C over a period of one day corresponds to [18.3°C −(−10°C)] × 1 day = 28.3 degree days (°C). For a house with a volume of 575 m^3 requiring 67 kJ/m^3 per degree day, the total energy required will be 67,000 J/m^3 × 575 m^3 × 28.3 = 1.09 × 10^9 J.

temperature of the water after the 1.09 GJ has been extracted, T_f. The mass of water can then be obtained by rewriting equation (8.12) as

$$m = \frac{Q}{c\Delta T}. \tag{8.13}$$

Minimizing the size of the tank means maximizing T_i and minimizing T_f. The maximum temperature is limited by the temperature of the water exiting the solar collector (in order to transfer heat from the solar collector water to the storage tank water). The minimum temperature is limited by the temperature of the house (in order to transfer heat from the heating system to the room). Although an absolute maximum temperature might be 90°C and an absolute minimum temperature would be about 18°C, a reasonable range of operating temperatures that provides good efficiency for heat transfer might be between $T_i = 65$°C and $T_f = 30$°C.

Example 8.8

Calculate the mass and volume of water needed to store 1.09 GJ of thermal energy if the maximum and minimum water temperatures are 65°C and 30°C, respectively.

Solution

Equation 8.13 gives

$$m = \frac{Q}{c\Delta T},$$

where Q = 1.09 × 10^9 J and ΔT = (65°C − 30°C) = 35°C. Table 8.6 gives c = 4186 kJ/(m^3·°C) for water, and, since water has a density of 10^3 kg/m^3, this is the same as 4186 J/(kg·°C). The mass of water is calculated to be

$$m = \frac{1.09 \times 10^9 \text{ J}}{4186 \text{ J/(kg} \cdot {}^\circ\text{C)}} \times 35{}^\circ\text{C)} = 7.4 \times 10^3 \text{ kg}.$$

This mass of water represents a volume of 7.4 m^3. This could be contained in a cubic tank about 1.9 m on a side.

8.7 Passive Solar Heating

All buildings are heated passively by solar radiation to some extent, but the design and placement of the building can maximize these benefits. In the design of a passively heated building, the steps taken to optimize passive solar heating must not ultimately require more heat. Thus, heat gain must be balanced against heat loss. A typical home designed to optimize passive solar heating incorporates large south-facing windows (Figure 8.11). Figure 8.12 shows how building design can be used to maximize passive solar heating during the winter and minimize excess heating during the summer.

Figure 8.11: Passive solar home showing large south-facing windows.

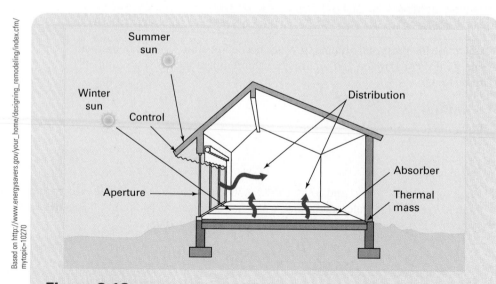

Based on http://www.energysavers.gov/your_home/designing_remodeling/index.cfm/mytopic=10270

Figure 8.12: Method of utilizing solar heating during the winter and reducing solar heating during the summer.

A massive stone floor absorbs heat during the day (especially if it is dark colored) and reirradiates the heat during the night to distribute the effects of passive heating over a 24-hour period.

Because passive heating results from energy entering through a window and being trapped as a result of the greenhouse effect, it is necessary to consider the heat gained in comparison with the heat lost through the window. It is easiest to consider the heat lost in the context of conductive heat transfer through the window. A standard double-pane window has an R-value of about 0.35, and a wall without a window that contains 15 cm. of insulation will have an R-value of about 3.5. Thus, it is important to ensure that, on average, the energy gain through windows from passive solar heating will outweigh the energy loss resulting from the low R-value of the windows. The effects of the R-values of windows will be discussed further in Chapter 17.

Example 8.9

Calculate the net energy gain (per day in January) from passive solar radiation incident on a double-pane window in a house at 40°N latitude. Consider a 1.5 m × 2.5 m south-facing (vertical) window and an average daily temperature of −5 °C with an inside temperature of 20°C. What is the efficiency of this heating method?

Solution

From Table 8.2, the average January daily insolation on a south-facing vertical surface is 11.2 MJ/m². For a (1.5×2.5)m² = 3.75 m² window, the total incident energy is $(11.2$ MJ/m²$) \times 3.75$ m² = 42 MJ per day. The energy loss is obtained from equation (8.5) by multiplying the power per unit area by the area and the time; that is,

$$E = \frac{(T_h - T_c)At}{R}.$$

For a standard double-pane window, the R-value is 0.35 (s·m²·°C)/J. Using area in square meters, time in seconds (86,400 s/d), and the temperature difference in °C yields energy in J. Thus

$$E = \frac{(20°C - (-5°C)) \times (3.75 \text{ m}^2) \times (86{,}400 \text{ s/d})}{0.35 \text{ (s·m}^2\text{·°C)/J}} = 23.1 \text{ MJ per day.}$$

The net energy gain from passive solar heating is 42.0 MJ − 23.1 MJ = 18.9 MJ per day. This is compared to the total incident energy to give an efficiency of $100 \times (18.9$ MJ$)/(42.0$ MJ$)$ = 45% efficiency.

The efficiency of solar energy collection under the conditions described in Example 8.9 is quite good. This calculation is based on clear sky conditions. The reduction of passive solar heating caused by clouds depends on local climate conditions. For example, if cloud cover reduces the insolation to 50% of its clear sky value, then the incident energy is 0.5×42 MJ = 21 MJ per day, resulting in a net energy loss through the window. This calculation also emphasizes the significance of a high R-value and the effects of lower outside temperatures.

8.8 Summary

The chapter has reviewed the properties of sunlight. Based on the total power radiated by the sun, the insolation at the surface of the earth, averaged over time and all locations, was found to be 168 W/m^2.

To describe how solar energy can be used for heating purposes, the chapter reviewed the three heat transfer mechanisms: conduction, convection, and radiation. The power transferred by conduction per unit area of a material is given as $P/A = \Delta T/R$, where ΔT is the temperature difference, and R is the R-value, defined as the thickness of the material divided by the thermal conductivity. R-values for insulating materials allow for the convenient calculation of heat transfer through walls, windows, and other building components. Radiative heat transfer is described by the Stefan-Boltzmann law and is proportional to the fourth power of the absolute temperature of an object.

A flat plate solar collector is a box in which solar energy is transferred to a working fluid for distribution to a building for space heating or hot water needs. The efficiency of the solar collector can be evaluated on the basis of the radiative and conductive heat mechanisms that transfer thermal energy into and out of the collector.

The chapter evaluates the usefulness of solar heating on the basis of heating requirements for a building. The energy needed for heating purposes is a function of the size of the building and the quality of its construction, as well as the local climate. The effects of climate are quantified using degree days, where one degree day is one day during which the outside temperature is one degree below a standard inside temperature. Annual heating needs are determined by the number of degree days integrated over the year. As illustrated in the chapter, the utilization of the roof area of a single-family home for solar collectors is sufficient to make a reasonable contribution to space heating needs.

The periodic nature of sunlight, on both a daily and a seasonal scale, as well as the variability of heating requirements, again on a daily and seasonal scale, means that the efficient utilization of solar energy requires some method of thermal energy storage. The chapter discusses the effectiveness of different materials for heat storage in terms of their heat capacity and the possible mechanisms for transferring heat into and out of the storage system.

The chapter describes the effectiveness of passive solar heating in terms of the trade-off between energy gain and energy loss through a window. Passive solar heating features can be incorporated into new building designs, often at little expense by utilizing specific design characteristics, building location, and/or orientation.

Problems

8.1 Calculate the power radiated by a woodstove of dimensions 65 cm high by 55 cm deep by 85 cm wide with a surface temperature of 120°C. Assume that heat is radiated from all surfaces of the stove and that the stove has an emissivity of 1. Note that woodstoves are painted black because black surfaces have high absorptance, and objects with high absorptance also have high emissivity.

8.2 Locate information about the current cost of home heating oil or natural gas in your area (whichever is in common use) and the cost of residential electricity.

Assuming an efficiency of 85% for an oil furnace and 100% for electric heat, calculate the relative cost of electric heat compared with oil heat or natural gas if both heating systems require the same net energy to heat a house.

8.3   Consider a vertical south facing window in a house at 40°N latitude. For an interior temperature of 20°C make a plot of the minimum R-value as a function of outside temperature from −30°C to 10°C for which the passive solar heating exceeds the heat loss though the window.

8.4   Compare the total solar energy received at the surface of the earth in one year to the total annual global energy requirements.

8.5   Compare the R-values of:
   a.   Two pieces of 3-mm thick glass in thermal contact.
   b.   Two pieces of 3-mm thick glass with a 1-cm air space between them.

8.6   Approximate a house as a cube with an edge length of 7 m. The house loses heat from the four walls and the roof (but not the floor). The average R-value for the walls and roof is $R = 1.2$ (this takes into account walls/windows/doors/etc.). Calculate the heat loss in MJ/m^3 per degree day (°C) and compare this to the estimated residential heating needs as discussed in this chapter.

8.7   A 300 liter electric hot water heater has provides 9000 W of power to heat water. If the heater is filled with water at an initial temperature of 10 °C, how long will it take for the water to reach 60°C? Assume there are no heat losses.

8.8   Compare the masses and volumes of water, concrete, sand and wood needed to store 1 GJ of heat if the operating temperatures are $T_c = 30$°C and $T_h = 80$°C. In each case calculate the edge length of a square storage unit with a height of 2.5 m.

Bibliography

B. Anderson. *Solar Energy: Fundamentals in Building Design.* McGraw-Hill, New York (1977).

B. Anderson and M. Riorden, *The New Solar Home Book* (2nd ed.). Brick House Publishing, Andover, MA (1996).

B. Anderson and M. Wells. *Passive Solar Energy: The Homeowner's Guide to Natural Heating and Cooling* (2nd ed.). Brick House Publishing, Andover, MA (1996).

A. K. Athienitis and M. Santamouris. *Thermal Analysis and Design of Passive Solar Buildings.* James & James, London (2002).

G. Boyle (Ed.). *Renewable Energy.* Oxford University Press, Oxford (2004).

D. Chiras. *The Solar House: Passive Heating and Cooling* Chelsea. Green Publishing, White River Junction, VT (2002).

J. A. Duffie and W. A. Beckman. *Solar Engineering of Thermal Processes* Wiley, Hoboken, NJ (2006).

J. Fanchi. *Energy: Technology and Directions for the Future.* Elsevier, Amsterdam (2004).

F. A. Farret and M. Godoy Simões. *Integration of Alternative Sources of Energy.* Wiley, Hoboken (2006).

D. Y. Goswami, F. Kreith, and J. F. Kreider. *Principles of Solar Engineering* (2nd ed.). Taylor & Francis, New York (2000).

D. Hawkes and W. Forster. *Energy Efficient Building: Architecture, Engineering and Environment.* W. W. Norton, New York (2002).

R. Hinrichs and M. Kleinbach. *Energy: Its Use and the Environment* (5th ed.). Brooks-Cole, Belmont, CA (2012).

R. W. Jones, J. D. Balcomb, C. E. Kosiewicz, et al. *Passive Solar Design Handbook, Volume Three: Passive Solar Design Analysis,* NTIS, U.S. Dept. of Commerce, Washington, DC (1982).

M. Kaltschmitt, W. Streicher, and A. Wiese (Eds.). *Renewable Energy: Technology, Economics and Environment.* Springer, Berlin (2007).

J. A. Kraushaar and R. A. Ristinen. *Energy and Problems of a Technical Society* (2nd ed.). Wiley, New York (1993).

F. Kreith and J. F. Kreider. *Principles of Solar Engineering.* McGraw-Hill, New York (1978).

F. Kreith and R. West. *Handbook of Energy Efficiency.* CRC Press, Boca Raton (1997).

G. Lof. *Active Solar Systems.* MIT Press, Cambridge, MA (1993).

P. J. Lunde. *Solar Thermal Engineering.* Wiley, New York (1980).

E. Mazria. *The Passive Solar Energy Book.* Rodale Press, Emmaus, PA (1979).

B. Norton. *Solar Energy Thermal Technology.* Springer-Verlag, London (1992).

T. Reddy. *The Design and Sizing of Active Solar Thermal Systems.* Clarendon Press, Oxford (1987).

S. Weider. *An Introduction to Solar Energy for Scientists and Engineers.* Kreiger Publishing, Melbourne (1992).

Electricity from Solar Energy

Learning Objectives: After reading the material in Chapter 9, you should understand:

- The generation of electricity from solar energy through heat engines.
- The basic physics of semiconducting materials and the properties of n-type and p-type materials.
- The construction of a semiconducting junction device.
- The production of electricity through the interaction of photons on semiconducting junctions.
- The sensitivity and efficiency of photovoltaic devices.
- The application of photovoltaic devices for electricity generation worldwide.
- The availability of solar energy and economic viability of its utilization.

9.1 Introduction

The extent of the solar energy resource was illustrated in the previous chapter. However, the applications presented thus far, which utilize thermal energy extracted from solar radiation, are most suitable for residential space heating. The conversion of solar energy into electricity allows for the wide-scale distribution of energy on the electric grid, in addition to local residential use. Two different approaches can be taken to the conversion of solar energy into electricity: (1) the conversion of solar radiation into heat, followed by the conversion of heat into mechanical energy by means of a heat engine, and finally the generation of electricity by means of a generator; (2) the direct conversion of radiation into electrical energy by means of a photovoltaic device. These two approaches are discussed in this chapter.

9.2 Solar Electric Generation

A relatively straightforward (at least conceptually) method of producing electricity from solar energy is to use the heat produced in a solar collector to generate electricity by means of a heat engine. This is the most easily accomplished when water is heated above its boiling point to produce steam that can then be used to drive a steam turbine. Focusing collectors are necessary for this purpose because flat plate collectors are not suitable for achieving the required temperatures. Several approaches to large-scale focusing collector designs have been taken in the past. The most notable are:

- Parabolic troughs.
- Parabolic dishes.
- Central receivers.

These three devices are now reviewed briefly.

9.2a Parabolic Troughs

Parabolic trough collectors heat a fluid that is flowing through a pipe located at the focus of a parabolic trough (Figure 9.1). The parabolic troughs rotate to track the sun to ensure that the radiation is properly focused on the fluid-carrying pipes. In this arrangement, the fluid is typically oil, and the heated oil transfers its thermal energy to water through a heat exchanger to produce steam. The only significant installation of this type is the Solar Electric Generating Station (SEGS) in Kramer Junction, California (Figure 9.1). This facility consists of several fields of collectors with over 2 million m^2 of solar collection area and a maximum output of 350 MW$_e$ of electricity. The overall efficiency of converting solar energy into electricity by this method is typically about 15% and is limited, in the best case, by the Carnot efficiency.

9.2b Parabolic Dishes

An alternative to the parabolic trough geometry is the parabolic dish geometry (Figure 9.2). Units may be either individual parabolic dishes or arrays of dishes. Instead of a line focus, as is the case for the parabolic trough, the parabolic dish has a point focus. These are tracking collectors and utilize one of two different approaches to converting solar energy into electricity. The Solar Total Energy Project (STEP) in Shenandoah, Georgia heated water to produce steam to generate electricity through conventional generators. This facility is no longer functional. Another approach is to directly convert the

Figure 9.1: Parabolic trough solar collectors at Kramer Junction, California.

Figure 9.2: Array of parabolic dish solar collectors at Maricopa Solar Project in Peoria, Arizona. The facility consists of a total of 60 parabolic dishes.

Figure 9.3 Parabolic dish solar collectors at Maricopa Solar Project in Peoria, Arizona showing the Stirling engine used to convert heat to mechanical energy.

energy content of hot gas produced by the absorption of solar radiation into mechanical energy using a Stirling engine. Figures 9.2 and 9.3 show the Stirling engine based parabolic dish array at the Maricopa Solar Project in Peoria, Arizona. The solar-to-electric efficiency for this system is around 26% and is limited primarily by the Carnot efficiency.

Example 9.1

Calculate the average power per unit area that is incident on a 5-cm diameter focal area for a 10-m diameter parabolic dish as illustrated in Figure 9.3. Assume a midday insolation of 650 W/m² on a sunny day.

Solution

A 10-m diameter dish has an area of

$$\frac{\pi(10 \text{ m})^2}{4} = 78.5 \text{ m}^2,$$

giving a total insolation on the dish of

$$(78.5 \text{ m}) \times (650 \text{ W/m}^2) = 51 \text{ kW}.$$

If this is focused on an area of

$$\frac{\pi(0.05 \text{ m})^2}{4} = 1.96 \times 10^{-3} \text{ m}^2,$$

then the power per unit area is

$$\frac{51 \text{ kW}}{1.96 \times 10^{-3} \text{m}^2} = 26 \text{ MW/m}^2.$$

9.2c Central Receivers

The final possibility for the conversion of solar energy to electricity by means of a heat engine is by means of a central receiver. This is analogous to a single large parabolic dish except that the parabolic dish is replaced with a planar array of computer-controlled mirrors, each of which tracks the sun and reflects the sunlight onto a single point. The individual mirrors are referred to as *heliostats*, and the central receiver is contained in the so-called *power tower*. An early example of this type of design is Solar One in California, which operated between 1982 and 1986. This was later redesigned to become Solar Two (Figures 9.4 and 9.5). Solar Two operated from 1995 to 1999 and achieved its rated output of 10 MW$_e$. A working fluid, circulated through the focal point at the top of the tower, is used to produce steam to either drive a turbine/generator to generate electricity directly or as a mechanism to store thermal energy for later use (Figure 9.6). Solar One used oil for heat storage, and Solar Two used a molten salt. An updated version of this type of facility, rated at 11 MW$_e$, is now operational in Spain (Figure 9.7).

9.3 Photovoltaic Devices

An alternative approach to the production of electricity from sunlight using a heat engine is *photovoltaics*. An attractive aspect of photovoltaics is the ability to implement this technology on a wide range of scales from milliwatt devices suitable for running

Figure 9.4: Solar Two in Daggett, California. The scale can be determined from the central tower that is about 90 m high.

Figure 9.5: One of Solar Two's heliostats. The Solar Power Tower is reflected in the mirror.

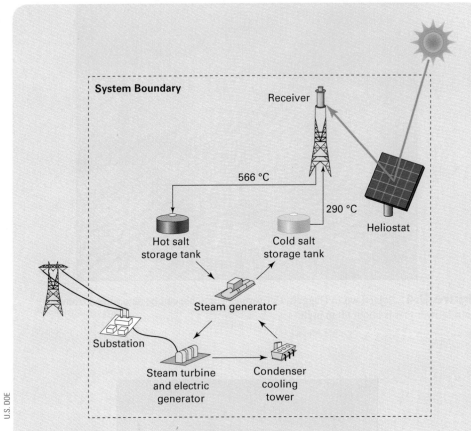

U.S. DOE

Figure 9.6: Schematic diagram showing the operation of Solar Two.

© Chris Sattlberger/Corbis

Figure 9.7: Aerial view of the PS10 solar generating station near Seville, Spain.

Figure 9.8: Photovoltaic device (black area above the model number) used to power a pocket calculator. The 1.5-cm² cell produces about 10 mW$_e$ in bright sunlight.

watches and pocket calculators (Figure 9.8) to multimegawatt installations (Figure 9.9). The operation of a photovoltaic device results from how light interacts with the electrons in a semiconducting material. The description of this behavior begins with an overview of some basic semiconductor physics.

Electrons that are associated with an atom may be described in terms of energy levels. For the hydrogen atom, the electron can occupy quantized levels with energies that are given by the expression

$$E = -\frac{2\pi e^4 m}{n^2 h^2},$$ **(9.1)**

Figure 9.9: The 40-MW$_e$ photovoltaic power plant Waldpolenz near Leipzig, Germany. The total length of the array in the photograph is about 2 km.

where e is the charge on the electron, m is the mass of the electron, h is Planck's constant, and n is an integer that defines the energy level. Increasing values of n correspond to increasing energy [Figure 9.10(a)]. In an atom that contains many electrons, the interactions between the electrons causes the energy levels to split into sublevels [Figure 9.10(b)]. The sublevels are labeled s, p, d, f, and so forth, as shown in the figure, and correspond to different values of the orbital angular momentum. An s-sublevel can hold 2 electrons, a p-sublevel can hold 6, a d-sublevel can hold 10, an f-sublevel can hold 14, and so on. Silicon (Si), which has 14 electrons, is a good example of a common semiconducting material. The electrons in their lowest energy configuration (ground state) can be expressed as

$$1s^2 2s^2 p^6 3s^2 3p^2 \qquad \textbf{(9.2)}$$

where 1s, 2s, 2p, and so on represent the energy sublevels, and the superscript after the level name gives the number of electrons in that level [Figure 9.11]. The two 3s electrons and two 3p electrons are not as firmly bound to the Si atom as the other electrons and are referred to as *valence electrons*.

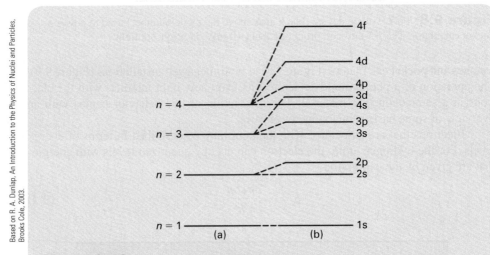

Based on R. A. Dunlap, An Introduction to the Physics of Nuclei and Particles, Brooks Cole, 2003.

Figure 9.10: Energy levels in (a) hydrogen and (b) a multielectron atom.

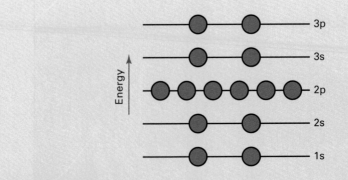

© Cengage Learning 2015

Figure 9.11: Occupation of electron energy levels in a Si atom in its ground state. The circles represent electrons.

If a large number of Si atoms are assembled to form a solid, then the interactions between the electrons associated with one atom and the electrons associated with another atom cause the energy levels to change a little—sometimes a bit lower, sometimes a bit higher. Thus, the energy levels get smeared out into *bands* (Figure 9.12). For a single Si atom, the 1s level can hold 2 electrons. For a piece of Si containing N Si atoms, the 1s band can hold $2N$ electrons. Similarly, in the piece of Si, the 2s band holds $2N$ electrons, the 2p band holds $6N$ electrons, and so on. It turns out that the interactions in the solid also cause the 2p and 3p bands to split into three sub-bands, each of which can contain $2N$ electrons. Thus, the electrons in a piece of Si fill up the 1s, 2s, 2p, 3s, and the first of the three 3p sub-bands (Figure 9.13). The $2N$ electrons in the 3s band and the $2N$ electrons in the lowest of the 3p sub-bands are the valence electrons in a piece of Si containing N Si atoms. The second 3p sub-band in Si is called the *conduction band* and, as shown in the figure, is empty in the ground state (i.e., contains no electrons). The spaces (in energy) between the bands are regions where no energy levels exist and are forbidden energy regions for the electrons. The forbidden space between the valence band and the conduction band is called the *energy gap* (or *band gap*) and has a width corresponding to an energy E_g. The size of the energy gap depends primarily on the kind of material; for Si, it is 1.1 eV.

The valence electrons contribute to the bonding between the atoms in a piece of solid Si, as represented by the simple two-dimensional picture of the atoms in a Si lattice in Figure 9.14. Each Si atom shares its four valence electrons with its four neighbors.

It is now possible to describe how light interacts with electrons in a solid. As described in Chapter 1, light is quantized in the form of photons. Each photon has an energy related to the frequency of the light (the greater the frequency, the greater the energy per photon [see equation (1.26)]. From equation (1.25), the energy per photon can be related to the wavelength of the light, λ, as

$$E = \frac{hc}{\lambda}. \tag{9.3}$$

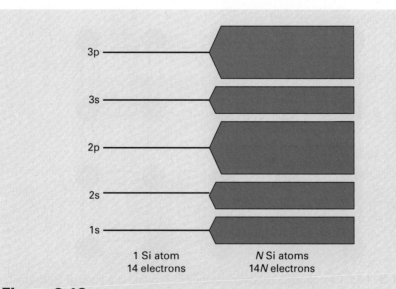

© Cengage Learning 2015

Figure 9.12: Energy bands in a piece of Si containing N Si atoms.

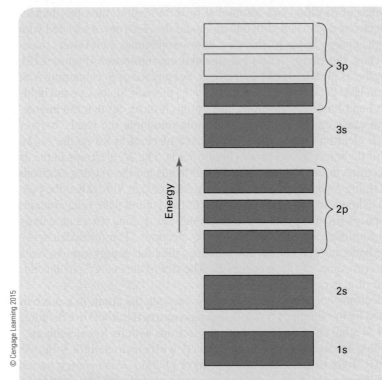

Figure 9.13: Band structure of a piece of Si. The solid boxes represent occupied bands, and the open boxes represent unoccupied bands.

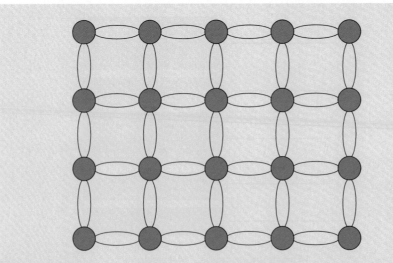

Figure 9.14: Two-dimensional representation of bonding in a piece of Si. Each of the lines between the atoms represents a 3s or 3p electron involved in bonding.

In customary units, this is

$$E = \frac{1240 \text{ eV} \cdot \text{nm}}{\lambda}, \qquad \textbf{(9.4)}$$

where E is in eV, and λ is in nm. If a photon of sufficient energy is incident on a piece of Si, it can impart that energy to an electron in the valence band of the material and cause it to move into an energy level in conduction band (Figure 9.15). When the electron moves to the conduction band, it leaves a vacant energy state in the valence band. This vacant state has an effective positive charge and is referred to as a *hole*. To form the electron-hole pair, the photon must have energy greater than the energy gap. In Si, the energy gap of 1.1 eV corresponds [according to equation (9.4)] to a wavelength of 1130 nm. From Figure 8.1, it can be seen that most of the solar spectrum (including the entire visible region) satisfies this condition. A simplified picture of the lattice, illustrating electron-hole pair formation, is shown in Figure 9.16. The electron formed in this

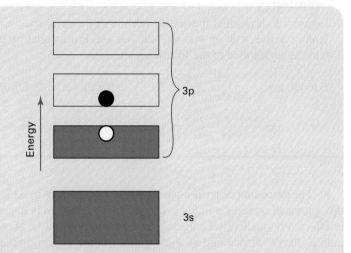

Figure 9.15: Band structure (upper portion only) of Si, showing the excitation of an electron from the valence band to the conduction band caused by the absorption of a photon of sufficient energy.

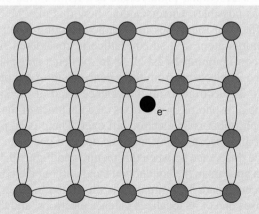

Figure 9.16: Portion of a Si lattice showing broken bond and resulting conduction electron.

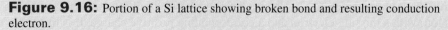

© Cengage Learning 2015

© Cengage Learning 2015

way is not bound to any particular atom and is free to move about in the material. Similarly, the hole can move about by exchanging places with one of the valence electrons associated with the other Si atoms in the material. If a number of photons are incident on a piece of Si and form a number of electron-hole pairs, then these electrons and holes can move through the material and constitute an electric current. In this way, the energy associated with the photons in the radiation can be converted into electrical energy.

Example 9.2

Estimate the range of photon energies that corresponds to the visible portion of the solar spectrum.

Solution

Figure 8.1 shows that the visible portion of the solar spectrum covers wavelengths between about 400 nm and 700 nm. From equation (9.4), these wavelengths correspond to the following energies:

For 400 nm: $\dfrac{1240 \text{ eV} \cdot \text{nm}}{400 \text{ nm}} = 3.1$ eV at the violet end of the visible spectrum.

For 700 nm: $\dfrac{1240 \text{ eV} \cdot \text{nm}}{700 \text{ nm}} = 1.8$ eV at the red end of the visible spectrum.

The preceding description may suggest a means of making a photovoltaic cell, but this approach presents a major problem. As the electron is negatively charged, the hole is effectively positively charged, and they can both move about in the material, these unlike charges attract and at some point are likely to combine and cancel each other out. This is called *recombination* and corresponds to a free electron in the conduction band (Figure 9.15), losing energy and falling across the energy gap to fill a hole in the valence band. This can be seen in the picture of the lattice in Figure 9.16, where an electron can fill a broken bond, thus eliminating both the free electron and the hole.

The design of a functional photovoltaic cell requires the elimination of recombination (as much as possible). Understanding how this may be done requires a consideration of the effects of impurities in a semiconducting material. The behavior of an impurity phosphorus atom in a silicon lattice is considered first. Phosphorus has 15 electrons in the configuration $1s^2 2s^2 2p^6 3s^2 3p^3$. When a phosphorus atom replaces a silicon atom, the two 3s electrons from the valence band of the phosphorus atom take the place of the two 3s electrons from the valence band of the silicon atom that was removed. Two of the phosphorus 3p electrons take the place of the two Si 3p electrons, but this leaves one phosphorus 3p electron left over. There is no place for this to go in the silicon valence bands because they are filled up, so it goes into the conduction band [Figure 9.17(a)]. The difference between this picture and Figure 9.15 is that, in the present case, an electron appears in the conduction band without the corresponding creation of a hole in the valence band. This situation is illustrated in Figure 9.18(a), where the phosphorus atom loses one of its localized valence electrons, and that electron is free to move about in the material. This leaves the phosphorus atom (now a positive phosphorus ion) with a missing electron and a corresponding positive charge. It is important to note that, as a whole, this material is electrically neutral because it is made up of atoms

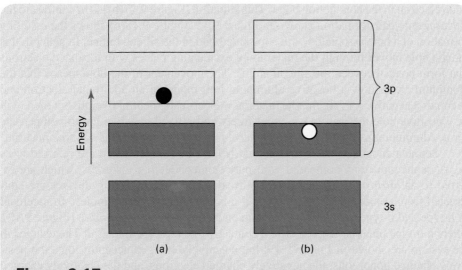

Figure 9.17: Occupied electron states in a piece of Si with (a) donor impurity and (b) acceptor impurity.

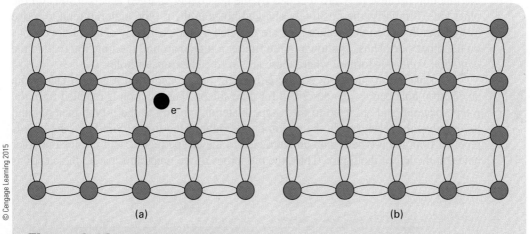

Figure 9.18: Portion of a Si lattice showing (a) conduction electron from donor impurity and (b) hole from acceptor impurity.

that were originally neutral. However, it is different than, say, a piece of pure silicon because a negatively charged electron is free to move about in the material, and a corresponding positively charged ion is fixed in its location in the lattice. The phosphorus atom is referred to as a *donor* because it gives up, or donates an electron to the conduction band, and the material is referred to as an *n-type* semiconducting material because there are free negatively charged electrons in the material.

In contrast, an aluminum impurity in silicon behaves differently. Aluminum has 13 electrons in the configuration $1s^2 2s^2 2p^6 3s^2 3p^1$. Thus there is one electron too few to fill up the 3s band and the first of the 3p valence bands. The lack of an electron results in the appearance of a hole in the 3p valence band without the corresponding appearance of an electron in the conduction band. This situation [Figure 9.17(b)] is in contrast to the electron-hole pair formation shown in Figure 9.15. Figure 9.18(b) shows the resulting formation of a hole associated with a missing bond in the silicon lattice. In general, this mobile hole moves through the material by exchanging places with localized electrons that form bonds between the silicon atoms. The movement of the hole means that the aluminum atom (now a negative aluminum ion) binds up an additional electron and becomes a negative ion. Again, the material is electrically neutral overall, but there will be a positive hole that can move about freely and a fixed negative ion to compensate for it. Aluminum in silicon is called an *acceptor*, and this material is a *p-type* material.

Semiconductors with impurities that behave as just described are referred to as *doped* semiconductors, compared with pure semiconducting materials, which are referred to as *intrinsic* semiconductors. Virtually all semiconducting devices are constructed from combinations of n-type and p-type materials. The simplest arrangement of n-type and p-type materials is the semiconducting junction (or *diode*) (Figure 9.19), where a n-type material is in electrical contact with an p-type material. The charges in the system are shown in Figure 9.19: the negatively charged acceptor ions, the positively charged donor ions, the negatively charged electrons, and the positively charged hole. The neutral silicon atoms are not shown. Recall that the charged electrons and holes are free to move around, and the charged acceptor and donor ions are fixed in the lattice. The negatively charged electrons in the n-type material tend to move away from the junction because they are repelled by the negatively charged acceptor ions on the other side. Similarly, the positively charged holes in the p-type material tend to move away from the junction because they are repelled by the positively charged donor ions on the other side. Thus, as shown in the figure, a region around the junction (called the *depletion region*) is formed where there are no free electrons or holes.

If a photon of sufficient energy is incident on the depletion region, it can create an electron-hole pair as just described for pure silicon. The electron is repelled from the p-type material and attracted to the n-type material, moving to the n-type material and joining the other electrons in the region on the right of the figure. Similarly, the hole is repelled from the n-type region and attracted to the p-type region, and it joins the other holes on the left of the figure. Thus, the properties of the junction separate the electrons

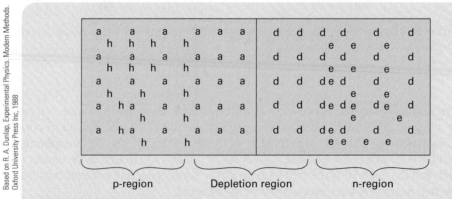

Based on R. A. Dunlap, Experimental Physics, Modern Methods. Oxford University Press Inc, 1988

Figure 9.19: Semiconducting junction formed from a piece of p-type and n-type semiconducting material showing acceptor impurities (a) holes (h), donor impurities (d) and electrons (e).

and the holes formed from incident electromagnetic radiation and reduce the probability that they will recombine. This movement of electrons and holes results in a net current flow across the junction, this electric current can be provided to an external device connected to the junction. This is the basis of operation for the *photodiode*, or *photovoltaic cell*. The spatial extent of the photosensitive region can be increased by creating the equivalent of a large depletion region (i.e., charge carrier–free region) by sandwiching an intrinsic (pure) semiconducting layer between the p-type and n-type regions.

9.4 Application of Photovoltaic Devices

Because the operation of the photovoltaic cell requires that the photons that are incident on it have sufficient energy to produce electron-hole pairs, there is, as described, a maximum wavelength of light that produces this effect. The spectral response for some photovoltaic cells is illustrated in Figure 9.20. The cutoff at long wavelengths is due to the fact that the photon energy for longer wavelengths is less than the energy gap and is insufficient to create electron-hole pairs. Figure 8.1 shows that about 23% of the photons received from the sun do not have sufficient energy to be converted to electricity by a Si photovoltaic cell. In addition, much of the electromagnetic energy is converted into heat rather than electricity. The spectral response of a photovoltaic cell can be improved by utilizing a semiconducting material with a smaller energy gap, thus enabling more of the solar spectrum to be effective in producing electricity. This is seen in Figure 9.20 by the longer wavelength corresponding to the cutoff for germanium ($E_g = 0.67$ eV) compared with Si ($E_g = 1.1$ eV).

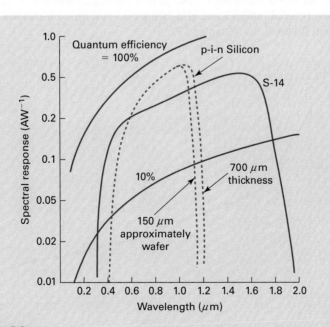

Based on RCA Solid State Division, Lancaster, Pennsylvania

Figure 9.20: Spectral response of silicon (p-i-n Si) and germanium (S-14) based photovoltaic cells.

Table 9.1: Energy gaps and maximum theoretical efficiency for solar radiation of some semiconducting materials for photovoltaic cell construction. Values are for room temperature. Some elemental components of these materials are toxic, such as arsenic and cadmium, and the world's resources of indium and gallium may be insufficient for large-scale use (Chapter 2).

material	E_g (eV)	maximum theoretical efficiency (%)
CdS	2.42	18
AlSb	1.6	27
CdTe	1.49	27
GaAs	1.43	26
InP	1.35	26
Si	1.11	25
Ge	0.67	13

© Cengage Learning 2015

The maximum theoretical efficiency of photovoltaic cells based on different semiconducting materials is summarized in Table 9.1. Typical efficiencies for Si-based cells are around 15–17%, with higher quality cells giving up to about 23%. Techniques such as concentrating the light and splitting the light into different spectral components that are incident on cells with specific spectral responses have yielded efficiencies in excess of 40%.

Inevitably, there is a trade-off between efficiency and economy. Although solar photovoltaic cells are becoming less expensive per W_e capacity, they are still more expensive than most other energy options. Figure 9.21 shows a comparison of cost per watt installed for photovoltaic installations in recent years. It should be realized that system costs include a suitable storage method (e.g., batteries) to make photovoltaic power practical, although local grid-connected systems can use the grid to buffer supply-demand fluctuations (Chapter 17).

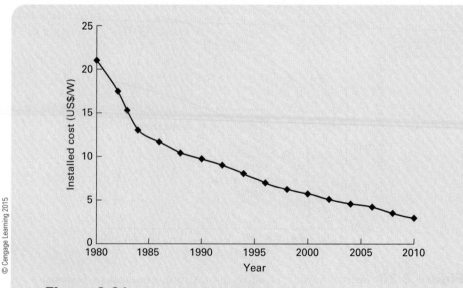

© Cengage Learning 2015

Figure 9.21: Cost per W_e installed for photovoltaic systems in recent years.

ENERGY EXTRA 9.1
Triple-junction photovoltaic cells

Most commercial photovoltaic cells have efficiencies for converting energy from solar radiation into electrical energy of 12–18%. A number of factors affect efficiency, but one of the most important is the spectral sensitivity of the semiconducting material used in the cell (Figure 9.20). The principle semiconductor property that governs the spectral response is the energy gap. Numerous semiconducting materials are known, and the energy gap (and hence the spectral response) of a material can be tuned by adjusting the composition of the material. The problem is that, if the energy gap is tuned to give the material a high efficiency for photons in the red end of the electromagnetic spectrum, then the efficiency at the blue end of the spectrum is sacrificed, and vice versa.

One approach to improving the efficiency of a photovoltaic cell over a broad spectral range is to make a sandwich (a multijunction) of materials that are sensitive over different portions of the spectrum. A triple junction consisting of one semiconducting material that is the most sensitive to the red photons, another that is the most sensitive to green photons, and a third that is the most sensitive to blue photons has been very successful in creating photovoltaic cells with overall high efficiencies to the solar spectrum.

An example of the triple-junction design is shown in the top right figure to the right. In this example, the top layer is made of amorphous silicon (the designation "i a-Si" in the figure refers to a layer of intrinsic amorphous silicon sandwiched between layers of p-type and n-type silicon). The i a-Si layer has an energy gap of about 1.8 eV and is most sensitive to photons in the blue region of the solar spectrum. The middle layer is intrinsic amorphous silicon-germanium (i a-SiGe), with about 80% Si and 20% Ge (again sandwiched between p-type and n-type regions). This material has an energy gap of about 1.6 eV and is the most sensitive to photons in the green portion of the spectrum. The bottom layer is also i a-SiGe with about 60% Si and 40% Ge. This composition reduces the energy gap to about 1.4 eV and makes the material the most sensitive to red photons. The ordering of the

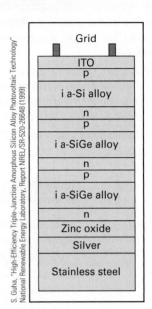

S. Guha, "High-Efficiency Triple-Junction Amorphous Silicon Alloy Photovoltaic Technology" National Renewable Energy Laboratory, Report NREL/SR-520-26648 (1999)

Geometry of triple-junction photovoltaic cell.

layers is important because the top layer utilizes the highest-energy photons and allows the lower-energy photons to propagate through to the layers beneath it, where they are utilized the most efficiently.

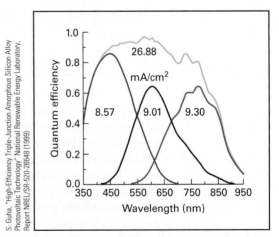

S. Guha, "High-Efficiency Triple-Junction Amorphous Silicon Alloy Photovoltaic Technology" National Renewable Energy Laboratory. Report NREL/SR-520-26648 (1999)

Spectral responses of the three layers of a triple-junction and the total response.

Continued on page 250

Energy Extra 9.1 continued

The bottom right figure on the previous page shows the spectral responses of the three layers; blue to red photons correspond to the horizontal axis from left to right in the figure. The total spectral response is illustrated by the total curve in the figure and shows a high efficiency over a wide range of photon energies (or wavelengths).

This approach has been successful in producing photovoltaic cells with overall efficiencies in the range of 40% (more than twice the efficiency of typical commercial silicon-based cells). Unfortunately, the cost of producing these cells, at least at the present,

makes them economically viable only for specific applications. These applications include space vehicles where weight is a more important consideration than cost and devices where radiation is focused onto a small area and the solar cell size is small compared to the total collector area.

Topic for Discussion

If triple-junction photovoltaic (PV) devices could be produced at the same cost as simple Si-based devices, how would the cost of PV-generated electricity compare to wind, nuclear, and fossil fuels?

Example 9.3

Calculate the cost per kWh averaged over an operational period of 10 years for a photovoltaic system with an installation cost of $3.00 per W if the system operates with a capacity factor of 8% (this is the net result of the photovoltaic efficiency and the fraction of time that is daylight). (Do not include the cost recovery factor.)

Solution

Ten years correspond to $(10 \text{ y}) \times (365 \text{ d/y}) \times (24 \text{ h/d}) = 87{,}600 \text{ h}$. Over this time period, one watt of photovoltaic installation, operating at an 8% capacity factor, produces a total of

$$87{,}600 \text{ h} \times 0.08 \text{ W/W} = 7008 \text{ Wh/W} = 7.0 \text{ kWh/W}.$$

At a total cost of $3.00 per W, this corresponds to

$$\frac{\$3.00/\text{W}}{7.0 \text{ kWh/W}} = \$0.43 \text{ per kWh}.$$

The cost of photovoltaics may be viewed as primarily a capital cost with little or no operating costs compared with, for example, coal power, which requires a continuous supply of fuel. However, it is necessary to amortize the capital costs of a photovoltaic system over its expected lifetime in order to determine the actual cost per unit energy (Chapter 2). The life expectancy of a photovoltaic cell is probably about 20 years, although the life expectancy for the batteries in the storage system might be only a third of that or less. The estimated cost per unit energy for photovoltaics compared with other energy production methods is summarized in Figure 9.22. Certainly, at present, other energy production methods are more economical than photovoltaics, but the trends shown in Figure 9.21 suggest that future developments may help alleviate this problem.

In addition to the financial cost to the consumer, it is necessary, as pointed out in earlier chapters, to consider the energy cost in producing energy. In the case of photovoltaics, the payback time may be defined as the amount of time needed for the system

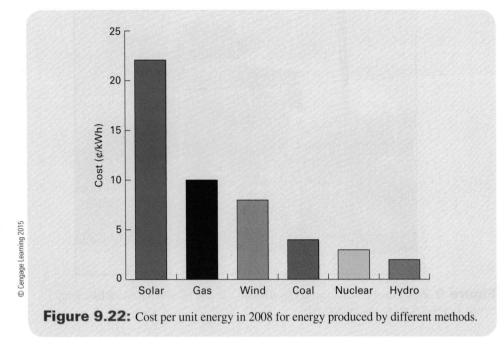

Figure 9.22: Cost per unit energy in 2008 for energy produced by different methods.

to produce as much energy as was required to manufacture and install it. Typical pay-back times for photovoltaic devices vary from a few months to a few years, depending on the type of cell and the details of the application. This is a relatively small fraction of the cell's life expectancy, so there is a net energy gain from manufacturing and using photovoltaic devices.

Photovoltaic cells have been implemented on a variety of size scales. These may be considered in two different categories: off-grid and on-grid. These designations refer to systems that are not connected to the power utility grid and those that are, respectively.

Off-grid photovoltaic cells typically fall into one of the following categories:

- Small units used to recharge batteries in portable electronic devices (Figure 9.8)
- Medium-sized units used for camping, emergency battery charging (Figure 9.23), portable road signs (Figure 9.24), or power sources in remote areas (Figure 9.25)
- Larger installations used for residential electric power (Figure 9.26)

Figure 9.23: Portable 1.8-W_e solar panel. The dimensions of the active area are approximately 9 cm × 30 cm.

Figure 9.24: Portable road sign at construction site that utilizes a photovoltaic array (about 1 m × 2.5 m) to charge batteries.

Figure 9.25: Photovoltaic array for providing power at a remote radio transmitter station.

Individual cells are fairly small (typically less than about 10 cm on a side), but they can be combined in series (to increase voltage) and/or in parallel (to increase current). Individual cells can be rectangular (Figure 9.23) or circular (Figure 9.25).

The Solar Settlement in Freiberg, Germany (Figure 9.26) is a sustainable model community designed to produce net zero carbon emissions. Many of the homes, as seen in the figure, incorporate photovoltaic arrays to provide electricity.

Experimental automobiles (Figure 9.27) and even airplanes (Figure 9.28) that are powered by photovoltaic cells have been constructed. Although these vehicles are

ENERGY EXTRA 9.2
Dye-sensitized solar cells

Clearly, the major problem with electricity generated by photovoltaic devices is the cost per kWh. One approach to dealing with this difficulty is to produce photovoltaic cell with higher efficiency. This, of course, has the advantage that the power generated per unit land area is greater. However, because more efficient photovoltaic cells are more expensive, this is not an obviously viable trade-off between productivity and price. The opposite approach is to design cells that are as inexpensive as possible without being so concerned with efficiency, thus taking a different approach to photovoltaic materials and photocell design. Dye-sensitized solar cells (DSSCs) are a type of organic photovoltaic cell that has attracted considerable interest in recent years. These cells utilize a very different method of cell construction and manufacture that may lead to innovative and practical devices.

A traditional photovoltaic cell consists of a sandwich of an n-type material and a p-type material between two conducting electrodes. When a photon of sufficient energy (i.e., greater than the energy gap) enters into the region near the semiconducting junction, an electron-hole pair can be formed, resulting in a current through an external circuit. The geometry is somewhat limited because the electron and hole must reach the electrode before recombining. However, the photosensitive region can be made larger and, in principle more economical, by using the heterojunction design. In a heterojunction device, the semiconducting region consists of a mixture of different materials as shown in the following figure. In a DSSC, the heterojunction is based on a piece of nonporous TiO_2. TiO_2 has a relatively poor efficiency for creating electron-hole pairs in response to photons because of its large energy gap. However, the pores in TiO_2 are filled with a very photosensitive organic dye. The top electrode is often made of a halide-doped tin oxide that is both optically transparent and conducting. This forms the electrical connection to the cell and allows light to enter into the photosensitive region.

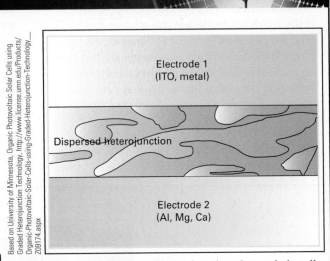

Based on University of Minnesota, Organic Photovoltaic Solar Cells using Graded Heterojunction Technology, http://www.license.umn.edu/Products/Organic-Photovoltaic-Solar-Cells-using-Graded-Heterojunction-Technology-Z09174.aspx

Schematic of dispersed heterojunction photovoltaic cell.

If the size of the dispersed grains in the heterojunction is comparable to the charge carrier diffusion length, then the charges formed in response to a photon have a good chance of reaching a semiconducting interface before recombination takes place.

The earliest DSSCs produced in the mid-1990s were primarily sensitive only to the high-energy (UV) part of the solar spectrum. In recent years, materials (dyes) that are sensitive to the entire optical spectrum have been developed. Because these materials are effective at absorbing all optical wavelengths, they appear black and are thus referred to as *black dyes*. The most common black dyes are based on organic ruthenium-containing complexes as shown in the next figure.

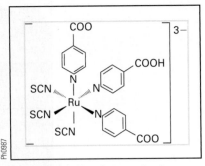

Structure of triscarboxy-ruthenium terpyridine (black dye).

Continued on page 254

Energy Extra 9.2 continued

DSSCs have been produced with efficiencies in the range of 11%. This is less than the efficiency for traditional Si-based semiconducting photovoltaic cells, which can be close to 20% efficient, and the element ruthenium is expensive. However, the hope is that new, economically viable materials (that do not include ruthenium) may be developed and that this approach to photovoltaic cell design may yield devices that are competitive in terms of cost per kW/m^2.

Topic for Discussion

The problem of the cost and availability of materials is prevalent when dealing with the development of new technologies. The use of ruthenium in DSSCs is only one example of a great need to develop new materials that have competitive properties but that are less expensive and more readily available. Find information about the market value of ruthenium and the quantity produced each year. An excellent website dealing with elemental resources that has useful information and links is http://minerals.usgs.gov/minerals/pubs/commodity/. Discuss the importance of considering how science, technology, and economics must all play a role in designing our future.

© imagebroker/Alamy

Figure 9.26: Photovoltaic solar panels on homes in the Solar Settlement in Freiberg, Germany.

Stefano Paltera, the NASC's official photographer

Figure 9.27: Photovoltaic-powered automobile.

Figure 9.28: Photovoltaic-powered airplane.

interesting engineering challenges, the power density in sunlight is insufficient to make them practical.

On-grid photovoltaic facilities exist in a number of countries and provide electric power for distribution through the power grid. A typical system for integrating a photoelectric array with the power grid is illustrated in Figure 9.29. The thermal generator (i.e., coal or nuclear generating facility) provides the primary AC power to the grid.

Example 9.4

Using the average (cloud-free) insolation on earth (Chapter 8), calculate the area of a horizontal photovoltaic array with an efficiency of 20% that would be needed to satisfy the residential electric needs of a city of 250,000 people with an average of 2.6 people per household.

Solution

This city would consist of $250,000/2.6 = 9.6 \times 10^4$ homes. As indicated, the average electricity power requirement per home is 1370 W. The total requirement for the city would be

$$(9.6 \times 10^4) \times (1370 \text{ W}) = 1.32 \times 10^8 \text{ W}.$$

The average cloud-free insolation averaged over the year and all locations on earth is (from Chapter 8) 168 W/m². Thus the city's electrical needs would require (at a 20% efficiency)

$$\frac{1.32 \times 10^8 \text{ W}}{0.2 \times 168 \text{ W/m}^2} = 3.9 \times 10^6 \text{ m}^2,$$

or a square array about 2 km on a side.

NASA photo by Nick Galante/PMRF

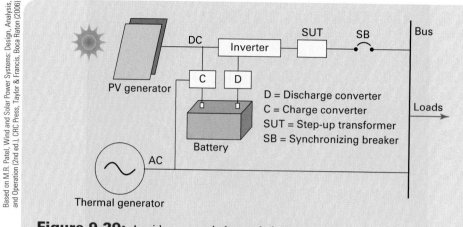

Based on M.R. Patel, Wind and Solar Power Systems: Design, Analysis, and Operation (2nd ed.), CRC Press, Taylor & Francis, Boca Raton (2006)

Figure 9.29: A grid-connected photovoltaic system.

U.S. Air Force photo by Airman 1st Class Nadine Y. Barclay

Figure 9.30: Photovoltaic arrays at Nellis Air Force Base, Nevada.

DC power from the photovoltaic array is stored in a battery system. DC power from the photovoltaic array and from the battery storage system is converted to AC by the inverter. Voltage is matched to the grid power from the thermal generator by a step-up transformer. The synchronizing breaker connects the power from the photovoltaic array to the grid when the voltages are matched in frequency, amplitude, and phase. Typical large photovoltaic facilities (Figure 9.9) produce a maximum output of tens of MW_e. The largest operational photovoltaic facility is currently Parque Fotovoltaico Olmedilla de Alarcon in Spain, with a peak output of 60 MW_e. Larger facilities are planned and are likely to be constructed in the near future. The largest photovoltaic installation in North America is the 14-MW_e facility at Nellis Air Force Base in Nevada (Figure 9.30). The photovoltaic panels employ an advanced solar tracking system and are spaced to minimize shadowing effects.

9.5 Global Use of Photovoltaics

The world use of photovoltaics is summarized in Table 9.2. A small number of countries (e.g., Germany, Japan, and Spain) have been very active in pursuing photovoltaic power. If solar energy is to be used on a large scale to satisfy our energy needs, then the economics and availability of this resource need to be considered. Figure 9.22 suggests that economically, solar energy is not as desirable as other energy sources. However, anticipated advances in photovoltaic cell design and manufacturing, as well as diminishing fossil fuel resources are likely to make solar energy an increasingly attractive resource. The longevity of sunlight as an energy resource is, for all practical purposes, infinite, but it is necessary to consider its availability. From equation (8.2), the total insolation on the outside of the atmosphere of the earth is 1.8×10^{17} W. On average, about half of this is transmitted through the atmosphere, giving a total insolation at the surface of 9×10^{16} W. Considering a modest photovoltaic efficiency of 15%, this gives the potential for 1.3×10^{16} W_e from photovoltaic generation worldwide. In the Preface, it was seen that the total primary energy use worldwide is 5.7×10^{20} J per year for an average power consumption of $(5.7 \times 10^{20}$ J/y$)/(3.15 \times 10^7$ s/y$) = 1.8 \times 10^{13}$ W. Thus, the utilization of only about 0.14% of the available solar energy would fulfill all of our energy needs. Certainly, the use of photovoltaic arrays would be impractical or uneconomical at some locations on earth. Terrestrial locations would probably be more practical than most of the oceans, while northern or southern latitudes would provide very low efficiency. The fraction of land area necessary for solar

Table 9.2: Peak photovoltaic capacity by country. Off-grid, on-grid, and total capacity, as well as capacity per capita are indicated, as of end of 2007. *Note:* Totals may have round-off error.

country	off-grid (MW$_e$)	on-grid (MW$_e$)	total (MW$_e$)	W$_e$/capita
Germany	35	3827	3862	46.8
Japan	90	1829	1919	15.0
United States	325	505	830	2.8
Spain	30	625	655	15.1
Italy	13	107	120	2.1
Australia	66	16	82	4.1
South Korea	6	72	78	1.6
France	23	53	75	1.2
Netherlands	5	48	53	3.3
Switzerland	4	34	36	4.9
Austria	3	24	28	3.4
Canada	23	3	26	0.8
Mexico	20	0	21	0.2
United Kingdom	1	17	18	0.3
Portugal	3	15	18	1.7
Norway	8	0	8	1.7
Sweden	5	2	6	0.7
Denmark	0	3	3	0.6
Israel	2	0	2	0.3

Table 9.3: Factors determining the fraction of land area needed to provide all primary energy needs by the use of photovoltaics in Canada and the United States. Photovoltaic efficiency of 15% has been assumed.

country	annual energy use per capita (J)	population	total annual energy use (J)	average power consumption (W)	typical insolation (W/m²)	land area needed (m²)	total land area (m²)	% land area needed
United States	3.3×10^{11}	3.06×10^{8}	1.01×10^{20}	3.21×10^{12}	200	1.07×10^{11}	9.63×10^{12}	1.11
Canada	4.0×10^{11}	3.34×10^{7}	1.34×10^{19}	4.24×10^{11}	150	1.88×10^{10}	9.98×10^{12}	0.19

energy to provide all energy needs can be considered on a country-by-country basis. Population density and per-capita energy use, as well as latitude and climate conditions, would be the relevant factors. A comparison between the United States and Canada can be considered as an example, and a summary of this analysis is given in Table 9.3. The insolation for the United States shown in Figure 8.2 has a typical value of about 200 W/m². For Canada, it is assumed that solar collectors would be located in fairly southern areas where the average insolation is in the range of about 150 W/m². The table shows that a relatively small amount of the land area of the United States and Canada would need to be dedicated to the production of electricity by solar photovoltaics in order to satisfy all primary energy requirements. For the United States, a square about 300 km on a side would be required; that is an area about the size of Ohio. For

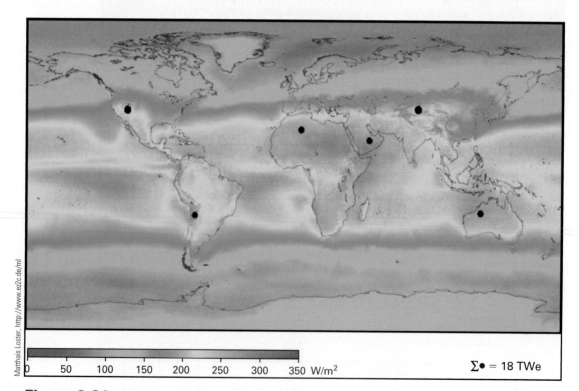

Matthais Loster, http://www.ez2c.de/ml

$\Sigma \bullet = 18$ TWe

Figure 9.31: Land area needed to supply the world's energy needs from solar. Average insolation is illustrated.

Canada, a square about 135 km on a side (about one third the area of Nova Scotia) would be required. A number of smaller facilities with an equivalent total area would likely be more practical than a single large facility. About six 300 km × 300 km facilities located worldwide in relatively sunny, low-population-density areas as shown in Figure 9.31 (or a comparable number of smaller facilities) would satisfy all of the world's energy needs.

Solar energy is the only single energy resource that has the capability to provide enough energy to fulfill all of our needs and that is indefinitely renewable. The technology for doing this is within our means would require that we adapt our energy infrastructure for the exclusive use of electricity as an energy source. The continued availability of (relatively) inexpensive fossil fuels does not provide the financial incentive for this development.

9.6 Summary

Solar radiation is the only source of nonfossil fuel energy that is plentiful enough to fulfill all of society's energy requirements, both for the present and for the foreseeable future. Although the direct use of solar energy for applications such as space heating contribute to energy needs on a residential heating scale, it is the production of electricity from solar radiation, as described in this chapter for distribution over the grid, that has the potential to satisfy the bulk of our energy requirements in the future. The ways in which solar energy can be used to generate electricity were described. There are basically two approaches: heat engines and photovoltaics. This chapter discussed systems for focusing sunlight onto a small area where it is used to heat a working fluid. The fluid is then used to run a heat engine to produce mechanical energy, which then generates electricity by means of a generator. Experimental facilities use one of three geometries: parabolic troughs, parabolic reflectors, or central receivers. The efficiency of these devices is ultimately limited by the Carnot efficiency.

This chapter provides an introduction to the basic semiconductor physics needed to describe the operation of a photovoltaic device. Both n-type and p-type semiconducting materials are formed by the inclusion of donor or acceptor impurities in a host material. These materials have either an excess electron in the conduction band or an excess hole in the valence band, respectively. A junction between an n-type and a p-type material is sensitive to solar energy and constitutes the basic design of a photovoltaic cell. Photons incident on the junction produce electron-hole pairs that provide a current through an external circuit, thus providing usable electrical energy.

Different semiconducting materials are sensitive to different regions of the electromagnetic spectrum, and this is a major factor in limiting the efficiency of photovoltaic devices. There is a trade-off between efficiency and economy. High-efficiency devices can have efficiencies of about 40% for conversion of solar energy to electrical energy, but they are expensive. More economical devices have efficiencies of about 18%.

Although the photovoltaic installations themselves would seem to be environmentally neutral, the overall environmental concerns of solar electricity are complex. The manufacturing processes needed to produce photovoltaic materials are intensive (per installed watt power capability) because the energy density of sunlight is low, and these processes must to be included in an overall evaluation of environmental impact. Some elements used in photovoltaic cells are toxic, and some, such as indium, are of limited worldwide availability. All of these factors contribute to the present high cost of solar electricity, have possible implications for future developments, and must be considered in any overall analysis of the viability of photovoltaics. However, solar photovoltaics are still an attractive alternative to diminishing fossil fuel supplies, and there has been a substantial increase in their use in recent years. Germany has a very ambitious development program in this area and has been the leader in implementing this technology for providing electricity to the grid. The development of photovoltaic cells with good efficiency that can be manufactured economically using low–cost, nontoxic components is an area of active research, and the future availability of such materials would benefit the growth of this renewable energy resource.

Problems

9.1 Consider a 1-m wide parabolic trough reflector with a 1-cm inside diameter pipe filled with oil at its focus. If all of the solar radiation is converted into heat, estimate the time required at midday on a sunny day to raise the temperature of the oil from 20°C to 100°C.

9.2 Assume that all of our energy is obtained from solar by means of photovoltaic devices with an efficiency of 18% using horizontal panels. If all energy is produced in the United States on a state-by-state basis, calculate the percent of land area needed to supply energy in Rhode Island and Idaho. Use an average annual per-capita energy requirement of 350 GJ. Comment on the suitability of this approach.

9.3 A house at 40°N latitude has a roof of dimensions 8 m by 15 m. This area is covered with photovoltaic panels at the optimal fixed angle with an efficiency of 17%. Assuming average sky conditions, what is the annual energy production from this installation? If this electricity were generated by a coal-fired thermal plant operating at 38% efficiency, how much coal would be needed? On the basis of this comparison, what is the annual reduction (in kilograms) of CO_2 that results from this use of photovoltaics?

9.4 If all of the energy for humanity is generated using horizontal solar photovoltaic arrays with an efficiency of 15% and which occupy 1% of the land area of the earth, what is the maximum population density that can be supported?

9.5 Estimate the diameter of a mirror array associated with a solar power tower, as illustrated in Figure 9.4, that would replace a 1-GW_e coal-fired generating station. Be sure to consider the method by which electricity is generated, as illustrated in Figure 9.6. Discuss the practicality of such a design. Consider the relationship between the height of the central tower and the arrangement of mirrors near the outer portions of the array. Each mirror must have a clear view of the central receiver that is not obscured by the mirror in front of it. Does this place a restriction on the size of the array in the context of the tower height?

9.6 What is the maximum wavelength of light (in nanometers) that would produce an output from a photovoltaic device manufactured from a semiconductor with an energy gap of 1.02 eV?

9.7   What is the maximum wavelength of the light that produces voltage in a CdS photovoltaic device? Discuss this in terms of the spectrum of sunlight as presented in Chapter 8.

9.8   Consider a typical family vehicle powered by photovoltaics similar to the vehicle shown in Figure 9.27 where the top surface is covered with 20% efficient cells. Compare the power available in such a vehicle on a sunny day in comparison with a typical gasoline-powered vehicle. Some typical vehicle characteristics are given in Table 19.3.

Bibliography

M. D. Archer and R. Hill. *Clean Energy from Photovoltaics*. Imperial College Press, London (2001).

E. G. Boes and A. Lague. "Photovoltaic concentrator technology." In *Renewable Energy*, T. B. Johansson, H. Kelly, A. K. N. Reddy, and R.H. Williams (Eds.). Island Press, Washington, DC (1993).

G. Boyle (ed.). *Renewable Energy*. Oxford University Press, Oxford (2004).

P. DeLaquil, D. Kearney, M. Geyer, and R. Diver. "Solar-thermal electric technology." In *Renewable Energy*, T. B. Johansson, H. Kelly, A. K. N. Reddy, and R. H. Williams (Eds.). Island Press, Washington, DC (1993).

R. A. Dunlap. *Experimental Physics—Modern Methods*. Oxford University Press, New York (1988).

M. Green. *Solar Cells*. Prentice-Hall, Englewood Cliffs, NJ (1982).

R. Hinrichs and M. Kleinbach. *Energy: Its Use and the Environment* (5th ed.). Brooks-Cole,

R. J. Komp. *Practical Photovoltaics*. Aatec Publications, Ann Arbor, MI (1995).

J. A. Kraushaar and R. A. Ristinen. *Energy and Problems of a Technical Society* (2nd ed.). Wiley, New York (1993).

T. Markvart. *Solar Electricity* (2nd ed.). Wiley, Hoboken, NJ (2004).

J. A. Merrigan. *Sunlight to Electricity* (2nd ed.). MIT Press, Cambridge, MA (1982).

J. Nelson. *The Physics of the Solar Cell*. Imperial College Press, London (2003).

M. R. Patel. *Wind and Solar Power Systems: Design, Analysis, and Operation* (2nd ed.). Taylor & Francis, Boca Raton, FL (2006).

S. Roberts. *Solar Electricity: A Practical Guide to Designing and Installing Small Photovoltaic Systems*. Prentice-Hall, London (1991).

J. N. Shive. *Physics of Solid State Electronics*. Merrill, Columbus, OH (1966).

F. C. Treble (Ed.). *Generating Electricity from the Sun*. Pergamon Press, Oxford (1991).

F. C. Treble. Solar Electricity: A Lay Guide to the Generation of Electricity by the Direct Conversion of Solar Energy (2nd ed.). The Solar Energy Society, Oxford (1999).

Wind Energy

Learning Objectives: After reading the material in Chapter 10, you should understand:

- The different designs of wind turbines.
- The physics of the energy content of wind.
- The efficiency of different wind turbine designs and the importance of the tip speed ratio.
- The geographic distribution of wind energy.
- The design of wind farms.
- The utilization of wind energy worldwide.
- Advantages of offshore wind farms.

10.1 Introduction

Along with solar energy, wind energy is the only alternative energy source that is prevalent virtually everywhere in the world. It is also one of the first sources of energy to be utilized by humanity. The earliest use of wind energy was in mechanical devices (e.g., for pumping water or grinding grain) and for transportation (e.g., sailing ships). Since the late nineteenth century, however, wind energy has also been utilized to generate electricity, and in recent years it has become a very attractive alternative to fossil fuels. Wind-generated electricity is based on a reasonably cost-effective, well established technology that has minimal environmental impact. Wind energy will certainly play an important role in the development of alternative energy strategies for the future. This chapter overviews the history, scientific foundations, and recent technological developments of wind energy.

10.2 Wind Turbine Design

Wind is moving air. The air moves because of pressure gradients caused by thermal gradients that result from the heating and cooling of the atmosphere. A typical example is the formation of winds along the shore of the ocean or a lake in the morning. The morning sun heats the land faster than it heats the water. The air above the land is heated more quickly and rises. The rising air results in a pressure gradient that pushes air from above the water onto the land. This is the typical onshore wind that is commonly seen in the morning. In the evening, the opposite happens, resulting in the typical offshore breeze.

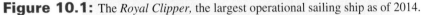

Figure 10.1: The *Royal Clipper,* the largest operational sailing ship as of 2014.

Like solar energy, wind energy is not constant over time. Although the average wind velocity at a particular location may be known, the variations are somewhat less predictable than the variations in solar insolation. Thus, the full utilization of wind energy must include some type of storage mechanism.

Wind energy has been harvested for literally thousands of years. One of the earliest uses of wind energy was for transportation, as shown in Figure 10.1. A large sailing ship of this type could produce close to 8 MW of power in a strong wind.

The use of wind energy for other purposes certainly dates back hundreds, if not thousands, of years. The use of the traditional so-called *Dutch windmill*, as shown in Figure 10.2, for the grinding of grain was widespread in Europe more than five

Figure 10.2: Traditional four-blade Dutch windmill in the Netherlands.

Figure 10.3: Post windmill in Britain.

centuries ago. Although many designs of traditional windmills suffered in their efficiency because they were in a fixed orientation, some designs incorporated a mechanism for rotating the mill so that it always faced into the wind. Such a design, known as a *post windmill*, is illustrated in Figure 10.3. This type of windmill was mounted on a central post, and the entire building could be rotated manually when the wind direction changed. Later windmills were used extensively for pumping water (for irrigation, etc.). The *American multiblade* design (Figure 10.4) was in common use for this purpose in North America during the nineteenth century.

Although there is current interest in the recreational use of wind energy (e.g., sailboats), the principal effort in wind energy development is in the production of electricity. Modern windmills (referred to as *wind turbines*) have been developed for this purpose. One modern design (developed in the 1930s) is the *Darrieus wind turbine*, or *Darrieus rotor* (Figure 10.5). These devices have two clear advantages over the more common designs discussed in this chapter: (1) The generator mechanism is at the bottom, making it easier to service, and (2) they do not have to be oriented to face into the wind. They have, however, some serious disadvantages: They are not self-starting, that is, if they stop due to the lack of wind, they need to be manually restarted when the wind returns; they also have a lower efficiency than other designs and are subject to excessive stresses that can result in mechanical failure. Despite numerous experimental devices of this type over the years, they are not as commercially attractive as other designs.

The modern *three-blade wind turbine* (Figure 10.6) is the most common design for electricity production. A *high-speed two-blade wind turbine* is shown in Figure 10.7. Although the two-blade design is intrinsically more efficient, it is more

Figure 10.4: American multiblade windmill in Maine.

Figure 10.5: Darrieus rotor on Îles de la Madeleine, Québec.

Figure 10.6: Modern three-blade wind turbine at Digby Neck Wind Farm, Nova Scotia. Turbine is a 1.5-MW GE 1.5sle with a rotor diameter of 77 m and a hub height of 80 m.

Figure 10.7: Modern high-speed two-blade wind turbine.

Figure 10.8: Nacelle of a 600-kW wind turbine.

Figure 10.9: Nacelle of a medium sized wind turbine.

difficult to balance, and, because it runs at a higher speed than the three–blade, it is more susceptible to failure. For this reason, most commercial wind turbines are of the three-blade design.

Typically in a wind turbine (Figure 10.6), the electric generator, along with associated drive systems and a gearbox to match the rotational speed of the wind turbine to the characteristics of the generator, are contained in a housing (called a *nacelle*) at the top of the wind turbine tower (Figures 10.8 and 10.9). Typically, a *yaw drive* is used to keep the wind turbine aligned into the wind.

10.3 Obtaining Energy from the Wind

The power available from the wind can be calculated by means of an analysis of the energy associated with a moving parcel of air. Figure 10.10 shows a cube of air with an edge length d moving with velocity v. The parcel of air has a mass $d^3\rho$, where ρ is the density of air (1.204 kg/m^3). Thus, the kinetic energy of the moving parcel of air is

$$E = \tfrac{1}{2} mv^2 = \tfrac{1}{2} d^3 \rho v^2. \tag{10.1}$$

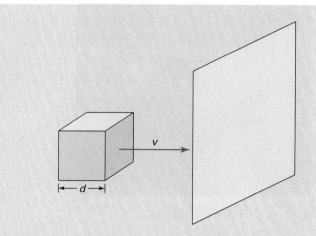

© Cengage Learning 2015

Figure 10.10: Cubic parcel of air with dimensions $d \times d \times d$ traveling with velocity v and passing through a plane parallel to one of its faces.

Example 10.1

Calculate the energy (in Joules) of 1 m³ of air moving at 4 m/s. How long could this energy be used to illuminate a 60-W light bulb if it could be converted to electricity with 40% efficiency?

Solution
From equation (10.1),

$$E = \tfrac{1}{2}d^3\rho v^2.$$

Using $d = 1$ m, $v = 4$ m/s, and $\rho = 1.204$ kg/m³ gives

$$E = (1/2) \times (1 \text{ m}^3) \times (1.204 \text{ kg/m}^3) \times (16 \text{ m}^2/\text{s}^2) = 9.6 \text{ J}.$$

Since 1 J = 1 W·s then at 40% efficiency,

$$(0.4) \times (9.6 \text{ W·s})/(60 \text{ W}) = 0.064 \text{ s}.$$

If this parcel of air passes through a plane, as shown in the figure, then the time required for the cube to pass through the plane is $t = d/v$. The power is then written as

$$P = \frac{E}{t} = \tfrac{1}{2}d^2\rho v^3. \qquad \textbf{(10.2)}$$

The power per unit area is

$$\frac{P}{A} = \frac{P}{d^2} = \tfrac{1}{2}\rho v^3. \qquad \textbf{(10.3)}$$

Using the density of air in kg/m³ and the velocity in m/s, this expression may be written as

$$\frac{P}{A} = (0.602 \text{ kg/m}^3)v^3, \qquad \textbf{(10.4)}$$

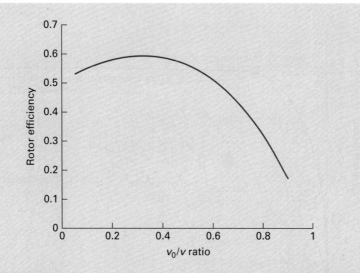

Figure 10.11: Wind turbine efficiency as a function of the relative air velocity after passing through the wind turbine.

where P/A is in W/m^2. Note that the units on the right-hand side are (kg/m^3) $\times$ (m^3/s^3) = kg/s^3. Since J = kg$\cdot$m^2/s^2 and W = J/s, then W/m^2 = (kg$\cdot$m^2/s^2)/(s$\cdot$m^2) = kg/s^3, as expected.

The objective of the wind turbine is to extract as much of the kinetic energy content of the wind as possible for conversion into mechanical energy and then, typically, into electricity. To extract all of the kinetic energy from the parcel of air, it would be necessary to stop the air. This would require a wind turbine design that did not allow the air to pass through the blades to the other side. If the wind turbine blocked the motion of the air entirely, the air would tend to flow around the wind turbine—not the ideal situation. On the other hand, if the wind turbine only very slightly reduced the velocity of the air, then only a small portion of the kinetic energy would be extracted. The best situation is somewhere in between, and a detailed analysis provides the results shown in Figure 10.11, where the efficiency is plotted as a function of the ratio of the air velocity after passing through the wind turbine, v_0, to the air velocity before the wind turbine, v. This figure shows that the maximum theoretical efficiency that can ever be achieved by a wind turbine, referred to as the *Betz limit*, is 59%.

Example 10.2

What is the velocity of the wind that provides ten times the power provided by wind with a velocity of 5 m/s?

Solution

Equation (10.4) shows that the power in a certain cross-sectional area is proportional to the wind velocity cubed. So, to increase the power by a factor of ten, the velocity has to be increased by a factor of the cube root of 10; that is, the velocity would be

$$(5 \text{ m/s}) \times (10)^{1/3} = 10.8 \text{ m/s}.$$

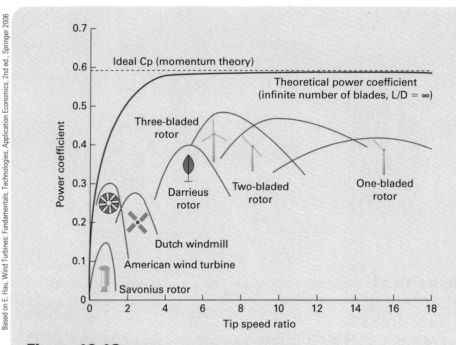

Based on E. Hau, Wind Turbines: Fundamentals, Technologies, Application Economics, 2nd ed., Springer 2006

Figure 10.12: Efficiency as a function of tip speed ratio for different wind turbine designs.

The efficiencies of different types of wind turbines are shown in Figure 10.12. The *tip speed ratio* is the ratio of the speed of the fastest moving tip on the blade or rotor to the actual wind speed. This ratio is determined by the design of the wind turbine and, in particular, by the details of the design of the blades. The best actual efficiencies are for the high-speed two-blade design and are about 45%. The modern three-blade wind turbine is slightly less efficient at around 40–42%. Thus, at best, actual wind turbines achieve about two-thirds to three-quarters of the theoretical maximum efficiency.

On the basis of equation (10.4) and Figure 10.12, the power output of a wind turbine of a given size can be calculated as a function of wind speed. For example, typical three-blade turbine will have an efficiency (η) of about 40% ($\eta = 0.40$). A turbine with a rotor diameter of 20 m in a wind of 5 m/s produces

$$P = (0.602 \text{ kg/m}^3)\eta A v^3$$
$$= (0.602 \text{ kg/m}^3) \times (0.40) \times (3.14) \times (10 \text{ m})^2 \times (5 \text{ m/s})^3 = 9450 \text{ W.} \quad \textbf{(10.5)}$$

This equation shows that the power is related to the cube of the wind velocity and increases substantially with increasing wind speed. For example, the wind turbine just described would produce 255 kW in a 15-m/s wind.

The power of an actual wind turbine differs from this simple analysis in several ways. A typical example of the output of a real wind turbine is shown in Figure 10.13. Below a certain speed (the *cut-in speed*), the wind turbine is inefficient and produces little power. From the cut-in speed to the rated speed, the behavior is fairly well described by the v^3 expression just derived. Above the rated speed, there is a risk of damage to the blades or the generator mechanism as a result of excessive speed. Thus, at this point

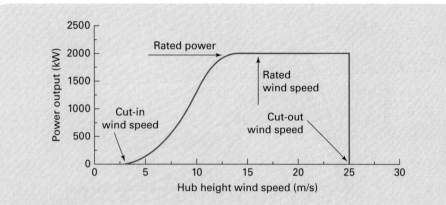

Figure 10.13: Power output for a 2-MW wind turbine as a function of wind speed.

Example 10.3

A large Dutch windmill can have blades 30 m in diameter. If such a windmill operates at optimal tip speed ratio in a wind with a velocity of 10 m/s, what is the mechanical power output in kW?

Solution

From equation (10.5), the power is

$$P = (0.602 \text{ kg/m}^3)\eta A v^3$$

The maximum efficiency of a Dutch windmill, as illustrated in Figure 10.2, is about 27%. The area of the rotor in square meters is

$$A = 3.14 \times (15 \text{ m})^2 = 707 \text{ m}^2.$$

The power in watts is then

$$P = (0.602 \text{ kg/m}^3) \times (0.27) \times (707 \text{ m}^2) \times (10 \text{ m/s})^3 = 115 \text{ kW}.$$

---

the speed of the wind turbine is regulated to avoid damage, typically by adjusting the angle (pitch) of the blades to make them intentionally less efficient. At some point (the *cut-out speed*), when the wind speed becomes too great, the wind turbine is shut down by applying a brake to avoid damage.

   The amount of power that can be obtained from a wind turbine depends, of course, on the size and design of the wind turbine and on wind conditions. Details of wind resources are discussed in the next section, but some general considerations for the design of wind energy systems are appropriate at this point. The design of the wind turbine

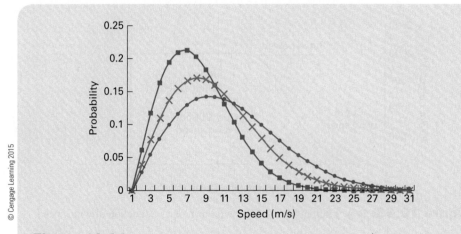

© Cengage Learning 2015

Figure 10.14: Rayleigh distribution of wind speeds for different average wind speed values.

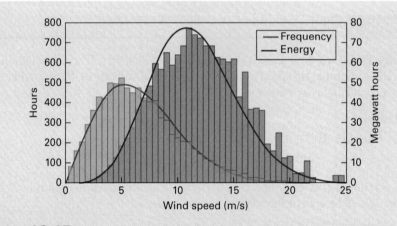

Based on Sandia National Laboratories, "New Mexico Wind Resource Assessment," Lee Ranch, http://windpower.sandia.gov/other/ LeeRanchData-2002.pdf

Figure 10.15: Comparison of speed and energy distributions for typical wind conditions. The energy is calculated for a 100-m-diameter circle normal to the wind direction with an efficiency equal to the Betz limit.

should match the wind conditions prevalent at the location. On one hand, the wind turbine should not operate below the cut-in speed too much of the time because it will not be producing power. On the other hand, the wind turbine should not be operating above the rated speed too much of the time because the efficiency will have to be limited to avoid damage. Although each location is associated with an average wind speed, it also has a well-defined distribution of wind speeds. Because the power output of the wind turbine goes up or down with the cube of the wind speed, it is important to know this distribution in order to obtain the maximum total energy production over a period of time. Typical wind speed distributions for different average wind speeds are shown in Figure 10.14. Because the energy is related to the cube of the speed, the peak in the energy distribution occurs at higher speeds than the peak in the speed distribution (Figure 10.15). A consequence of this is that the greatest portion of energy produced over a

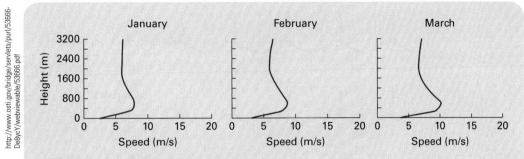

Figure 10.16: Typical variation of wind speed with altitude.

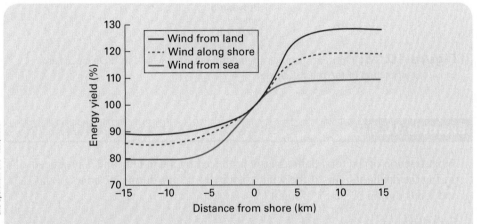

Figure 10.17: Variation of wind energy near seacoast.

period of time is generally produced in a relatively small fraction of the time when the wind velocity is well above average.

A factor that can play an important role in the power output of a wind turbine is altitude. Typical wind speed variations with altitude are shown in Figure 10.16. Wind speed is seen to increase substantially with increasing altitude. Thus, locating the turbine at the top of a tall tower is advantageous. However, there is certainly a trade-off between power output and the cost and serviceability of the unit. A consideration of the local terrain is important because it is generally beneficial to locate a wind turbine at the top of a hill rather than at the bottom of a valley.

For installations that are close to the seacoast, the wind is stronger offshore than on land (Figure 10.17). Placing wind turbines a few kilometers offshore can provide more energy and also help to minimize the environmental impact of the facility. This is further discussed later in the chapter.

For installations containing multiple wind turbines (generally referred to as *wind farms*), the relative placement of the wind turbines is important. If they are too far apart, then the facility will occupy more land area than necessary. If they are too close together, each wind turbine will not achieve its optimal output. Figure 10.18 shows the optimal spacing of wind turbines in a wind farm.

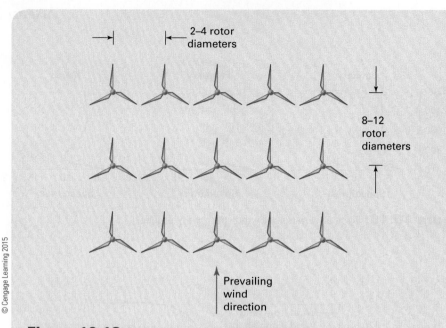

2–4 rotor diameters

8–12 rotor diameters

Prevailing wind direction

© Cengage Learning 2015

Figure 10.18: Optimal spacing of wind turbines in a wind farm. Distances are shown as a function of the rotor diameter.

Example 10.4

What fraction of the total daily energy produced by a wind turbine is produced during the daylight hours if the wind velocity is 12 m/s during 12 hours of day and 3 m/s during 12 hours of night?

Solution

The energy produced by the turbine over a time t is given by equation (10.5):

$$E = (0.602 \text{ kg/m}^3)\eta A v^3 t.$$

Thus for a specific wind turbine, the energy produced during a specific time interval is

$$E = (\text{constant})v^3 t.$$

So the energy produced during the day is

$$E = (\text{constant})(12 \text{ h})v_{\text{day}}{}^3,$$

and during the night it is

$$E = (\text{constant})(12 \text{ h})v_{\text{night}}{}^3,$$

for a total energy production of

$$E = (\text{constant})(12 \text{ h})(v_{\text{day}}{}^3 + v_{\text{night}}{}^3).$$

Thus, the ratio of fraction of energy produced during the day is

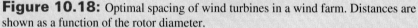

$$\text{fraction} = \frac{v_{\text{day}}{}^3}{v_{\text{day}}{}^3 + v_{\text{night}}{}^3} = \frac{(12 \text{ m/s})^3}{(12 \text{ m/s})^3 + (3 \text{ m/s})^3} = 0.98.$$

This shows that the majority of the energy is produced when the wind is blowing the strongest.

10.4 **Applications of Wind Power**

Like solar power, wind power has an essentially infinite longevity. It is important, however, to consider how much power is available. The average wind energy density for various locations in the United States is shown in Figure 10.19. There are regions of high wind energy along the Atlantic and Pacific coasts, in the mountainous regions (Rockies and Appalachians), and in the Western Plains. These are the regions in which the development of wind energy would be the most viable.

Wind power installations may be small residential facilities for generating electricity in individual homes (Figure 10.20), or they may be large stand-alone wind turbines (Figure 10.21) or wind farms (Figure 10.22) for the production of electricity for the grid. The former may actually be connected to the grid so that excess energy can be sold back to the public utility when production exceeds demand. A typical grid connection for a wind turbine is shown in Figure 10.23. The details of this system follow very much along the lines of the grid-connected photovoltaic system shown in Chapter 9.

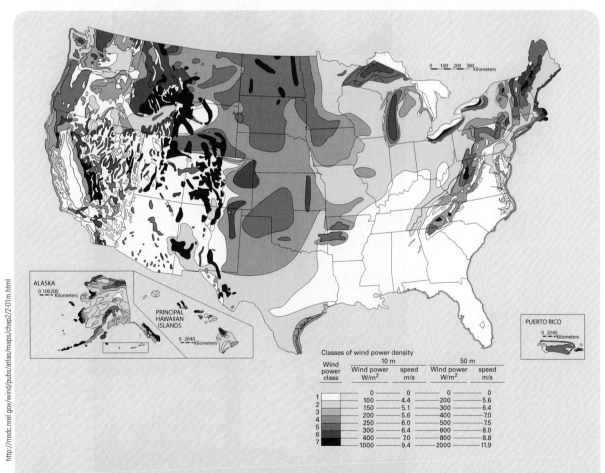

http://rredc.nrel.gov/wind/pubs/atlas/maps/chap2/2-01m.html

Figure 10.19: Average wind power (in W/m^2) for the United States.

Figure 10.20: Small residential wind turbine.

Figure 10.21: 600-kW wind turbine in Goodwood, Nova Scotia.

Richard Thornton/Shutterstock.com

Figure 10.22: Tehachapi wind farm in California.

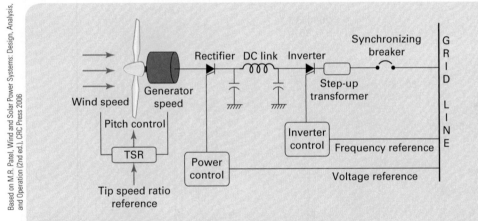

Based on M.R. Patel, Wind and Solar Power Systems: Design, Analysis, and Operation (2nd ed.), CRC Press 2006

Figure 10.23: Schematic of grid-connected wind turbine system.

We can make a rough estimate of the amount of energy available from the wind. As an example, for a typical value of 350 W/m² (corresponding to regions where it would be beneficial to develop wind energy), the power produced by a single wind turbine with 50-m diameter blades operating at 40% efficiency is

$$P = 350 \text{ W/m}^2 \times 3.14 \times (25 \text{ m})^2 \times 0.4 = 275 \text{ kW}. \qquad \textbf{(10.6)}$$

Using the mean spacing shown in Figure 10.18 of 3 rotor diameters in one direction and 10 rotor diameters in the other direction gives an area of 150 m × 500 m (or 0.075 km²) per wind turbine. This means that about 275 kW/0.075 km² = 3.6 MW$_e$/km² can be generated from wind energy. Note that, to a first approximation and within certain limits, the power produced by a wind turbine is proportional to the square of the rotor diameter. The optimal spacing of wind turbines in a wind farm is proportional to the rotor diameter in both north-south and east-west directions (Figure 10.18). Thus,

ENERGY EXTRA 10.1
Limits to wind turbine size

The increase in the utilization of wind energy is most apparent in Denmark, Spain, and Germany, but active development is taking place in many areas throughout the world, including North America, India, and China. These developments have motivated designs for larger and larger wind turbines. A simple analysis of Figure 10.18 might suggest that a wind farm consisting of several large wind turbines would produce the same energy as one consisting of a larger number of small turbines as long as the total area of the rotors is the same. However, large wind turbines are advantageous for two basic reasons. One is that larger turbines are taller, and, as Figure 10.16 suggests, the power density in the wind increases with height (at least up to a point). Another reason is that the infrastructure cost includes not only the cost of the turbine but also the cost of the control systems and grid connections. These costs are minimized if the number of wind turbines is smaller.

The size of the largest common wind turbines as a function of time is illustrated in the next figure:

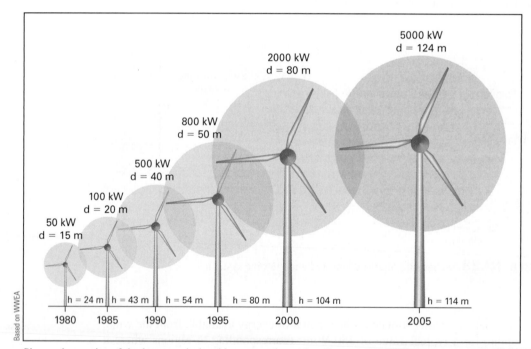

Size and capacity of the largest wind turbines over time (d = diameter, h = height).

Currently, the largest wind turbine is the Enercon E-126 (introduced in 2007) with a rated capacity of 7.58 MW and a total height of 198 m, more than half the height of the Empire State Building. As the capacity of a wind turbine is scaled up, the height increases and the mass of the nacelle increases. This requires the tower to be not only taller but also more massive to support both itself and the weight of the nacelle. The size of the Enercon E-126 is probably the limit of wind turbine size that can be constructed with current designs and materials technology.

One approach to scaling up wind turbine size is to reduce the mass of the generator by using superconducting winding (see Chapters 2 and 18 for more information on superconductors). The American Superconductor Corporation has collaborated

Continued on page 279

Energy Extra 10.1 continued

with the U.S. Department of Energy to investigate the design of a 10-MW superconducting wind turbine. The proposed design would have a mass of about 120 t, in comparison to a nonsuperconducting 10-MW turbine mass of around 300 t. It is possible that this approach could be scaled up to produce a 20-MW wind turbine.

Another approach to higher-capacity wind turbines is to abandon the currently adopted horizontal axis geometry. Vertical axis wind turbines may be an approach that will allow the generator to be in a stationary structure below the rotor and may help to alleviate some of the design limitations of horizontal axis turbines. Some examples of experimental horizontal axis turbines are shown in the photograph below.

Topic for Discussion

In addition to structural considerations, the transport and assembly of components can also limit turbine size, particularly for the turbine blades that are manufactured and transported as single units to the turbine site. A standard transport trailer in North America is 14.6–16.2 m in length; in the United Kingdom, it is about 13.7 m. What capacity wind turbine could be constructed from components transported on the highway using standard trucking methods? The transport of blades for today's larger wind turbines requires special consideration. You can usually find some interesting videos if you search for "wind turbine blade transport" on YouTube. Discuss the compatibility of the highway system with potentially ideal wind farm locations.

Martin Bond/Science Source

Examples of experimental vertical axis wind turbines.

the number of wind turbines that can be accommodated within a given area of land is inversely proportional to the rotor diameter. The total power is therefore roughly independent of the rotor diameter. However, when other factors, such as the typical altitude dependence of wind velocity, are considered, larger wind turbines have clear advantages.

As an example of land requirements for wind power, consider the region that is most likely the best candidate for large-scale wind production in the United States, the Great Plains. A strip through the central United States that includes North Dakota, South Dakota, Nebraska, Kansas, Oklahoma, and Texas has an area of about 1.7×10^6 km^2. At a power density of 3.6 MW$_e$/km^2, this gives a total power capacity for the Great Plain area of $(1.7 \times 10^6$ km$^2) \times (3.6 \times 10^6$ W$_e$/km$^2) = 6.1 \times 10^{12}$ W$_e$. A comparison with the estimate of the total power requirement for the United States from Table 10.3 of 3.2×10^{12} W indicates that about half of this portion of the country would need to be covered with wind turbines at their optimum spacing to fulfill all of the country's energy needs. Because many locations would be inappropriate for wind turbines, the possibility of supplying all U.S. energy needs from the wind is very questionable. However, the current cost of wind power (Figure 10.22) makes it an attractive alternative that can supplement other energy sources.

Because wind conditions in all locations are somewhat unpredictable and cannot be forecast with any certainty more than a few days in advance, it is difficult to provide a stable base load of power to the grid from wind energy. To ensure an adequate supply of energy at all times, wind energy needs to be combined with other sources. As with photovoltaics, it is necessary to integrate a suitable energy storage system with any large-scale wind facility. Energy storage will be discussed in Chapter 18.

The current status of wind power development worldwide is summarized in Table 10.1. A few countries have been very serious about developing wind power;

Table 10.1: Installed wind power as of 2007.		
country	installed capacity (MW$_e$)	capacity per capita (W$_e$)
Germany	22,247	270.6
United States	16,818	55.0
Spain	15,145	328.5
India	8000	7.0
China	6050	4.6
Denmark	3129	568.9
Italy	2726	45.7
France	2454	38.1
United Kingdom	2389	39.0
Portugal	2150	202.8
Canada	1856	55.6
Netherlands	1747	105.9
Japan	1538	12.0
Austria	982	118.3
Greece	871	77.8
Australia	824	38.3
Ireland	805	182.9
Sweden	788	85.7

Tony Moran/Shutterstock.com

Figure 10.24: Danish offshore wind farm.

Germany and Spain are notable in this respect. However, on a per-capita basis, Denmark has been the most active, and wind power presently accounts for more than 20% of their electricity generation. Offshore wind farms in Denmark (Figure 10.24) are particularly noteworthy. These take advantage of the greater offshore wind energy and help (perhaps) to alleviate some of the environmental concerns.

Wind energy has some (although perhaps not serious) environmental concerns. Like any so-called clean energy source, it is truly clean only if the manufacturing of its equipment and its construction and maintenance all use energy that comes from environmentally friendly sources. The environmental impact of wind energy production itself is fairly minimal. The major concerns that have been expressed are:

1. Noise.
2. Effect on wildlife.
3. Effect on land use.
4. Aesthetics.

All wind turbines make some noise. Small wind turbines are the type most commonly used for residential electricity production, and they make the most noise. This is because of the greater rotational frequencies that are necessary in order to achieve a tip speed ratio compatible with efficient operation. Large wind turbines are typically quieter and produce noise at lower frequency. This can be a nuisance for residents living near these facilities, and the possible long-term health effects of the noise need to be carefully considered.

Birds and bats are at risk from wind turbines, and this has been a public concern for some time. All relevant studies have shown that, even if the use of wind power increases substantially, the risk to birds, and probably to bats, from plate glass windows, vehicles, and power lines is several orders of magnitude greater than it is from wind turbines.

The deforestation of wilderness areas for the development of wind power is an important concern, but the use of agricultural land can be compatible with wind power. The spacing between rows of wind turbines in a wind farm along the prevaling wind direction can be close to a kilometer (Figure 10.25). An attractive option is the dual use of land in the Midwest for wind power generation and agriculture.

Figure 10.25: San Gorgonio Pass wind farm in California. The rows of wind turbines are clearly seen.

Wind farms certainly change the appearance of the landscape (Figure 10.22), and this may be considered a relevant factor in assessing the effects of different energy production methods. Offshore wind farms influence the aesthetics of the ocean (Figure 10.24) and, of course, must be located so as to minimize possible adverse effects on navigation.

10.5 Summary

Wind energy has attracted much interest as an alternative to fossil fuel energy sources. Wind has a number of important advantages over some other alternatives and relatively few disadvantages. The attractiveness of wind as an environmentally conscientious energy source has resulted in its growth in recent years. Wind resources that are suitable for development are available in nearly all regions of the world. The basic technology for the utilization of wind energy is reliable and well established and is fairly cost-effective. Wind energy is relatively safe, and overall environmental concerns are minimal.

This chapter began with an overview of the different types of wind turbines. These include vertical axis turbines, such as the Darrieus rotor, as well as the more common horizontal axis turbines.

The chapter presented a basic development of the physics of wind energy and shows that the energy content of moving air is proportional to the cube of the velocity. Wind turbines extract energy from the wind by slowing it down but they cannot extract all the available energy because they cannot stop the air altogether. The maximum amount of energy that can be extracted from moving air is about 59%, as given by the Betz limit. Actual wind turbines more commonly have efficiencies up to about 40%. Efficiency varies with the tip speed ratio, that is, the ratio of the speed of the blade tip to the speed of the wind. Each wind turbine's geometry has an optimal tip speed ratio.

The chapter reviewed wind properties and their relationship to turbine operation. The normal distribution of wind speeds is given by the Rayleigh distribution. Because the energy is proportional to the cube of the velocity, the majority of the energy is available during brief periods of high wind speed. Up to some point, wind speed increases with increasing altitude. This suggests clear advantages to locating turbines atop high towers, although there are trade-offs with structural design and maintenance costs. Wind energy is also typically less over land and increases with distance up to about 5 km over the ocean. Over land, wind energy resources are related to a variety of geographical and climatic features. In the United States, the greatest wind resources are available over the mountainous regions and across the central plains.

The design of wind farms was described in this chapter. To extract the optimum amount of wind energy, turbines should be spaced in accordance with the rotor diameter. The low density of energy in the wind means that fairly large land areas are needed to accommodate wind farms. Offshore wind farms have the advantage of eliminating the need to utilize land resources, as well as taking advantage of the higher energy density of offshore winds.

There have been ambitious efforts in recent years to utilize wind energy. Substantial development has occurred in the European Union, particularly in Denmark, Spain, and Germany. India also has an active wind energy program and has seen significant increases in capacity in recent years.

Problems

10.1 Using Figure 10.16 (for March), estimate the relative power provided by a wind turbine at an altitude of 600 m compared with one at ground level.

10.2 a. Consider a wind farm consisting of 10-m diameter wind turbines with efficiencies of 40% on a grid given by the average spacing shown in Figure 10.18. For a wind velocity of 6 m/s, what is the power output in kW per km^2

   b. Compare the result of part (a) with the 24-hour average output of a 1-km^2 photovoltaic array with an efficiency of 18%. Assume average (worldwide) insolation conditions.

10.3 A wind turbine with a 40-m diameter rotor produces 287-kW output in a 10-m/s wind. What is its efficiency?

10.4 A wind turbine with an efficiency of 42% produces 1-MW output at a wind velocity of 13 m/s. What is the turbine rotor diameter?

10.5 Wind velocities are not constant throughout the day. The daily average power produced by a wind turbine is the power averaged over the wind velocity for the day. Calculate the average power for a turbine with a diameter of 20 m and an efficiency of 37% if, during a 24-hour period, the wind velocity is

- 2 m/s for 4 hours.
- 8 m/s for 16 hours.
- 14 m/s for 3 hours.
- 17 m/s for 1 hour.

10.6 A home owner installs a wind turbine with a rotor diameter of 2 m to supplement electricity from the public utility. The cost of the turbine, the associated electronics, and energy storage system (batteries) is $10,000. If the turbine has an efficiency of 35% and the energy is utilized and or stored at an efficiency of nearly 100%, what is the payback period for the investment? Assume that maintenance costs are minimal, the cost recovery factor is unity, electricity from the public utility costs $0.10 per kWh, and the wind velocity is constant at 12 m/s.

10.7 A Darrieus rotor has an area of 1500 m^2 and operates with the optimal tip speed ratio. What is its power output in a wind with a velocity of 20 m/s?

10.8 A country with good wind resources decides to make a national commitment to the development of wind energy. It is decided that 10% of the land area will be devoted to wind energy with a goal of 1.5 kW$_e$ per-capita wind capacity. This is roughly the electricity used by one person in an industrialized nation (Chapter 2). Two-megawatt wind turbines with rotor diameters of 80 m (Energy Extra 10.1) are placed at the average spacing shown in Figure 10.18. Average output based on wind conditions is 650 kW$_e$ per turbine. Estimate the maximum population density that is consistent with this plan. Locate geographical information about Germany, Spain, India, and Denmark (all of whom have active wind power programs), and discuss the possibility of achieving this goal in each of these countries.

Bibliography

G. Boyle (Ed.). *Renewable Energy*. Oxford University Press, Oxford (2004).

T. Burton, D. Sharpe, N. Jenkins, and E. Bossanyi. *Wind Energy Handbook*. Garrad Hassan and Partners, Bristol (2003).

D. M. Eggleston. "Wind power." In *CRC Handbook of Energy Efficiency*, F. Kreith and R. E. West (Eds.). CRC Press, Boca Raton, FL (1997).

D. M. Eggleston and F. S. Stoddard. *Wind Turbine Engineering Designs*. Van Nostrand Reinhold, New York (1987).

F. R. Eldridge. *Wind Machines* (2nd ed.). Van Nostrand Reinhold, New York (1980).

F. A. Farret and M. Godoy Simões. *Integration of Alternative Sources of Energy*. Wiley, Hoboken, NJ (2006).

L. L. Freris (Ed.). *Wind Energy Conversion Systems*. Prentice-Hall, London (1990).

P. Gipe. *Wind Energy Comes of Age*. Wiley, New York (1995).

P. Gipe. *Wind Power: Renewable Energy for Home, Farm, and Business* (rev. ed.). Chelsea Green Publishing, White River Junction, VT (2004).

E. W. Golding. *Generation of Electricity by Wind Power*. E. & F. N. Spon, London (1955).

S. Heier. *Grid Integration of Wind Energy Conversion Systems*. Wiley, New York (1998).

R. Hinrichs and M. Kleinbach. *Energy: Its Use and the Environment* (5th ed.). Brooks-Cole, Belmont, CA (2012).

G. L. Johnson. *Wind Energy Systems*. Prentice Hall, London (1985).

J. F. Manwell, J. G. McGowan, and A. L. Rogers. *Wind Energy Explained: Theory, Design, and Application*. Wiley, Chichester, U.K. (2002).

J. Park. *The Wind Power*. Cheshire Books, Palo Alto, CA (1981).

M. R. Patel. *Wind and Solar Power Systems: Design, Analysis, and Operation* (2nd ed.). Taylor & Francis, Boca Raton, FL (2006).

P. C. Putnam. *Power from the Wind.* Van Nostrand Reinhold, New York (1948).

J. Reynolds. *Windmills and Watermills.* Hugh Evelyn, London (1970).

R. A. Ristinen and J. J. Kraushaar. *Energy and the Environment.* Wiley, Hoboken (2006).

D. Spera (Ed.) *Wind Turbine Technology: Fundamental Concepts of Wind Turbine Engineering.* ASME Press, New York (1994).

R. Wolfson. *Energy, Environment, and Climate* (2nd ed.). W. W. Norton, New York (2011).

World Energy Council. *Survey of Energy Resources* (20th ed.). Elsevier, Amsterdam (2004).

A. J. Wortman. *Introduction to Wind Turbine Engineering.* Butterworth, Boston (1983).

Hydroelectric Energy

Learning Objectives: After reading the material in Chapter 11, you should understand:

- The potential energy associated with water.
- The kinetic energy associated with water.
- The different types of water turbines and their applications.
- The design and properties of high head hydroelectric systems.
- The design and properties of low head and run of the river hydroelectric systems.
- The availability and utilization of hydroelectric energy worldwide.
- The effects of hydroelectric energy on the environment and risks to society.

11.1 Introduction

Hydropower has been used for many centuries. Like wind power, it was first used as a source of mechanical power and was typically used for grinding grain and for sawing wood. Figure 11.1 shows a reconstruction of a waterwheel used in the nineteenth century to obtain mechanical energy. During the past century, hydropower has been primarily used as a source of electricity. Hydroelectric power has been and continues to be the most prevalent method of generating electricity that is generally referred to as renewable and carbon free. The degree to which this description of hydroelectric power is accurate depends largely how this resource is utilized and includes factors such as geography and climate. The scientific and technological aspects of hydroelectric power are presented in this chapter. The chapter also overviews the extent of hydroelectric resources, their utilization, and the associated risks and environmental impact.

11.2 Energy from Water

The energy associated with water running downhill can be harnessed in two ways: (1) The potential energy of water confined behind a dam can be used to run turbines at the bottom of the dam, or (2) the kinetic energy of flowing water in a river can be used to operate turbines. The first case is generally used for the construction of large

© WoodyStock/Alamy

Figure 11.1: Waterwheel used to convert the kinetic energy of moving water to mechanical energy for sawing wood (reconstruction of nineteenth-century facility in Nova Scotia, Canada).

hydroelectric generating facilities (Figure 11.2). These typically have capacities of hundreds of megawatts electrical but can be up to 10,000 MW$_e$ or more. They are referred to as *high head* hydroelectric facilities.

Facilities that use the kinetic energy of the water are referred to as run-of-the-river systems, although many are associated with small (i.e., *low head*) dams. They typically have a capacity of a few megawatts electrical (up to about 10 MW$_e$), although small versions can be less than 100 kW$_e$ (Figure 11.3). Medium head systems are intermediate between these two extremes, although, in practice, the distinction between these designations is not always well-defined.

Figure 11.2: Three Gorges hydroelectric facility in China. This is the world's largest hydroelectric facility, currently rated at 22,500 MW$_e$. The environmental consequences of the Three Gorges Dam are discussed in Energy Extra 11.2.

Richard A. Dunlap

Figure 11.3: Small run-of-the-river hydroelectric facility on the Musquash River in New Brunswick, Canada, with output of about 7 MW$_e$. The cylindrical tank is a surge tank used for evening out fluctuations in water flow to the turbines.

A high head hydroelectric dam blocks a river to create a reservoir. Water is allowed to flow through turbines at the bottom of the dam, and the height of the water in the reservoir above the height of the turbines is called the *head*. The potential energy of water near the surface falls through that distance and is converted into kinetic energy as it flows through the turbines that converts it into electrical energy. The potential energy of the water is

$$E = mgh. \tag{11.1}$$

where m is the mass of the water, g is the gravitational acceleration (9.8 m/s^2), and h is the head, or height, of the water above the turbine. The power generated is determined by the rate at which energy is generated:

$$P = \frac{E}{t} = \frac{m}{t}gh. \tag{11.2}$$

If the rate at which water flows through the turbine (i.e., volume per unit time) is φ, expressed in cubic meters per second (m^3/s), then $(m/t) = \rho\varphi$, where ρ is the water density. Thus

$$P = \rho\varphi gh. \tag{11.3}$$

For equation (11.3), the power is expressed in watts when the density is in kg/m^3, the flow rate is in m^3/s, g is in m/s^2, and the height is in m.

In a run-of-the-river system, water with a mass, m, flowing in a river at a velocity, v, will have kinetic energy

$$E = \frac{1}{2}mv^2. \tag{11.4}$$

This represents a power-generating capacity of

$$P = \frac{1}{2}\left(\frac{m}{t}\right)v^2, \tag{11.5}$$

Example 11.1

One cubic meter of water is dropped from a vertical height of 100 m. If the change in potential energy is converted into electricity with an efficiency of 85%, how long would this output fulfill the electrical needs of a typical North American single-family home? (See Chapter 2, and do not include electric heating.)

Solution

One cubic meter of water has a mass of 1000 kg, so the energy given by equation (11.1) is

$$E = (1000 \text{ kg}) \times (9.8 \text{ m/s}^2) \times (100 \text{ m}) = 980,000 \text{ J}.$$

At 85% efficiency, this represents a total of $(980,000 \text{ J}) \times (0.85) = 833,000 \text{ J}$ of electrical energy. From Chapter 2, the average annual residential electric use is approximately 1.2×10^4 kWh $= 4.32 \times 10^{10}$ J. This represents an average power of $(4.32 \times 10^{10} \text{ J/y})/(3.15 \times 10^7 \text{ s/y}) = 1370$ W. Thus, 833,000 J will last for

$$\frac{833,000 \text{ J}}{1370 \text{ J/s}} = 608 \text{ s}.$$

Example 11.2

What is the flow rate in terms of volume per unit time and in terms of mass per unit time that would be required for a 90% efficient hydroelectric facility with a head of 50 m to satisfy (on average) the electrical needs of a typical single-family home?

Solution

Following from Example 11.1, the average power requirement is 1370 W. At 90% efficiency, this represents $(1370 \text{ W})/0.9 = 1522$ W total. From equation (11.3), the flow rate (in m³/s) can be found to be

$$\varphi = P/(\rho g h).$$

Substituting appropriate values gives

$$\varphi = \frac{1522 \text{ W}}{1000 \text{ kg/m}^3} \times (9.8 \text{ m/s}^2) \times (50 \text{ m}) = 3.11 \times 10^{-3} \text{ m}^3/\text{s}.$$

Multiplying by the density of water gives the flow rate in mass per unit time:

$$(3.11 \times 10^{-3} \text{ m}^3/\text{s}) \times (1000 \text{ kg/m}^3) = 3.11 \text{ kg/s}.$$

or

$$P = \frac{1}{2}\rho\varphi v^2.$$
(11.6)

For water flow through an opening with a cross-sectional area A, equation (11.6) reduces to the form given in equation (10.3) for the power per unit area for moving air:

$$\frac{P}{A} = \frac{1}{2}\rho v^3.$$
(11.7)

Numerically, equation (11.7) differs substantially from equation (10.3) because the density of water is much greater than the density of air.

Example 11.3

A small river, 40 m wide and 6 m deep, flows at a velocity of 2 m/s. If 20% of the flow of the river is diverted through a run-of-the-river hydroelectric system that generates electricity at an efficiency of 90%, what is the output in watts electrical?

Solution

The cross-sectional area of the river is (40 m) × (6 m) = 240 m². The volumetric flow rate is

$$\varphi = (240 \text{ m}^2) \times (2 \text{ m/s}) = 480 \text{ m}^3/\text{s},$$

and 20% of this flow corresponds to

$$(480 \text{ m}^3/\text{s}) \times (0.2) = 96 \text{ m}^3/\text{s}.$$

From equation (11.6), the power output is

$$P = \frac{1}{2}\rho\varphi v^2 = (0.5) \times (1000 \text{ kg/m}^3) \times (96 \text{ m}^3/\text{s}) \times (2 \text{ m/s})^2 = 192,000 \text{ W}.$$

At 90% conversion efficiency to electricity, the output is

$$(192,000 \text{ W}) \times (0.9) = 173 \text{ kW}_e.$$

The following sections describe the details of the design of the turbine and specific design considerations for high head and low head systems.

11.3 Turbine Design

To produce electricity from the energy content of water, a turbine is used to produce rotary mechanical energy, and this, in turn, is used to produce electricity by means of a generator. There are two basic types of turbines for use with water: *reaction turbines* and *impulse turbines*. In the first case, the flowing water is contained in an enclosure around the turbine and experiences a pressure drop as it passes through the turbine. In the second case, a free jet of water is incident on the turbine and does not experience a

Table 11.1: Some important water turbine designs and their operating conditions.			
name	type	suitable head (m)	maximum power (MW$_e$)
Kaplan	reaction	2–40	200
Francis	reaction	10–350	800
Pelton	impulse	50–1500	400
Turgo	impulse	50–250	5

© Cengage Learning 2015

pressure drop. There are numerous designs of both of these types of turbines. The most common are summarized in Table 11.1.

Perhaps the most obvious design for a water turbine is one that resembles a wind turbine. The *Kaplan turbine* (Figure 11.4) is a reaction turbine and is the closest to this design. Figure 11.5 shows a close-up of the rotating hub and blades of the turbine (i.e., the *runner*). The runner must be enclosed in a cylindrical tube, through which the water flows, in order to prevent the water from being diverted around the blades and thus reducing the turbine's efficiency.

U.S. Army

Figure 11.4: A Kaplan turbine being assembled.

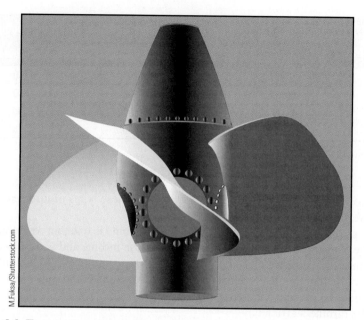

M.Fuksa/Shutterstock.com

Figure 11.5: Kaplan turbine runner.

The *Francis turbine* is a reaction turbine and probably the most commonly used type of turbine in the electric power industry (Figure 11.6); the details of the runner are shown in Figure 11.7. In this type of turbine, the water enters radially from the sides and is guided through the blades, causing them to rotate, and then exits axially from the center of the turbine.

Impulse turbines are suitable for high head applications where the water at the base of the dam is at high pressure. The water is allowed to exit through a small opening, forming a free jet of water at high velocity. This high-velocity water jet is incident

U.S. Bureau of Reclamation photo archives

Figure 11.6: Francis turbine at the Grand Coulee Dam. The inlet scroll is clearly seen in the photograph.

Figure 11.7: Francis turbine runner.

on the runner of the turbine. An impulse turbine runner has what look like buckets along the outer edge that catch the high-velocity jet of water, deflecting it and imparting momentum to the runner. In the *Pelton turbine*, the buckets are U-shaped [Figure 11.8(a)]. Water is ejected through a nozzle and is incident on the runner [Figure 11.8(b)]. The *Turgo impulse turbine* is basically one-half of a Pelton turbine. The water jet enters into the half-U-shaped bucket at an angle and exits in a different direction thus transferring momentum to the runner (Figure 11.9).

A summary of the operating ranges of the various types of turbines is shown in Figure 11.10. Although the efficiency of turbines and the associated generators depends

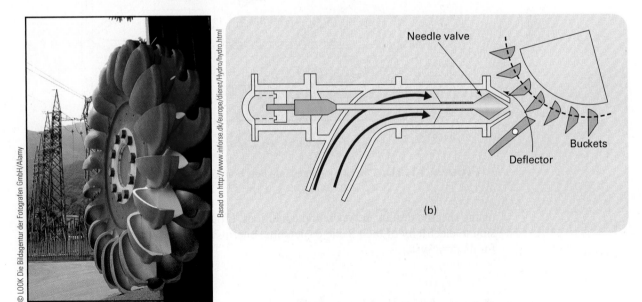

(a)

(b)

Figure 11.8: (a) Photograph of a Pelton runner; (b) diagram of a Pelton turbine.

Joseph Hartvigsen

Figure 11.9: A micro-Turgo runner.

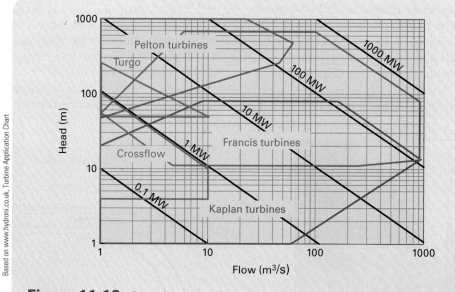

Based on www.hydroni.co.uk, Turbine Application Chart

Figure 11.10: Operating ranges for some common water turbine designs.

on the specific design and operating conditions, efficiencies of 85–90% for the conversion to electricity of the kinetic or potential energy associated with water are common for modern systems.

11.4 High Head Systems

For the present discussion, *high head systems* are defined as those that impound water behind a dam for the purpose of producing a reservoir to create a head of water. Most high-capacity hydroelectric facilities are of this type.

U.S. Bureau of Reclamation photo archives

Figure 11.11: Grand Coulee Dam in Washington State, United States.

One of the best known examples of a high head system is Grand Coulee Dam in Washington State (Figure 11.11). The dam is 1592 m in length and 168 m in height. The size of this structure is indicated by the fact that the volume of concrete in the dam is sufficient to construct a road 2.5 m wide and 10 cm thick around the equator of the earth.

Typically, in a high head system there is an underwater water intake in the dam. The water flows through a pipe (the *penstock*) to the turbine near the bottom of the dam (Figure 11.12). The general features of a typical high head hydroelectric dam are shown in Figures 11.11 and 11.13. The portion of the dam on the right side of both photographs is the *spillway*, which allows excess water to bypass the turbines. On the left side of the photographs, the structure near the bottom of the

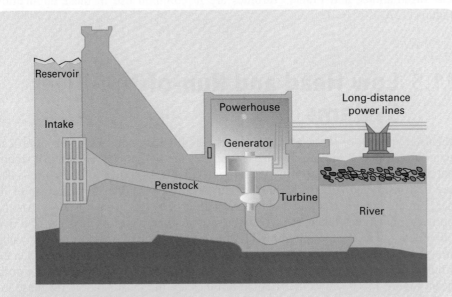

Based on http://www.window.state.tx.us/specialrpt/energy/renewable/hydro.php

Figure 11.12: Schematic of a typical high head hydroelectric system showing dam, penstock, and turbine.

U.S. Army Corps of Engineers

Figure 11.13: The Libby Dam in Montana.

dam houses the turbine/generator assemblies. A vertical axis configuration (more on this in the next section) is the most common arrangement for turbines in high head installations, and Francis turbines are in common use in these applications (Figure 11.6).

11.5 Low Head and Run-of-the-River Systems

Low head and *run-of-the-river* hydroelectric facilities do not impound a significant quantity of water. While most of these facilities are low power, some have significant capacity, like the facility shown in Figure 11.14. As shown in the figure, a dam may be constructed across the entire width of a river in order to make use of the energy of the flowing water without the formation of a significant reservoir. Another approach, sometimes taken, is a diversion system, where part of the flow of a river is diverted into a penstock (Figures 11.15 and 11.16) and allowed to flow downhill through the penstock to the generating system. After passing through the generator turbines, the water is returned to the river or, in some cases, output into a lake or the ocean. Figure 11.10 shows that Kaplan turbines are the most suitable for these low head applications. The turbine and generator arrangement may be on a horizontal axis (Figure 11.17) or on a vertical axis, as is customary for high head applications (Figure 11.18).

U.S. Army Corps of Engineers

Figure 11.14: Chief Joseph Dam in Washington, a low head hydroelectric facility with a capacity of 2620 MW$_e$.

Erick Margarita Images/Shutterstock.com

Figure 11.15: Penstocks used to divert part of a river's flow through a generating station.

Figure 11.16: Small hydroelectric facility in Head of St. Margaret's Bay, Nova Scotia, Canada, showing the penstock that supplies the water to the turbines.

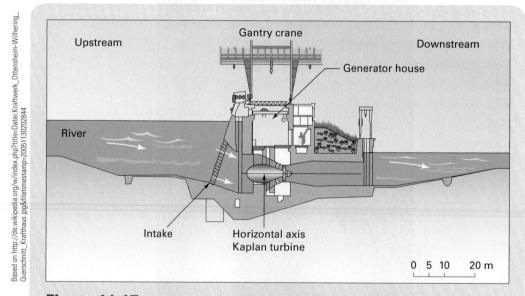

Figure 11.17: Low head or run-of-the-river hydroelectric generating facility utilizing a horizontal axis bulb (Kaplan-type) turbine.

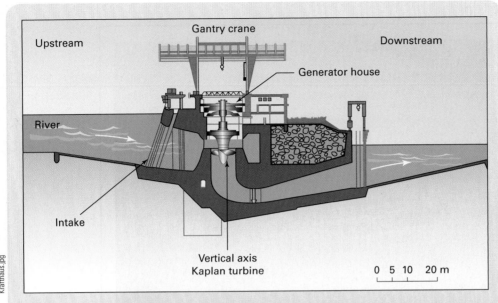

Figure 11.18: Low head or run-of-the-river hydroelectric generating facility utilizing a vertical-axis Kaplan turbine.

Based on http://en.wikipedia.org/wiki/File:Kraftwerk_Wallsee-Mitterkirchen_Querschnitt_Krafthaus.jpg

ENERGY EXTRA 11.1
Niagara Falls: Energy and tourism

Panoramic view of Niagara Falls.

Niagara Falls is one of the best known waterfalls in the world. It is also a major source of hydroelectric power. The falls are on the Niagara River on the border between the United States and Canada and have a vertical drop of 51 m. The falls, shown in the photograph, consist of the American Falls on the left of the image, the (small) Bridal Veil Falls just to the right of the American Falls, and the Horseshoe Falls (also known as the Canadian Falls) at the right in the photograph. The falls have been used as a source of mechanical power since 1759 and as a source of hydroelectric power since 1881.

Because Niagara Falls is a popular (and economically important) tourist attraction, its utilization for hydroelectric generation is managed in order to minimize its impact on the appearance of the falls. Water

Continued on page 300

Energy Extra 11.1 continued

Busfahrer

The Robert Moses Niagara Falls Generating Station.

from the Niagara River is diverted upstream from the falls and directed through canals or underground tunnels that are up to 10 km or more in length to hydroelectric generating stations and then back to the river downstream from the falls. Eight hydroelectric generating stations are associated with Niagara Falls, with a total generating capacity of about 5 GW. The largest of these is the Robert Moses Niagara Generating Station, shown in the above photograph, which has a maximum generating capacity of about 2.5 GW.

The use of the Niagara Falls hydroelectric resources is governed by a 1950 treaty between the United States and Canada. The United States can utilize a maximum of 920 m³/s of water from the Niagara River for hydroelectric generation; Canada can utilize a maximum of 1600 m³/s of water. The inequality of these amounts compensates for other hydroelectric

utilization agreements for resources along the border that favor the United States. These amounts are subject to the condition that at least 50% of the flow of the Niagara River must go over the falls during the daylight hours between April and October (the prime tourist season). At other times, a minimum of 25% of the river's flow must go over the falls. Niagara's hydroelectric stations are an interesting example of the management of a significant energy resource in a way that optimizes its economic benefits.

Topic for Discussion

Estimate the financial cost (per year) of maintaining Niagara Falls as a natural attraction. Specifically, what would be the value of the additional electricity that could be generated by diverting all of the flow of the Niagara River through turbines?

11.6 Utilization of Hydroelectric Power

Of all the sources of energy that are generally considered to be renewable, hydroelectricity is the only one that has seen widespread use. A summary of the major producers of hydroelectric power by country is shown in Table 11.2. The installed capacity is the maximum power output from all hydroelectric facilities in each country. These facilities

Based on "Binge and purge". The Economist. 2009-01-22. Retrieved 2009-01-30; "Indicators 2009. National Electric Power Industry". Chinese Government. Retrieved 18 July 2010.

Table 11.2: Capacity and production of major hydroelectric power producers (2009).

country	installed capacity (GW$_e$)	actual annual production (TWh$_e$)	capacity factor	% domestic electricity production
China	197	652	0.38	22
Canada	89	369	0.47	61
Brazil	69	364	0.60	86
United States	80	251	0.36	6
Russia	45	167	0.42	18
Norway	28	141	0.57	98

produce energy, on average, at a rate that is less than their rated maximum capacity. In general, the output is determined by rainfall, and, because of variability in conditions, hydroelectric power is generally used in conjunction with other energy sources in order to provide a reliable and consistent energy supply. Energy storage methods (discussed in Chapter 18) are also useful in this respect. The capacity factor is the fraction of maximum capacity that is actually produced on the average. Typical capacity factors for hydroelectric facilities are between 0.4 and 0.6 (or 40–60%). Table 11.2 also gives the percentage of the national electricity use that is produced by domestic hydroelectric power. Certainly, a few countries (e.g., Canada, Brazil, and Norway) rely heavily on hydroelectric power. Canada and Norway have traditionally had substantial hydroelectric facilities, while Brazil has installed considerable capacity in the latter part of the twentieth century. Overall, hydroelectric energy production accounts for about 16% of the world's electricity, or about 6% of the world's total energy.

The trends in hydroelectric energy production as a function of time are interesting to consider. Figure 11.19 shows the total hydroelectricity production worldwide for

Based on http://openlearn.open.ac.uk/file.php/4184/!via/oucontent/course/479/s278_8_f024hi.jpg

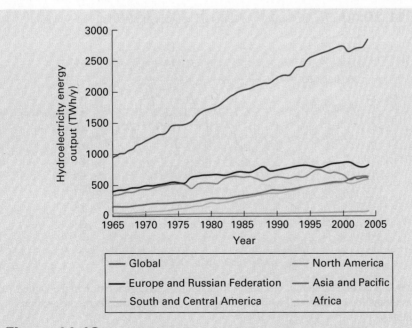

Figure 11.19: Annual production of hydroelectric power worldwide.

the past 30 years. There is a fairly consistent increase in hydroelectric use over time, although the trends in the past few years are not as clear. The general increase illustrated in the figure is largely due to the development of hydroelectric resources in developing countries, particularly China. However, much can be understood about the future of hydroelectricity from trends in countries where this industry is much more mature. The trends in Canada are shown in Figure 11.20, and the trends in the United States are shown in Figure 11.21. For the past half century or more, the fraction of electricity coming from hydroelectricity in these two countries has been declining. Although the actual amount of electricity from hydroelectric sources has not changed significantly since about 1970, increased demand, due to both a growing population and changes in per-capita electricity consumption, has been met primarily by increased production from fossil fuel and nuclear sources. The graphs show that in the United States hydroelectricity accounts for about 6% of the electrical supply, whereas in Canada it is currently around 60%. Traditionally, most of Canada's electricity has come from hydroelectric

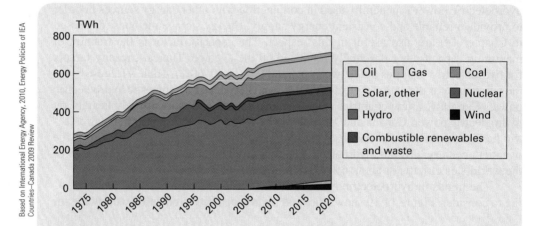

Figure 11.20: Relative proportions of different sources of electricity in Canada.

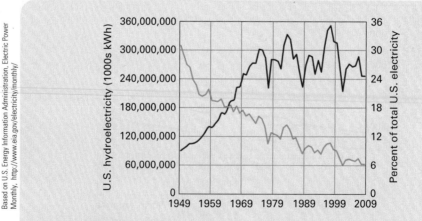

Figure 11.21: Fraction of electricity from hydroelectric generation in the United States as a function of year (yellow line), along with the total hydroelectric generation (blue line).

Based on U.S. Energy Information Administration, http://www.eia.doe.gov/cneaf/electricity/epm/table1_1.html

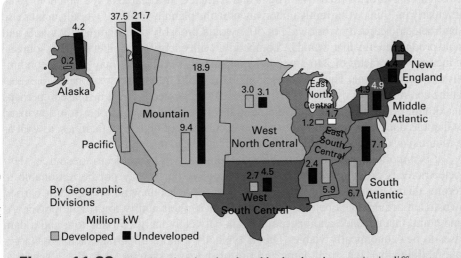

Figure 11.22: Developed and undeveloped hydroelectric capacity in different regions of the United States.

generation. In fact, in Canada the term *hydro* is synonymous with electricity supplied by a public utility regardless of how it is generated, and the majority of provincial electric companies have names like Québec Hydro and Manitoba Hydro, even though the electricity is produced by a variety of methods.

The reasons for these trends can be readily understood in terms of the availability of hydroelectric power. Hydroelectric power is not an unlimited resource because the locations where it can be developed on a large scale in an economically viable and environmentally acceptable way are limited. (See the next section for environmental considerations.) In fact, in the more industrialized regions of the world, such as North America and Europe, the largest portion of the potential for hydroelectric power has already been developed. Figure 11.22 show the developed and undeveloped hydroelectric potential in different regions of the United States. On average, over half of the hydroelectric potential is already in use. In Canada, the fraction is even greater. Worldwide, the potential for use of hydroelectricity is illustrated in Figure 11.23. The greatest possibility

Based on 2010 World Energy Council, Survey of Energy Resources, http://www.worldenergy.org/documents/ser_2010_report_1.pdf

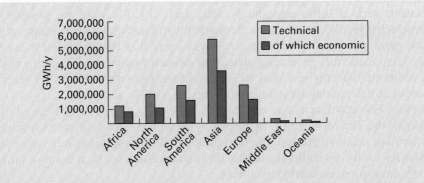

Figure 11.23: Technically viable and economically viable hydroelectric capacity in different parts of the world.

for increased hydroelectric use is in Asia and Africa. Considerable development in Asia, particularly in China, is currently underway or in the planning stages. Overall, an increase in hydroelectric capacity worldwide of 50% would probably be economically viable and would produce energy that would be competitive with energy from other existing sources. An increase by a factor of 2 to 3 would be technically feasible but may not be economically viable at this time. Thus, at some point in the future, it might be reasonable to expect that the contribution of hydroelectricity to our electricity needs could increase to perhaps a maximum of 10–15%. Much of the new development is likely to be from low head installations due to the cost and environmental concerns of high head facilities, as well as the minimal availability of viable undeveloped high head resources.

It is important, as well, to consider the longevity of hydroelectric resources. Hydroelectricity is promoted as a renewable resource, but it may not be reasonable to interpret this classification in the same way as for solar or wind energy. Silt carried downstream accumulates behind dams and reduces their ability to provide power. At some point, maintenance and operating costs outweigh energy production, and a dam ceases to be economically viable. This is typically a more important factor for large-scale facilities than for small dams and run-of-the-river systems. It is also a sensitive function of the geography of the region. Overall, the longevity of some high head hydroelectric facilities may be limited to somewhere in the range of 50 to 200 years, making them clearly not indefinitely renewable.

11.7 Environmental Consequences of Hydroelectric Energy

Although hydroelectric power is often considered an environmentally friendly source of energy, it is important to consider the details of its impact on the environment. Large-scale high head dams are the most concern, whereas run-of-the-river facilities are not as invasive for many reasons. The construction of a large dam is a major undertaking (Figure 11.24), and the dam can produce a very sizable reservoir upstream. For example, the reservoir associated with the Three Gorges Dam in China extends 600 km upriver. The replacement of land, which in most cases supported vegetation (e.g., trees, etc.), with a reservoir has important implications. Trees sequester carbon and help to reduce greenhouse gases (i.e. CO_2). Vegetation that has been flooded decays and produces greenhouse gases, notably CO_2 and methane. Although other pollutants such as sulfur compounds, NO_x, and particulates are not produced by a hydroelectric facility, studies have shown that in tropical regions the release of greenhouse gases associated with reservoirs can be more than that produced by a comparably sized fossil-fuel generating plant. In temperate regions, this is not the case, and emissions are typically less than 10% of that from equivalent fossil-fuel generation. Changes in habitat may have adverse effects on wildlife and particularly on wildlife diversity. Fish mobility is one of the most obvious factors to be influenced by dam construction. The replacement of forests by an aquatic environment may have both positive and negative effects. The disruption of silt transport may have adverse consequences for the agricultural industry downstream from the dam because farmlands benefit from nutrients carried by a river.

In addition to environmental effects, hydroelectric dam construction generally has social and cultural consequences. People living in areas that are flooded by dam construction need to be relocated. It is estimated that, to date, many tens of millions

ENERGY EXTRA 11.2
Three Gorges carbon payback period

Although alternative energy sources are attractive as replacements for fossil fuels and as a mechanism for reducing the adverse environmental effects, carbon release is always associated with energy production. Large-scale hydroelectric power is an interesting example of the carbon cost associated with renewable energy. Although an analysis of the environmental impact of changes in ecology due to land reallocation is generally not simple, a life cycle analysis of the materials involved in dam construction is reasonably straightforward.

Hydroelectric dams are constructed largely of concrete, which is comprised of three components—cement (CaO), sand, and filler (typically gravel)—in a ratio of about 1:2:3, respectively. While sand and gravel have a relatively small carbon cost, cement contributes carbon to the environment for three reasons: (1) CaO is made by the decomposition of limestone ($CaCO_3$) by heating ($CaCO_3$ + heat → CaO + CO_2): (2) furnaces used to heat the limestone use primarily fossil fuels; and (3) carbon emissions from transportation, etc. The last factor is relatively unimportant, but the first two, together, contribute to an emission of about 0.15 kg CO_2 per kg of concrete. Consider as an example the construction of the Three Gorges Dam in China. This dam used 6.5×10^{10} kg of concrete and emitted $(6.5 \times 10^{10}$ kg$) \times (0.15) \approx 10^{10}$ kg of CO_2 to the atmosphere. The payback period for this carbon release to the environment can be calculated by equating the electric generation from the dam to the generation from a coal-fired plant that would produce the same emissions.

The burning of coal produces 0.11 kg CO_2/MJ. At a Carnot efficiency of about 40% a coal fired station produces about (0.11 kg/MJ)/0.40 = 0.28kg CO_2/MJ_e or about 1 kg CO_2/kWh_e. The release of 10^{10} kg of CO_2 would therefore correspond to the generation (by coal) of 10^{10} kWh_e.

The following graph shows the actual average monthly power output of the Three Gorges Dam in

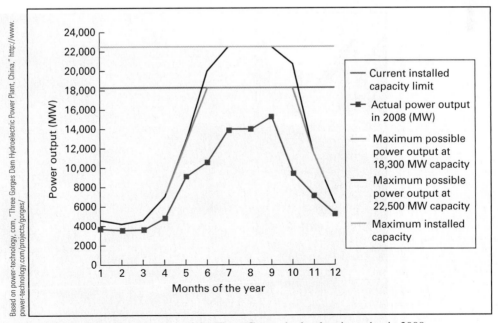

Monthly average power output of the Three Gorges hydroelectric station in 2008.

Continued on page 306

Energy Extra 11.2 continued

2008. Integrating data over the year yields the total energy generated during 2008: 7.4×10^{10} kWh$_e$. Thus, the carbon payback period for the concrete used in the construction of the Three Gorges Dam is about $[10^{10} \text{ kWh}_e]/(7.4 \times 10^{10} \text{ kWh}_e/\text{y}) = 0.14$ years (about a month and a half). Other material used in dam construction (e.g., steel) would need to be considered in a similar way in a more detailed analysis. This type of analysis assumes that the dam is carbon-neutral once it is constructed.

This analysis shows that the carbon payback period for hydroelectric dam construction can be quite reasonable. It is the long-term environmental effects of large scale hydroelectric generation that need to be carefully assessed.

Topic for Discussion

The Three Gorges Dam contains 4.6×10^8 kg of steel. Estimate the additional carbon payback period for this component of the dam.

of people have been affected in this way. It is a matter not only of relocating people but also, in most cases, of dealing with cultural or historical sites and cemeteries. A recent example of social and cultural effects of hydroelectric power is the construction of the Three Gorges Dam in China. The project required the relocation of 1.24 million residents. The relocation costs were approximately the same as the actual construction costs for the hydroelectric facility. In addition, the project affected approximately 1300 archeological sites. Although many were preserved and relocated, other known, as well as yet undiscovered, sites were destroyed.

A final point to consider in the implementation of hydroelectric power is safety. A large dam can fail with serious consequences. Dams may be constructed for a number of reasons other than for the production of hydroelectricity, such as water storage, irrigation, flood control, etc., and many dams are considered multipurpose. All have at least some risk of failure. The severity of such a failure is a function of many variables,

Figure 11.24: Construction of the Three Gorges hydroelectric facility.

U.S. Bureau of Reclamation

Figure 11.25: Failure of the Teton Dam in Idaho in 1976.

such as the volume of water impounded by the dam, the population distribution living downstream, and the warning time before failure. The most notable dam failure was the Banqiao Reservoir Dam and 61 other associated dams in China in 1975 during Typhoon Nina. The Banqiao Dam was constructed on the Ru River in the 1950s. Excessive rain associated with the typhoon increased water flow and levels in the river and ultimately caused a series of dam failures. The Banqaio Dam, at a height of 118 m, was the largest of these. A design that did not include an adequate number of sluice gates to release excess water was a factor in the failure of the dam. Several other dams were intentionally destroyed in an attempt to control water levels. The flooding caused by these failures caused 26,000 deaths, and the resulting epidemics and famine in the region led to an estimated additional 145,000 deaths. Although such catastrophic events are rare, there have been more than 30 dam failures in the past century that have resulted in more than 100 fatalities. A well-known dam failure that resulted in 11 deaths was the Teton Dam in Idaho. The dam had existed for less than a year before failure resulted from poor design and unstable geological conditions in the area. Figure 11.25 shows the reservoir emptying through the failed dam within a few hours of the catastrophic event.

Despite notable dam failures, hydroelectric power is, on average, a relatively safe option overall, as indicated in Table 6.8. However, dam failure, although unlikely, has the greatest risk for large-scale disaster of any of the anthropogenic causes presented in Figure 6.27.

11.8 **Summary**

In this chapter, the basic physics of the potential and kinetic energy associated with water were presented. The potential energy associated with water as a result of the gravitational interaction is given as $E = mgh$, where m is the mass, g is the gravitational acceleration, and h is the height. The kinetic energy of moving water is analogous to that of moving air, as discussed in the previous chapter, and is related to the cube of the velocity.

This chapter also reviewed the various designs of turbines. Different runner geometries are appropriate for different applications on the basis of flow rate and pressure. Generally, Kaplan turbines are most suitable for low head systems, whereas Francis turbines are most suitable for high head systems.

High head hydroelectric systems are typically constructed using a dam to create a large reservoir. Generating capacity can exceed 20 GW_e, compared with about 1 GW_e for a typical coal-fired or nuclear generating station.

Low head and run-of-the-river systems typically do not incorporate a reservoir to create a significant head but primarily utilize the kinetic energy associated with the flow of the river. These systems typically have a much smaller capacity than high head system, often in the 10–100 MW_e range.

The availability and use of hydroelectric power were discussed. Hydroelectric power comprises about 16% of the world's electricity generation. It is a major source of energy and, together with nuclear power (which has a similar share), contributes the majority of the world's non–fossil fuel electricity. Although much of the practical hydroelectric capacity in North America and Europe has already been developed, extensive new development continues in many other regions of the world. Hydroelectric power has some very positive attributes. It is low cost, and the facilities are relatively low maintenance. It comprises a significant enough resource that it can make a major contribution to the world's electrical needs. Thus far, the safety record of hydroelectric facilities has been quite good, and the average overall risk is probably lower than for most large-scale generating techniques. Small-scale hydroelectricity, specifically run-of-the-river facilities, has relatively low environmental impact

On the negative side, hydroelectric power, at least large-scale high head installations, may not be as environmentally neutral as other alternative energy technologies, particularly in warmer climates. Hydroelectric installations may contribute to greenhouse gas emissions in addition to having adverse effects on agriculture and wildlife. The longevity of hydroelectric dams is related to geographical conditions, and, in the long term, some hydroelectric resources may have limited renewability. Finally, there is always a risk of catastrophic dam failure, which carries the possibility of substantial human causalities.

Problems

11.1   One cubic meter of water at 20°C is dropped from a vertical height of 100 m. If the change in potential energy is converted into heat with an efficiency of 100% and this thermal energy is used to heat the water, what is its final temperature?

11.2   The Three Gorges Dam is rated at 22.5 GW_e maximum output. If its generating efficiency is 90% and its head is 81 m, what is the rate of flow (in kg/s) of water through its turbines at maximum output?

11.3   Water falls from a head of 100 m at a rate of 1 m^3/s, and this energy is converted into electrical energy at an efficiency of 90%. How many typical single-family homes can this provide with electricity?

11.4   Consider the run-of-the-river system described in Example 11.3. What is the most reasonable turbine runner design for such a system and why?

11.5 Use the total annual U.S. hydroelectric production and the information in Chapter 2 to confirm the percentage of U.S. electricity that comes from hydroelectric facilities as shown in Table 11.2.

11.6 The reservoir created by the Three Gorges Dam has an area of 1045 km^2. It is anticipated that, in the long term, the capacity factor of this generating facility will be around 0.5. Consider an alternative situation where the land area utilized by the Three Gorges reservoir was utilized instead for a photovoltaic array with an average efficiency of 20%. Would one of these situations be significantly more advantageous than the other in terms of energy production?

11.7 A reservoir is 1 km wide and 10 km long and has an average depth of 100 m. Every hour, 0.1% of the reservoir's volume drops through a vertical height of 100 m and passes through turbines to produce electricity with an efficiency of 92%. What is the electrical power output of this facility?

11.8 The term *micro hydro* refers to very small-scale hydroelectric power and includes installations that can supplement other sources of electricity for single-family residences. Typical efficiencies are less than commercial hydroelectric installations and may be in the range of 75%. A home owner plans to install a micro hydro generator on a stream that has an average flow of 1200 L/min and a vertical drop of 5 m. What is the average electric power that can be expected from this facility? Should this project be seriously considered if the construction cost is $3000.

Bibliography

G. Boyle (Ed.). *Renewable Energy.* Oxford University Press, Oxford (2004).

R. Hinrichs and M. Kleinbach. *Energy: Its Use and the Environment* (5th ed.). Brooks-Cole, Belmont, CA (2012).

M. Kaltschmitt, W. Streicher, and A. Wiese (Eds.). *Renewable Energy: Technology, Economics and Environment.* Springer, Berlin (2007).

R. A. Ristinen and J. J. Kraushaar. *Energy and the Environment.* Wiley, Hoboken (2006).

C. Simeons. *Hydropower: The Use of Water as an Alternative Source of Energy.* Pergamon Press, Oxford (1980).

N. Smith. "The origins of the water turbine." *Scientific American* **242** (1980): 138.

J. Tester, E. Drake, M. Driscoll, M. W. Golay, and W. A. Peters. *Sustainable Energy: Choosing Among Options.* MIT Press, Cambridge, MA (2006).

C. C. Warnick. *Hydropower Engineering.* Prentice-Hall, Englewood Cliffs, NJ (1984).

World Energy Council. *2010 Survey of Energy Resources*, available online at http://www.worldenergy.org/documents/ser_2010_report_1.pdf.

Wave Energy

- The availability of wave energy worldwide.
- The relationship between wave properties and energy.
- The types of wave energy devices.
- The design of oscillating water columns and the properties of the Wells turbine.
- The use of floating and pitching devices for wave energy generation.
- The use of wave-focusing devices.

12.1 Introduction

Energy can be obtained from the ocean either from temperature or salinity gradients or from the movement of water. The former is discussed in Chapter 14. The latter may be from tides, ocean currents, or waves. Tidal energy, which is the result of the gravitational interaction, is discussed in the next chapter. Ocean currents may have components that result from temperature gradients, winds, and/or tides. The possibility of obtaining energy from currents that arise primarily from tidal motion is discussed in the next chapter. This chapter discusses energy from ocean waves.

Ocean waves are primarily the result of the interaction between wind and the sea surface. Like wind energy, wave energy is primarily a manifestation of solar energy. Waves occur everywhere in the oceans, but some locations on earth are more likely to have larger waves than others. Figure 12.1 shows the average wave power available at various locations worldwide. Typically, the larger wave energies are over 50 kW/m averaged over the year. Areas with wave energies much below this value are probably not economically viable. For a region with an average wave energy of 50 kW/m, a facility that extracts the energy from 20 m of wave front could theoretically have an average output of as much as 1 MW. The more industrialized regions where wave power is attractive are the southern coast of Australia and the Atlantic coast of Norway, the United Kingdom, Ireland, and Portugal. Much of the activity in wave energy development occurs in these countries, although there are also significant activities elsewhere, such as Canada, the United States, China, Denmark, and Japan.

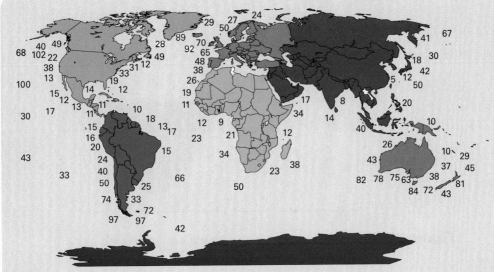

Figure 12.1: Wave power at various locations. Powers are average annual values in kW/m of wave front.

This chapter begins with a review of the basic physics of wave energy. This is followed by an overview of some of the technologies that have been developed to extract useful energy from waves and a summary of the utilization of these technologies.

12.2 Energy from Waves

A schematic of an ocean wave in cross section is shown in Figure 12.2. The wave amplitude is A, and the wavelength is λ. A water wave has energy associated with it for two reasons: (1) It is moving, so it carries kinetic energy, and (2) the surface of the water is not flat, so that potential energy is associated with the raising of some of the water above the flat surface and the lowering of some water below the surface. It is easiest to consider the potential energy in detail.

As illustrated in Figure 12.2, the wave is formed by raising the surface on the left side of the figure and depressing the surface on the right side. If the wave is modeled as an ideal sine wave, it can be created by merely flipping the water in the darker blue portion of the wave above the zero level to form the lighter blue portion. The area of the portion of the wave above the zero level is found by integrating a sine wave of wavelength λ and an amplitude over half a wavelength. This area is $A\lambda/\pi$. Thus, the mass of this half wave per unit length in the direction orthogonal to the plane of the figure (and the direction of propagation of the wave) is

$$\frac{m}{l} = \frac{\rho A \lambda}{\pi},$$ **(12.1)**

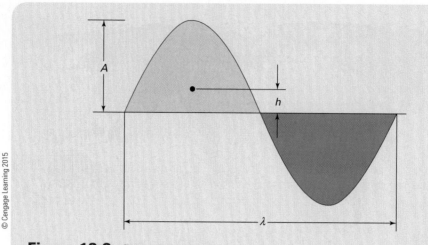

Figure 12.2: Cross section of a wave on the surface of the ocean.

where ρ is the water density. The center of gravity of this half wave is a distance h above the zero level (Figure 12.2). From some simple geometry, h can be calculated to be

$$h = \frac{\pi A}{8}.$$ (12.2)

Thus the formation of the wave from a flat surface by flipping over the lower portion of the wave to form the upper portion of the wave corresponds to moving a mass, as given in equation (12.1), vertically upward by a distance of $\Delta h = 2h$. The potential energy (per unit length of the wave front) associated with this process is

$$\frac{E}{l} = \frac{m}{l}g\Delta h = \left(\frac{\rho A \lambda g}{\pi}\right)\left(\frac{\pi A}{4}\right) = \frac{1}{4}\rho A^2 \lambda g.$$ (12.3)

It can be determined (although not very easily) that the kinetic energy of a water wave per unit length is equal to its potential energy per unit length. Thus the total energy associated with the wave is equal to twice that given in equation (12.3):

$$\frac{E}{l} = \frac{A^2 \lambda \rho g}{2}.$$ (12.4)

This is the energy content of one full wavelength of the wave. If a device is utilized to convert this wave energy into a usable form (e.g., electricity), then the energy content of one wavelength of the wave is incident on the device during one period of the wave. It can be shown that the period of a water wave, T, is related to its wavelength by the expression

$$\lambda = \frac{gT^2}{2\pi}.$$ (12.5)

Substituting this expression into equation (12.4) gives

$$\frac{E}{l} = \frac{A^2 g^2 \rho T^2}{4\pi}.$$ (12.6)

Since we are primarily concerned with the power generated, which is the energy per unit time, then the left-hand side of equation (12.6) is divided by the wave period to give the power per unit length of the wave front as

$$\frac{P}{l} = \frac{A^2 g^2 \rho T}{4\pi}. \qquad (12.7)$$

Note that in this expression the amplitude A, as shown in Figure 12.2, is one-half of the crest to trough height of the wave, H, i.e. $H = 2A$. Equation (12.7) is, therefore, often written in terms of the wave height as

$$\frac{P}{l} = \frac{H^2 g^2 \rho T}{16\pi}. \qquad (12.8)$$

Using the density of seawater ($\rho = 1025 \text{ kg/m}^3$) and appropriate values for the constants in equation (12.8), the power in kilowatts per meter is related to the wave height in meters, and the period in seconds and is expressed as

$$\frac{P}{l} = [1.96 \text{ kW/(m}^3\text{·s)}]H^2 T. \qquad (12.9)$$

Example 12.1

An ocean wave has a height of 2 m and a period of 10 seconds. What is the power available in 1 km of wave front?

Solution

From equation (12.9), the power in kilowatts per meter of wave front is

$$\frac{P}{l} = [1.96 \text{ kW/(m}^3\text{·s)}] \times (2 \text{ m})^2 \times 10 \text{ s} = 78.4 \text{ kW/m}.$$

For 1 km of wave front, the total power available is

$$P = 78.4 \text{ kW/m} \times (1000 \text{ m/km}) = 78.4 \text{ MW}.$$

Equation (12.5) also gives the wave's velocity. As it travels a distance of λ in time T, its velocity is

$$v = \frac{\lambda}{T} = \frac{gT}{2\pi}. \qquad (12.10)$$

Example 12.2

An ocean wave has a period of 8 seconds. What are its velocity and wavelength?

Solution

From equation (12.10), the velocity is given in terms of the period as

$$v = \frac{gT}{2\pi} = \frac{9.8 \text{ m/s}^2 \times 10 \text{ s}}{2 \times 3.14} = 15.6 \text{ m/s}.$$

From equation (12.5), the wavelength is

$$\lambda = \frac{gT^2}{2\pi} = \frac{9.8 \text{ m/s}^2 \times (10 \text{ s})^2}{2 \times 3.14} = 156 \text{ m}.$$

12.3 Wave Power Devices

Devices to extract energy from ocean waves can be located onshore, where they obtain energy from breaking waves, or they can be offshore and extract energy from waves as they propagate in the ocean. From a practical standpoint, facilities are best located on the shore or at least not far offshore. Facilities near shore can be connected via transmission cables, as is the case for offshore wind farms, and this has been found to be fairly practical (see the information later in the chapter about the Pelamis). Connection of onshore facilities is straightforward. Facilities that are far offshore need an appropriate mechanism for transporting the generated electricity back to shore. Energy storage mechanisms, such as batteries (Chapter 19) or hydrogen (Chapter 20), are either expensive or have low efficiency. Unfortunately, the efficiency of wave power devices is fairly low, and the actual (electrical) energy output is substantially less than the theoretical wave energy. For an offshore device, the situation is somewhat like that for a wind turbine; to extract all of the kinetic energy from a wave, the wave has to be stopped. For an onshore device, the problem is that waves lose energy as they approach shore due to interaction with the sea floor. At a water depth of 20 m, a typical wave has lost over 60% of the energy it had in deep water.

There are three main approaches to the development of devices for harnessing wave power: oscillating water columns are primarily onshore devices while floats and pitching devices, as well as wave focusing devices, are typically located offshore. These three technologies are discussed in detail below.

12.3a Oscillating Water Columns (OWCs)

Oscillating water columns are devices that are typically permanently attached to the shoreline or anchored close to shore. Onshore units have been constructed in Scotland, and near-shore units have been developed in Australia (Figures 12.3 and 12.4, respectively).

The basic design of the OWC is shown in Figure 12.5. The details of the generator assembly are shown in Figure 12.6. The device is positioned onshore or near-shore, and, because waves are incident on the device, the level of the water in the wave

Figure 12.3: Illustration of LIMPET the 500 kW oscillating water column constructed by Wavegen in Scotland.

Figure 12.4: Oscillating water column developed by Energetech in Australia that is anchored near shore, rated at 500 kW.

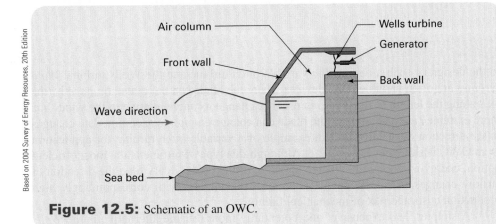

Figure 12.5: Schematic of an OWC.

chamber rises. When the height of the water column rises, then air is pushed out through a turbine, which drives a generator to produce electricity. When the water level in the wave chamber falls, then air is pulled in through the turbine. A typical wind turbine is designed so that, if air flows in one direction, the turbine rotates in one direction, and, if the air flows in the opposite direction, the turbine rotates in the opposite direction. This situation is acceptable for a traditional wind turbine because, if the direction of the wind changes, then this is a slow process, and the wind turbine can be rotated to ensure that the turbine always faces into the wind. The problem with the OWC is that the direction of air flow changes with the periodicity of the ocean waves, which might be about 10 seconds.

The practical solution to this problem is to design a turbine that always rotates in the same direction regardless of the direction of the air flow. There are two approaches

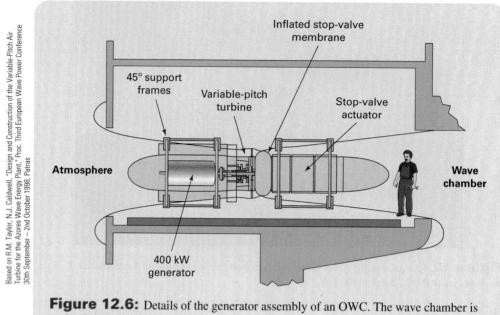

Based on R.M. Taylor, N.J. Caldwell, "Design and Construction of the Variable-Pitch Air Turbine for the Azores Wave Energy Plant," Proc. Third European Wave Power Conference 30th September – 2nd October 1998, Patras

Figure 12.6: Details of the generator assembly of an OWC. The wave chamber is to the right in the figure.

to the design of such a device: the variable pitch turbine and the Wells turbine. *Wind turbines* are often variable pitch devices where the blade can be rotated on its axis (the axis along the length of the blade) in order to change its angle relative to the wind. In a more extreme case, the angle of the blade can change enough to accommodate changes in the direction of the airflow. An example of a variable pitch turbine for application in an OWC is shown in Figure 12.7. Although this type of turbine is the most efficient option, energy is required to change the pitch of the blades every time the direction of airflow changes. Also, the device is somewhat complex, with correspondingly high production costs and risk of mechanical failure.

A simpler, less expensive, and potentially more reliable alternative is the *Wells turbine*, which was developed by Alan Wells of Queen's University, Belfast, in the 1980s. The trade-off is a potentially lower efficiency. In a Wells turbine (Figure 12.8), the blades are fixed and symmetric (somewhat almond-shaped in cross section, as shown in the figure), so that they appear the same from either air flow direction. The net resulting force on the blade causes a rotation in the same direction regardless of the airflow direction.

12.3b Floats and Pitching Devices

Devices of several different geometries fall into this category, and two characteristic designs are described in the section. These are anchored near shore, and the electricity that is generated can be transferred to shore through appropriate power lines. The most common of these devices is the *Pelamis* (pronounced pel-AH-mis), which is named after a genus of sea snake (because of its geometry). (*Note:* The genus Pelamis contains a single species, *P. platura*, which is highly venomous and

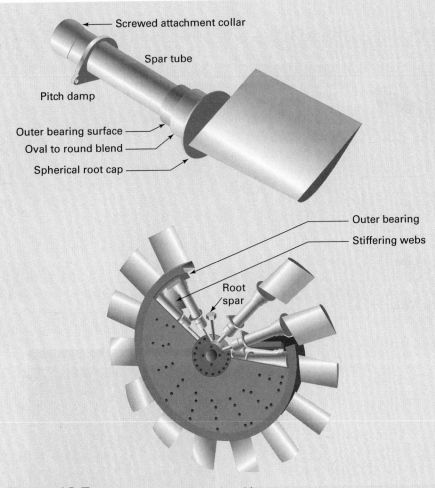

Based on R.M. Taylor, N.J. Caldwell "Design and Construction of the Variable-Pitch Air Turbine for the Azores Wave Energy Plant" Proc. Third European Wave Power Conference, 30th September–2nd October 1998, Patras

Figure 12.7: Design of a variable pitch turbine.

inhabits coastal regions in much of the Pacific and Indian Oceans.) An example of a Pelamis (the energy generator) is shown in Figure 12.9. These devices have been developed in Scotland and Portugal. They are typically about 3.5 m in diameter and 120 m long and are made in sections (typically four) that are connected by hinges. They are anchored in water that is typically about 50 m deep and are allowed to rotate so that they face in the direction of the waves. As the waves pass the sections of the Pelamis, it flexes at the hinges, giving rise to a snake-like appearance. Hydraulic cylinders associated with the hinges are used to pump hydraulic fluid through hydraulic turbines, which drive electric generators. Typical Pelamis generators are rated at 750-kW$_e$ output.

Other devices that have been tested in recent years are buoys that ride up and down in the waves. (Figure 12.10). These typically generate electricity either by pumping fluid through a turbine that is connected to a generator or by means of coils that move up and down in a magnetic field produced by permanent magnets.

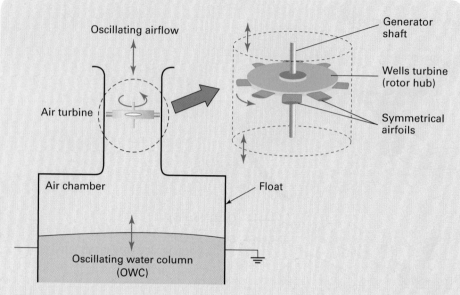

Based on M. Mamun et al., Ocean Engineering 31 (2004) 1423–1435

Figure 12.8: Design of a Wells turbine.

Figure 12.9: Pelamis prototype machine on-site at the European Marine Energy Centre off Orkney.

© David Fleetham/Alamy

Figure 12.10: Experimental wave energy buoy off Kaneohe Bay, Oahu, Hawaii. Portions of the device above and below the surface are shown.

ENERGY EXTRA 12.1
The design of a Pelamis

The Pelamis has been one of the most successful devices for harvesting wave energy. Its efficiency in converting wave energy into electricity depends on its ability to couple the wave motion to the mechanical motion of device. This coupling is optimized when the frequency response of the Pelamis (that is, its resonant frequency) is close to the frequency of the waves. Although the Pelamis can be designed so that its resonant

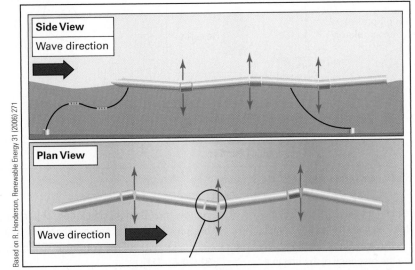

Based on R. Henderson, Renewable Energy 31 (2006) 271

Movement of a Pelamis in the vertical (top) and horizontal (bottom) directions.

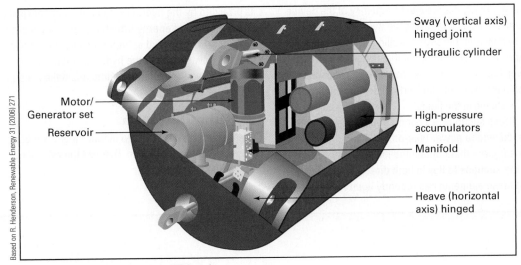

Based on R. Henderson, Renewable Energy 31 (2006) 271

Internal design of the Pelamis PTO unit.

Continued on page 320

Energy Extra 12.1 continued

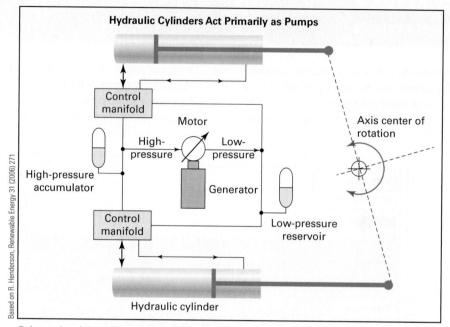

Based on R. Henderson, Renewable Energy 31 (2006) 271

Schematic of the PTO unit in a Pelamis.

frequency is near the average wave frequency, it is necessary to adjust the Pelamais frequency response to account for variations in the wave frequency. It is also necessary to vary the response of the unit in a cyclic way through the wave cycle in order to maximize energy transfer. Because the Pelamis generates electricity by using hydraulic cylinders to pump fluid through a turbine, the resonant frequency of the device can be adjusted by controlling the hydraulic pressure in the system and thereby controlling the resistance of the joints. A detailed picture of the motion of a Pelamis in waves is shown in the first diagram. The sections of the Pelamis flex in both the vertical and horizontal directions, as shown in the figure.

These sections are connected by shorter elements, called the power take-off (PTO) mechanisms (see the figure), that contain the hydraulic systems and allow the sections to flex in both directions. The design of these elements has recently been reviewed by

Henderson "Design, simulation, and testing of a novel hydraulic power take-off system for the Pelamis wave energy converter" [*Renewable Energy* **31** (2006): 271].

A simplified schematic of the hydraulic control system is shown in the above figure. The control manifolds ensure that optimal resistance to wave motion is provided by the hydraulic cylinders and maximize the efficiency of energy transfer. High-pressure accumulators store fluid at high pressure and provide a smooth flow of fluid to the hydraulic motor to even out any cyclic variability and random anomalies in electricity generation.

Topic for Discussion

The Pelamis is constructed in four hinged sections and has a total length of 120 m. Discuss the relevance of this geometry and size.

12.3c **Wave-Focusing Devices**

The Wave Dragon (Figure 12.11) is a wave-focusing device. The device is moored offshore, typically a few kilometers from land. Two arms act as curved reflectors to gather waves, which are directed to the sloping ramp at the far end of the device (Figure 12.12). The water collected in the reservoir runs back into the sea through turbines, which are used to run generators. Several prototype devices have been constructed off the coasts of the United Kingdom, Portugal, and Denmark. Because the device can be large, it harvests energy from a substantial length of wave front. Prototype and planned devices have outputs in the range of tens of megawatts electrical for moderate sea states.

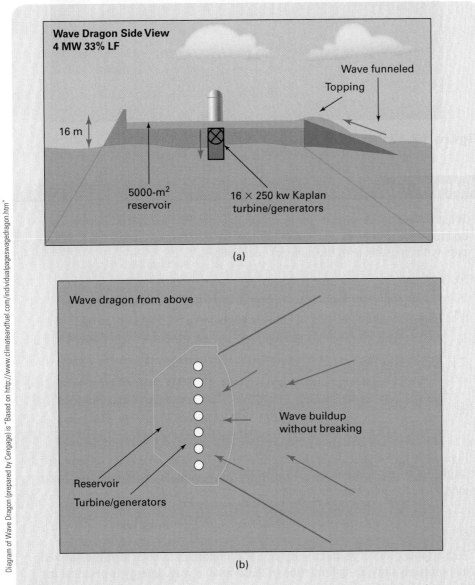

Diagram of Wave Dragon (prepared by Cengage) is "Based on http://www.climateandfuel.com/individualpageswagedragon.htm"

Figure 12.11: Schematic of the Wave Dragon: (a) side view and (b) top view.

Erik Friis-Madsen/Wave Dragon

Figure 12.12: Photograph of the Wave Dragon.

Example 12.3

What is the total average wave power available in Portugal? The country has a continental coastline of about 940 km. How does the annual wave energy compare with an estimate of the national energy needs?

Solution

Figure 12.1 shows an average wave power in the region around Portugal of 48 kW/m. For a total coastline of 940 km, the total power is

$$P = (48 \text{ kW/m}) \times (1000 \text{ m/km}) \times (944 \text{ km}) = 4.53 \times 10^7 \text{ kW}.$$

The per-year energy is

$$E = (4.53 \times 10^7 \text{ kW}) \times (8760 \text{ h/y}) = 3.97 \times 10^{11} \text{ kWh},$$

or

$$E = (3.9 \times 10^{11} \text{ kWh}) \times (3.6 \times 10^6 \text{ J/kWh}) = 1.43 \times 10^{18} \text{ J}.$$

Figure 2.4 shows an average per-capita energy use of around 150 GJ for most southern European countries. Portugal, with a population of 10.6 million, has a total energy requirement of about

$$(10.6 \times 10^6) \times (150 \times 10^9 \text{ J}) = 1.59 \times 10^{18} \text{ J}.$$

These values are very similar, although it is important to note, as just discussed, that probably only a small fraction of the available wave energy is practical for utilization.

12.4 **Wave Energy Resources**

At present all wave power devices are prototypes or have had limited commercialization. The future development of this energy resource requires a consideration of the extent of wave energy. Certainly, it is a renewable resource with a lifetime that is, for all practical purposes, infinite. However, as shown in Figure 12.1, it is a resource that is best exploited in a limited number of locations around the world. It is estimated that the average deepwater wave power worldwide is in the range of 10^{12}–10^{13} W. This is similar to, or somewhat less than the current world power requirements. Since all of this wave power cannot be exploited, it cannot meet our total needs. An ambitious effort to utilize wave power with current or foreseeable technology could make up to perhaps a few percent of the available wave power commercially viable. Because wave energy resources occur in specific geographical areas, it is a resource that can be locally significant. An analysis of wave power places production costs in the 5–10 cents per kilowatt-hour range, which, according to Figure 9.22, make it reasonably competitive with other options.

Like wind power, wave power is subject to daily and seasonal fluctuations, as well as random fluctuations due to changing meteorological conditions. Thus, this is best used as a source of energy in conjunction with storage systems and/or other energy resources, as discussed in Chapter 18.

Wave power has minimal environmental impact. The Wells turbines used in most onshore OWCs tend to be noisy and need to be acoustically insulated, particularly in populated areas. Offshore wave farms, consisting of devices like the Pelamis, Aquabuoy or Wave Dragon must be cognizant of navigational considerations. Aesthetically, wave farms are probably less conspicuous than offshore wind farms.

12.5 **Summary**

Wave energy is an extensive resource, comparable to our total energy needs. However, only a small fraction of wave energy can be extracted in a viable manner. This chapter described the geographic distribution of wave resources that may be of commercial interest. If viable resources are fully utilized, wave energy could account for a few percent of our energy requirements. Because waves are variable and somewhat unpredictable, wave energy is most appropriately utilized in conjunction with other resources and a suitable energy storage system.

This chapter considered the basic physics of the power available from ocean waves and demonstrated that the total power per unit length (l) of a wave is $\dfrac{P}{l} = \dfrac{H^2 g^2 \rho T}{16\pi}$, where H is the wave height, and T is the wave period. The dependence of power on the square of the wave height indicates that the majority of energy available occurs during periods of above-average wave conditions.

The chapter described the three current technologies for harnessing wave energy: oscillating water columns, floats and pitching devices, and wave-focusing devices. The oscillating water column uses wave motion to force air through a turbine to drive a generator to produce electricity. The most common turbine for this purpose is the bidirectional Wells turbine.

Floating and pitching devices, such as the Pelamis, have probably been the most extensively developed of the wave energy–harvesting technologies. These devices use the mechanical movement of the wave to operate turbine/generators to produce electricity. Wave-focusing devices, such as the Wave Dragon, use ramps to convert the kinetic energy of waves into gravitational potential energy, which can then be used to drive turbines.

Although the technology for the conversion of wave energy to electricity is relatively straightforward, the development of commercial devices is still in its early stages. Portugal and Scotland have been the most active countries in developing this resource. In the long term, wave energy may be economically competitive with other alternative energy technologies. Overall, its environmental impact is relatively low, and safety and security are not significant concerns.

Problems

12.1 North Carolina has some of the highest average wave energies in the United States along its roughly 500 km of coastline (Figure 12.1). Would it be reasonable for North Carolina to utilize wave energy for all of its electricity? Assume that a total annual per-capita electricity requirement of 15 MWh$_e$ and that 5% of the coastline is used for wave electricity generation at an average efficiency of 20%.

12.2 A small boat with two occupants and a 15-kW outboard motor (total mass = 500 kg) moves up and down in a wave with a height of 1 m and a period of 10 seconds. Calculate the ratio of the average wave power lifting the boat from the trough of the wave to the crest of the wave to the power available from the motor.

12.3 An ocean wave has a velocity of 8.5 m/s. What are its period and wavelength?

12.4 An ocean wave has a height of 2.5 m and a period of 10 s. What is the total power available from 20 m of wave front for this wave?

12.5 The average wave power available in Hawaii (Figure 12.1) is 100 kW/m. For waves with a period of 10 s, what is the average wave height?

12.6 In a storm, waves may have a height of 12 m and a period of 15 s. Compare the power per meter of wave front to the average values shown in Figure 12.1.

12.7 In deepwater, tsunamis have a relatively small height (typically 2 m) and a very long period (typically 30 minutes). The amplitude becomes larger as they pile up when they reach shallow water. For a tsunami in deep water, calculate the energy per meter of wave front (in J/m). Compare this with the energy per meter of width of a 2-m wide, 1500-kg automobile traveling at 120 km/h. This comparison emphasizes the damage that can be caused by a tsunami.

12.8 A wave travels at 10 m/s and has a height of 2 m. If it is incident on a wave energy device that generates electricity from wave energy with an average efficiency of 20%, how large would the device have to be to generate 1 MW$_e$?

Bibliography

G. Boyle (Ed.). *Renewable Energy.* Oxford University Press, Oxford (2004).

R. Cohen. "Energy from the ocean." *Philos. Trans. Royal Soc. London* **A307** (1982): 405.

A. Goldin. *Oceans of Energy—Reservoir of Power for the Future.* Harcourt, Brace, Jovanovich, New York (1980).

M. Knott. "Power from the waves." *New Scientist* **179** (2003): 2473.

M. McCormick . *Ocean Wave Energy Conversion.* Wiley, New York (1981).

T. R. Penney and D. Bharathan. "Power from the sea." *Scientific American* **256** (1987): 86.

D. Ross. *Energy from the Waves.* Pergamon Press, Oxford (1979).

R. J. Seymour. *Ocean Energy Recovery: The State of the Art.* American Society of Civil Engineers, New York (1992).

World Energy Council. *2010 Survey of Energy Resources*, available online at http://www.worldenergy.org/documents/ser_2010_report_1.pdf.

Tidal Energy

Learning Objectives: After reading the material in Chapter 13, you should understand:

- The reasons for tidal motion and resonance effects in enclosed basins.
- The energy associated with tidal movement.
- The design of barrage systems.
- The availability and utilization of tidal energy based on barrage systems.

- Environmental and other factors that affect the viability of barrage systems.
- The design of tidal current energy systems.
- Experimental and commercial tidal current systems.

13.1 Introduction

In addition to waves, energy is associated with the movement of water in the ocean because of tides. Tidal motion is much more predictable than wave activity, and the realization that the energy associated with this movement could be harnessed dates back to well before 1000 CE. In medieval Europe, barrages were constructed to trap tidal water and subsequently release it through waterwheels. Like hydromechanical energy harnessed from flowing rivers, this tidal energy was used to perform tasks such as grinding grain. The interest in using tidal energy to generate electricity on a commercial scale originated in the 1960s as part of the growing desire to develop nonfossil fuel energy sources. The first (and still functional) commercial tidal electricity generating facility grew out of this interest and opened in 1966. Although tides exist everywhere in the world's oceans, a sufficiently large tidal range is necessary to make tidal energy practical. These conditions exist only in a few locations on earth. This is particularly true in the United States where tidal energy opportunities are very scarce. Nonetheless, interest in tidal energy as a alternative to fossil fuels has grown, and there are now several new tidal energy initiatives in North America and elsewhere. This chapter reviews the historical, scientific and engineering aspects of tidal power development and the progress that has been made in this area in recent years.

13.2 Energy from the Tides

The ocean tides are the result of the gravitational interaction of the ocean's water with the sun and the moon. This source of energy, along with nuclear and geothermal, are the only energy sources available that are not in some way a direct manifestation of solar radiation. Tidal energy may be available in the form of gravitational potential energy resulting from the raising and lowering of the ocean water level, or it may be kinetic energy associated with currents flowing through channels as a result of changing water levels. There has been increased interest during the past half century or so in the development of tidal power. It has, however, been used since the eleventh century in England and France, where the use of watermills, powered by tidal water motion paralleled the use of traditional waterwheels utilizing flowing rivers.

The tides arise primarily from the gravitational force acting between the oceans and the moon. Water on the side of the earth facing the moon is closer to the moon than the water on the other side of the earth, and is therefore more strongly attracted to the moon. This causes the water to form a bulge toward the moon. The water on the side of the earth away from the moon is less strongly attracted to the moon and bulges in the opposite direction. This behavior is illustrated in Figure 13.1. As the earth rotates, this tidal bulge moves around the earth with a period of half the *lunar day* (that is the time it takes the earth to return to the same position relative to the moon). This effect gives rise to the *diurnal period* of the tides, which is about 12 hours and 25 minutes.

Tidal behavior is complicated by the presence of the sun, which also has a gravitational effect on the oceans. This effect is significant, although smaller than that of the moon. When the moon is in the new phase or full phase, the earth, moon, and sun align in a single line (Figure 13.1). In this case, the solar tide adds to the lunar tide, giving rise to an increased tidal range. This situation is referred to as a *spring tide*. When the moon is in its quarter phase, it lies at right angles to the line between the earth and the sun (Figure 13.1). In this case (known as a *neap tide*), the lunar and solar effects do not

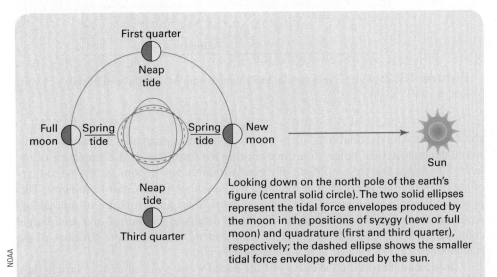

Looking down on the north pole of the earth's figure (central solid circle). The two solid ellipses represent the tidal force envelopes produced by the moon in the positions of syzygy (new or full moon) and quadrature (first and third quarter), respectively; the dashed ellipse shows the smaller tidal force envelope produced by the sun.

Figure 13.1: The solar and lunar components of the tides for different phases of the moon.

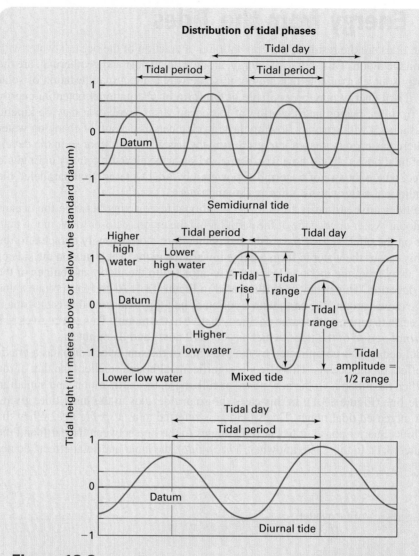

Figure 13.2: Water levels for diurnal (bottom) semidiurnal (top) and mixed (center) tides.

work together and the tidal range is not as great. During a neap tide, the earth's oceans would experience tidal bulges four times during the lunar day. If the amplitude of the lunar tidal bulge and the amplitude of the solar tidal bulge are similar, the periodicity of the tide would be one-fourth of the lunar day, or about 6 hours and 13 minutes. This is referred to as a *semidiurnal tide*. In most cases, tides are a mixture of diurnal and semidiurnal, as shown in Figure 13.2. The exact details of the diurnal and semidiurnal components of the tide at a particular location depend on the relative positions of the sun and the moon and on the specific coastal geography.

Interest in harnessing the energy of the tides for the production of electricity has, until fairly recently, dealt with the utilization of the gravitational potential energy associated with the rising and falling water level. This approach has been primarily aimed at using enclosed basins where resonance conditions (as described later in the chapter) make the tidal range particularly large. Particular sites around the world that satisfy this

© Cengage Learning 2015

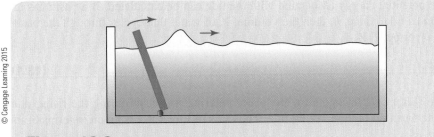

Figure 13.3: Generation of waves in a tank of water.

Richard A. Dunlap

Figure 13.4: Low tide at Parrsboro, Nova Scotia, on the Bay of Fundy, illustrating the extreme tide conditions experienced at this location.

condition are described in detail in the next section. The tidal range in an enclosed basin results from the tidal period and the geometry of the basin. Increased tidal range occurs as a result of resonance effects. This behavior can be illustrated using the properties of water in a tank with a mechanism (a paddle) for making waves. If a wave is produced as shown in Figure 13.3, it will travel to the far end of the tank and reflect back to the paddle. The period of time it takes to travel this distance depends on the properties of the water and the geometry (e.g., length) of the tank and is referred to as the *resonant period*. If a second wave is created just as the first wave is returning to the paddle, the wave amplitudes will add together. Thus, if waves are produced at the resonant period of the tank, then a resonance effect occurs, and the wave amplitude is very large. If the resonant period of an ocean basin is near the tidal period, then this effect gives rise to an increased tidal amplitude. This is what happens, for example, in the Bay of Fundy where the resonant period of the bay is about 13.3 hours compared to the tidal period of about 12.4 hours, leading to a very large tidal range of about 13 m. These extreme tides are illustrated in Figure 13.4.

The potential energy associated with the tide can be calculated. If a basin has an area, A, and a tidal range, h, then the volume, V, of water that cycles through the basin every tidal period, T, is

$$V = 2Ah. \tag{13.1}$$

where the factor of 2 accounts for the water entering the basin during the rising tide leaving the basin during the falling tide. The mass of this water is given in terms of its density, ρ, as

$$m = 2Ah\rho. \tag{13.2}$$

The potential energy available as a result of the tidal movement is

$$E = 2Ahg\rho\left[\frac{h}{2}\right] = Ah^2g\rho, \tag{13.3}$$

where $h/2$ is the change in height of the center of mass of the water in the basin between high tide and low tide, and g is the gravitational acceleration. If this energy is available over the tidal period, it will correspond to an average power of

$$P = \frac{Ah^2g\rho}{T}. \tag{13.4}$$

It is seen in this equation that having a large area is important for the production of power, but having a large tidal range is even more important.

Example 13.1

Calculate the average tidal power that is available from a basin that is 10 km wide and 25 km long if the tidal range is 10 m.

Solution
The average available power is given by equation (13.4) as

$$P = \frac{(10 \times 25 \times 10^6\,\mathrm{m}^2) \times (10\,\mathrm{m})^2 \times (9.8\,\mathrm{m/s}^2) \times (1025\,\mathrm{kg/m}^3)}{(12.4\,\mathrm{h} \times 3600\,\mathrm{s/h})} = 5.6\ \mathrm{GW},$$

equivalent to several large coal-fired or nuclear power plants (Chapters 3 and 6).

For reasons to be described, however, only about a quarter of this is actually accessible; that is a capacity factor of about 25%.

13.3 Barrage Systems

Barrage systems make use of tidal energy in the manner just described. The barrage is basically a dam that closes off a basin (sometimes called a *headpond*). *Sluice gates* in the barrage are doors that can be opened or closed-to allow the water to enter or leave the basin. In a simple depiction (Figure 13.5), the sluice gates are opened as the tide

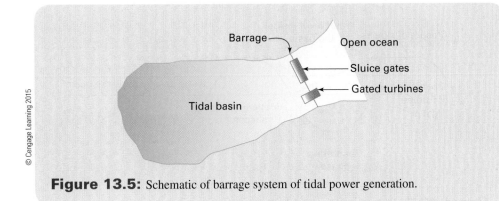

Figure 13.5: Schematic of barrage system of tidal power generation.

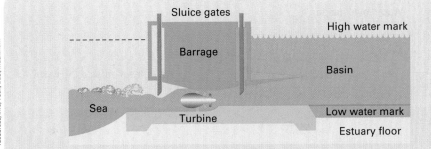

Figure 13.6: Schematic of barrage design showing turbines. Gates are used to allow water in and out of the basin and to direct water through the turbine.

is rising to allow the basin to fill from the ocean. When the tide is high and the basin is filled, the sluice gates are closed. At this point, the turbine gates are closed, and the water level in the ocean drops as the tide goes out. When the tide is low, the turbine gates are opened to allow the water to flow through the turbines and back into the ocean, thereby generating electricity. The output of the turbines depends on the head of water enclosed in the basin, and the turbines are designed so that the water flow rate through the turbines (and hence the electric power generated) is compatible with the timescale of the tidal period. A simplified diagram of the barrage system in cross section is shown in Figure 13.6.

The main problem with this simple scheme is that, as the water is flowing out of the basin, the water level in the ocean is rising due to the rising tide. Thus, as the tide rises, the effective head of the water in the basin is decreasing, and when the water levels become equal on the two sides of the barrage, no further power is generated.

One way of improving on this simple scheme is shown in Figure 13.7. This is known as the *ebb generation scheme*. In this scheme, the basin is allowed to fill, and then the turbine gates are opened when the tide in the ocean is still on the way down, and electricity is generated until the tide rises again to the level of the falling water in the basin. This scheme, shown in the figure, illustrates that electricity is generated from a point where the tide is about half low until it is half high again.

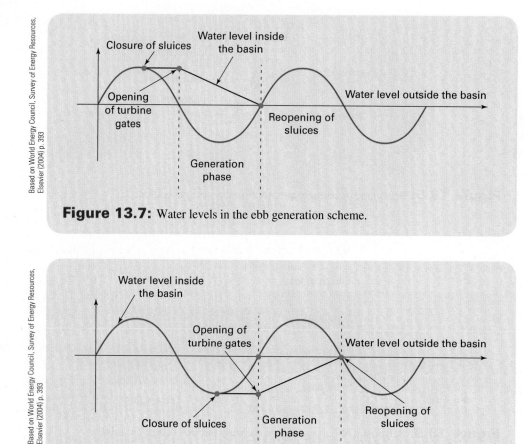

Figure 13.7: Water levels in the ebb generation scheme.

Figure 13.8: Water levels in the flood generation scheme

An alternative approach is to close the sluice gates at low tide and allow the water to rise in the ocean before opening the turbine gates to allow the water into the basin. This situation (Figure 13.8) is referred to as the *flood generation scheme*. In general, the ebb generation scheme is more efficient than the flood generation scheme because, as a result of the typical shape of a basin, more water is contained in the top half than in the bottom half. Also, any rivers that flow into the basin will augment ebb generation but reduce flood generation. A simple analysis of Figures 13.7 and 13.8 provides a rough estimate of the capacity factor. Both illustrations show that power generation occurs over only one-half of the total tidal period; that is, water is flowing through the turbines only when the water level in the basin is changing linearly. In addition, it is seen that the total range of the water level inside the basin is one-half of the total range of the water level in the ocean. Thus, combining these two factors of one-half yields a typical capacity factor of 25%.

The implementation of barrage systems worldwide has been minimal. At present there are two major operating facilities, although a few other small prototype systems have also been built. The largest operational tidal power plant (Figure 13.9) is the Rance River Tidal Generating Station in Bretagne, France. Located on the estuary of the Rance River at its outflow into the English Channel between Dinard and Saint-Malo (Figure 13.10), it became operational in 1966. The barrage is 750 m in length

Figure 13.9: Barrage at the Rance River Tidal Generating Station in France.

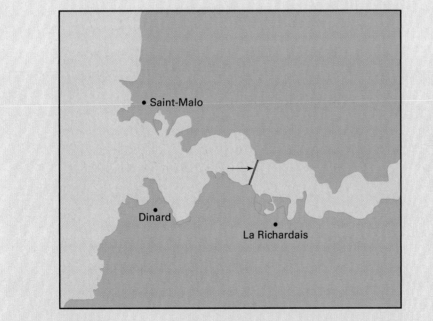

Figure 13.10: Location of the Rance River Tidal Power Station in France.

and accommodates a roadway across the river. It encloses a basin of 22.5 km² with a mean tidal range of about 8 m. The barrage contains 24 *bulb turbines* of the type shown schematically in Figure 13.11. These are similar to Kaplan turbines (Figures 11.4 and 11.5). The maximum capacity of the station is 240 MW$_e$ but the actual average output is 68 MW$_e$, consistent with about a 25% capacity factor.

The second largest tidal power plant, at Annapolis Royal in Nova Scotia, Canada, became operational in 1984. This location is shown on the maps in Figures 13.12 and

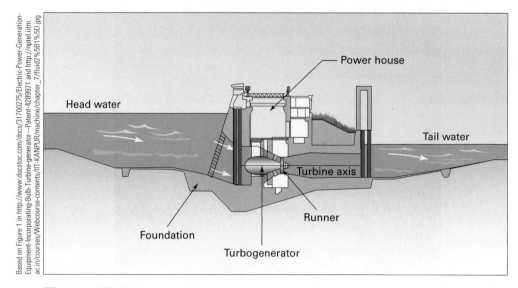

Figure 13.11: Bulb turbine used for a tidal power plant.

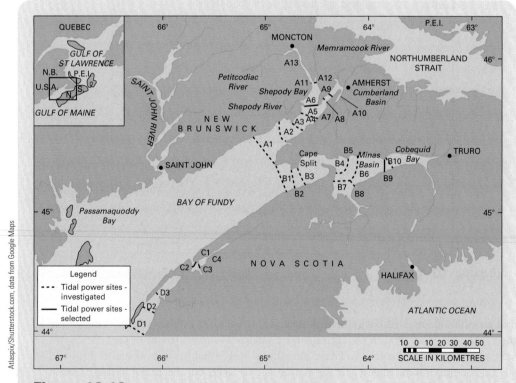

Figure 13.12: The Bay of Fundy showing the location of the tidal generating station at Annapolis Royal (site C4) and other possible tidal power sites.

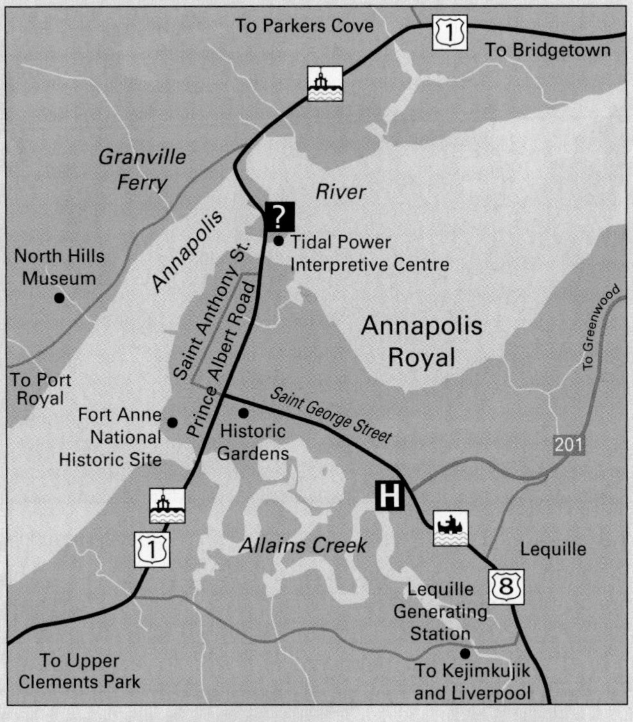

Figure 13.13: Detailed map showing location of the Annapolis Royal tidal generating station on the Annapolis River.

13.13. The basin or headpond is formed upstream by the barrage. The facility is shown in the photograph in Figure 13.14, and the details of the sluice gates (which are large concrete doors that can be lowered and raised) are shown in Figure 13.15. The barrage also serves as a causeway for vehicular traffic. The generating station has a maximum capacity of 20 MW$_e$, and typically generates electricity for about 5 hours during each tidal cycle. A *rim turbine* design (Figure 13.16) is used in this facility and reduces maintenance costs compared with bulb turbines.

At present, the fraction of tidal power resources that have been utilized is very small. Certainly there exists the possibility of implementing large-scale systems similar to that in France (or larger) at a number of locations worldwide. If tidal range and basin area are considered, then a dozen or more clearly identifiable sites are available for possible exploitation (Table 13.1). Even considering a capacity factor of about 25%, most of the sites in Table 13.1 would provide as much power as a major nuclear or fossil fuel generating station. In fact, some would provide substantially more. The Penzhinsk location has the potential to satisfy the electricity needs of a population of 6 million or more (at typical North American usage).

It is interesting to consider why development of these resources has not been pursued. The basic technology is straightforward, and in most ways, is similar to that used for hydroelectric power. However, large-scale tidal barrage systems are a substantial engineering challenge (as explained later in this chapter). Also, the longevity of tidal power resources and their environmental impact need to be investigated thoroughly.

Figure 13.14: Annapolis Royal, Nova Scotia tidal generating station. The photograph was taken from the north side of the Annapolis River looking southwest (see Figure 13.13). The headpond is in the foreground, and the exit to the ocean is behind the barrage.

Figure 13.15: Sluice gates at the Annapolis Royal tidal power station.

Constructing a barrage across an ocean basin has some similar effects to constructing a major hydroelectric installation. The natural flow of water is disrupted. Consequences can include effects on navigation, fish and mammal movements, and the transportation of sediments. In the case of the Rance River station, locks have been constructed to allow small craft to pass from one side to the other, although changes in the marine ecology have been observed since the facility was constructed. The effects on the transport of sedimentation are analogous to the collection of silt behind

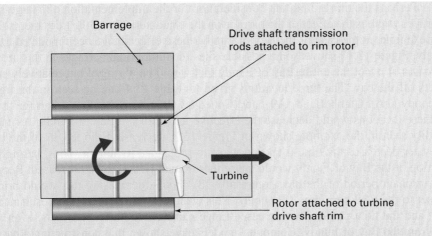

Figure 13.16: Schematic diagram of a rim turbine as used at the Annapolis Royal tidal power station. In this arrangement, the rotating turbine shaft is connected by rods to the generator, which is in the form of a ring outside the region that is exposed to the marine environment. This facilitates maintenance and reduces adverse environmental effects.

Table 13.1: Possible sites for the development of tidal power using a barrage system. The table is not necessarily comprehensive but includes all major possibilities for barrage systems with a maximum installed capacity of 1000 MW or greater.

country	location	mean tidal range (m)	basin area (km²)	maximum capacity (MW)
Argentina	San Jose	5.8	778	5040
	Golfo Nuevo	3.7	2376	6570
	Santa Cruz	7.5	222	2420
	Rio Gallegos	7.5	177	1900
Australia	Secure Bay	7.0	140	1480
	Walcott Inlet	7.0	260	2800
Canada	Cobequid	12.4	240	5338
	Cumberland	10.9	90	1400
	Shepody	10.0	115	1800
India	Gulf of Khambhat	7.0	1970	7000
United Kingdom	Severn	7.0	520	8640
Russia	Mezen	6.7	2640	15,000
	Tuigar	6.8	1080	7800
	Penzhinsk	11.4	20,530	87,400

hydroelectric dams, as discussed in Chapter 11. The Bay of Fundy is an interesting case in this respect because the region is very muddy and sedimentation collection on the basin side of the barrage may substantially reduce the lifetime of the resource and contribute to ecological changes.

Perhaps the most important consideration for the implementation of a large-scale barrage system is the effect of the barrage on the characteristics of the tides themselves. For example, in the case of the Bay of Fundy, these effects have been modeled extensively. Figure 13.12 shows some possible sites for future tidal barrages in this area. In all cases, it is not the entire Bay of Fundy that would be barraged but relatively small inlets off the bay. The three locations given in Table 13.1 can be seen in the figure; Shepody (A6), Cumberland (A9), and Cobequid (B9). Constructing a barrage at any of these locations would decrease the effective length of the Bay of Fundy. As can be readily seen in the example shown in Figure 13.3, decreasing the length of the body of water decreases the time it takes a wave to propagate its length. Thus, barraging a portion of the Bay of Fundy would cause its resonant period to be decreased. Because the resonant period of the bay is currently 13.3 hours, decreasing this would bring it closer to the tidal period of 12.4 hours. This change would increase the resonance effect and the tidal range. The increased range can have significant effects, even very far from the Bay of Fundy. Computer models have shown that constructing a barrage across Cumberland Basin (site A8) would give rise to an increased tidal range of about 3 cm in Boston. Constructing a barrage across Cobequid Bay (site B9) would result in a corresponding 13-cm increase in tidal range in Boston. Even very moderate changes in mean ocean water level can have severe consequences for coastal environments and marine structures.

Finally, the length of the necessary barrage needs to be considered as an important factor in the economic viability of this approach. For example, a barrage across Cobequid Bay would be 8 km in length, while a barrage across the Severn Estuary in England would be 17 km in length. The Gulf of Khambhat in India would require a barrage of 25 km long. Feasibility studies of the Severn barrage system have estimated the infrastructure cost to be around US$30 billion. Certainly these are major undertakings from the standpoints of both construction and maintenance.

For a variety of reasons, barrage systems are less appealing than they may have been in the past. The focus on tidal energy is now aimed at systems that do not involve barrages. Some of these approaches are discussed in the next section.

Example 13.2

For a generation efficiency of 25%, calculate the total tidal energy available during one tidal cycle for Cobequid Bay in Canada.

Solution

Use equation (13.3),

$$E = Ah^2g\rho,$$

and the following values:

$$A = 240 \text{ km}^2 = 2.4 \times 10^8 \text{ m}^2.$$
$$h = 12.4 \text{ m}.$$
$$g = 9.8 \text{ m/s}^2.$$
$$\rho = 1025 \text{ kg/m}^3.$$

Continued page 339

Example 13.2 continued

The total energy is

$$E = (2.4 \times 10^8 \text{ m}^2) \times (12.4 \text{ m})^2 \times (9.8 \text{ m/s}^2) \times (1025 \text{ kg/m}^3) = 3.7 \times 10^{14} \text{ J}.$$

At an efficiency of 25%, this provides $(0.25) \times 3.7 \times 10^{14} = 9.3 \times 10^{13}$ J. Dividing by 3.6×10^6 J/kWh gives about 2.6×10^7 kWh. The average power can be determined from the energy by dividing the tidal period as

$$(3.7 \times 10^{14} \text{ J}) \times \frac{10^{-6}\text{MJ/J}}{12.4 \text{ h} \times 3600 \text{ s/h}} = 8290 \text{ MW}.$$

This calculated output power may be compared with the actual expected value of 5338 MW from Table 13.1.

13.4 Nonbarrage Tidal Power Systems

Several approaches to the utilization of tidal power that do not require a barrage system have been investigated. The more significant options include tidal lagoons, which are artificial enclosures that utilize the same principles as barrage systems, and underwater turbines and tidal fences, which make use of the kinetic energy associated with tidal currents.

13.4a Tidal Lagoons

A tidal lagoon is an artificial enclosure located up to a couple of kilometers offshore in a region where there is a large tidal range. Although, in principle, the operation of a tidal lagoon is similar to that of a barrage system, it avoids many of the potential environmental consequences that result from blocking off an ocean basin with a barrage. It does not significantly affect the resonant period of the basin, and adverse effects due to sedimentation are minimized. Also, navigation and aquatic animal movements are not greatly influenced. Bidirectional flow generation (i.e., generation during both ebb and flood cycles) can be much more easily implemented than in traditional barrage systems. This ability is largely due to the more regular shape of the lagoon compared with a natural basin. Typically, the design also eliminates the need for separate sluice and turbine gates. The water levels in the ocean and inside the lagoon, as a function of tidal period, are shown in Figure 13.17, where periods of ebb and flood generation are illustrated. Bidirectional generation is expected to give about a 35% capacity factor and is predicted to be a cost-effective improvement over single-direction ebb generation. This improvement in capacity factor can be roughly seen by combining part of the ebb generation period in Figure 13.7 with part of the flood generation period in Figure 13.8. An illustration of a typical (proposed) tidal lagoon installation is shown in Figure 13.18, where possible turbine locations allowing energy generation during water flow into or out of the lagoon are indicated. The calculated power output during the tidal cycle is shown in Figure 13.19. A prototype tidal lagoon is in the planning stages at Swansea off the coast of Wales. The mean annual tidal range in this region is 6.3 m, and the proposed facility would have a lagoon area of about 5 km^2. Twenty-four 2.5-MW$_e$ turbines

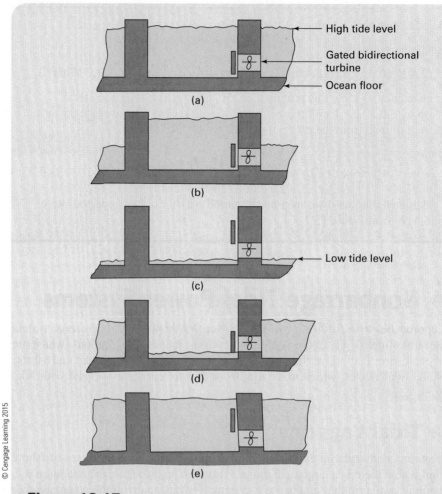

Figure 13.17: Simple bidirectional energy generation scheme for a tidal lagoon.
(a) High tide–turbine gates closed.
(b) Falling tide–head created inside lagoon. Turbine gates opened when tide is about half
    way down to generate electricity.
(c) Low tide–turbine gates closed as tide begins to rise.
(d) Rising tide–head created outside lagoon. Turbine gates opened when tide is about
    half way up to generate electricity.
(e) High tide–turbine gates closed when tide is high to return to diagram (a).

would give a maximum output of 60 MW$_e$. A larger, 300-MW$_e$, tidal lagoon system is
in the planning stages in China.

13.4b Underwater Turbines

Underwater turbines make use of the kinetic energy associated with tidal currents and
rely more on current velocity and volume than actual tidal range. The same technology is applicable to the harvesting of energy associated with any type of ocean current.
The basic design follows along the lines of the design of a wind turbine. Analogous to

Figure 13.18: Illustration of typical proposed tidal lagoon power generation system.

Courtesy of Aqua-RET, www.aquaret.com

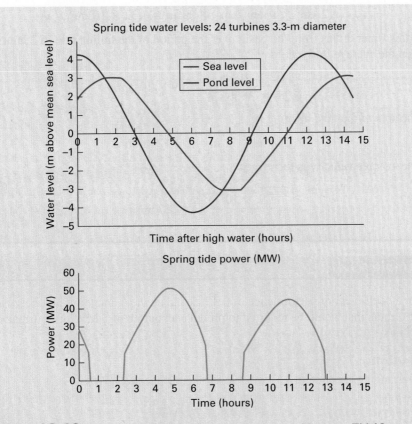

Based on Tidal Lagoon Power Generation Scheme in Swansea Bay. April 2006. A report on behalf of the Department of Trade and Industry and the Welsh Development Agency

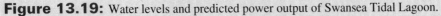

Figure 13.19: Water levels and predicted power output of Swansea Tidal Lagoon.

equation (10.3) for wind energy, the power per unit area (P/A) generated by a water turbine is given as

$$\frac{P}{A} = \frac{1}{2}C\rho v^3,$$

where C is referred to as the coefficient of performance and is a measure of the turbine efficiency. Typical C values are in the range of 0.3 to 0.4. ρ is the water density, and v is the current velocity. Current velocities are typically less than wind velocities, and velocity appears as v^3 in the equation. However, the difference between the density of air (1.204 kg/m^3) and the density of seawater (1025 kg/m^3) is the most significant difference between the analysis of the energy content of currents and wind. Even relatively small (5- to 10-m diameter) water turbines are rated in the megawatt range for moderate tidal currents.

Example 13.3

A basin of area 12 km^2 with a tidal range of 11 m drains through an opening 200 m wide. What is the average water velocity during the falling tide cycle?

Solution

The total volume of tidal water in the basin is

$$(12 \text{ km}^2 \times 10^6 \text{ m}^2/\text{km}^2) \times 11 \text{ m} = 1.32 \times 10^8 \text{ m}^3.$$

If this water drains from the basin over the period of the falling tide (6.2 hours), then the average flow rate is

$$\frac{1.32 \times 10^8 \text{ m}^3}{6.2 \text{ h} \times 3600 \text{ s/h}} = 5.9 \times 10^3 \text{ m}^3/\text{s}.$$

The area of the tidal opening is

$$200 \text{ m} \times 11 \text{ m} = 2.2 \times 10^3 \text{ m}^2.$$

The average velocity is therefore

$$\frac{5.9 \times 10^3 \text{ m}^3/\text{s}}{2.2 \times 10^3 \text{ m}^2} = 2.7 \text{ m/s}.$$

Example 13.4

For a coefficient of performance of 0.35, calculate the diameter of an underwater turbine that would be required to produce an output of 5 MW in a current of 3.5 m/s.

Solution

Rearranging equation (13.5) to solve for the turbine area gives

$$A = (2P)/(C\rho v^3)\frac{2P}{C\rho v^3}.$$

Continued on page 343

Example 13.4 continued

Using values of

$$P = 5 \times 10^6 \text{ W},$$
$$C = 0.35,$$
$$\rho = 1025 \text{ kg/m}^3, \text{ and}$$
$$v = 3.5 \text{ m/s}$$

gives an area of

$$\frac{2 \times 5 \times 10^6 \text{ W}}{0.35 \times 1025 \text{ kg/m}^3 \times (3.5 \text{ m/s})^3} = 650 \text{ m}^2,$$

giving a diameter of

$$d = (4 \times 650 \text{ m}^2/\pi)^{1/2} = 29 \text{ m}.$$

A number of devices along these lines have been tested. The simplest, perhaps, are bladed devices that are similar to modern wind turbines (Figures 13.20 to 13.22). With these devices, it is straightforward to track the direction of the current, as is done for the wind direction with wind turbines by rotating the rotor assemblies about a vertical axis to ensure that the blades always face into the direction of the current. Other turbine geometries that are similar to rim turbines (Figure 13.23) are also being investigated. It

Figure 13.20: Underwater turbine constructed by Seagen at Strangford Lough. Turbine blades are in their operating position below the water and the support tower is shown above the surface. Compare with Figure 13.21.

Bloomberg via Getty Images

Figure 13.21: Illustration of the Seagen underwater turbine with rotors raised above the water's surface for maintenance.

Courtesy of Atlantis Resources Corporation

Figure 13.22: Atlantis' AR 1000 turbine.

is possible that *shrouded turbines*—that is, turbines with an enclosure to direct water toward the blades—may be effective for water devices. In a suitable location, a number of individual tidal turbines of, say, 1-MW output could be combined into a tidal energy farm, much as wind farms have been implemented.

Figure 13.23: The rotor for the OpenHydro system being transported. Note the workers in red to the right of the rotor for scale.

ENERGY EXTRA 13.1
The SeaGen tidal turbine in Strangford Lough

The Seagen tidal turbine (Figures 13.20 and 13.21) was installed in the 500-m wide mouth of Strangford Lough in Northern Ireland (see the following map) in 2008. The twin rotor turbine was constructed by Marine Current Turbines (MCT) at a cost of about £8.5 million and has a total rated output of 1.2 MW. Each of the 16-m diameter rotors is attached to a drive system with a mass of 27 t at the end of a transverse arm. The arm assembly is mounted on a 3-m diameter tower, which is attached to the seabed by a single piling. The total device has a mass of about 1000 t.

During the maximum tidal current of about 3.6 m/s, the rotors rotate at about 14 rpm, corresponding to a tip speed of about 12 m/s. Because of the difference in density of air and water, the tidal turbine is most efficient at a tip speed of about one-third that of a typical modern wind turbine (Chapter 10). The rotor assembly can be rotated on a vertical axis around the tower by 180 degrees so that the turbine is equally functional for both tidal current directions. Because of the predictable nature of the tides, this 1.2-MW tidal current generator actually produces about the same total energy output integrated over time as a wind turbine rated at 2.5 MW.

During normal operation (Figure 13.20), the water surface is about 3 m above the tips of the rotors, even at low tide. Thus, the turbine has no effect on navigation for small boats. The twin rotor assembly can be raised on the central tower (Figure 13.21), making maintenance straightforward. Studies have suggested that Seagen has minimal impact on marine animals because of the relatively low speed of the rotors.

The Seagen project is the first successful commercial tidal current generator, and the electricity that it generates is sufficient to fulfill the needs of over 1000 homes. It has led to the planning of further tidal current initiatives in the United Kingdom over the next few years, including a proposed tidal current farm off the coast of Scotland. It has also renewed interest in this technology in other parts of the world.

Continued on page 346

Energy Extra 13.1 continued

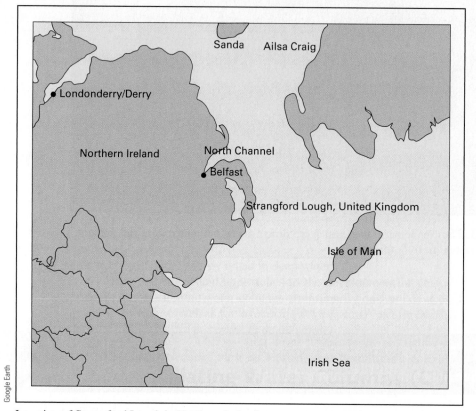

Google Earth

Location of Strangford Lough in Northern Ireland.

Topic for Discussion

A public utility determines its cost ($/kWh) for generating electricity at a particular facility by dividing the total cost of the facility over its lifetime by the total number of kilowatt-hours the facility produces. The latter includes capital infrastructure costs, operation and maintenance costs, fuel costs, and financing costs (if any). The latter is based on rated capacity and the overall capacity factor. This analysis is well established for coal-fired stations or nuclear plants, for example. Discuss how this analysis might be the same or different for a tidal turbine farm and where there may be different certainties and uncertainties in the various factors.

There has been significant activity in the development of tidal energy in the United Kingdom, Australia, and Canada, and several tidal turbines have been constructed or are in the planning stages. The facility shown in Figure 13.20 is located off the coast of Northern Ireland and has operated commercially since 2008 and is rated at 1.2-MW_e output. Canadian prototype systems are planned for the near future for British Columbia, and some prototype testing has been undertaken in the Bay of Fundy. The environmental impact of water turbines is expected to be minimal.

ENERGY EXTRA 13.2
Tidal energy in the United States

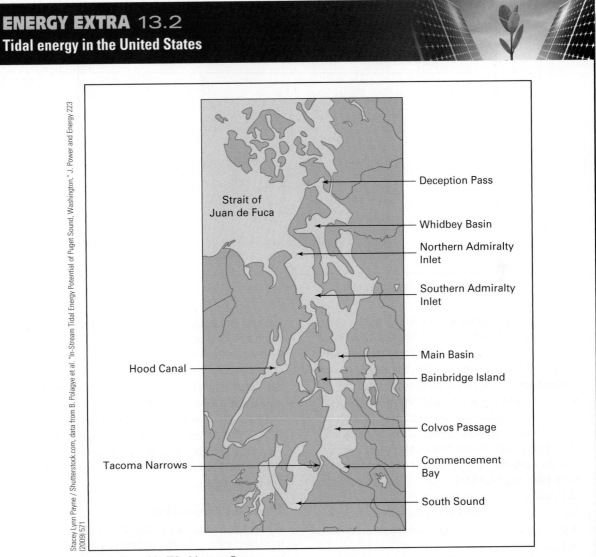

Stacey Lynn Payne / Shutterstock.com, data from B. Polagye et al. "In-Stream Tidal Energy Potential of Puget Sound, Washington." J. Power and Energy 223 (2009) 571

Puget Sound in Washington State.

While the most significant tidal energy resources worldwide are located outside the United States, a few sites within the United States are under consideration for development of this energy resource. The three locations that have attracted the most interest are Puget Sound in Washington State, Cook Inlet in Alaska, and the Bay of Fundy in Maine.

Puget Sound is a complex estuarine system that connects to the Pacific Ocean through the Strait of Juan de Fuca, as shown in the following map. Several organizations have investigated possible tidal

generation in Puget Sound. Tacoma Power has looked at installing an in-stream turbine at Tacoma Narrows but has concluded that this is not commercially viable at the present. The U.S. Navy has shown interest in tidal power in Puget Sound, but their investigations are at the very early stages. The most advanced investigations are by the Snohomish County Public Utility District and have considered sites at Admiralty Inlet and Deception Pass (where tidal currents can exceed 4 m/s) and plan to deploy two underwater turbines in the near future.

Continued on page 348

Energy Extra 13.2 continued

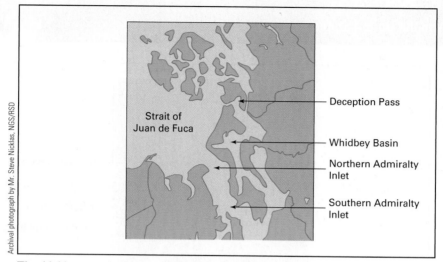

Archival photograph by Mr. Steve Nicklas, NGS/RSD

The tidal bore on the Turnagain Arm of Cook Inlet.

Cook Inlet has the second largest tidal range (after the Bay of Fundy) in North America. It has been estimated that 90% of the tidal energy resources in the United States occur in Alaska. The Turnagain Arm on Cook Inlet is one of only a few dozen locations worldwide where a tidal bore occurs. A tidal bore is a wave-like structure that is formed in a tidal river or narrow inlet by the leading edge of the incoming tide (see photograph). Ocean Renewable Power Company is in the early stages of projects at two locations in Cook Inlet, one near Anchorage and one near the town of Nikiski. If these projects move to commercialization, the electricity generated will help to offset Alaska's dependence on electricity generated from local fossil fuel resources.

The most advanced tidal energy efforts in the United States are in Maine. Maine borders the Bay of Fundy, along with New Brunswick and Nova Scotia, Canada, and experiences tidal ranges that can be as much as 8 m. Since 2006, pilot projects run by Ocean Renewable Power Corporation have been in place in Cobscook Bay near the Maine–New Brunswick border. These projects have involved tidal stream generators, and it is expected that up to 3 MW of commercial generating facilities could be installed by around 2014–2015. Additional collaborative projects are underway in collaboration with the Canadian company, Fundy Tidal Power, to utilize the energy potential of this border region.

Topic for Discussion

Would it be reasonable to expect that a tidal energy farm consisting of 1-MW_e turbines located in Cook Inlet could provide all the necessary electric power for Anchorage, Alaska? How many turbines would be needed?

13.4c Tidal Fences

Water turbines are horizontal axis machines. Like the Darrieus rotor used for wind power generation, water turbines can also be vertical axis machines. An advantage of this type of design over the horizontal axis machine is that the gearbox and generator can easily be located above water, where it is much less susceptible to adverse environmental conditions. It is anticipated that a number of such turbines could be combined to span the mouth of an estuary or basin. Such a construction, referred to as a *tidal fence*, is shown in Figure 13.24. Although prototype vertical axis machines have been built, full-scale implementation of a tidal fence is still in the future. The Dalupiri Passage in

Blue Energy

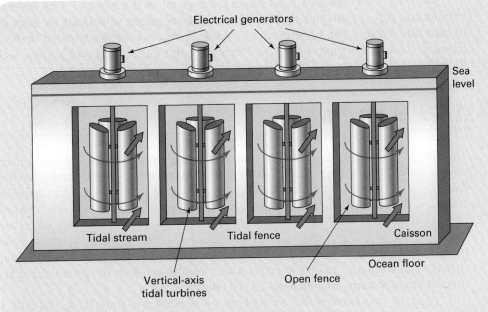

Figure 13.24: Illustration of proposed tidal fence.

Courtesy of Alternative Energy Tutorials, www.alternative-energy-tutorials.com

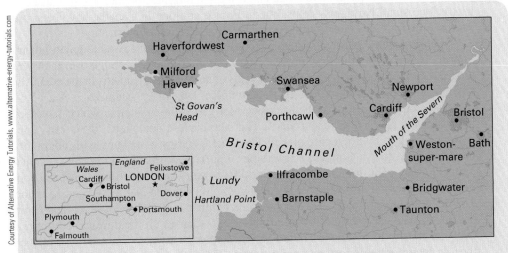

Figure 13.25: Location and geography of the Mouth of Severn in England.

the Philippines is a possible site for such development and could have a maximum capacity of 2200 MW_e. Capacity factors can be as high as about 50% for such installations as power generation is bidirectional and it is not necessary to wait for a tidal basin to fill and/or empty before generation begins. The Severn Estuary (or Mouth of Severn) in England (Figure 13.25) is another site where a tidal fence may be a viable alternative to a traditional barrage system. The tidal fence can also serve as a causeway for vehicles. Environmental and navigational concerns for tidal fences may be more serious than for horizontal axis water turbines, either isolated or in tidal power farms, and careful consideration of any possible adverse environmental impacts is necessary.

If properly implemented, tidal power can be a renewable resource with low environmental impact. Although the tidal energy in the oceans is substantial, the number of locations where it can be utilized as a viable energy source is somewhat limited. However, new, developing technologies have met with success and may lead the way to further implementation of this energy source.

13.5 Summary

The tides in the earth's oceans are the result of the combined gravitational forces from the sun and the moon. This chapter has reviewed the properties of tidal energy and has discussed the possible approaches to harnessing this energy. Although the tidal amplitude in most locations on earth is relatively small, the tidal range becomes amplified in certain enclosed basins as a result of resonance effects. These resonance effects are caused by the similarity of the tidal period and the natural oscillation period of the basin. In these areas, the possibility of utilizing tidal energy is the greatest. This chapter showed that, analogous to hydroelectric power from rivers, both the potential and kinetic energy associated with the tides can be utilized.

A barrage system uses the potential energy associated with the rising and falling tide to generate electricity. For example, at high tide, the water inside an enclosed basin is trapped by a gate. At low tide, when the sea level outside the basin has dropped, the water inside the basin can be released and can drive turbine/generators to produce electricity. Two small barrage systems, one in France and one in Canada, have been used successfully for many years for the commercial generation of electricity.

Unfortunately, barrage systems can have significant undesirable environmental impact. Manipulating the flow of tidal water can affect the tidal range over a very extended region and can also alter the life cycles of biological organisms in the area. Also, silt deposits can adversely affect the functioning of a barrage system.

This chapter also described tidal current generating systems, which have been more actively pursued in recent years. This approach is analogous to the run-of-the-river approach to hydroelectric power. The kinetic energy associated with tidal movement is harnessed using underwater turbines. This type of system minimizes the impact on the environment. Experimental devices have been tested, and there has been some limited commercial utilization of this technology in Canada and the United Kingdom. The tidal current approach also increases the number of possible locations that can be developed. The harvesting of tidal current energy focuses on the development of relatively small individual devices, rather like the approach toward wind energy, which can then be incorporated into tidal energy farms. Tidal energy development in the future is likely have fairly local, rather than global, influences on energy production. This is because, like wave energy, it is confined to a fraction of the coastal areas worldwide.

Problems

13.1 Calculate the total kinetic energy (in MJ and in kWh) of a 1-m^3 parcel of seawater moving with a velocity of 1 m/s.

13.2 What is the total tidal energy available (at 100% efficiency) during the falling tide from a basin of area 100 km^2 with a tidal range of 8 m?

13.3 Using the tidal range and basin area for Penzhinsk inlet given in the chapter, estimate the monetary value of the electrical energy that can be generated (at 25% efficiency) during one falling tide cycle. Assume electricity has a value of $0.10/kWh.

13.4 A water turbine 15 m in diameter is placed in a channel with a tidal current moving with a velocity of 3.5 m/s. Estimate the power produced by the turbine.

13.5 What is the power available from an underwater turbine with a coefficient of performance of 0.4 that is 10 m in diameter in a current of seawater traveling at a velocity of 2.0 m/s?

13.6 For a tidal current with a velocity of 1.5 m/s in a channel that is 0.5 km wide and 28 m deep (on average), calculate the mass of water per unit time moving through the channel.

13.7 Calculate the average current velocity that is necessary for an 8-m diameter water turbine with a coefficient of performance of 0.3 to generate 1 MW.

13.8 The capacity factor of a barrage system may be defined in terms of the actual maximum power output compared with the theoretical maximum, as given by equation (13.4). Using data from Table 13.1, show that the four proposed barrage generating facilities proposed for Argentina would operate at similar capacity factors.

Bibliography

A. C. Baker. *Tidal Power*. Peter Peregrinus, London (1991).

G. Boyle (Ed.). *Renewable Energy*. Oxford University Press, Oxford (2004).

R. Charlier. *Tidal Energy*. Van Nostrand Reinhold, New York (1992).

R. Charlier. *Ocean Energies: Environmental, Economic, and Technological Aspects of Alternative Power Source*. Elsevier, New York (1993).

R. H. Clark. "Tidal power." *Annual Review of Energy* (1997): 2648.

R. H. Clark. "Tidal power." In *Wiley Encyclopedia of Energy Technology and the Environment*, Vol. 4. A. Bisio and S. Boots (Eds.). Wiley, New York (1995).

R. H. Clark (Ed.). *Tidal Power: Trends and Developments*. Thomas Telford, London (1992).

R. Cohen. "Energy from the ocean." *Philos. Trans. Royal Soc. London* **A307** (1982): 405.

M. W. Conley and G. R. Daborn (Eds.). *Energy Options for Atlantic Canada*. Formac Publishing, Halifax (1983).

A. Goldin. *Oceans of Energy—Reservoir of Power for the Future*. Harcourt, Brace, Jovanovich, New York (1980).

D. A. Greenberg. "Modeling tidal power." *Scientific American* **257** (1987): 128.

T. J. Hammons. "Tidal power." *Proc. IEEE*, **8** (1993): 419.

T. R. Penney and D. Bharathan. "Power from the sea." *Scientific American* **256** (1987): 86.

R. J. Seymour. *Ocean Energy Recovery: The State of the Art*. American Society of Civil Engineers, New York (1992).

World Energy Council. *2010 Survey of Energy Resources*, available online at http://www.worldenergy.org/documents/ser_2010_report_1.pdf

Ocean Thermal Energy Conversion and Ocean Salinity Gradient Energy

Learning Objectives: After reading the material in Chapter 14, you should understand:

- The distribution of thermal energy in the oceans.
- The use of a heat engine to extract energy from the oceans and convert it into electricity.
- The design of the different types of OTEC systems.

- Experimental OTEC facilities and their performance.
- The basic principles of osmotic energy production.
- Experimental facilities for osmotic energy production.

14.1 Introduction

Several options for obtaining energy from the kinetic or potential energy associated with the movement of water in the oceans have been discussed in the past two chapters. Two additional methods of extracting energy from the oceans are discussed in this chapter. The first, *ocean thermal energy conversion* (OTEC), is a method of making use of the thermal energy in the ocean. It is basically a heat engine, like the heat engine used to convert the thermal energy of steam produced by burning fossil fuels or from nuclear reactions into mechanical energy, which can be used to generate electricity. Another option is to make use of energy associated with the chemical composition of seawater. The removal of salt from seawater to produce freshwater (i.e., desalination) requires energy. Conversely, energy can be extracted when seawater and freshwater are mixed. Energy obtained by this method is referred to as salinity gradient energy, or *osmotic energy*.

14.2 Basic Principles of Ocean Thermal Energy Conversion

The conversion of thermal energy in the ocean to mechanical energy depends on the availability of a cold reservoir into which excess heat can be transferred. This situation exists in the ocean because water temperature is a function of depth. In deep water, the depth dependence of the temperature is shown in Figure 14.1.

The figure shows that, below about 1000 m, the temperature is fairly constant at about 4°C. This is independent of the location on earth, the time of the year, and the air temperature. In temperate and colder regions, particularly in the winter, the temperature difference between deep water and surface water is relatively small. In tropical regions, particularly in the summer, the temperature difference between deep water and surface water is large. The OTEC process relies on a temperature gradient between the surface and the very deep water in the ocean and is suited only to tropical regions. Figure 14.2 shows the mean annular temperature difference between the ocean surface and water at 1000 m depth for various locations on earth. A temperature difference of at least 20°C (and preferably as large as possible) is necessary to make use of OTEC.

In an OTEC facility, cold water is pumped from deep in the ocean. Heat may then be transferred from the warm surface water to the cold water obtained from the ocean's

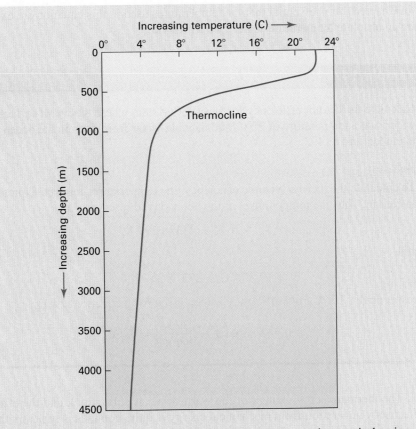

Based on http://www.windows2universe.org/earth/Water/temp.html

Figure 14.1: Typical depth variation of the ocean temperature in a tropical region.

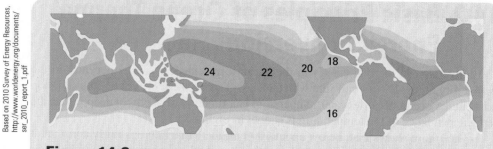

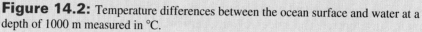

Figure 14.2: Temperature differences between the ocean surface and water at a depth of 1000 m measured in °C.

depths. According to Figure 1.4, the warm surface water acts as the hot reservoir, and the cold water acts as the cold reservoir, and the transfer of heat allows for the extraction of mechanical energy. This process, as in all heat engines, is governed by the laws of thermodynamics, and the maximum efficiency that can be obtained is limited by the Carnot efficiency (Chapter 1):

$$\eta = 100\left(1 - \frac{T_c}{T_h}\right). \tag{14.1}$$

As in Chapter 1, it is essential that the temperatures used in this equation are expressed using an absolute temperature scale, such as Kelvin.

Example 14.1

Calculate the Carnot efficiency for an OTEC system where the warm surface water is at a temperature of 22°C, and the cold water from deep in the ocean is at a temperature of 4°C.

Solution

To calculate the thermodynamic efficiency, the temperatures must be expressed in Kelvin. Thus the temperatures of the hot and cold reservoirs are

$$T_h = 26°C + 273 = 299 \text{ K}$$

and

$$T_c = 4°C + 273 = 277 \text{ K},$$

respectively. The Carnot efficiency is then given from equation (14.1):

$$\eta = 100 \times \left(1 - \frac{277 \text{ K}}{299 \text{ K}}\right) = 7.4\%.$$

The thermal energy content of the warm water is substantial, and the efficiency of electricity generation from mechanical energy is quite good, as is typically the case for a generator. However, the low Carnot efficiency, as calculated in Example 14.1, is a factor that must be considered in the design and implementation of an OTEC system.

Example 14.2

Calculate the maximum amount of energy that can be extracted from 1 m^3 of seawater by cooling from 20°C to 12°C. Note the specific heat of seawater is about 3930 J/(kg·°C) and its density is 1025 kg/m^3.

Solution

From equation (8.12), the relationship between the change in temperature, ΔT, and heat, Q, is

$$Q = \Delta TCm,$$

where C is the specific heat, and m is the mass. In this case,

$$\Delta T = (20°C - 12°C) = 8°C, C = 3930 \text{ J/(kg·°C)}, \text{ and } m = 1025 \text{ kg.}$$

Thus

$$Q = (8°C) \times [3930 \text{ J/(kg·°C)}] \times (1025 \text{ kg}) = 32.2 \text{ MJ/m}^3.$$

The possibility of extracting useful energy by the OTEC process was first proposed in 1881 by the French physicist Jacques Arsene d'Arsonval. The first operational OTEC system was constructed by Georges Claude in Mantanzas Bay, Cuba, in 1930. It produced 22 kW of electricity, but unfortunately, due to its low thermodynamic efficiency, it consumed more than that to operate. It is important to understand the operation of an OTEC plant, as described in the next section, in order to assess the possibilities of improving on this design.

14.3 OTEC System Design

14.3a Open Cycle Systems

There are basically three types of OTEC systems: *open-cycle systems, closed-cycle systems, and hybrid systems.* The open-cycle system (Figure 14.3) is the simplest, and uses seawater itself as the working fluid.

Warm water from near the ocean surface enters the system on the left side of the diagram and is vaporized in an evaporator by lowering the vapor pressure with a pump. The details of this process become evident in the phase diagram of water in Figure 14.4. At a pressure of 1 atm, the transition line from the liquid (water) region to the vapor (steam) region (that is, the boiling point) occurs at 100°C. If the pressure is lowered, the transition line from liquid to vapor, as illustrated in the figure, shows that the boiling point occurs at a lower temperature. For a sufficiently low pressure, the water boils at (or below) room temperature. So when the warm seawater enters the evaporator (Figure 14.3), the pressure is lowered sufficiently to cause it to become vapor. This vapor drives a turbine (which, in turn, drives a generator to produce electricity) and then enters into a condenser where it is turned back into a liquid. The condenser is cooled by the cold seawater pumped from deep in the ocean, and this lowers the temperature of the vapor to a temperature below the liquid–vapor transition line. Generally, such systems utilize equal quantities of warm surface water and cold deep-ocean water. Ideally,

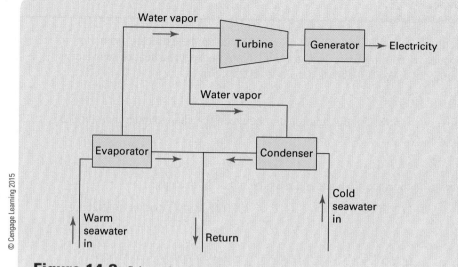

Figure 14.3: Schematic of an open-cycle OTEC plant.

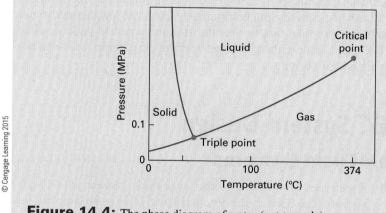

Figure 14.4: The phase diagram of water (not to scale).

the energy produced is determined by the energy extracted from the vapor by the turbine, which is related to the temperature difference between the evaporator and the condenser, and by the thermodynamic efficiency of the system, which is determined from the difference between the hot reservoir (warm surface water) and the cold reservoir (cold deep-ocean water). However, efficiencies are never ideal, and, in addition, pumps to bring water from deep in the ocean to the surface require substantial energy input.

A detailed diagram of the operation of an open-cycle OTEC is shown in Figure 14.5. As the figure shows, a by-product of energy production is desalinated water. This is produced by the condensation of water vapor and can be used to provide freshwater, which is often a valuable resource in locations where OTEC plants may be viable.

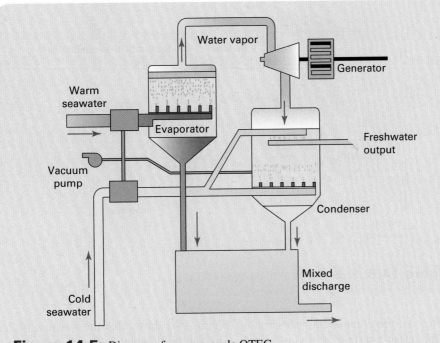

Figure 14.5: Diagram of an open-cycle OTEC.

Example 14.3

Consider an OTEC system that operates in a region where the warm surface water is at 26°C and the cold reservoir is at 4°C (Example 14.1). The temperature of the evaporator is at 20°C, and the condenser is at 12°C. Calculate the net (ideal) energy gain from 1 m³ of water. Make the simple approximation that 1 m³ (1025 kg) water input yields 1025 kg vapor through the turbine.

Solution

From Example 14.2, it is seen that the energy available from cooling 1 m³ (or 1025 kg) of water from 20°C to 12°C is 32.2 MJ. From Example 14.1, it is seen that the ideal Carnot efficiency for the reservoir temperature difference given is 7.4%. The net energy extracted is therefore

$$(32.2 \text{ MJ/m}^3) \times (0.074) = 2.4 \text{ MJ/m}^3.$$

For reasons described in the text, this simple calculation significantly overestimates the energy produced by an OTEC system.

14.3b Closed-Cycle Systems

A schematic diagram of a closed-cycle OTEC system is shown in Figure 14.6. In this system, the working fluid is a substance with a phase diagram that is compatible with the temperatures and pressures present in the system. These substances, such as ammonia,

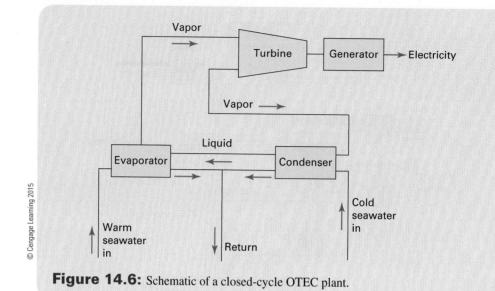

© Cengage Learning 2015

Figure 14.6: Schematic of a closed-cycle OTEC plant.

are often used in commercial refrigerators. Figure 14.6 shows that the warm seawater from the ocean's surface is used to heat the working fluid, which is in a closed-system, to a temperature above its boiling point. The vaporized fluid is used to drive the turbine, after which it is condensed back into a liquid by cooling it with cold seawater pumped from deep in the ocean. A more detailed diagram of a closed-cycle OTEC is shown in Figure 14.7. Closed-cycle systems have the disadvantage that leaks may release hazardous materials to the environment.

14.3c Hybrid Systems

The operation of a hybrid system (Figure 14.8) follows along the lines of the operation of a closed-cycle system. However, the warm seawater used to vaporize the working fluid is itself vaporized. The vaporized working fluid runs the turbine, as in the closed-cycle system, and the vaporized water is recondensed to produce desalinated water. Thus the system works as a closed-cycle system with the added value of producing freshwater.

14.4 Implementation of OTEC Systems

OTEC systems can be constructed in three types of locations: onshore, mounted to the ocean floor, or floating.

In the case of onshore installations, the facility is located on land or in very shallow water close to shore in a region where appropriate ocean thermal gradients exist. Warm surface water is supplied from the region near the facility, while a supply pipe carries cold water from deeper water offshore. This type of facility has clear advantages and disadvantages. On the plus side, the plant itself is easy to maintain. It is less susceptible to the adverse marine environment, and it is simple to connect the electric output to the power grid. In general, it is important to avoid mixing output water with the warm water supply. However, for land-based systems, the freshwater produced from the condensed working fluid can be readily used as a supply of quality drinking water,

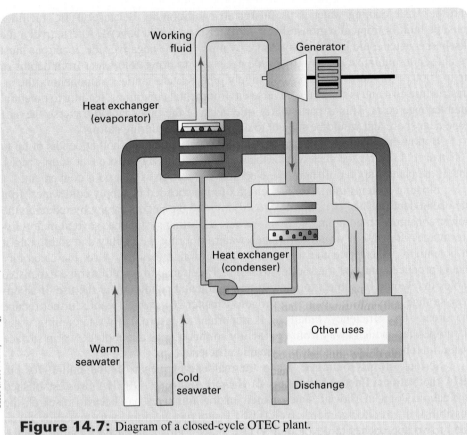

Figure 14.7: Diagram of a closed-cycle OTEC plant.

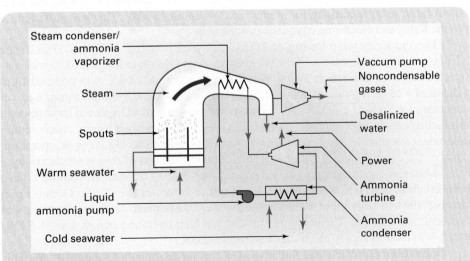

Figure 14.8: Schematic diagram of hybrid OTEC plant.

and the waste cooling water is useful for air conditioning. Because OTEC facilities must be built in tropical regions, this cold water is an added benefit. Alternatively, the discharge water could be carried offshore by pipe. On the negative side, locations must be chosen to minimize the length of pipe necessary to bring cold water from depths of up to 1000 m (or more) offshore. The supply pipe is a substantial component of the infrastructure cost and is subject to adverse environmental conditions, leading to potential maintenance costs. This is particularly true because the pipes would traverse the surf zone near shore and would be subject to extreme stresses during storms.

Bottom-mounted systems can be seated on the continental shelf at depths of up to 100 m or so. Locating the plant outside the surf zone minimizes stress on the supply pipes, and, if the plant is not far offshore, the electrical connection to the grid is convenient.

Floating systems minimize the length of pipes needed to supply cold water from the ocean depths. This, in principle, would improve the net efficiency by reducing the power consumed in pumping water. However, a floating system has several major disadvantages. Because it would be in deep water, mooring the facility and stabilizing it in potentially adverse weather conditions is not straightforward. Also, the electricity that is produced must be transported to shore. This transport would involve expensive cables that would be subject to adverse environmental conditions or the use of an energy storage mechanism [such as hydrogen (Chapter 20)], which would further reduce the already low efficiency. Finally, it is important to ensure that waste cooling water is pumped far enough away from the facility so that it does not cool the warm surface water, thereby reducing the thermodynamic efficiency.

Serious efforts to utilize ocean thermal energy began in the mid-1970s. In 1974, the National Energy Laboratory of Hawaii Authority (NELHA) was established at Keahole Point in Hawaii. One of their major initiatives has been to research the possibility of producing energy by OTEC. Prototype facilities have generally taken one of two approaches to developing OTEC power; either an onshore facility, which minimizes operational and maintenance costs, or portable ship-based plants. The latter optimizes efficiency without incurring substantial infrastructure costs.

In 1979, NELHA launched the *MiniOTEC*, a 50-kW$_e$ OTEC facility mounted on a barge. The barge was moored about 2 km off the coast of Hawaii near Keahole Point. The closed-cycle OTEC system produced a gross electric output of 52 kW$_e$ but a net output (after using electricity for pumps, etc.) of about 15 kW$_e$. This was sufficient to run the lights and electronic equipment on the ship.

The most extensive testing of OTEC occurred between 1992 and 1998 at the NELHA facility in Hawaii. A 250-kW$_e$ open-cycle OTEC plant (Figure 14.9) was constructed on shore at Keahole Point. Water at a temperature of 6°C was pumped from a depth of 820 m at a rate of about 400 L/s (0.4 m^3/s = 6400 gal/min) through a 1-m diameter pipe. In May 1993, 50 kW$_e$ of net power was produced by this plant. Although this facility is no longer operating as an OTEC plant to produce electricity, it does pump cold water on shore. This is used locally for air conditioning and, at peak operation, replaces the equivalent of traditional air conditioning requiring 200 kW of electricity.

Although Japan has no OTEC possibilities itself, Japanese researchers have been involved in OTEC activities on the island of Nauru in the Pacific Ocean. From October 1981 to September 1982 a closed-cycle OTEC plant, rated at 100 kW$_e$, was operated on the island. The net power production after operating pumps was 31.5 kW$_e$ which was used to supply electricity to a local school.

It has been estimated that globally OTEC resources amount to about 10^{13} W. This is close to our total world power requirements. However, only a small fraction of this resource is economically viable or practical. Onshore or close-to-shore sites are the

Figure 14.9: Land-based 250-kW open-cycle OTEC demonstration plant at Keahole Point, Hawaii.

easiest to implement. There are probably a few hundred sites, at most, that fulfill the necessary criteria for exploitation. The principal locations with cold water resources within 10 km or so of land are summarized in Table 14.1. Many of these locations are islands where the use of OTEC could provide a substantial contribution to local energy needs and reduce the need for energy importation.

Certainly this discussion illustrates the problems facing the development of OTEC as an energy resource. Thus far, prototype plants have produced a small net energy output. However, infrastructure and operating costs, at this time, make OTEC commercially unattractive. Technical difficulties in the efficient implementation of OTEC are significant. Fundamental problems that need to be dealt with include the intrinsically low efficiency of the process and the difficulties associated with the marine environment. These difficulties include the corrosive nature of seawater and the probability of biofouling in the system. The transport of very large quantities of seawater is also a challenge. A full-scale plant producing, say, 100 MW_e would require pumping something in the order of 1000 m^3 of seawater per second through the system.

However, OTEC energy production does provide a number of advantages. It has relatively low environmental impact and produces freshwater as a by-product as well as cold water for air conditioning use. Aquaculture is also an added benefit of OTEC. Cold water from the ocean depths has a large concentration of organic nutrients that are depleted in the surface water by biological processes, and this water may be used effectively for aquaculture. Also, the low temperature of this water opens up the possibilities for aquaculture in tropical regions that include the growth of fish (e.g., salmon) and invertebrates (e.g., lobsters) that are normally native to temperate regions—rather the opposite of geothermal aquaculture discussed in the next chapter. In some ways, OTEC may be considered as a viable resource for reasons other than energy production, with any generated electricity a bonus.

Predictions for the utilization of OTEC have often been far from accurate. A forecast made in 1980 (Figure 14.10) illustrates the optimism with which this resource was

Table 14.1: Principal locations for the utilization of on shore ocean thermal energy conversion.

area	country	temperature difference (°C)
Africa	Sao Tome	22
Latin America/Caribbean	Barbados	22
	Cuba	22–24
	Dominica	22
	Dominican Republic	21–24
	Grenada	27
	Haiti	21–24
	Jamaica	22
	Saint Lucia	22
	Saint Vincent	22
	Trinidad/Tobago	22–24
	U.S. Virgin Islands	21–24
Indian Ocean/Pacific	Comoros	20–25
	Cook Islands	21–22
	Fiji	22–23
	Guam	24
	Kiribati	23–24
	Maldives	22
	Mauritius	20–21
	New Caledonia	20–21
	Philippines	22–24
	Samoa	22–23
	Seychelles	21–22
	Solomon Islands	23–24
	Vanuatu	22–23

© Cengage Learning 2015

viewed. The actual cumulative capacity at a point close to the right-hand side of the graph is identically zero. Overall OTEC is an interesting, low-environmental-impact energy technology that unfortunately suffers form low efficiency. It is unclear whether the complex technological challenges facing OTEC can be overcome to make this resource economically viable.

14.5 Ocean Salinity Gradient Energy: Basic Principles

In 1784, the French physicist Jean-Antoine Nollet placed a pig's bladder filled with wine in a barrel of water. Over time, the bladder swelled and eventually burst. This behavior is the result of the osmotic pressures (Figure 14.11) of freshwater and saltwater

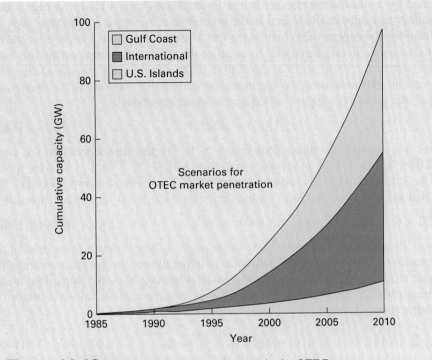

United States Department of Energy

Figure 14.10: Predicted production of electricity by OTEC.

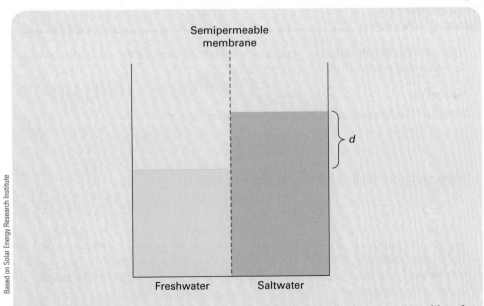

Based on Solar Energy Research Institute

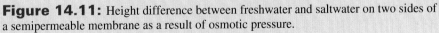

Figure 14.11: Height difference between freshwater and saltwater on two sides of a semipermeable membrane as a result of osmotic pressure.

separated by a suitable membrane. In the figure, freshwater is placed in one side of a container and saltwater is placed in the other side of a container. The two sides are separated by a semipermeable membrane that allows water molecules to pass but not Na or Cl ions. A driving force tries to equalize the salt concentration on the two sides of the membrane, causing water to move from the fresh side to the salt side to dilute the saltwater, thereby increasing the height and pressure on the salt side of the membrane. The osmotic pressure is approximated by the Morse equation:

$$p = iRMT, \tag{14.2}$$

where i is a constant (the dimensionless van 't Hoff factor) that depends on the solute ($i = 2.0$ for NaCl), R is the universal gas constant ($8.315 \text{ J} \cdot \text{K}^{-1} \cdot \text{mol}^{-1}$), M is the molarity of the solution, and T is the absolute temperature. For seawater at, say, 10°C (i.e., 283 K) with a molarity of 600 mol/m^3 (about average for the oceans), the osmotic pressure is

$$p = 2.0 \times (8.315 \text{ J} \cdot \text{K}^{-1} \cdot \text{mol}^{-1}) \times (600 \text{ mol/m}^3) \times (283 \text{ K}) = 2.82 \times 10^6 \text{ N/m}^2 \tag{14.3}$$

or about 28 atm. Thus, in Figure 14.12 the pressure difference pushes the level of the seawater above the freshwater by a distance (d in the figure), so that the pressure in the seawater at a height level with the top of the freshwater is 2.82×10^6 N/m^2. That means that the osmotic pressure is equal to the mass per unit area of the saltwater above the freshwater times the gravitational acceleration ($g = 9.8$ m/s^2), or

$$p = mg/A = \frac{dA\rho g}{A} = d\rho g, \tag{14.4}$$

where ρ is the density of water on the right side of the membrane.

Example 14.4

Calculate the height difference between a column of freshwater and a column of seawater that results from osmotic pressure.

Solution

Equation (14.4) gives an expression for the height, d, as

$$d = \frac{p}{\rho g}.$$

Substituting equation (14.4) for the pressure gives

$$d = \frac{iRMT}{\rho g}.$$

We use the values $i = 2.0$ (for NaCl), $R = 8.315 \text{ J} \cdot \text{K}^{-1} \cdot \text{mol}^{-1}$, and $\rho = 1025$ kg/m^3, and $g = 9.8$ m/s^2. The molarity of seawater is given $M = 600$ mol/m^3, and we assume a temperature of 10°C + 273 = 283 K. The height is then calculated to be

$$d = \frac{2.0 \times (8.315 \text{ J} \cdot \text{K}^{-1} \cdot \text{mol}^{-1}) \times (600 \text{ mol} \cdot \text{m}^{-3}) \times (283 \text{ K})}{(1025 \text{ kg} \cdot \text{m}^{-3}) \times (9.8 \text{ m} \cdot \text{s}^{-2})} = 281 \text{ m}.$$

Note: The units in this expression reduce to $\text{J}/(\text{kg} \cdot \text{m} \cdot \text{s}^{-2}) = \text{m}$.

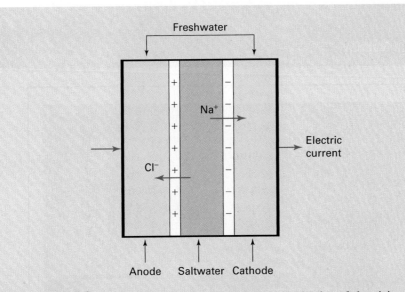

Figure 14.12: Schematic of reverse electrodialysis production of electricity.

The potential energy associated with the height difference calculated in Example 14.2 can be converted to kinetic energy and used to run turbines to generate electricity. In practice, only a small fraction of this energy could actually be extracted, because the membrane, which allows water (but not salt) to pass, will not withstand the osmotic pressure caused by the head (Figure 14.11). (This is why the pig's bladder burst.) A customary approach is to utilize the increased pressure on the saltwater side of the membrane to push water through a turbine to generate electricity. This approach to the production of useful energy is referred to as *pressure-retarded osmosis* (PRO).

Another method of producing electricity from the energy associated with salt concentration differences between freshwater and saltwater is *reverse electrodialysis* (RED). This method produces electricity directly, without first producing mechanical energy, by utilizing the charges associated with the dissolved ions in the water. A schematic diagram of an electrochemical cell that can be used for this purpose is illustrated in Figure 14.12. Pairs of membranes (+ and − in the figure) separate the negative (Cl^-) and positive (Na^+) ions, respectively, in a salt solution. Freshwater and saltwater are circulated through the device. Cl^- ions diffuse in one direction (from the saltwater to the freshwater), and this flow of charged ions gives rise to a current in an external circuit as shown. To increase the voltage produced, several RED cells can be connected in series, or the cell can have several layers of alternating salt and freshwater channels.

14.6 Applications of Ocean Salinity Gradient Energy

The utilization of salinity gradient energy is possible at locations where both saltwater and freshwater are available in large quantities and in close proximity. This situation occurs most commonly at the outflow of rivers into the ocean. The world potential for viable power from salinity gradients at river outflows is estimated at about 150 GW_e.

ENERGY EXTRA 14.1
Pressure-retarded osmosis in Norway

Pressure-retarded osmosis generating station in Norway.

The world's first fully functional osmotic power facility opened in 2009 in Tofle, on the Oslo Fjord in Norway (see the above photograph). The facility utilizes the pressure-retarded osmosis method to produce electricity and was constructed by the Norwegian energy company Statkraft at a cost of $7–8 million. The facility is located where a river flows into the ocean, providing a source of both freshwater and saltwater.

The Statkraft facility is designed for a total electrical output of 2 to 4 kW (on average, about enough for two or three single-family homes) using 2000 m^3 of membrane. It is intended as a three-year study to assess the viability of a large-scale facility. Scaling up such a facility is not necessarily straightforward. To make a large-scale commercial plant economically viable will require the improvement of membrane characteristics from its current output of about 1 W/m^2 to about 5 W/m^2. Even under these

conditions, a small 25-MW$_e$ plant (small compared to a coal or nuclear plant) would require 5×10^6 m^2 of membrane. This area is about 10% of the current total world production of such membranes (largely for desalination facilities). The 25-MW$_e$ osmotic generating station would be about the size of a football stadium and would require a flow of 25 m^3/s freshwater and 50 m^3/s saltwater. Statkraft anticipates that the construction of a commercial osmotic power plant would take place around 2015–2017 and that this technology could be economically competitive with other alternative energy generation methods by around 2030.

The general design of a PRO generating station is shown in the figure. Osmosis of freshwater across a membrane in order to equalize the salt concentration gradient, produces excess pressure of the seawater side of the membrane and this drives the turbine/generator to produce electricity.

Continued on page 367

Energy Extra 14.1 continued

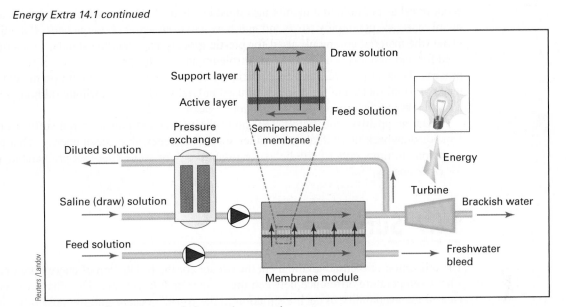

Schematic of pressure-retarded osmosis generating system.

Topic for Discussion

Consider the merits of the following energy genera- tion proposal: An onshore OTEC facility brings in cold seawater and produces electricity. It also produces desalinated water. A PRO facility could be located next to the OTEC facility and could pump in additional sea- water and use the desalinated water from the OTEC plant to produce more electricity.

About half of this capacity is from the world's 50 largest rivers. Inland salt lakes are also a resource that could be utilized. These are highly saline and have a large osmotic pressure relative to freshwater. For example, the Dead Sea has a salinity that corre- sponds to a head of over 5000 m.

Although it is in the early stages of development, salinity gradient research is ac- tive in several places. The Netherlands has a research program to investigate electricity generation by RED. Freshwater that collects behind dykes is normally pumped to the sea, and this proximity of freshwater and saltwater (rather than at river outflows) makes salinity gradient power of interest. In Russia, a prototype RED plant at Vladivostok has been operational for several years. Norway has the capacity for about 3 GW_e of salinity gradient power from rivers flowing into the ocean and has the most advanced program to utilize this resource. Since 2009, a fully functional PRO facility has been operat- ing outside of Oslo, Norway. Until fairly recently, the cost of osmotic membranes has been prohibitive. These membranes have a finite lifetime and must be replaced. Recent developments in membrane production have done much to alleviate this problem.

Generally, power produced from salinity gradients is environmentally friendly. However, there are some environmental concerns. The mixing of saltwater and fresh- water produces brackish water. This happens naturally at the outflow of rivers, but if this process is altered by the utilization of salinity gradient power, then the envi- ronmental consequences of these changes must be considered. In PRO methods, the membranes tend to get clogged with impurities and particulate matter from the water. To optimize the lifetime of these membranes and to make salinity gradient power as economical as possible, the membranes must be cleaned. The chemicals used to clean membranes are potentially toxic and can have environmental consequences. Polyethylene (the material commonly used for plastic grocery bags and the standard material used for membranes) is derived from petroleum and is a by-product of the oil refining process. Thus, at present, salinity gradient energy seems to be dependent on the existence of fossil fuels, and the disposal of used polyethylene will contribute (although in a small way) to the production of greenhouse gases.

On the positive side, both OTEC and salinity gradient power do not suffer from a major drawback that affects most other sources of renewable ocean energy. That is, variations in time, either periodic as for tidal energy or somewhat less predictable as for wave energy.

14.7 Summary

The utilization of thermal gradients in the oceans for the production of energy requires a large temperature difference between the surface and deep water. This chapter overviewed the availability of such a situation in the world's oceans and the possibility of constructing an OTEC generating station. As the temperature of water deep in the ocean is relatively constant worldwide, a large gradient is characteristic of tropical regions where the surface temperature is high.

This chapter discussed the availability of a large thermal difference between surface and deep water as a mechanism for generating electricity using a heat engine. Warm surface water forms the hot reservoir, and cold water pumped from the oceans depths is the cold reservoir. The basic principle of operation follows the general design of a heat engine and allows for the transfer of thermal energy to produce mechanical energy, which can then be used to drive a generator to produce electricity.

The chapter reviewed the three possible designs of an OTEC system: open-cycle, closed-cycle, and hybrid systems. The open-cycle uses seawater as the working fluid in the heat engine. The closed-cycle system uses heat exchangers to transfer thermal energy from the seawater to a closed-system containing a working fluid such as ammonia. The hybrid system is similar to the closed-cycle system except that, in the process of transferring heat from the working fluid, the seawater is vaporized and freshwater, which can be used for domestic purposes, is produced.

The principles of OTEC have been known for well over a century, and experimental work has been conducted for over 80 years in the warmer regions of the world. The overall efficiency of these systems is very low because the temperature difference between the hot and cold reservoirs running the heat engine is very small (compared to the situation for a coal-fired or nuclear power plant). As a result of this low efficiency, the economic viability of OTEC-generated electricity is questionable.

This chapter also described another method of extracting energy from the oceans based on the salinity gradient that exists where freshwater from rivers mixes with seawater. This approach is based on the osmotic energy that results from two liquids with different solute concentrations. Two technologies allow for the utilization of osmotic energy: pressure-retarded osmosis (PRO) and reverse electrodialysis (RED). The chapter overviewed the basic principles of osmotic pressure and the methods for using osmotic pressure to drive a turbine/generator by the PRO technique. RED uses the charges associated with ions in solution to generate electricity directly in an electrolytic cell. The principles of salinity gradient energy have also been known for a number of years, but technical developments for its use as a source of electricity have not been undertaken until relatively recently. At present, a pilot PRO facility in Norway has shown encouraging results. While the cost of electricity generated by this method is still high, the hope is that new developments in membrane technology may make PRO- or RED-generated electricity a small but viable component of future energy production.

Problems

14.1 A 1 m^3 parcel of seawater is cooled from 24°C to 8°C in 1.2 seconds. Calculate the average thermal power that is available during that time.

14.2 An ideal OTEC system operates with a warm reservoir of 23°C and a cold reservoir of 5°C. The temperature of the evaporator is 22°C, and the temperature of the condenser is 11°C. Calculate the mass of water (vapor) flowing through the turbine needed to generate 1 MWh of electricity. Assume a 90% conversion from mechanical to electrical energy.

14.3 An ideal OTEC system has a flow rate of 100 m^3/s of seawater. The warm reservoir is at 26°C, and a cold reservoir is at 6°C. The temperature of the evaporator is 20°C, and the temperature of the condenser is 12°C. For a conversion efficiency for mechanical to electrical energy of 90%, what is the facility's output in MW$_e$.

14.4 It has been speculated that an OTEC facility could use wave energy to offset its low efficiency by providing electrical energy for operating the plant. Where would the most advantageous location(s) be for such a facility? If the facility could intercept 200 m of wave front for your chosen location and had a total output of 100 MW$_e$, what fraction of this total output would be from waves? Assume the efficiency of wave-generated electricity to be 35%.

14.5 A 20 MW$_e$ OTEC facility operates at a net thermal efficiency of 5.4%. The warm surface water is cooled by 12°C while passing through the system. Calculate the flow rate of warm water in cubic meters per second (m^3/s). Assume that the conversion from mechanical to electrical energy is 86% efficient.

14.6 Make a plot of the height of a column of water (in meters) that the osmotic pressure between freshwater and saltwater can support as a function of salinity between 0 and 10% (weight of salt per weight of solution) of NaCl in water. Assume a constant density of 1025 kg/m^3.

14.7 The salinity (mostly NaCl) in some parts of the Great Salt Lake (Utah) is 27%; this is sometimes expressed as 270 parts per thousand, meaning 270 g of salt per liter of solution. Estimate the height of a column of water that can be supported by the osmotic pressure between Great Salt Lake water and freshwater. The density of a 27% salt solution is $\approx$ 1200 kg/m^3.

14.8 Consider the experiment of Jean-Antoine Nollet as described in the text, which illustrates the osmotic pressure between water and wine. Assume wine is 12% (by volume) of ethanol in water, and estimate, as in Example 14.4, the height of a column of water that the osmotic pressure between water and wine can support. Note that the van 't Hoff factor depends on the nature of the solvent and that a value of $i \approx 1.0$ is appropriate for ethanol in water.

Bibliography

W. H. Avery and C. Wu. *Renewable Energy from the Ocean: A Guide to OTEC.* Oxford University Press, New York (1994).

G. Boyle (Ed.). *Renewable Energy.* Oxford University Press, Oxford (2004).

R. Charlier. *Ocean Energies: Environmental, Economic, and Technological Aspects of Alternative Power Source.* Elsevier, New York (1993).

R. Cohen. "Energy from the ocean." *Philos. Trans. Royal Soc. London* **A307** (1982): 405.

A. Goldin. *Oceans of Energy—Reservoir of Power for the Future.* Harcourt, Brace, Jovanovich, New York (1980).

B. K. Hodge. *Alternative Energy Systems and Applications.* Wiley, Hoboken (2010).

J. A. Kraushaar and R. A. Ristinen. *Energy and Problems of a Technical Society*, 2nd ed. Wiley, New York (1993).

C. Ngô and J. B. Natowitz. *Our Energy Future: Resources, Alternatives and the Environment.* Wiley, New York (2009).

T. R. Penney and D. Bharathan "Power from the sea." *Scientific American* **256** (1987): 86.

R. J. Seymour. *Ocean Energy Recovery: The State of the Art.* American Society of Civil Engineers, New York (1992).

World Energy Council. *2010 Survey of Energy Resources*, available online at http://www.worldenergy.org/documents/ser_2010_report_1.pdf

Geothermal Energy

Learning Objectives: After reading the material in Chapter 15, you should understand:

- The reasons for geothermal energy.
- The types and distribution of geothermal energy resources.
- The direct use of geothermal heat.
- The types of geothermal electric generating facilities.

- The use of geothermal energy worldwide.
- The sustainability and environmental consequences of geothermal energy use.

15.1 Introduction

Geothermal energy refers to energy that is extracted from the interior of the earth. It is most convenient to access this energy in locations where hot regions are close to the surface. This source of energy has been used for more than 2300 years. The oldest known use of geothermal energy was the utilization of hot springs for bathing purposes in China. The Roman Empire also used geothermally heated water for bathing in England in the first century CE. The first extensive use of geothermal energy for heating purposes occurred in France in the fourteenth century when the city of Chaudes-Aigues developed a district heating system. The first commercial geothermal electric generating station was constructed in Italy in 1911. Since then, the use of geothermal energy for electricity generation has grown worldwide. The current chapter reviews the properties of geothermal energy, how it can be used for electricity generation, and its prospects for the future.

15.2 Basics of Geothermal Energy

The earth consists of an outer layer called the *crust*, an intermediate layer called the *mantle*, and an inner layer called the *core*. The crust consists primarily of rock and is typically about 30 km thick but can range from about 3 km to about 60 km in different locations. The mantle is about 2900 km thick and consists of rock that, because of the

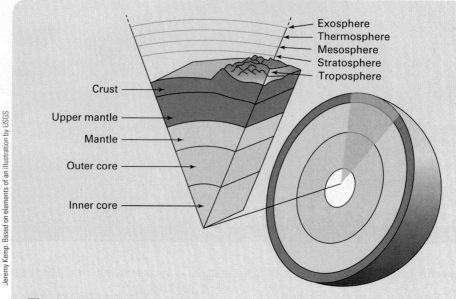

Figure 15.1: The internal structure of the earth.

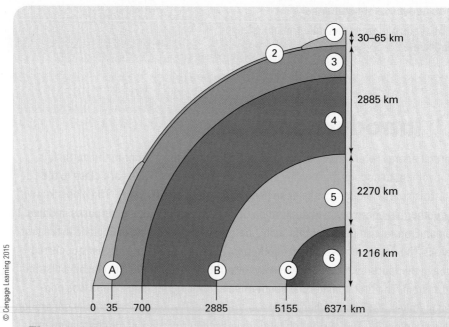

Figure 15.2: Properties of the crust, mantle, and outer and inner cores of the earth.
1. Continental crust; 2. Oceanic crust; 3. Upper Mantle; 4. Lower Mantle;
5. Outer Core; 6. Inner Core; A: Crust/Mantle Discontinuity (Mohorovicic
Discontinuity); B: Mantle/Core Discontinuity (Gutenberg Discontinuity);
C: Outer Core/Inner Core Discontinuity (Lehmann Discontinuity)

temperature and pressure present, is very plastic and made up largely of Mg, Fe, Al, Si, and O. The core is comprised primarily of Fe and is divided into an outer core (about 2300 km thick), which is liquid, and an inner core (about 2400 km diameter), which is solid (Figures 15.1 and 15.2).

The interior of the earth is hot. This heat is believed to have three causes:

1. The primordial heat associated with the material that condensed to form the earth during the early history of the solar system
2. The decay of radioactive nuclides inside the core
3. Friction in the liquid outer core that results from the tidal forces between the earth's core and the moon and sun

Large planets, like Jupiter, cool slowly, and their interior temperature is dominated by primordial heat. The earth has lost a larger fraction of its primordial heat than Jupiter, and its interior temperature is dominated largely by the rate of radioactive decay of nuclides of uranium, thorium, and potassium. It is estimated that approximately 75% of the heat in the interior of the earth comes from this source. The heat from the interior of the earth, flowing outward (toward the cooler crust), amounts to about 0.087 W/m^2 at the surface. This is small (less than 0.1%) of the heat flow into the earth from solar radiation (average of 168 W/m^2; see equation 8.3).

Example 15.1

Calculate the total geothermal heat flow through the surface of the earth.

Solution

The heat flow from the earth's interior averages 0.087 W/m^2. The total surface area of the earth is $4\pi r^2$, where the mean radius of the earth is $r = 6371$ km $= 6.371 \times 10^6$ m (Figure 15.2). Thus the total heat flow is

$$4 \times 3.14 \times (6.371 \times 10^6 \, \text{m})^2 \times (0.087 \, \text{W/m}^2) = 4.4 \times 10^{13} \, \text{W}.$$

The crust of the earth consists of separate regions referred to as *tectonic plates*. The heat transfer from the hot interior of the earth toward the crust generates convection currents in the mantle. These convection currents drive the movement of tectonic plates, referred to as *continental drift*. Near the middle of the oceans, the oceanic plates move apart, allowing molten rock (*magma*) from deep within the mantle to push upward and to form a rift. At continental boundaries, plates collide. As this occurs, one plate (the oceanic plate) is pushed below the other (or *subducted*). This causes the subducted plate to heat (because it is pushed into hotter regions of the earth) and sends plumes of magma up toward the surface.

The possibility of utilizing geothermal energy is the greatest at or near plate boundaries because the thermal gradient below the surface is the largest in these regions. Figure 15.3 shows the tectonic plates on the surface of the earth. The locations of volcanoes are also shown in the figure and, in most cases, are seen to fall on or near plate boundaries. These regions correspond to the subduction of oceanic plates by continental plates, as along the Pacific coast of North and South America or at midoceanic ridges as in Iceland (Figure 15.4). The most active regions of the earth occur along the edges of the Pacific Plate, where the greatest number of volcanoes occurs. For this reason, the edge of the Pacific Plate is sometimes referred to as the *Ring of Fire*. This ring offers the greatest potential for the use of geothermal energy

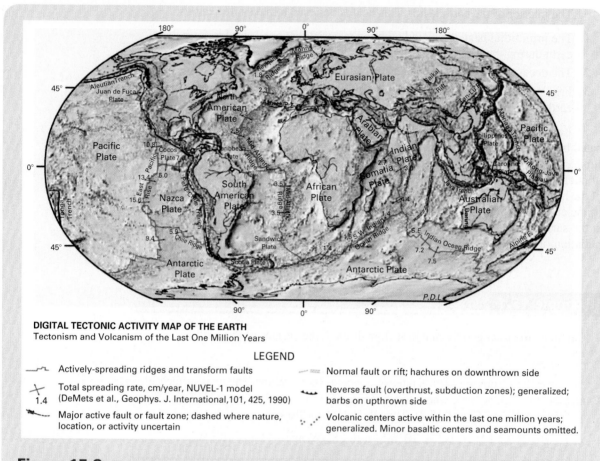

DIGITAL TECTONIC ACTIVITY MAP OF THE EARTH
Tectonism and Volcanism of the Last One Million Years

LEGEND

Actively-spreading ridges and transform faults		Normal fault or rift; hachures on downthrown side
Total spreading rate, cm/year, NUVEL-1 model (DeMets et al., Geophys. J. International,101, 425, 1990) 1.4		Reverse fault (overthrust, subduction zones); generalized; barbs on upthrown side
Major active fault or fault zone; dashed where nature, location, or activity uncertain		Volcanic centers active within the last one million years; generalized. Minor basaltic centers and seamounts omitted.

Figure 15.3: The earth's tectonic plates. The locations of volcanoes that have erupted during historic times are shown by the red dots.

(although other active regions, e.g., in the Caribbean and in Iceland, have important resources as well).

The temperature inside the earth increases as a function of depth. The temperature gradient in normal cases is in the range of 17–30°C per km. In active regions, the temperature gradient can be much greater. The thermal gradient is a measure of the usefulness of geothermal energy in a particular region. The measured temperature gradient for different regions of the United States is shown in Figure 15.5 and suggests that the western parts of the country are the most suitable for the development of geothermal energy resources.

Geothermal resources are roughly divided into six categories:

1. Normal geothermal gradient
2. Hot dry rock
3. Hot water reservoirs
4. Natural steam reservoirs
5. Geopressurized regions
6. Molten magma

The possibilities of utilizing these resources are now discussed.

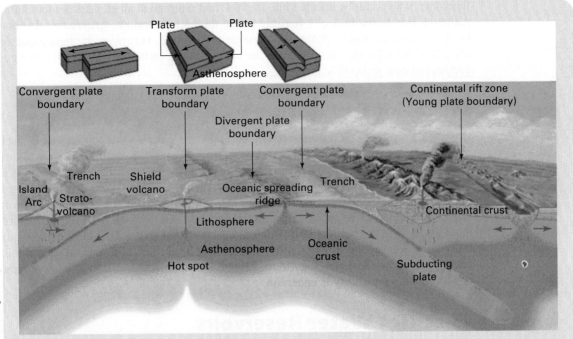

U.S. Geological Survey

Figure 15.4: Movement of tectonic plates resulting in the formation of hot regions close to the surface of the earth at midoceanic ridges and at continental shelf boundaries.

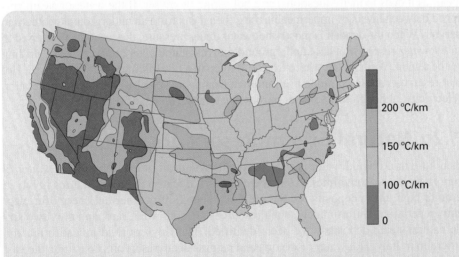

U.S. DOE

Figure 15.5: Geothermal temperature gradient for the United States in °C/km.

15.2a Normal Geothermal Gradient

Even in regions where the normal geothermal gradient exists, the rock becomes hot enough to provide useful energy if one drills deep enough. For a gradient of 30°C/km, a well drilled to a depth of 6 km will access a temperature of about 200°C. A working fluid (e.g., water) can be injected into the well and returned at an elevated temperature. Although it is technologically feasible to use this as a method of generating electricity, this approach is not economically viable at this time. However, the heat capacity associated with the soil and rock near the surface of the earth can be exploited using a geothermal heat pump (Chapter 17).

15.2b Hot Dry Rock

Hot dry rock refers to a resource that is basically the same as the normal geothermal gradient except that the temperature increases at a greater rate (i.e., more than about 40°C/km). This feature increases the possibility of utilizing geothermal energy and is a resource that is estimated to be capable of providing up to 200 GW of thermal energy in the United States alone. The basic approach would be to drill a well into the region of elevated temperature and then, using hydraulic or explosive techniques, to fracture the rock to form a reservoir into which a working fluid (water) could be injected to artificially replicate the situation found in hot water reservoirs, as discussed later in this chapter. Thus far, the technology to implement this approach is still in its early stages, and its economic viability is uncertain.

15.2c Hot Water Reservoirs

Surface water from rain, melting snow, and other sources penetrates the earth's surface in regions where the crust has faults and cracks. If this occurs in a region where there is an anomalously large geothermal gradient, then the water can be heated underground to form hot water and/or steam. If the water or steam has a clear path back to the surface, it can appear, as it does in many locations, as a hot spring or geyser. If the water or steam gets trapped between layers of impermeable rock, then it can form an underground geothermal reservoir. When the deposit is mostly hot water under pressure, the resource is referred to as a *hot water reservoir*. These underground sources of hot water are accessed by drilling into the region. Wells of anywhere from 100 m up to a few kilometers deep are used. Hot water reservoirs are in common use for direct heating and can be used for the generation of electricity, although for the latter purpose, natural steam deposits are preferred.

15.2d Natural Steam Reservoirs

When little or no liquid water is associated with a deposit, it is referred to as a *natural steam reservoir* or sometimes a *dry steam deposit*. Because the temperatures involved are quite high, these deposits are the preferred source of geothermal energy for electricity generation. Unfortunately, such resources are somewhat rare, and only two sizable natural steam deposits have been identified: The Geysers field in California and Larderello in Italy. Hot water reservoirs and natural steam reservoirs constitute the vast majority of geothermal resources that have been exploited commercially.

15.2e Geopressurized Regions

Geopressurized resources are reservoirs of pressurized hot water containing dissolved methane gas. They typically have temperatures in the range of 90–200°C and occur at depths of 3–6 km. The only extensive deposits of this type that have been identified occur along the Gulf of Mexico coast. This resource can provide energy by three mechanisms: heat associated with the water, combustion of the methane, and direct use of the pressure to drive turbines. Test facilities were constructed in the 1980s and 1990s in the Gulf region. Thermal, as well as chemical energy, was extracted, although the mechanical energy associated with the pressure was not utilized. It was concluded that extraction of all forms of available energy was not economically feasible at that time. It is probable that some aspects of the energy associated with geopressurized regions will be developed in the future in conjunction with natural gas recovery.

15.2f Molten Magma

Regions where molten magma appears at or near the surface of the earth (e.g., active volcanoes) are too unstable and unpredictable to provide a reliable and safe means of extracting geothermal energy.

15.3 Direct Use of Geothermal Energy

Geothermal energy can be used in two ways: (1) direct use of the heat for space heating and similar applications or (2) use in a heat engine (e.g., turbine/generator) to produce electricity. The heat associated with geothermal resources has been utilized directly in several ways (Figure 15.6), and the more important ones will be described in some detail.

15.3a Bathing

Perhaps the oldest use of geothermal energy is for bathing (or *balneology*). In regions where geothermally heated hot water appears at the surface, these so-called hot springs

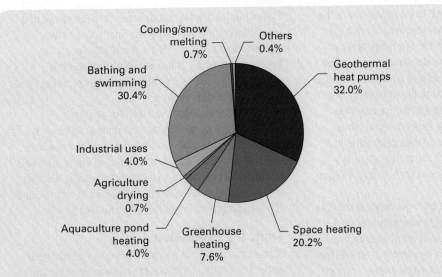

Figure 15.6: Breakdown of the direct use of geothermal energy.

Based on http://climatelab.org/Geothermal_Energy

ENERGY EXTRA 15.1
Geothermal aquaculture

Geothermal Education Office, Tiburon, California

Alligators raised in a geothermally heated pond in Idaho.

Aquatic farming in ponds heated by geothermal energy has increased the growth rate and diversity of animals that can be raised. Fish, shellfish, and reptiles have been produced by aquafarmers in geothermal regions, often in locations where they might otherwise not survive.

Geothermal aquaculture can nicely complement other geothermal energy uses. In locations where the temperature of the geothermal resource is not high enough to provide an acceptable thermodynamic efficiency for electricity generation, aquaculture is still a possibility. Also, in locations where geothermal energy is used for electric generation or direct heating, the runoff from such systems is still at sufficiently high temperatures to be put to use in geothermal farms. Geothermal water can be mixed with cold water in a controlled manner to regulate the temperature in a pond. Circulation is typically necessary to avoid excessive thermal gradients.

The availability of heat from geothermal sources provides reasonably constant temperature ponds for aquafarming. Because most species of aquatic creatures grow fastest and remain healthiest in an environment where the temperature remains within a certain range, productivity in geothermally heated aquatic farms can exceed that in unheated farms, and such facilities can function economically and without the carbon footprint of farms that utilize artificial heating methods.

Notable geothermal aquafarming activities in the United States exist in Arizona, California, Colorado, Idaho, and Nevada. Total heat utilization in the United States for geothermal aquaculture is in the range of a few million gigajoules per year. Commonly grown organisms are catfish, tilapia, trout, sturgeon, and freshwater prawns. Other activities include the growth of tropical fish for aquaria and tropical reptiles, such as alligators in Idaho.

In other locations worldwide, geothermal aquaculture is becoming more common. Some activities include raising eels in Slovakia, fish and shellfish (particularly abalone) in Iceland, fish in China, and eels and alligators in Japan. In addition to aquatic animals, aquatic plants that are useful for human or animal food can also be grown in geothermally heated ponds.

Topic for Discussion

Discuss the design of a system using hot geothermal water along with a cold water supply to regulate the temperature of a pond or tank for aquaculture.

have been used since ancient times for bathing and therapy. Today, these resources form an important tourist industry in the United States, Japan, Mexico, New Zealand, and elsewhere.

15.3b Space Heating

Space heating is another traditional use of geothermal energy where hot water (typically at 60°C or hotter) from geothermal reservoirs is pumped through buildings to provide heating. After passing through a heat exchanger to heat the building, the water is reinjected into the geothermal reservoir for reheating. This may be done on an individual building basis but is often done collectively for large residential and commercial districts. The largest development of this type is in Reykjavik, Iceland, although the resource is used substantially in France, the United States, Turkey, Poland, and Hungary.

15.3c Agriculture

Along the lines of space heating, hot water from geothermal reservoirs can be used to provide heat to greenhouses to improve growing conditions. This approach is in common use in Italy and in Hungary, where it satisfies about 80% of the greenhouse heating needs.

15.3d Industrial Uses

Heat from geothermal sources has been used worldwide for a variety of applications that would otherwise use electric heat or heat from fuel combustion. Some industrial uses that have been found for geothermal heat are drying of food and wood products.

15.3e Snow Melting

Geothermally heated water is used in a variety of locations to melt snow and ice in the winter. A common approach is to pump hot water from geothermal reservoirs through pipes embedded in sidewalks and roads to keep them free from snow and ice in cold weather. A geothermally heated sidewalk is shown in Figure 15.7.

15.3f Geothermal Heat Pumps

The temperature of the ground at a depth of about 3 m remains very constant throughout the year and ranges from about 10°C to about 16°C depending on the local climate. This very constant temperature reservoir can be utilized by a heat pump (Section 1.5) for space heating purposes. For traditional heat pumps, as discussed in more detail in Section 17.6, the outside air is typically used as the cold reservoir. However, geothermal heat pump systems offer the advantage that the cold reservoir is at a much more stable temperature. This type of system utilizes the heat capacity of the earth as a heat source and does not depend on the availability of enhanced geothermal activity. As such, it can be implemented in any location.

Christopher Porter/Flickr/Getty Images

Figure 15.7: Construction of a sidewalk in Reykjavik, Iceland, utilizing heating pipes carrying geothermally heated water to prevent icing.

15.4 Geothermal Electricity

Geothermal energy is converted into electricity by extracting hot geothermal fluid (water or steam or a mixture of the two) from a *production well* drilled into the geothermal reservoir. Hot water, extracted from the well, is converted into steam (as described in Section 15.4b). The steam is then used to run a turbine, which in turn drives a generator to produce electricity. The fluid is then injected back into the geothermal reservoir through an injection well in order to replenish the supply of water in the reservoir and to minimize the distribution of geothermal water on the earth's surface (Section 15.5). The general layout of a geothermal power plant is shown in Figure 15.8.

There are several methods by which the hot fluid is actually utilized to drive the turbine, and the choice of method depends to some extent on the nature of the geothermal resource. Several possibilities are now described.

15.4a Dry Steam Plants

A few geothermal reservoirs provide essentially pure steam with little hot water. This simplifies the utilization of geothermal fluids for driving a turbine. Basically, the steam can be input directly into the turbine as it comes out of the well (after passing through a filter to remove debris). After driving the turbine, the steam is condensed and reinjected into the reservoir (Figure 15.9). The Geysers reservoir in Northern California is currently the world's largest known dry steam resource. One of the 22 power plants associated with the Geysers resource is shown in Figure 15.10. Typical facilities of this type have capacities in the range of 50 MW$_e$.

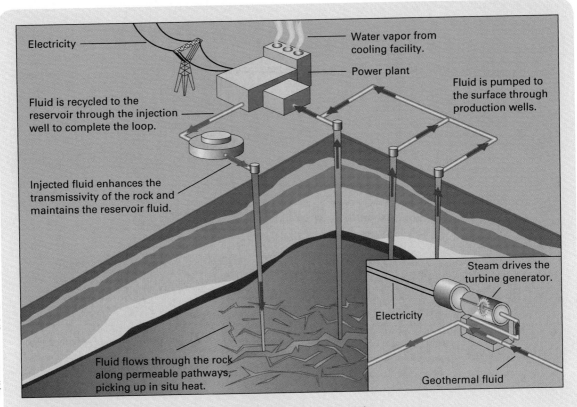

Figure 15.8: General schematic of a geothermal generating station.

Example 15.2

Calculate the ideal Carnot efficiency of a geothermal generating station if the temperature of the geothermal fluid is 210°C and the temperature of the cold reservoir is 85°C.

Solution

From equation (1.34), the ideal Carnot efficiency of a heat engine is

$$\eta = 100\left(1 - \frac{T_c}{T_h}\right).$$

Converting temperatures to Kelvin gives

$$T_h = 210°C + 273 = 483 \text{ K}$$

and

$$T_c = 85°C + 273 = 358 \text{ K}.$$

The efficiency is therefore

$$\eta = 100 \times \left(1 - \frac{358 \text{ K}}{483 \text{ K}}\right) = 25.9\%.$$

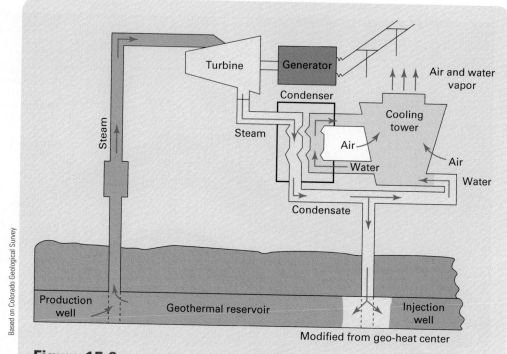

Based on Colorado Geological Survey

Figure 15.9: Schematic of a dry steam geothermal generating station.

Bureau of Land Management, U.S. Department of Interior

Figure 15.10: A power generating station at The Geysers in California.

15.4b Flash Steam Plants

Flash steam power plants are the most common type of geothermal generating plants in operation today. This is because most reservoirs contain hot water at an elevated pressure, and flash steam plants are the most convenient technology for utilizing this type of resource. When hot water under pressure is removed from its underground reservoir,

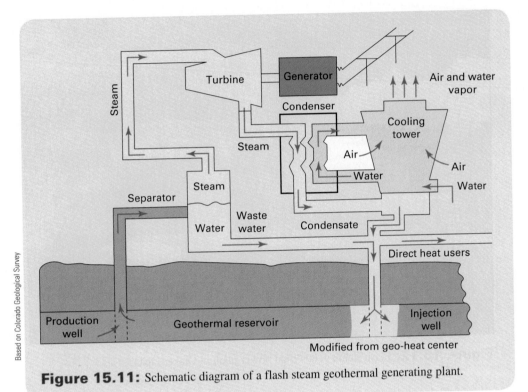

Based on Colorado Geological Survey

Modified from geo-heat center

Figure 15.11: Schematic diagram of a flash steam geothermal generating plant.

the pressure drops, and, according to the water phase diagram in Figure 14.4, this lowering of the pressure causes the water to vaporize, producing steam. This steam is then used to run a turbine, and after going through a condenser, it is reinjected into the reservoir. A general schematic of a flash steam power plant is shown in Figure 15.11, and a photograph of a plant in Japan is shown in Figure 15.12.

15.4c Binary Power Plants

In both the dry steam and flash steam generating systems, the hot water and/or steam from the geothermal reservoir are used to drive the turbine. In a binary cycle power plant (Figure 15.13), the thermal energy of the hot water and/or steam is transferred to a second working fluid through a heat exchanger. This working fluid is in a closed-system and is vaporized in the heat exchanger. The vapor is used to run the turbine and is then returned as a liquid to the heat exchanger. This system has the advantage that a working fluid with a boiling point that is lower than that of water can be used, thus allowing geothermal reservoirs with lower temperatures to be used for electricity generation. A binary cycle plant in Nevada is shown in Figure 15.14.

15.4d Hybrid Power Plants

Hybrid power plants combine a binary cycle power plant with another method of utilizing geothermal energy. Figure 15.15, for example, shows a combined flash steam and

Geothermal Education Office, Tiburon, California

Figure 15.12: Flash steam geothermal power plant in Otake, Japan.

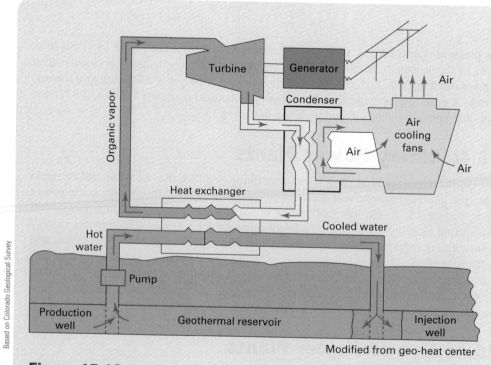

Based on Colorado Geological Survey

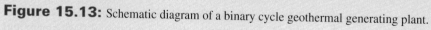

Figure 15.13: Schematic diagram of a binary cycle geothermal generating plant.

Figure 15.14: Binary cycle geothermal plant in Soda Lake, Nevada.

Figure 15.15: Hybrid flash/binary cycle geothermal plant on the Big Island of Hawaii.

binary cycle plant. Flash steam is produced from geothermal hot water and is used to run a turbine. The condensed steam (which is still quite hot) can be used in a binary cycle system to extract more of the heat content of the water.

15.5 Utilization of Geothermal Resources and Environmental Consequences

The installed capacity for geothermal energy use in various countries is summarized in Tables 15.1 and 15.2 for electricity generation and direct use, respectively. Capacity factors are typically about 0.7 for electricity generation and around 0.3 for direct use, but they can vary considerably, particularly for direct use where demand (e.g.,

Table 15.1: Electricity generation from geothermal sources as of 2008 for the world's principal producers and total world capacity.

country	installed capacity (MW$_e$)
United States	3277
Philippines	1958
Indonesia	1054
Mexico	958
Italy	810
New Zealand	585
Iceland	573
Japan	535
El Salvador	204
Kenya	163
Costa Rica	162
Other	377
world total	**10,656**

© Cengage Learning 2015

Table 15.2: Installed capacity for direct use of geothermal power in 2008.

country	installed capacity (MW)
United States	12,037
China	8898
Sweden	4460
Norway	3300
Japan	2100
Turkey	2084
Iceland	1826
Germany	1640
France	1607
Netherlands	1410
Canada	1126
Austria	1080
Switzerland	1054
other	6888
world total	**49,636**

Based on data from World Energy Council

for heat throughout the year) is not constant. The data for direct use are also subject to substantial uncertainty. Resources are utilized in specific locations where geothermal reservoirs occur. The nature of the resource is a major factor in the development of geothermal energy for electricity generation or direct use. The distribution of electricity-generating facilities that utilize geothermal energy is illustrated in Figure 15.16. This

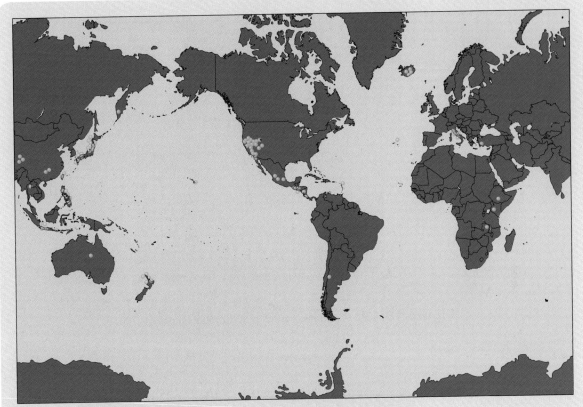

Geothermal Education Office, Tiburon, California

Figure 15.16: Distribution of geothermal electricity generating plants.

correlates well with the availability of geothermal energy, as shown in Figure 15.4, and the population density. The United States has the largest installed geothermal electric generating capacity. Roughly two-thirds of this capacity is in California where geothermal electricity accounts for about 4.5% of the state's total electric use. Although there have been increases in geothermal energy production in some developing countries over the past few years, the overall growth worldwide has been minimal since the late 1980s (Figure 15.17). This is perhaps due to the continued low prices of fossil fuels and the economic and environmental advantages of pursuing other renewable energy sources, such as wind. Since the nuclear accident at the Fukushima reactor in Japan in 2011, there has been renewed interest in developing that country's geothermal resources for electricity production. At present, Japan has a geothermal electric capacity of 536 MW_e, or about one-half of the capacity of a medium-sized nuclear reactor. The Geothermal Research Society of Japan has estimated that 1.5 to 2.4 GW_e of Japan's 23 GW_e geothermal potential could be developed within the next 40 years. These numbers can be viewed in the context of Japan's current nuclear generating capacity of about 47 GW_e.

Table 15.3 gives predictions from around 1980 for U.S. geothermal electricity capacity. A comparison with Figure 15.17 shows that the business-as-usual prediction for 1990 turned out to be relatively accurate but that the predicted growth beyond 1990 did not occur. The national commitment prediction far exceeds reality.

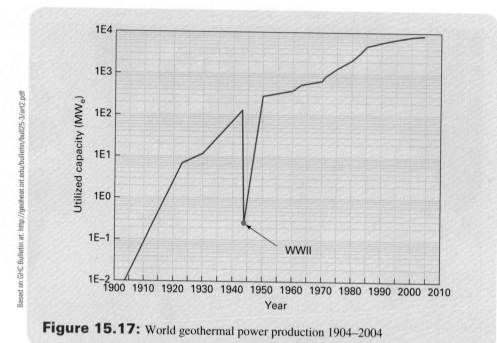

Based on GHC Bulletin at: http://geoheat.oit.edu/bulletin/bull25-3/art2.pdf

Figure 15.17: World geothermal power production 1904–2004

Based on data from CONAES 1980

Table 15.3: Predicted growth in geothermal electricity capacity in the United States in MW$_e$.		
scenario	1990	2010
business as usual	2930	18,870
national commitment	8270	60,900

15.5a Sustainability of Geothermal Resources

The total worldwide geothermal capacity is difficult to assess, and estimates have varied greatly. In an analysis of the future use of geothermal energy, it is important to consider the lifetime of these resources. Although geothermal energy is considered a renewable resource and one may have the impression that its lifetime is more or less infinite, this is not generally the case. Because of the thermal gradient in the earth, heat is constantly flowing from the inside toward the surface. When heat is removed from a reservoir beneath the surface, it is being removed faster than it is being replenished from deeper within the earth. Reinjecting water into the reservoir provides a source of fluid to carry the heat, but it does not help replace the extracted heat. If we stop removing heat from a reservoir, then the temperature increases due to the heat flowing from the interior. This process is, however, very slow and occurs on a timescale of probably hundreds of thousands of years. How long the geothermal energy in a particular reservoir will last depends on how much thermal energy is in the reservoir and how fast it is removed. In general, lifetimes can be as short as a few years, or as long as a few hundred.

The environmental impact of geothermal power is generally minimal, but some concerns should be mentioned. The water or steam extracted from a geothermal reservoir contains a number of impurities. Among these are dissolved gases such as carbon dioxide, nitrogen compounds, and sulfur compounds. These can be released to the atmosphere, particularly in dry steam and flash steam plants (less in binary cycle plants). These emissions typically amount to about 5% of those released by a coal-fired generating station (per unit energy produced), and appropriate methods of removing at least some of these pollutants can be implemented. Water from underground reservoirs often contains anomalously high levels of toxic elements, such as arsenic, mercury, lead, antimony, and the like. Environmental consequences that could result from the release of these toxins are minimized by the common practice of reinjecting geothermal fluids back into the deposit. An undesirable side effect of this practice, however, is the possibility of introducing geological instabilities, and these effects may preclude the possibility of future development of certain geothermal resources.

A final point in geothermal energy's favor is land utilization. Compared with renewable energy sources that have a very low energy density on the earth's surface, such as solar or wind, or compared with coal, which requires substantial mining activities, geothermal energy utilizes land area very effectively. Per unit of power produced, geothermally generated electricity occupies only about 15–30% of the land area compared with most other energy sources.

15.6 Summary

This chapter described the origins of geothermal energy. This energy is believed to be associated with three factors: the primordial heat from the material that formed the earth, radioactive decay in the earth's interior, and tidal friction in the earth's molten core. Geothermal energy resources may be categorized in a number of ways. The major types of geothermal energy resources are the normal geothermal gradient, hot dry rock, hot water reservoirs, natural steam reservoirs, geopressurized regions, and molten magma. Hot water reservoirs and natural steam reservoirs are the most useful for wide-scale utilization of geothermal energy.

This chapter presented the ways in which geothermal energy resources may be utilized. These uses fall into two general categories: direct use of the thermal energy and conversion of the thermal energy into electricity. Typical direct uses of geothermal energy include bathing, space heating, agriculture, and aquaculture. Hot water reservoirs are typically the most suitable for these applications.

Geothermal electricity-generating facilities extract energy in the form of hot water or steam from a production well. This fluid is used to generate electivity using a turbine/generator and is then reintroduced into the reservoir through an injection well. Hot dry steam is the most convenient resource for this application and is the most commonly found in the geothermal areas of California. Most geothermal electricity generation worldwide uses hot water as a source of geothermal energy.

The United States has the greatest utilization of geothermal energy resources both for direct use and for electricity generation. While some further development is possible in the United States, the greatest potential for additional geothermal energy use is probably in countries, such as Japan, that lie in geothermally active areas but that have not been as active in pursuing the use of this resource.

As noted in the chapter, geothermal resources are not indefinitely renewable because utilization on any reasonable commercial scale for electric generation depletes the resource faster than it is replenished by heat flowing outward from the earth's interior. Lifetimes of geothermal resources depend on the nature of the resource and the manner of its use. Typically, with extensive use, the lifetime of a resource may be in the order of a century.

Problems

15.1 Calculate the ideal Carnot efficiency for a turbine operating with a hot reservoir at a temperature of 225°C and a cold reservoir at a temperature of 75°C.

15.2 Assuming that the geothermal heat flux in the United States is (on average) typical of that which occurs worldwide, calculate the fraction of total U.S. primary energy use that could be provided by geothermal energy if this energy could be utilized in its entirety.

15.3 Consider the total energy use for a typical North American single-family home, as discussed in Chapter 2. If the house has a footprint of 160 m^2, what fraction of its energy needs could be met by the average geothermal heat flow from the earth's interior?

15.4 Heat is extracted from hot rock at a temperature of 250°C and used to produce electricity with the ideal Carnot efficiency. If the temperature of the cold reservoir is 75°C, what mass of rock would be needed to yield 1 GWh$_e$? See Chapter 8 for the thermal properties of typical rock, and assume a generator efficiency of 90%.

15.5 Heat is extracted from geothermal pressurized water at a temperature of 225°C with the ideal Carnot efficiency. If the cold reservoir is at 80°C and the generator efficiency is 90%, what is the flow rate of geothermal water (in m^3/s) needed to yield 20 MW$_e$ output?

15.6 Assume that geothermal heat transfer (at least near the surface of the earth) is by conduction through the crust rocks. For an average geothermal heat flow of 0.087 W/m^2 and typical thermal gradient of 100°C/km, calculate the thermal conductivity of the rock. Compare with the known thermal conductivities of similar materials given in Chapter 8.

15.7 It is clear that geothermal generating stations extract energy from a resource more rapidly than it is replenished from the interior of the earth. Assume that a resource has 10 EJ of energy that can be extracted economically. If this resource is used to generate 250 MW$_e$ of electricity, how long will the resource last? It is necessary to know the efficiency of electric generation. Assume an efficiency of one-third the ideal Carnot efficiency (Example 15.2). This is typical of the actual average operational efficiency of a geothermal generating station.

15.8 A geothermal electrical generating station has an actual efficiency of one-third of the ideal Carnot efficiency. For a cold reservoir temperature of 70°C, calculate and plot the efficiency of the station as a function of the geothermal fluid temperature from 100°C to 300°C. If a minimum efficiency of 5% is necessary to make the utilization of the resource economical, what minimum geothermal fluid temperature is needed?

Bibliography

H. Armstead. *Geothermal Energy*, 2nd ed. E&FN Spon, London (1983).

G. Boyle (Ed.). *Renewable Energy*. Oxford University Press, Oxford (2004).

M. Dickson and M. Fanelli. *Geothermal Energy: Utilization and Technology*. Earthscan Publishers, London (2005).

R. Harrison, N. D. Mortimer, and O. B. Smarason. *Geothermal Heating: A Handbook of Engineering Economics*. Pergamon Press, Oxford (1990).

B. K. Hodge. *Alternative Energy Systems and Applications*. Wiley, Hoboken (2010).

M. Kaltschmitt, W. Streicher, and A. Wiese (Eds.). *Renewable Energy: Technology, Economics and Environment*. Springer, Berlin (2007).

R. A. Ristinen and J. J. Kraushaar. *Energy and the Environment*. Wiley, Hoboken (2006).

World Energy Council. *2010 Survey of Energy Resources*, available online at http://www.worldenergy.org/documents/ser_2010_report_1.pdf

Biomass Energy

Learning Objectives: After reading the material in Chapter 16, you should understand:

- The properties of wood and its uses for energy.
- The production and use of ethanol and methanol.
- A comparison of the Brazilian and U.S. ethanol programs.
- Biodiesel production.
- Environmental consequences of biofuel utilization.
- The use of municipal waste for energy production.

16.1 Introduction

Biomass energy refers to energy extracted from recently grown biological matter. It is renewable (as compared with fossil fuels) because, as it is used, new material can be grown to replace it. Biofuels are the oldest source of energy used by humans; the use of wood for heating and cooking dates back to prehistoric times. Because of their renewability, biofuels are also a source of energy that has generated substantial interest in recent years. On a global scale, biofuels are currently the largest contributor to renewable energy. However, to be utilized in a renewable manner and remain carbon neutral, new biomaterial must be grown to replace material that has been used in order to sequester the carbon that has been produced. Biofuels may be materials, such as wood, that are utilized directly in the form in which they are found in nature. They may also be materials that are produced by processing naturally grown materials into a form of fuel that is more readily utilized (e.g., the production of ethanol from plant matter). Also, biofuels include societal waste products (e.g., municipal waste), which are either burned directly or processed to produce a more convenient form of fuel. In this chapter, biofuels are placed in four basic categories: wood, bioalcohols, biodiesel, and municipal waste.

16.2 Wood

Wood has long been used as a source of energy, and until the late 1800s, was the major source of energy worldwide, after which it was replaced (for some time at least) by coal. Over the years, wood has also been an important source of material for industry and building (and continues in this role). In North America, more than 70% of wood that is harvested is used for industry (largely the paper industry) and building construction. Less than 30% is used as a source of energy. In Bangladesh, 98% of the harvested wood is used for energy. Today one-third of world's population (mostly in developing countries) relies on wood as a major source of energy. This discussion is confined to the use of wood as a source of energy.

Wood that is used for the production of energy is used in one of three forms:

1. *Firewood*: Wood that is used directly
2. *Charcoal*: Primarily carbon that is produced by heating wood
3. *Black liquor*: A combustible oil-like substance that is a by-product of the pulp and paper industry

The combustion of wood is basically the oxidation of carbon or hydrocarbons and is, from a chemical standpoint, very much like the combustion of a fossil fuel. The energy content of wood is typically about 14 MJ/kg compared with about 31 MJ/kg for bituminous coal. The main difference between burning wood and burning coal is that trees (or other living organic matter) sequester carbon dioxide so that the CO_2 released by the combustion of wood can be eliminated from the environment if the trees that are burned are replaced by an equal number of new trees. In this way, wood can be used as a renewable resource that does not have a net contribution to the greenhouse gas in the environment. In this case, it is referred to as being *carbon neutral*. Unfortunately, much of the wood used for both energy production and industrial purposes is not used in a renewable way, and the resulting deforestation means that the greenhouse gases released into the environment by the burning, decomposition, or decay of wood are not sequestered by the growth of new trees. Efforts exist in many areas to alleviate this problem (Figure 16.1).

Figure 16.1: Sign promoting reforestation activities in New Brunswick, Canada.

Example 16.1

How many kilograms of wood have the same energy as one tonne of bituminous coal?

Solution

From Appendix IV 1 tonne of bituminous coal has an energy content of

$$(3.10 \times 10^7 \text{ J/kg}) \times (1000 \text{ kg/t}) = 3.10 \times 10^{10} \text{ J/t}.$$

From Appendix IV, 1 kg of wood has an energy equivalence of 1.4×10^7 J. Thus

$$(3.10 \times 10^{10} \text{ J/t})/(1.4 \times 10^7 \text{ J/kg}) = 2.21 \times 10^3 \text{ kg wood is}$$
equivalent to 1 tonne of coal.

Much of the wood that is used as a source of energy is used on a residential basis, rather than for generating electric power for the grid. Difficulties with the widespread use of wood for commercial energy production stem largely from problems with processing and transportation costs. A 1500 MW$_e$, wood-fired generating station operates in Sweden, and there is a 500-MW$_e$ facility in Austria. In North America, a few small wood-fired commercial electric generating facilities exist. Even so, wood remains a major contributor to renewable energy production (Figure 16.2). The use of wood resources for energy production worldwide is summarized in Table 16.1. The black liquor that is produced by the pulp and paper industry is used to meet most of the heating and much of electricity needs for this industry.

Although the use of wood in a renewable way is carbon neutral, the burning of this resource is not without other environmental consequences. Wood contains much less sulfur than coal, so SO$_x$ emissions are not a concern. However, burning wood produces substantial quantities of NO$_x$ and particulate matter. Because wood use is not generally a large-scale commercial operation, control of emissions is more difficult. Also, the combustion of wood releases benzo(a)pyrene, a known carcinogen.

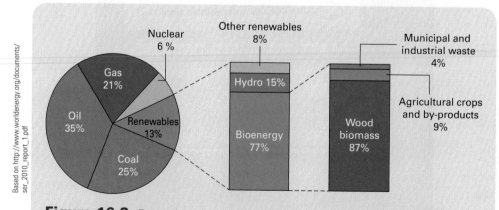

Based on http://www.worldenergy.org/documents/ser_2010_report_1.pdf

Figure 16.2: Proportion of renewable energy production worldwide (in 2007) and a breakdown of the various renewable sources.

Table 16.1: Energy content of wood used annually as an energy source (in 2002) in different regions of the world.

region	fuelwood (10^{15} J)	charcoal (10^{15} J)	black liquor (10^{15} J)	total (10^{15} J)
Africa	6088	453	~0	6541
North/Central America	1673	64	1599	3335
South America	1528	211	601	2341
Asia	9254	145	414	9812
Europe	806	23	592	1420
Oceania	86	1	29	115
world total	**19,458**	**897**	**3234**	**23,589**

Based on data from World Energy Council

The consequences of leaving dead trees to decay is also a concern, and it has been suggested that clearing deadwood from forests for fuel use is the lesser of two evils. Decaying wood emits carbon mostly in the form of CH_4, while burning wood emits carbon mostly in the form of CO_2. Although both CH_4 and CO_2 are greenhouse gases (Table 4.6), methane is about 25 times more effective at absorbing infrared radiation than carbon dioxide.

16.3 Ethanol Production

Light hydrocarbons such as methane and ethane are gaseous at room temperature (Table 3.2), and hydrated light hydrocarbons are typically liquid at room temperature. Some of these are alcohols, such as methanol, ethanol, and the like, are represented by the formula $C_nH_{2n+1}OH$. The properties of the light alcohols are shown in Table 16.2. These alcohols may be produced as a by-product of the distillation of petroleum or by the fermentation of sugar containing biological materials. Bioalcohols can be used as a source of energy and, because they are liquids at room temperature, provide a convenient replacement for petroleum-based transportation fuels such as gasoline. This discussion focuses on ethanol because this has been shown to be the most practical in terms of ease of production and suitability for direct replacement of gasoline in internal combustion engines. Methanol has found some use as a fuel for fuel cells and is discussed further in Chapter 20.

Table 16.2: Properties of the light alcohols of the composition $C_nH_{2n+1}OH$. Heat of combustion is the HHV (Chapter 1).

n	name	formula	molecular mass (g/mol)	density (g/cm³)	boiling point (°C)	heat of combustion (MJ/L)
1	methanol	CH_3OH	32.04	0.792	64.7	17.9
2	ethanol	C_2H_5OH	46.07	0.789	78.4	23.5
3	propanol	C_3H_7OH	60.10	0.785	82.3	26.3
4	butanol	C_4H_9OH	74.12	0.810	117.7	29.7

© Cengage Learning 2015

Example 16.2

Calculate the mass of CO_2 (in kilograms) produced by the combustion of 1 kg of methanol.

Solution

Methanol combines with oxygen to yield CO_2, H_2O, and energy. Balancing the chemical formula shows that the reaction is

$$2CH_4O + 3O_2 \rightarrow 2CO_2 + 4H_2O + energy.$$

Therefore, 2 moles of methanol yields 2 moles of carbon dioxide, or a 1:1 molar ratio. Using approximate molar weights of methanol (32 g/mol) and carbon dioxide (44 g/mol) means that 1 g of methanol produces (1 g) $\times$ (44 g/mol)/(32 g/mol) = 1.37 g of CO_2. Thus 1 kg of methanol produces 1.37 kg of carbon dioxide.

Currently about 95% of all ethanol is bioethanol; the remainder is produced from petroleum. Ethanol is the alcohol found in alcoholic beverages and is also used in various industrial processes. The majority of ethanol produced at present is used for the production of energy. Energy from bioalcohol is a manifestation of solar energy. Light from the sun produces glucose ($C_6H_{12}O_6$) by means of photosynthesis using water and CO_2. The process is

$$6CO_2 + 6H_2O + light\ energy \rightarrow C_6H_{12}O_6 + 6O_2. \tag{16.1}$$

Living organisms can modify glucose to form fructose (same chemical formula as glucose but different structure), bond glucose together in strings to form starch or cellulose, or bond glucose to fructose to form sucrose (normal table sugar). The fermentation of simple glucose produces ethanol, along with CO_2 and heat, according to the process

$$C_6H_{12}O_6 \rightarrow 2C_2H_5OH + 2CO_2 + heat. \tag{16.2}$$

The combustion of ethanol produces CO_2 and water along with heat:

$$C_2H_5OH + 3O_2 \rightarrow 2CO_2 + 3H_2O + heat. \tag{16.3}$$

The photosynthesis process is inherently inefficient (perhaps about 2%), meaning that the production of electricity by burning ethanol in a heat engine has an overall efficiency that is considerably lower than that achieved by photovoltaics. However, as the technologies involved in these two approaches are quite different, a more detailed analysis of biofuel use is needed.

The steps in the commercial production of ethanol from organic matter are: fermentation, distillation and dehydration. The details of these are as follow:

16.3a Fermentation

Traditional fermentation processes convert simple sugars (including glucose, fructose, sucrose, and starches) to ethanol according to equation (16.2). More complex processes (discussed later in the chapter) are required to break down cellulose prior to fermentation. Some plant material (e.g., sugarcane) contains a significant proportion of simple sugar, whereas others, such as corn, contain much less. For example, the current production of ethanol from corn utilizes only about 50% of the dry corn kernel [although the remainder of the plant may be used for other purposes (e.g., livestock feed, etc.)].

16.3b Distillation

The distillation of the fermentation product is normally required to remove water from the ethanol. Traditional distillation techniques yield an *azeotropic* mixture of ethanol with about 4% water. Although this mixture can, in principle, be used directly to produce energy by combustion, the presence of water makes it immiscible with gasoline, limiting its use for fuels that are a gasoline-ethanol mixture (see section 16.3d). It is therefore generally necessary to use dehydration methods to remove the remaining water.

16.3c Dehydration

Traditional dehydration techniques mix, say, benzene with the azeotropic ethanol-water mixture produced by distillation. The water preferentially mixes with the benzene, and the ethanol can be separated. While this technique allows for the extraction of high-purity dehydrated ethanol, the waste material contains benzene, a known carcinogen. To eliminate the health and environmental hazards associated with carcinogenic materials, new benzene-free techniques have been developed that use molecular sieves to absorb water and allow the ethanol to be extracted. These techniques are now in common use and can take the place of both the distillation and dehydration processes.

16.3d Use of Ethanol

Ethanol can be used in its pure form as a fuel in an internal combustion engine, or it may be blended with gasoline in various proportions. Fuel mixtures are designated by the percentage (by volume) of ethanol where, for example, E10 represents a mixture of 10% ethanol and 90% gasoline. The mixing of ethanol with gasoline as a transportation fuel dates back to the 1920s. It became popular in the 1970s during the oil crisis when E10 fuel was commonly marketed under the name of *gasohol*. Normal gasoline internal combustion engines can use ethanol-gasoline mixtures up to about E10, or in some cases slightly higher, without any modification. In some countries, low-ethanol blends have been mandated by government regulation (e.g., E10 in Sweden and E5 in India). As of 2008, nine states in the United States have mandated the use of E10. In other states, the addition of ethanol to gasoline is a common practice, and ethanol in some amount is present in more than 60% of the gasoline sold in the United States. In some cases (e.g., see Figure 16.3), this is indicated on the gasoline pump, although this is not always the case.

The use of higher percentages of ethanol in gasoline requires minor modifications to the engine design. Most modifications are related to the increased reactivity of polymers to ethanol containing fuels and require the replacement of some plastic components in the fuel delivery system. Typical modifications to an engine that will allow the use of E85 cost less than US$100. However, retrofitting older engines can be problematic because the ethanol dissolves accumulated organic deposits in the fuel system (i.e., those that are relatively insoluble in gasoline) and causes clogging in the engine. Many vehicles sold in North America (referred to as *flex fuel* vehicles) come with the ability to utilize ethanol mixtures up to E85 (Figure 16.4). In considering fuels

Figure 16.3: Gasoline pump in Maine indicating 10% ethanol content (E10).

Figure 16.4: Nameplate on vehicle designed to run on gasoline with up to 85% ethanol (E85).

with higher ethanol content, it is important to realize the lower energy content (per unit volume) of ethanol compared with gasoline when analyzing fuel consumption, fuel cost, and driving range. Ethanol (Table 16.2) has an energy content of 23.5 MJ/L whereas gasoline has 34.8 MJ/L, although increased engine efficiency may partially offset these differences. The utilization of ethanol-containing fuels in the gasoline engines of recreational vehicles has been somewhat controversial, particularly for marine engines. In addition to the potential clogging problems, the presence of ethanol in fuel stored in a humid marine environment can promote the condensation of moisture from the air, resulting in performance loss and mechanical difficulties. This problem is often exacerbated by the fact that boaters often replace fuel less frequently in their boats than drivers do in their automobiles.

Example 16.3

For vehicles with the same range, calculate the fuel tank volume and equivalent fuel mass for vehicles using methanol and ethanol compared to a gasoline-powered vehicle with a fuel tank volume of 70 L. Assume that the engine efficiencies are the same for all three fuels. (*Note:* The density of gasoline depends on the exact ratio of the different hydrocarbons present but has an average value of about 0.72 kg/L.)

Solution

If the energy content of gasoline is 34.8 MJ/L, a 70 L tank would have a total energy content of 34.8 MJ/L × 70 L = 2436 MJ. Using the energy content per unit volume given in Table 16.2 for methanol and ethanol (17.9 MJ/L and 23.5 MJ/L, respectively), the volumes of these two biofuels necessary to produce the same energy as 70 L of gasoline are

$$\text{Methanol:} \qquad \frac{2436 \text{ MJ}}{17.9 \text{ MJ/L}} = 136 \text{ L}$$

and

$$\text{Ethanol:} \qquad \frac{2436 \text{ MJ}}{23.5 \text{ MJ/L}} = 104 \text{ L.}$$

Using the densities from Table 16.2, the masses of these two biofuels are

$$\text{Methanol:} \qquad (136 \text{ L}) \times (0.792 \text{ kg/L}) = 108 \text{ kg}$$

and

$$\text{Ethanol:} \qquad (104 \text{ L}) \times (0.789 \text{ kg/L}) = 82 \text{ kg.}$$

By comparison, 70 L of gasoline have a mass of (70 L) × (0.72 kg/L) = 50 kg. This shows that biofuel-powered vehicles must accommodate both larger volumes and masses of fuel.

The breakdown of world ethanol production is shown in Table 16.3. Currently, about 90% of ethanol production is for fuel purposes. The increase in fuel ethanol production worldwide since the mid-1970s is shown in Figure 16.5. From Table 16.3, it is seen that almost 70% of the ethanol produced worldwide is produced in either the

| Table 16.3: Ethanol production (for 2009). ||
country	ethanol production 10⁶ L
United States	40,121
Brazil	24,898
European Union	3936
China	2051
Thailand	1646
Canada	1098
Columbia	314
India	348
Australia	216
Other	1749
world total	**73,940**

© Cengage Learning 2015

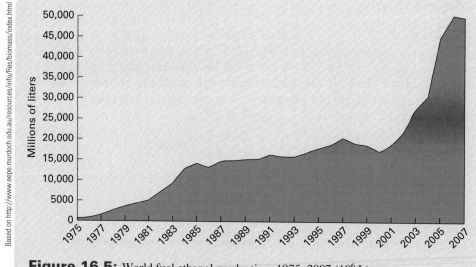

Based on http://www.eepe.murdoch.edu.au/resources/info/Res/biomass/index.html

Figure 16.5: World fuel ethanol production, 1975–2007 (10^6 L).

United States or Brazil. It is interesting to compare the approaches taken to fuel ethanol production in these two countries.

In the United States, virtually all fuel ethanol production is produced from corn. As previously explained, only a fraction of the corn plant can be utilized in ethanol production using current technology. Celluosic ethanol production (i.e., the use of plant cellulose rather than just the fermentation of sugars) would provide a substantial increase in opportunities for ethanol production in the United States and elsewhere by allowing for greater utilization of plants like corn and also allowing for the utilization

Table 16.4: A comparison of ethanol production and use in the United States and Brazil.

property	United States	Brazil
major crop	corn	sugarcane
total ethanol production (2007)[10^6 L]	24,600	19,000
total arable land [10^6 km^2]	2.70	3.55
land used for ethanol production [10^5 km^2]	1.0	0.36
percent arable land used for ethanol production	3.7%	1.0%
productivity [L/km^2]	380,000–400,000	680,000–800,000
energy balance	~1.4	~9.2
ethanol fueling stations	1700	33,000
percent fueling stations selling ethanol	1.0	100
ratio of fuel ethanol to gasoline used	0.04	1.0

of other (low-sugar-containing) plant species such as grasses. At present, this is not technologically viable but may be developed at some point in the future. The utilization of sugarcane as a source material for ethanol production has recently been initiated in some states with warmer climates (i.e., Louisiana, Hawaii, Texas, and Florida), but in most parts of the country, sugarcane is not a viable crop. Some of the properties of ethanol production in the United States are summarized in Table 16.4.

In Brazil, sugarcane is used almost exclusively for ethanol production. Brazil has seriously pursued the use of ethanol as a fuel for more than 30 years; some of the characteristics of their program are summarized in Table 16.4. The improvement in ethanol production per area of farmland over the years in Brazil is illustrated in Figure 16.6. While normal U.S. passenger vehicles typically tolerate up to 10% ethanol in gasoline

José Goldemberg. "The Brazilian biofuels industry." Biotechnology for Biofuels 2008, 1:6. © 2008 Goldemberg; licensee BioMed Central Ltd.

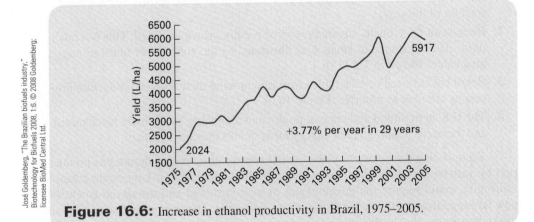

Figure 16.6: Increase in ethanol productivity in Brazil, 1975–2005.

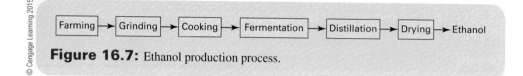

Figure 16.7: Ethanol production process.

mixtures and flex fuel vehicles can use up to E85, vehicles in Brazil are designed to operate on a much wider variety of fuels. Most run on any mixture up to E100. In fact, flex fuel vehicles in Brazil can utilize hydrated ethanol (that is the azeotrope consisting of about 96% ethanol with 4% water), thus eliminating the need for additional dehydration after distillation.

A comparison of the information in Table 16.4 for the United States and Brazil shows that, although the total ethanol production in the two countries is similar, the program in the United States is a minor addition to gasoline use, while in Brazil it fulfils the major portion of transportation fuel needs. This comparison is clear from the fraction of fueling stations in the two countries that sell ethanol-rich fuel mixtures: 1% in the United States and 100% in Brazil. It is also seen in the table that U.S. ethanol production utilizes a larger fraction of available arable land than is the case in Brazil. To understand why ethanol production and use seem to be much more successful in Brazil than it is in the United States, it is important to look at some of the details of the ethanol production process. As described, plant material is grown, ground, and fermented and the resulting ethanol is distilled and possibly dehydrated (Figure 16.7). At present, each of these steps, at least in the United States, is accomplished by machines that are fueled by fossil fuels or utilize electricity generated largely by burning coal. This will continue to be the situation until ethanol production becomes self-sustaining. Even so, energy input is required to produce ethanol, and the ratio of energy input to the energy content of the ethanol produced is the important factor. In the United States, this ratio is about 1.3, meaning that 1 J of energy input produces ethanol with an energy content of 1.3 J; that is a gain of 0.3 J of energy. Although there is a net energy gain in the process, the numbers indicate that the 0.3 J of energy gained is expensive in terms of both energy and dollars. In Brazil, the ratio is about 9.2, meaning that 1 J of energy input into ethanol production yields a net gain of 8.2 J—clearly a much more advantageous situation.

Several factors contribute to this situation:

1. Brazil has a climate that is much more conducive to plant growth than much of the United States has. Thus, plants grow faster and can be grown over a greater fraction of the year.

2. Because of the climate, sugarcane can be readily grown in Brazil. This is a much more efficient source of ethanol, as illustrated by the volume-per-unit-farming-area values shown in Table 16.4.

3. Brazil has invested considerable effort into making their ethanol production process as efficient as possible (Figure 16.6).

4. The U.S. agricultural and ethanol production technologies are more mechanized and therefore require greater energy input than is the case in Brazil.

There is also the question of the overall contribution to greenhouse gas production by various energy production methods. In the United States, large-scale ethanol utilization would require the development of new farmland and the development of new farming infrastructure (e.g., tilling of new soil, etc.). The carbon payback period in the United States, that is, the time required to compensate for the carbon released to the atmosphere during the development of ethanol production (recall the carbon payback period for hydroelectric discussed in Energy Extra 11.2), has been estimated to be about 93 years. A similar period for Brazil is about 17 years. This difference results from the greater difficulty in growing crops with a greater ethanol return in a more temperate climate coupled with the greater degree of mechanization (and associated carbon emissions) in the U.S. agricultural system.

A final point that needs to be considered is the impact of fuel ethanol production on food production. The extensive use of farmland for the production of plant matter for ethanol production would clearly impact the ability to produce food crops. Such food crops are essential both for direct human consumption and for use as feedstock for animals, although some use of waste material from ethanol production from corn for livestock feed is possible. The extent of the conflict between ethanol production and food production in the United States is emphasized by the fact that, if all corn currently grown in the United States were used for ethanol production, it would replace only about 12% of the gasoline used. Thus, the large-scale use of ethanol will require considerable new farming activity or a decrease in food productivity or (most likely) both. A decrease in food production will lead to a decrease in food supply and an increase in prices. Agricultural methods that are the most effective at maximizing ethanol production will likely lead to a decline in soil fertility, the increased use of pesticides and fertilizers (which require energy for their production), increased deforestation (to make additional land available), and a reduction in the availability of water for irrigation. Thus, in addition to the reallocation of existing farmland, increased ethanol production can have adverse indirect consequences on food production and the environment. Cellulostic ethanol production will increase the productivity of ethanol from corn and open up the possibility of producing ethanol from other crops. This could include the use of grasses that can be grown on land that is less desirable for food production. Switchgrass (*Panicum virgatum*) is a rapidly growing grass that thrives in most parts of the United States and in a wide variety of habitats. The cellulostic ethanol industry is in its very early stages of development, and further technological advances, as well as careful integration into the agricultural system, will be needed to benefit fully from this approach.

Overall, a comparison of the Brazilian and American approaches to ethanol production emphasizes the importance of climate, as well as the development of efficient production technologies. It also makes it clear that, from a climatic and geographical perspective, ethanol production is not a viable energy alternative in all parts of the world.

16.4 Biodiesel

Biodiesel fuel is comprised of short chain alkyl esters and is similar in most ways to traditional petroleum-derived diesel fuel. It is made by the transesterification of vegetable oils or animal fats. Biodiesel is distinguished from unprocessed vegetable oil (often called straight vegetable oil, SVO), which is sometimes used as a fuel. The latter may be nonfood-grade vegetable oil or waste vegetable oil from the food industry. Although SVO may be an attractive alternative to diesel fuel, engines require significant modifications to utilize this fuel. Biodiesel, on the other hand, can readily replace petroleum-derived diesel fuel, much as ethanol can be used to replace gasoline. Minor engine modifications, as for ethanol use, are required because the natural rubber components often used in diesel engines may degrade upon exposure to biodiesel. Also, engines that have previously run on petroleum diesel typically have deposits left behind from the fuel, and these are soluble in biodiesel fuel, possibly leading to clogging of filters. In addition to use in vehicles, biodiesel is a direct replacement for domestic or commercial heating fuel, subject to the solubility issues described. It may also find use as a fuel for trains or aircraft.

Biodiesel may be blended with petroleum diesel, and the resulting fuels are designated by the percentage of biodiesel (e.g., B5 for 5% biodiesel, 95% petroleum diesel etc.), up to B100 for 100% biodiesel. B99 is a common fuel in some parts of the world. The 1% petroleum diesel is added to biodiesel to retard mold growth. The compatibility of biodiesel blends with unmodified diesel engines is somewhat unclear, although blends up to B5 are generally felt to be acceptable.

While waste vegetable oil and waste animal fat may seem like an attractive source material from which to make biodiesel, it is unlikely to be a major component in the production of this fuel. The collection of waste oil may be inconvenient or difficult, as well as being expensive. The major problem, however, is availability. In the United States, approximately 1.8×10^{11} L of diesel fuel are used for transportation and heating annually. Estimated primary production of vegetable oil is about 1.3×10^{10} L per year, and the production of animal fat is about half that amount. Thus, waste from the food industry could account for, at most, about 10% of the diesel needs in the United States It is therefore necessary to utilize predominately source material that is grown for the specific purpose of producing biodiesel. In the United States, most current biodiesel production uses oil extracted from soy. Supplying the U.S. needs for diesel fuel with biodiesel produced from soy would require about 2×10^6 km^2 of farmland. This is the majority of all arable land available (Table 16.4). Clearly this is not a viable situation; biodiesel would replace food production (and ethanol production as well). Other sources of materials for biodiesel need to be found if this energy source is going to make a significant contribution to our energy needs. Table 16.5 shows the productivity for various plants that can be used for biodiesel production. Productivity from all common terrestrial oil-producing plants that are compatible with a temperate climate is similar. Tropical plants (palm and coconut) are somewhat more productive. However, algae is clearly the best choice for producing biodiesel by far. An additional feature of algae is that it can be grown in marine environments and in ponds on land that is not otherwise suitable for farming. Total U.S. diesel needs could be satisfied by algae grown in an area of about 100,000 km^2, or about the land area of Iceland or South Korea. Certainly future developments in biodiesel production must consider algae as an important possibility.

Table 16.5: Typical annual productivity of biodiesel from different plant materials per unit farming area.

plant	annual biodiesel production 10^3 L/km^2
algae	1700
palm oil	475
coconut	215
rapeseed	95
soy	55–91
peanut	84
sunflower	77

Example 16.4

The actual ocean coast line of a country is a somewhat ambiguous quantity, depending on the scale on which harbors, islands, estuaries, and other features are included. For example, the coastline of the continental United States has been estimated at a minimum of about 8000 km or a maximum of about 80,000 km, if all small-scale features are included. Assuming, from a practical standpoint, that a maximum of 16,000 km of the U.S. coastline to a distance of 100 m off shore is utilized for algae production to make biodiesel, what fraction of the total U.S. diesel needs could be met?

Solution

The total available area for algae production is

$$(16000 \text{ km}) \times (0.1 \text{ km}) = 1600 \text{ km}^2.$$

The annual productivity of biodiesel from algae is given in Table 16.4 as 1.7×10^6 L/km^2. The available coastal area would allow for the production of

$$(1600 \text{ km}^2) \times (1.7 \times 10^6 \text{ L/km}^2) = 2.7 \times 10^9 \text{ L of biodiesel per year.}$$

If the total diesel use in the United States is 1.8×10^{11} L, the amount that can be produced from algae grown under the conditions specified in this example would amount to $(2.7 \times 10^9 \text{ L})/(1.8 \times 10^{11} \text{ L}) = 0.015$, or 1.5% of the requirement. This calculation emphasizes the magnitude of the agricultural task of producing biofuels.

In recent years, biodiesel production has increased substantially, particularly in Europe where more than 85% of the world production occurs (Figure 16.8). World production in 2005 was about 4×10^9 L, with Germany the clear leader in the production of biodiesel.

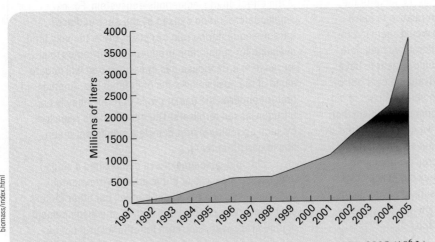

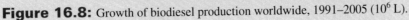

Figure 16.8: Growth of biodiesel production worldwide, 1991–2005 (10^6 L).

ENERGY EXTRA 16.1
Environmental effects of biofuels

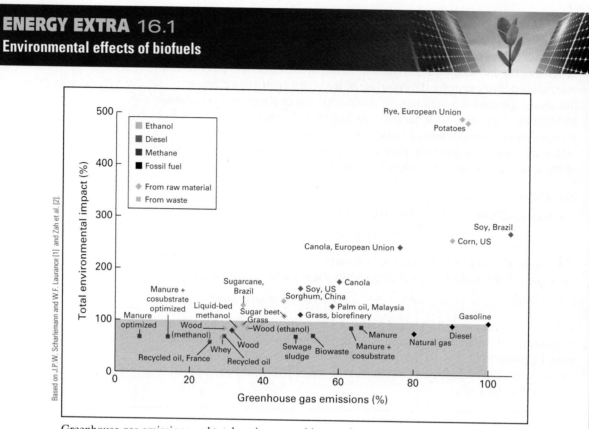

Based on J.P.W. Scharlemann and W.F. Laurance [1] and Zah et al. [2].

Greenhouse gas emissions and total environmental impact factors for some biofuels.

As illustrated by the comparison in the text between ethanol production from Brazilian sugarcane and from U.S. corn, it is clear that all biofuels do not offer the same degree of carbon reduction relative to fossil fuels. While Brazilian sugarcane would seem to provide substantial benefits as an alternative fuel, corn produced in the United States would appear to have only a marginal environmental advantage over petroleum. The question of the environmental impact of biofuels, however, is a much more complex problem than merely a comparison of the relative carbon footprint. A recent analysis by Zah and colleagues[1] has provided a quantitative analysis of the overall environmental consequences of the use of various biofuels by establishing a single quantitative measure of environmental impact. Scharlemann and Laurance[2] have provided further clarification of this approach.

One of the factors that contribute to the overall environmental impact of a particular fuel assesment is the effect on the indigenous ecosystem. For example, deforestation caused by the destruction of carbon-sequestering rain forests in Brazil to make land available for sugar cane production contributes to an increase in greenhouse gas emissions that will negate some of the positive benefits of ethanol use. Another factor concerns the use of crops that rely heavily on nitrogen-based fertilizers. The nitrous oxide released by such agricultural practices is a significant greenhouse gas (Chapter 4).

Zah and colleagues have established a single quantitative measure of the overall environmental consequences of a particular biofuel, but other factors are difficult to quantify. For example, the displacement of food crops with, say, corn for ethanol production

Continued on page 407

will likely affect market prices for certain crops, in turn, possibly altering the approach to agricultural land utilization. However, Zah and colleagues' analysis is a huge step forward in quantifying the benefits of biofuels.

A summary of Zah and colleagues' analysis of biofuels is shown in the figure. All data are normalized to the values for gasoline; to be environmentally advantageous to gasoline, a fuel must have both of these factors less than 100%. Very nearly all biofuels are acceptable in terms of greenhouse gas emissions. However, consistent with the discussion in the text, it is seen that Brazilian sugarcane does quite well in this area, whereas U.S. corn is fairly marginal. A large fraction of biofuels fail to meet the criterion for total environmental impact. Those that fail include both Brazilian sugarcane and U.S. corn (which is one of the worst performers in this respect). The analysis by Zah and colleagues did not include crops, such as switch grass, which may, at some point in time,

provide a suitable material for cellulosic ethanol production.

A careful consideration of the type of analysis present by Zah and colleagues is crucial in establishing policies for biofuel production.

1. R. Zah et al., *Ökobilanz von Energieprodukten: Ökologische Bewertung von Biotreibstoffen* (Empa, St. Gallen, Switzerland, 2007).
2. J. P. W. Scharlemann and W. F. Laurance, How green are biofuels?" *Science* **319** (2008): 43–44.

Topic for Discussion

Discuss the relative merits (in the United States) of the following two possibilities:

- Converting land used for growing corn as a food crop to land for growing corn for ethanol production
- Erecting wind turbines in corn fields (grown for food) and use as much of the land as possible for dual purposes

The question of the environmental impact of biodiesel use is important. Compared with petroleum-derived diesel, biodiesel produces less sulfur compounds, less hydrocarbon emission, and fewer particulates. However, it produces somewhat more NO_x. Clearly, the combustion of a hydrocarbon releases carbon to the atmosphere in the form of CO_2. Although carbon is absorbed from the atmosphere by the growth of plant material (such as soy), a detailed and reliable evaluation of the net carbon contribution from biodiesel use is difficult at best. As with ethanol use, many factors must be considered, including the carbon payback period for infrastructure development, and this depends greatly on crop selection and production methods.

16.5 Municipal Solid Waste

Humans produce considerable quantities of waste. This includes household waste, as well as commercial and industrial waste. Municipal solid waste refers to waste from all sources but does not include chemically or biologically hazardous waste or radioactive waste because these must be disposed of by special means according to the hazards they pose. It also does not include sewage and other nonsolid waste. Municipal solid waste does include durable goods such as old appliances and furniture, as well as nondurable goods, such as newspapers and food waste. The most appropriate method of disposing of municipal solid waste depends on the nature of the waste. Methods include recycling (as is appropriate for metals, glass, and plastics), composting (as is appropriate for

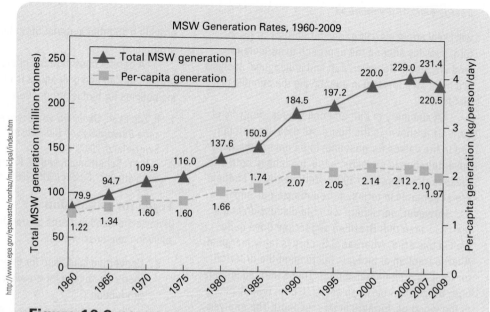

Figure 16.9: Municipal solid waste production in the United States. The left axis is total annual production for the U.S. and the right axis is the daily per capita production.

biodegradable materials), landfilling, and waste-to-energy incineration (as is appropriate for combustible materials).

The amount of waste produced per person has increased over the years. An example for the United States is shown in Figure 16.9. Per capita, the United States has one of the highest rates of waste production. Other similarly industrialized countries typically produce less waste per capita; for example, values for Canada and New Zealand are about half those shown for the United States in Figure 16.9. Over the years, the ways in which waste has been treated have changed. The percentages of waste disposal by various methods in the United States are shown in Figure 16.10. While composting, incineration, and recycling have increased over the past 45 years, landfilling is still clearly the major method of waste disposal. Overall, waste management is an extensive and complex problem; however, the use of waste for the production of energy is relevant to this discussion.

The utilization of municipal solid waste to produce energy can be approached in several ways. Noncombustible materials, such as glass and metal, can first be removed, and the remaining combustible material can be burned directly to produce heat, which can then be used to generate steam to run a turbine and generator. This is the most common approach in the United States. Another simple approach is to collect gas from the decomposition of organic material at landfill sites (mostly methane) and burn it to produce heat and subsequently electricity. A more sophisticated approach is to shred the combustible material to produce refuse-derived fuel (RDF). RDF material can then be formed into pellets for combustion, or it can be heated to produce gas (i.e., gasification). At low cooking temperatures, mostly methane is given off, whereas at higher

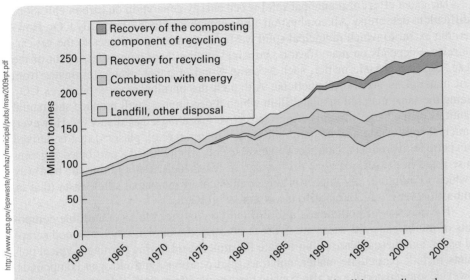

http://www.epa.gov/epawaste/nonhaz/municipal/pubs/msw2009rpt.pdf

Figure 16.10: Relative importance of various municipal solid waste disposal methods in the United States.

temperatures the total gas production increases, and the gas is largely hydrogen. Methane can be used for the thermal generation of electricity, and hydrogen is suitable for combustion or fuel cell utilization (Chapter 20).

Society produces substantial quantities of waste, and there are benefits to the proper disposal of this waste, both from an energy standpoint and an environmental standpoint. It is important, however, to realize that the possible contribution of RDF to our overall energy needs is quite small.

Example 16.5

Estimate the percentage of the total energy needs of the United States that could be provided by MSW if the total content of this waste is 10 MJ/kg and it is utilized with an efficiency of 30%.

Solution

From Figure 16.9, the daily average per-capita waste production in the United States is about 2 kg. This has an energy content of 20 MJ per day or 7.3 GJ per year. From Figure 2.4, the per-capita primary energy use in the United States is about 330 GJ per year. Thus, RDF can supply a maximum fraction of 7.3/330 = 0.022, or about 2% of the energy needs. This is a fairly optimistic estimate because it does not take into account the energy required to collect and process the MSW.

The exact effect of municipal solid waste energy generation on carbon emissions is difficult to determine. All combustion emits greenhouse gases, primarily CO_2. However, the extent to which municipal solid waste combustion contributes to the net carbon release depends on many factors, principal among them being the content of the waste. Burning organic matter, such as scrap wood or paper, which originates from plant material, is generally beneficial. Although the burning process produces CO_2, placing the same material in a landfill in which it decomposes will produce substantial quantities of methane, which is a more effective greenhouse gas than CO_2. However, municipal solid waste also contains substantial material (e.g., plastics) that is derived from petroleum products. This is not renewable and, when burned, contributes to greenhouse gases in the same way as the burning of fossil fuels. Determining the best way in which to minimize the impact of this component of municipal solid waste (that is, combustion, recycling, or landfill) is a complex problem.

The question of pollution is also difficult to analyze. Most combustible components of municipal solid waste are hydrocarbon-based materials, such as wood scraps and paper, and these release pollution into the atmosphere during burning. Typical pollutants include particulate matter, unburned hydrocarbons, and nitrogen compounds. However, some material may also contain toxic components such as lead, mercury, and cadmium. It is important from an environmental standpoint to remove these materials prior to combustion in order to avoid their release into the atmosphere or into groundwater supplies.

The overall effectiveness in producing energy from municipal waste, compared to the benefits of recycling and/or composting, needs to be established.

16.6 Summary

Wood is the most traditional biomass-based energy, and has been a component of our energy production for many years. As discussed in this chapter, a portion of the use of wood may be considered carbon neutral (or at least low carbon) because it includes an awareness of the renewability of the resource. This, however, is not true of wood use in many parts of the world where this resource is used to produce energy at the expense of deforestation.

This chapter described the straightforward method of producing ethanol from organic matter that contains glucose. Ethanol production has been successful in Brazil but less so in countries with more temperate climates. The economic and environmental factors related to ethanol production in, for example, the United States are not simple to evaluate. The relatively high cost from both an economic and an energy standpoint, coupled with the agricultural demands on farm land, make the wide-scale conversion from fossil fuels to biofuels for transportation use a difficult task. Certainly, the development of efficient celluosic ethanol production methods would make the situation much clearer, although much research is necessary to make this a reality.

Biodiesel is derived from vegetable oils or animal fats and is a direct replacement for petroleum-based diesel fuel, either alone or as a mixture. The production of biodiesel is subject to considerations similar to those of ethanol production. The availability of arable land resources is a major concern, and current technologies may be viable only in regions with tropical climates. The ability to utilize plant material grown in an aquatic environment or on land that is otherwise not agriculturally useful would be a significant factor.

The chapter showed that the environmental consequences of biofuel utilization are not straightforward to assess. At present, fossil fuel energy is a major component of the energy used to produce biofuels, and as a result, the net reduction of greenhouse gas emissions may be less than perceived. Also, the interrelationship of agriculture for biofuel production, and agriculture for food production, needs to be carefully analyzed.

The use of waste material (e.g., municipal solid waste or agricultural by-products) for the production, of heat/electricity or for conversion into appropriate transportation fuels, may form a small component of our future energy mix. Although it is not always obvious that significant net economic and environmental benefits are associated with the utilization of refuse-derived fuels, the alternatives for waste management may be less attractive. The recovery of some of the energy content of waste material has advantages over the mere use of energy for waste collection and processing, followed by landfill greenhouse gas emissions, without the energy benefits.

Problems

16.1 How many tonnes of wood per day would be needed for a wood-fired generating station that produces 1000 MW_e continuously (typical of a coal-fired facility) at 35% efficiency?

16.2 Calculate the mass of CO_2 (in kilograms) produced by the combustion of 1 kg of ethanol.

16.3 Calculate the mass of CO_2 (in kilograms) per MJ of energy produced for the hydrocarbons with $n = 1$ through 4 shown in Table 16.2.

16.4 Following Example 16.3, estimate the necessary fuel tank volume and fuel mass for a vehicle operating on E85 as compared to a gasoline vehicle in order to achieve the same driving range.

16.5 Photosynthesis is a very inefficient process for converting sunlight into usable chemical energy. While the theoretical efficiency is 25%, the actual efficiency is affected by the wavelength distribution of the light, the absorptance of the plant matter, and various other factors and may typically be in the range of about 1%. Using the energy content of wood as a guide and the average solar insolation of 168 W/m^2 (Chapter 8), how much land area would be needed to fulfill an average person's energy needs (in the United States)? Assume that biomass energy content can be converted into end user energy with the same efficiency as current energy sources (Chapter 2).

16.6 A large maple tree collects sunlight over an area that is 8 m in diameter. The average solar radiation (Chapter 8) is 168 W/m^2. The tree grows for 8 months of the year and is dormant for 4 months of the year. After 10 years, the mass of the tree has increased by 540 kg. What is the efficiency of converting sunlight into chemical energy?

16.7 Using information from Figure 16.9 and an energy content of MSW of 10 MJ/kg, estimate the number of 35% efficient 1000 MW_e coal-fired power plants that could be eliminated if 50% of MSW were burned to produce electricity.

16.8 Two methods are used to generate 1 GWh$_e$ of electricity: (1) burning coal with a generation efficiency of 30% and (2) burning MSW with an efficiency of 25%. For MSW generation, assume an energy content of 10 MJ/kg and a total net energy requirement of 5000 MJ/tonne for collection and transportation, which is provided by burning fossil fuels at 20% efficiency. The MSW consists of 50% carbon-neutral organic material and 50% fossil fuel–derived material. Compare the relative contributions to greenhouse gas emissions for these two methods. Ignore any effects of possible carbon sequestration.

Bibliography

G. Boyle (Ed.). *Renewable Energy*. Oxford University Press, Oxford (2004).

R. C. Brown. *Biorenewable Resources: Engineering New Products from Agriculture*. Iowa State Press, Ames (2003).

H. S. Geller. "Ethanol fuel from sugar cane in Brazil." *Annual Review of Energy* **10** (1985) 135–164.

R. Hinrichs and M. Kleinbach. *Energy: Its Use and the Environment*. Brooks-Cole, Belmont, CA (2002).

B. K. Hodge. *Alternative Energy Systems and Applications*. Wiley, Hoboken, NJ (2010).

A. Nag. *Biofuels Refining and Performance*. McGraw-Hill, New York (2008).

L. Olsson (Ed.). *Biofuels (Advances in Biochemical Engineering/Biotechnology)*. Springer-Verlag, Berlin (2007).

G. Pahl. *Biodiesel: Growing a New Energy Economy*. Chelsea Green Publishing, White River Junction, VT (2005).

D. Pimentel and T. Patzek. "Ethanol production using corn, switchgrass, and wood; Biodiesel production using soybean and sunflower." *Natural Resources Research* **14** (2005): 65.

P. Quack. *Energy from Biomass: A Review of Combustion and Gasification Technology*. World Bank, New York (1999).

V. Quaschning. *Renewable Energy and Climate Change*. Wiley, West Sussex, United Kingdom (2010).

R. A. Ristinen and J. J. Kraushaar. *Energy and the Environment*. Wiley, Hoboken, NJ (2006).

J. Tester, E. Drake, M. Driscoll, M. W. Golay, and W. A. Peters. *Sustainable Energy: Choosing Among Options*. MIT Press, Cambridge, MA (2006).

F. M. Vanek and L. D. Albright. *Energy Systems Engineering: Evaluation and Implementation*. McGraw Hill, New York (2008).

PART V

Energy Conservation, Energy Storage, and Transportation

In this section of the book, mechanisms for energy conservation, energy storage, and the ways in which energy can be used for transportation purposes are described. Energy conservation must go hand in hand with the development of sustainable energy development. The reduction of energy needs and the efficient use of the energy produced will help to maximize the effectiveness of any energy strategy. Energy conservation efforts may be based on improved energy utilization technologies, but they may also be based on changes in how we view energy use. Options for energy conservation may be things that we can do as individuals with the products that we purchase, or with our energy use in our homes and for transportation. Energy conservation may also be viewed on a much larger scale in terms of the development of new energy-efficient products and the implementation of proactive national energy policies. These factors are considered in Chapter 17.

Energy utilization is efficient and effective only if the energy is available when and where it is needed. The need for energy is not always readily compatible with the way in which it is produced, particularly for many of the nontraditional energy technologies. Wind energy is available only when the wind blows, and solar energy is available only during the daytime. Our energy needs, however, are not necessarily coupled to the behavior of nature. Most alternative energy technologies produce electricity. This is readily distributed on the electric grid for residential or commercial use, but appropriate technologies are needed to use electricity for transportation.

The availability of energy when and where it is needed depends on the implementation of appropriate energy storage mechanisms, as well as an effective distribution technology. These aspects of efficient energy use are often not given the serious consideration that they deserve. Inevitably, the storage of energy involves the conversion of one type of energy to another (and generally back again), and these conversions are always less than 100% efficient. The most appropriate energy storage mechanism for a particular application depends on a number of factors: the form of end-use energy, the amount of energy that needs to be stored, the rate at which energy needs to be stored and extracted (i.e., the power), and any space or weight constraints (i.e., portability).

Technologies that are potentially viable for large-scale energy storage to even out supply and demand variations on the electric grid are considered. For the most part, these systems are not portable, and size and weight considerations are secondary.

Reliability, efficiency of energy storage, and conversion, as well as economy of infrastructure and operation, are generally the first concerns (Chapter 18).

Finally, in Chapters 19 and 20, technologies for portable energy storage are discussed. These technologies are appropriate for application to transportation. High-energy-density storage technologies, which mean small and lightweight devices, are a crucial factor for providing energy for vehicles. The two major competing approaches to this problem, batteries and hydrogen, are discussed in detail.

The photograph at the beginning of this part of the text shows a Tesla roadster being charged. This battery electric vehicle is produced by Tesla Motors in California and was first sold in 2008. It is the first production electric vehicle to utilize Li-ion batteries and the first to have a range of more than 320 km. ■

Energy Conservation

Learning Objectives: After reading the material in Chapter 17, you should understand:

- How government energy policies deal with conservation matters.
- How combined electricity and heat production can make the best use of resources.
- Approaches to electricity distribution and how the smart grid can be used to integrate different energy sources and regulate the use of electricity.
- Energy conservation in the community through the use of LED streetlights.

- The efficient operation of residential HVAC systems.
- The application of heat pumps to space heating needs.
- Reducing heat transfer in and out of buildings.
- High-efficiency lighting technologies.
- Vehicle fuel efficiency and government standards.
- The viability of hybrid vehicles as a means of fossil fuel conservation.

17.1 Introduction

While most of this text has dealt with methods of producing energy (i.e., extracting energy from our environment and converting it to a form that suits our needs), it is also important to make efficient use of the energy that we produce. Energy resources are limited, and the more efficiently they are utilized, the more available they will be in the future. The *conservation of energy* (Chapter 1) is a basic principle of physics. The term *conservation of energy*, or, more commonly, *energy conservation*, can also refer to actions taken to best minimize the quantity of energy that we use to fulfill our needs. Actions to minimize our energy use can occur on several levels, from the scale of the individual to a global scale. For example, we can control energy conservation measures: turning lights off when they are not being used or driving more fuel-efficient vehicles. On a somewhat larger scale, energy can be conserved by public utilities by improving the efficiency of electricity-generating stations or by cities by using LEDs (light emitting diodes) for street lighting. On a national scale, government energy policies may include guidelines that help to promote efficient utilization of energy in addition to directing the methods by which energy is produced. Clearly, the topic of energy conservation is very diverse and may be approached in a number of ways. In many ways, individuals control their own energy use. They can make choices in their lifestyles and in how they satisfy their transportation needs that can contribute to more effective utilization of energy resources. Individuals and organizations may, to the limit of their ability, influence corporate and/or government policy in order to optimize energy use on a local, regional, national, or even global scale.

In this chapter, the topic of energy conservation is approached from several viewpoints, beginning with the ways in which the efficient energy use can be promoted by policies and incentives at the national or international levels. Energy conservation measures that are applicable on a regional or local level are then presented. Finally, the things we can do as individuals to conserve the energy that we use in our own lives are discussed.

17.2 Approaches to Energy Conservation

Environmental concerns are integrated with energy production methods and conservation methods. Utilizing energy production technologies that are less harmful to the environment has a positive impact, although all energy technologies have some impact, as discussed throughout this text, and the severity of this impact is not always obvious. Using less energy has a positive environmental impact, whether the energy comes from traditional fossil fuel resources or from alternative technologies, and this emphasizes the need to actively pursue appropriate approaches to conservation in addition to developing new resources.

Actions to conserve energy may be made by individuals, businesses, or governments at the national, regional, state/provincial, or local levels. As individuals, our approach to energy conservation includes things that we can do in our own home, the ways that we satisfy our transportation needs, and the ways we interact with local, regional, or national representatives and government agencies or organizations to promote energy conservation measures. Business can contribute to energy conservation efforts through proactive approaches to reducing energy use, the utilization of suppliers who are energy conscious, utilizing environmentally aware manufacturing processes, and producing energy-efficient products. In the case of governments, the development of energy policies, as well funding and incentives to other levels of government, industry, business, and/or individuals for the implementation of conservative approaches to energy use, can have a positive effect on the efficient utilization of energy resources.

Motivations for energy conservation may include concerns for the environment, long-term energy security and independence, but ultimately a significant component of the motivation for energy conservation is economic; if we save energy, we save money. The most appropriate approach to energy conservation varies considerably in different parts of the world. The availability of energy resources, climate, and the distribution of energy requirements between, for example, transportation, electricity, heating, industry, and so on, are all important factors in determining the best approach to energy conservation. Economy, social factors, and politics also play important roles in implementing conservation measures. In most countries, a national energy policy guides the use of energy resources and typically involves a consideration of such factors as:

- The assessment of energy needs.
- The availability of energy resources.
- The possibility of energy self-sufficiency.
- The environmental consequences of energy use.
- The promotion of energy-efficient products.
- The promotion of energy conservation measures.

- Interaction with energy-related activities at state/provincial and/or municipal levels.
- The development of mechanisms to implement energy policy, including incentives, subsidies, and the like.

Such policies involve various aspects of energy conservation, as well as the development of energy resources. Although it is difficult to generalize, some examples of the approach to energy conservation from diverse locations will be shown in this chapter, to help provide insight into the ways of dealing with this issue.

17.2a Energy Conservation in the United States

Energy policy in the United Sates is largely implemented by the Department of Energy (DOE). The U.S. DOE divides energy use in the United States into four broad categories: residential, commercial, industrial, and transportation. According to DOE statistics, these sectors account for 21%, 17%, 33%, and 28%, respectively, of the primary energy use nationwide. The approaches to energy use and conservation in these four sectors are now outlined.

Heating and/or cooling accounts for the largest component of residential energy use, about 42% on average. About 13% of energy use is for water heating, about 10% for lighting, and the remainder mostly for electronics and appliances. The exact breakdown of residential energy use varies considerably in different regions of the United States and is directly related to climate. In colder regions, heating is a major component of energy use, while in warm regions, air conditioning is a substantial factor. In regions with a moderate climate, lighting and appliances may account for the largest component.

Historically, changes in the overall energy consumption are determined by two competing factors. Energy consumption is reduced by the use of more efficient appliances and measures to reduce energy loss, whereas energy consumption is increased by greater energy needs of consumers.

To promote more efficient energy use, the Department of Energy, as authorized by Congress, established the 1987 National Appliance Energy Conservation Act. This Act specifies minimum efficiencies for residential energy consuming devices, including appliances, electronics, and heating and cooling equipment. Regulations are upgraded on the basis of a yearly assessment of what is technologically feasibly and economically justifiable. Exceptionally efficient devices are awarded the ENERGY STAR approval (Energy Extra 17.2). Because many appliances consume substantial electricity while in their standby mode, improvements in energy consumption when the appliance is in off mode are an important consideration for energy conservation. The ENERGY STAR program includes not only devices that consume energy but also construction materials and construction techniques that are a factor in determining energy needs.

Consumer energy demands have increased as a result of lifestyle changes. This has been particularly significant since the 1970s with increases in average home size, in the utilization of electronic devices, and in the prevalence of central air conditioning systems. These have all put a greater strain on energy requirements in the United States.

Both federal and state governments have promoted more efficient residential energy use (Section 17.6) by offering tax credits for home improvements that decrease energy consumption. In many areas, public utilities also offer subsidies for improvements leading to more efficient energy use. The possibility of the future implementation of a smart grid (Section 17.4) will require input from both utilities and various levels of government. A step in this direction may be the use of real-time energy (e.g., electricity) monitors to provide consumers with feedback on their energy use and to promote energy-conscious behavior.

The commercial sector includes retail businesses, restaurants, schools, and other facilities. Much of the energy use in the commercial sector follows along the lines of the residential sector in its energy use for heating, cooling, hot water, and lighting. However, the commercial sector typically uses a larger fraction of energy for lighting (up to about 25%) compared to the residential sector. As a result, fluorescent lighting is used for nearly all lighting in this sector. Centralized control of building energy use, such as programmed heating and lighting systems, provides an effective means for the implementation of energy conservation measures. Efficient energy use in the commercial sector is strongly motivated by economic factors. In most areas, municipal building codes for new commercial structures require designs that comply with energy efficiency standards. U.S. government policy specifies that federal government buildings must conform to certain energy efficiency requirements. Fifteen percent of existing buildings are required to meet minimum energy requirements by 2015, and all new federal government buildings must be *zero net energy* (*ZNE*) buildings by 2030. *ZNE building* is a term used somewhat ambiguously, and usage in North America is often different from its usage elsewhere. In the strict sense, a ZNE building is a building in which all required energy is produced on-site from renewable sources such as solar or wind. The official definition of a ZNE building used in U.S. government policy is a building whose construction and operation contribute zero net greenhouse gas emissions. This is sometimes referred to as a zero-net-carbon building.

Energy conservation in the industrial sector, which includes all manufacturing, mining, construction, and agricultural activities, is motivated by both environmental and economic issues. The steel industry and the paper industry are major energy users and have made substantial reductions in their energy use in the past 30–40 years. An important factor in this reduction has been the utilization of cogeneration (Section 17.3). Agricultural activities have become more energy efficient in recent years, and, as seen in Chapter 16, this increased efficieny will be observed directly as a reduction in the ratio of primary energy input to the caloric content of the food produced. In many regions, the energy used in the treatment and distribution of freshwater to consumers is a significant fraction of total energy use. Thus, the availability of freshwater is a long-term environmental concern and an integral component of our future energy concerns.

The total energy use in the industrial sector in the United States has decreased in recent years. This is, in part, due to more energy-efficient technologies but reflects changes in the types of industrial activities in the United States and the increase in the outsourcing of many manufacturing processes.

Transportation (Section 17.8) is a major component of energy use. In the United States, 65% of transportation energy use is for gasoline vehicles (mostly privately owned vehicles), 20% for diesel vehicles (mostly commercial trucks, trains, and ships), and 15% for aircraft. U.S. government energy policy has dealt with the improvement in fuel consumption by vehicles since the 1970s. These actions have primarily been through federal, or in some cases, state regulations (California most notably) aimed at automobile manufacturers. As a result of the energy crisis in the early 1970s, the U.S. government implemented the *CAFE* (*Corporate Average Fuel Economy*) policy in 1975 to pressure automobile manufacturers to improve the fuel efficiency of their vehicles. In 1978, the *Gas Guzzler Tax* was introduced to penalize drivers of vehicles that consumed excessive amounts of fuel. Although the tax is still in effect, few vehicles are subject to the tax as a result of general improvements in the fuel efficiency of passenger vehicles.

One approach to reducing the environmental impact of gasoline-powered vehicles has been the introduction of flex fuel vehicles. These vehicles (Chapter 16) can operate on various mixtures of gasoline and ethanol, typically up to 85% ethanol (E85) in the United States, although in other countries flex fuel vehicles may be able to use up to 100% ethanol or various mixtures of gasoline and methanol. Over 25 million flex fuel vehicles are on the roads worldwide. Most are in Brazil, but the United States is next with about 40% of the total. However, despite the substantial number of flex fuel vehicles on the roads in the United States, the availability of flex fuels (e.g., E85) is minimal. Many owners of flex fuel vehicles are not aware that their vehicles can run on high ethanol-gasoline mixtures, and so more than 90% of these vehicles are operated exclusively on gasoline. Part of the motivation for automakers to produce these vehicles [which are only minimally different than non–flex fuel vehicles (Chapter 16)] has been a government fuel economy credit for each flex fuel vehicle sold (even if owners run them on gasoline). This credit helps manufacturers meet CAFE guidelines without improving fuel efficiency.

Other activities that have promoted improved efficiency of transportation have included subsidized public transportation. Such subsidies are typically provided at the municipal level. Also, at the municipal or state level, the ride-sharing or carpooling of private vehicles has been encouraged by designating special lanes in urban areas for vehicles with more than a certain number of occupants.

Incentives are available for purchasers of plug-in hybrids (Section 17.8) and pure battery electric vehicles, BEVs (Chapter 19). These are in the form of tax credits and/or rebates and are offered by the federal government and a number of state governments.

17.2b Energy Conservation in Canada

According to the Constitution of Canada, provinces and territories have jurisdiction over natural resources, which include energy resources in the form of both nonrenewable resources such as oil and natural gas and renewable energy such as wind, solar, and tidal. In terms of energy, the federal government deals largely with interprovincial and international matters. Most of the initiatives for energy development and many of the programs dealing with energy conservation come at the provincial level. As presented in Chapter 2, Canada has among the highest per-capita energy consumption in the world. This is due to factors such as GDP, climate, industrial diversity, and population density. Although Canada is a net energy exporter (producing more energy than it consumes), conservation is an important concern. Heating needs (due to the climate) and transportation (due to the large land area) are of particular interest, and a number of initiatives and incentives are available in these areas.

A number of incentive programs for energy conservation vary from province to province depending on the specific energy needs and concerns. A number of these are aimed specifically at the home owner and include actions such as:

- Distributing free compact fluorescent lamps (Section 17.7) at major building supply and department stores.
- Free removal and a rebate for home owners who replace old (energy-inefficient) appliances with new energy-efficient ones.
- Free energy-efficient upgrades such as blown-in insulation, energy-efficient lighting, programmable thermostats, and air leak reduction for low income families.

- Provincial and federal rebates for energy-efficient upgrades such as ENERGY STAR windows or improved insulation.
- Rebates for the development of residential renewable energy, such as solar collectors or residential wind turbines.
- Rebates for supplementing or modifying heating systems to use lower greenhouse gas emission options.
- Rebates for new home construction that meets certain energy efficiency standards.

The Office of Energy Efficiency of Natural Resources Canada offers a number of programs dealing with energy conservation that are aimed at business and industry. These include distribution of information on energy efficiency of appliances and equipment, guidelines for energy-efficient construction, and cost sharing programs to undertake energy assessments and retrofits for existing structures to improve energy efficiency.

Following the energy crisis of the early 1970s, the Canadian government introduced measures to improve the fuel efficiency of vehicles in that country. In 1975, the Joint Government–Industry Fuel Consumption Program (FCP) was introduced. Since that time, Transport Canada has collected fuel consumption data for vehicles, and this information is made available through Natural Resources Canada. The purpose of this program was to encourage the introduction of fuel-efficient vehicles in Canada and to promote public awareness of energy conservation. Annual voluntary CAFE guidelines, the equivalent of the U.S. CAFE, were established to encourage manufacturers and importers of automobiles and light-duty trucks to strive for improved fuel efficiency. This program was ended in 2010 with the introduction of new *Passenger and Light Truck Greenhouse Gas Emission Regulations*, which form part of the *Canadian Environmental Protection Act*. Motor vehicle manufacturers now submit data on vehicles sold in the country to Environment Canada.

Flex fuel vehicles are readily available in Canada, and, in 2008, about 600,000 such vehicles were on the road. Because only three service stations that were open to the public in Canada in 2012 sold E85, virtually all flex fuel vehicles in that country operated exclusively on gasoline.

At the provincial and municipal levels, carpool lanes are common in a number of locations. The governments of Ontario and Quebec have initiated a rebate program for purchasers of plug-in hybrids and BEVs. Ontario also issues Green Licence Plates to BEVs and plug-in hybrids, which allow these vehicles to travel in carpool lanes regardless of the number of occupants.

17.2c Energy Conservation in the European Union

A common energy policy for the European Union (EU) was approved by the European Council in 2007. Key points of this policy are:

- Reducing greenhouse gas emissions.
- The development of renewable energy technologies.
- Energy conservation.
- Development of advanced nuclear reactors.
- Carbon sequestration.

While conservation is a specific goal of the policy, substantial emphasis in the EU has been placed on the reduction of greenhouse gas emissions, and conservation is an important component of this reduction. The EU energy policy is implemented through the *Strategic Energy Technology (SET) Plan*, which includes initiatives in wind energy, solar energy, bioenergy, carbon sequestration, smart grid, and nuclear fission. The plan coordinates EU programs with national programs of the member states. Much of the energy conservation effort in the European Union falls under the *SAVE (Specific Actions for Vigorous Energy Efficiency)* Programme. Originally, SAVE was established in 1991 and ran until 1995. It was followed by an updated version, SAVE II, from 1996 to 2002, and then SAVE III, now part of *IEE (Intelligent Energy–Europe)*. These programs deal specifically with energy conservation in the residential, commercial, and industrial sectors.

Various approaches to residential energy conservation in the EU parallel efforts in North America. The goal of these programs is both to reduce greenhouse gas emissions and to conserve energy resources. Long-term goals include requirements for all new homes to be ZNE buildings, meaning zero-net-carbon buildings, and the installation of real-time electricity monitors in all homes.

In the commercial sector, building efficiency is an important concern. Common methods for the implementation of efficiency standards for buildings in the EU are that:

- Requirements that all large new buildings must meet certain efficiency standards.
- Large existing buildings must undergo major renovations and comply with minimum requirements.
- All large buildings require energy certification before sale.
- Furnaces, boilers, and air conditioning systems must have regular inspections to ensure efficient operation.

EU energy policy includes a voluntary agreement with the *ACEA (Association des Constructeurs Européens d'Automobiles*, or, in English, *European Automobile Manufacturers' Association)* to achieve specified reductions in CO_2 emissions from automobiles. Improved fuel efficiency is an important component in greenhouse gas reduction.

Most EU countries have incentives for the use of biofuels for transportation. Flex fuel vehicles, typically operating on E85, are promoted in many locations, with E85 fueling stations becoming more common. Incentives in terms of reduced taxes on fuel cost or vehicle taxes help to offset the lower energy content of E85 compared with gasoline. Germany, in particular, has placed an emphasis on biodiesel vehicles. Incentives include a fuel tax waiver on biogenetic fuels. Because biodiesel is a 100% nonfossil fuel, there are no fuel taxes, whereas for ethanol-gasoline mixtures, the tax is prorated by the concentration of fossil fuel present.

The EU is also a member of *International Partnership for Energy Efficiency Cooperation (IPEEC)* along with Australia, Brazil, Canada, China, India, Mexico, Russia, South Korea, and the United States. This organization enhances global cooperation on energy policies with a focus on energy efficiency. Many of the initiatives are led by one of the member states. Some of these initiatives are as follows:

- *IPEEI (Improving Policies through Energy Efficiency Indicators)*, led by France, develops and implements methodologies to assess energy efficiency.
- *SBN (Sustainable Buildings Network)*, led by Germany, investigates policies dealing with building efficiency. Guidelines for new construction, as well as approaches to the upgrade of existing buildings, are being explored.
- *WEACT (Worldwide Energy Efficiency Action through Capacity Building and Training)*, led by Italy, facilitates policy creation and implementation related to energy efficiency in developing countries.

17.2d Energy Conservation in India

Energy use is growing rapidly in India as a result of its significant economic growth. At present, more than 30% of the population has no access to electricity, and this percentage is greater than 50% in rural areas. Thus, the needs that must be addressed by energy policy in India are more complex than in many other nations. Currently, about 70% of India's energy generation is from fossil fuel sources, and expanding efforts in alternative renewable sources is a major priority. In fact, there is active development in rural electrification, using renewable energy sources such as hydroelectric energy, wind energy, and solar energy. Due to the need for more energy generation in India, conservation of the available energy is an important component of energy policy. Several government agencies are involved in conservation efforts.

The *Petroleum Conservation Research Association* (*PCRA*) is an Indian government body established in 1977 for the purpose of promoting energy efficiency, specifically in the area of fossil fuel conservation measures. Media campaigns to promote awareness of the need for energy efficiency to conserve resources and reduce environmental pollution have been undertaken in recent years and have been effective in bringing information about these issues to the public. The association deals with energy conservation in four broad areas of national importance: residential (or domestic) energy use, industrial energy use, agricultural energy use, and transportation.

Public education is a major focus of the PCRA in the residential sector and involves programs aimed at promoting efficient energy use in the home. The PCRA is also involved in the development of fuel-efficient stoves, the use of energy-efficient lighting by compact fluorescent lamps, and the introduction of alternative renewable energy sources. In the industrial sector, the PCRA has been active in conducting energy audits for industries and providing recommendations for more efficient energy use. Operational adjustments that can save energy without capital investment are an important focus of these activities. Interactions with industry include seminars, clinics, and workshops, as well as follow-ups to assess actual energy savings realized. Agricultural initiatives include educational programs for farmers to promote the proper maintenance of equipment, such as tractors and pumps, to optimize energy efficiency. Finally, the transportation sector is the focus of driver education programs to promote fuel-efficient driving habits.

In 2001, the Indian Parliament passed the *Energy Conservation Act*. This Act places constraints on large energy consumers to adhere to energy utilization guidelines, requires that new construction follow energy-efficient building codes and requires that appliances follow energy-efficient guidelines. As part of the implementation of this Act, the *Bureau of Energy Efficiency* (*BEE*) was established with the mandate to develop programs to increase the efficiency of energy use in India. Several focus sectors for BEE have been identified:

- Replacement of incandescent lamps with compact fluorescent lamps (Section 17.7)
- Marketing of high-efficiency appliances and providing consumers with energy consumption information
- Replacing municipal streetlights with LED lamps (Section 17.5)
- Improving energy efficiency in municipal water treatment and distribution systems
- Establishment of voluntary energy efficiency standards for new building construction
- Replacement of inefficient agricultural equipment with energy-conserving devices

The BEE has been instrumental in coordinating a wide variety of conservation efforts in India and has established systems and procedures to monitor the effectiveness of implemented measures.

In the transportation sector, there is a major effort in India to produce biodiesel derived from Jotropha plants as a means of reducing dependence on fossil fuels. Numerous projects producing and using biodiesel from this source have been undertaken. The area around the railroad lines between Mumbai and Delhi is used to grow Jotropha, and the train itself operates on a biodiesel mixture containing Jotropha oil. This program has been a major success for biodiesel development in India.

India is also a member IPEEC (Section 17.2c). One of India's roles in this organization is to lead the *AEEFM* (*Assessment of Energy Efficiency Finance Mechanisms*) initiative. This program identifies and documents methods to overcome barriers to the successful financing of energy efficiency projects worldwide. This includes assessing effective mechanisms for utilizing financing to undertake energy efficiency programs and how tax initiatives and subsidies can best be used to benefit these activities.

17.3 Cogeneration

All methods of producing electricity that involve a heat engine generate excess heat in accordance with the thermodynamic efficiency of converting thermal energy into mechanical energy. These methods include thermal generation using nuclear fuel, coal, and natural gas. On the one hand, this excess heat is an undesirable consequence of heat engines because energy is not converted into the desired form and because the excess heat must be removed. Another approach to this situation may be to view the excess heat as a resource. The heat may be used locally to provide heating and/or hot water to the community. This approach is referred to as *cogeneration*, or *combined heat and power* (*CHP*). As efficiencies of generating stations are typically less than about 40%, the majority of energy produced is available for use as heat rather than as electricity. A major difficulty in implementing CHP is to match the demand for electricity and the demand for heat. Electricity may be widely distributed via the grid, but thermal energy must, for practical reasons, be utilized locally. A cogeneration plant may be primarily designed to meet the base load requirements for electricity or the base load requirements for heat/hot water. In either case, other sources of energy may be required, particularly in periods of high demand for one or the other resource.

Although it might seem that the utilization of any excess heat for a useful purpose might be more beneficial than merely dumping it into the environment, a certain degree of efficiency must be achieved to justify (economically) the necessary infrastructure and maintenance costs. Distribution of electricity via connection to the electrical grid is relatively straightforward, but the distribution of heat and hot water requires a system of insulated pipes.

Cogeneration has been utilized more effectively for natural gas than for coal-fired plants, but it has also been successfully implemented for stations burning a variety of alternative fuels, including wood and agricultural waste. The greater need for heat in cooler climates has made the implementation of cogeneration popular and often economical in such regions. In fact, it has become an important component of the energy economy in Denmark, the Netherlands, and Finland. In warmer climates, the utilization of excess heat from power plants may provide heat/hot water, as well as cold water for air conditioning purposes via an *absorptive chiller*. Such an approach to the use of energy from a thermal generating station is referred to as *trigeneration*.

Figure 17.1: Cornell University cogeneration plant.

Cogeneration has been used very effectively where a large amount of heat is used in a fairly small area. This is the case, for example, in Manhattan where approximately 100,000 buildings receive heat through cogeneration. Cogeneration is also commonly used in decentralized facilities such as businesses, manufacturing facilities, and hospitals. The needs of those small communities can be met through small generating stations. The natural gas–fired combustion turbine cogeneration facility at Cornell University is an example of this approach (Figure 17.1). The 30-MW generating facility provides the majority of energy needs on campus. The general design of such a system is illustrated in Figure 17.2.

Example 17.1

A natural gas-fired generating station, rated at 200 MW_e with an efficiency of 39%, operates with a capacity factor of 68%. Hot water for heating use in the local community is produced by cogeneration. Calculate the heat energy available per year.

Solution
At the rated output, the total power produced is

$$\frac{200 \text{ MW}_e}{0.39} = 513 \text{ MW}.$$

Thus, the excess heat produced is

$$513 \text{ MW} - 200 \text{ MW} = 313 \text{ MW}.$$

At a capacity factor of 68%, the total energy produced per year is

$$313 \text{ MW} \times 0.68 \times 3.15 \times 10^7 \text{ s/y} = 6.7 \times 10^9 \text{ MJ}.$$

Overall, cogeneration can, under the right circumstances, be an effective method of utilizing waste energy and improving the efficiency of our energy use. It has been more extensively used in Europe (particularly Northern and Eastern Europe) than in North America, but its use has begun to become more widespread in recent years.

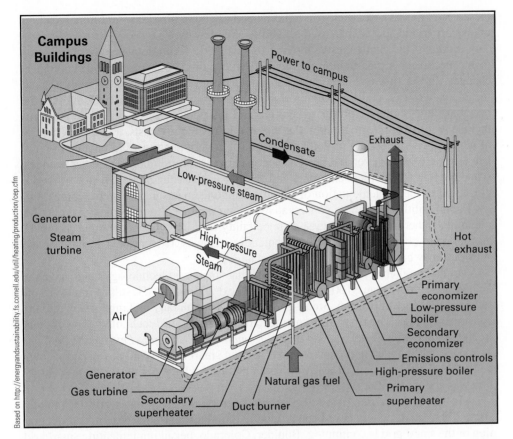

Based on http://energyandsustainability.fs.cornell.edu/util/heating/production/cep.cfm

Figure 17.2: Schematic diagram of Cornell University cogeneration plant.

17.4 Smart Grid

The first electric distribution grids were established in the 1890s. Over the next 70 years or so, the electric grid system expanded and grew into the large-scale system that effectively supplied our electric needs in the 1960s and 1970s through a one-way interconnected system. During this time, the tendency was to construct large, centralized generating facilities (i.e., 1 GW_e and larger) in geographically appropriate locations (e.g., hydroelectric dams on rivers, nuclear stations near cooling water supplies, coal plants near railway connections, etc.). One-way transmission of electricity to users via an interconnected grid minimized the possibility for large-scale interruptions.

In recent years, the desire to integrate alternative energy sources into the grid has made the previous approach to electricity distribution less effective than it had previously been. This is because many alternative energy facilities are smaller (e.g., tens-of-MW_e wind farms), are intermittent in their output (e.g., wind, solar, tidal, etc.), and require the integration of energy storage systems (e.g., pumped hydroelectric, batteries, etc.). There is also a move for users to become suppliers by selling excess electricity generated by wind and/or solar back to the utility, as well as an incentive to shift usage to off-peak hours by real-time metering. The so-called *Smart Grid* would deal with these issues to make electricity distribution and use more efficient. Additional features of a smart grid would be increased reliability and security.

Some of the challenges to moving to smart grid technology are the installation of the necessary infrastructure and dealing with public concerns over technological changes. In the first instance, each household would have to be fitted with a smart meter for two-way, real-time monitoring and control of electrical use, along with the necessary integration of power generation and distribution systems. Public concerns include security of information on the Internet, infrastructure costs, and loss of control of personal energy use. In principle, utilities (or governments) would be able to regulate electricity use to even out peak use periods by, say, turning off unused appliances, scheduling electric vehicle charging, and so on. Rapidly changing appliance technologies requires the implementation of suitable standards to ensure long-term compatibility. Public utilities tend to be conservative in their approach to technology, aiming for reliability rather than some degree of uncertainty, and they view economic factors as an important consideration when implementing changes. Government incentives and subsidies may be important in promoting change.

In the long term, it has been estimated that smart grid technology could save the United States more than $100 billion over the next 20 years as a result of improved efficiency of energy utilization. Smart grid technology is already making its way into the consumer market. The first full implementation of smart grid technology was in Italy in 2005 by the utility Enel S.p.A. (*Ente Nazionale per l'Energia eLettrica, Società per Azioni*). Enel designed and manufactured its own smart meters and developed its own computer control systems. It was estimated that the payback period for the infrastructure was about four years.

In North America, Austin, Texas, has been working on smart grid technology since 2003. Smart meters have been gradually introduced into the system, and integration of the smart grid is continuing. Boulder, Colorado, began implementing smart grid technology in 2008. Hydro-One in Ontario, Canada, is in the process of introducing smart grid technology to over 1 million customers. In 2009, however, electricity regulators in Massachusetts rejected plans to deploy smart grid technology out of concerns for lack of protection for low-income customers.

Smart grid technology is certainly an important component of a future electrical system that includes the effective integration of nontraditional energy generation technologies and energy storage methods. As evidenced by previous efforts, however, the implementation of this approach involves technical innovation but also must deal with political and economic challenges in order to function efficiently.

17.5 Energy Conservation in the Community—LED Streetlights

On a municipal level, many approaches can be taken to reduce energy consumption. Because up to 40% of municipal energy budgets are used for lighting streets, this is an area where significant energy savings can be realized. It is also an area where a clear, well-defined, and cost-effective approach can be taken. The replacement of traditional sodium vapor or metal halide streetlights with high-efficiency light emitting diode (LED) lamps typically reduces energy consumption by around 60%.

The physics of the operation of an LED follows closely from the discussion of photovoltaics in Chapter 9. A photovoltaic cell is a semiconducting junction (diode) that produces an electric current in response to incident light. Electron-hole pairs are formed in the depletion region (Figure 9.19) as a result of energy deposited there by photons. The electron-hole pairs contribute to a current that results from the electric field across the device. A light emitting diode operates in basically just the opposite manner. A voltage is applied across the junction, causing electron-hole pairs to combine and produce photons. Electrons the in the n-type region on one side of the junction (Figure 9.19), can move toward the junction and recombine with holes from the p-type region that are moving in the opposite direction. The energy that results from this recombination produces photons. The energy of the photons is governed by the difference in the energy of the electrons near the bottom of the conduction band and the energy of the holes near the top of the valence band. The energy is therefore determined by the size of the energy gap and is fairly narrowly defined. Semiconducting materials with different energy gaps give rise to photons of different energies with small-energy gap materials, producing small-energy, long-wavelength light toward the red end of the spectrum and large-energy gap materials producing large-energy, short-wavelength light toward the blue end of the spectrum. Table 9.1 gives a general idea the energy gaps associated with different common semiconductors. Figure 1.1, along with the relationship between energy (in eV) and wavelength (in nm),

$$E = \frac{1240 \text{ eV} \cdot \text{nm}}{\lambda}, \qquad \textbf{(17.1)}$$

provides an idea of the colors of the light that can be produced. Early LEDs produced red light, whereas modern devices can be tuned by adjusting the composition of the semiconductors, and hence the energy gap, to produce virtually any color desired.

For lighting purposes, it is desirable (for good color rendition) to have a broad spectrum of wavelengths giving rise to light that, to the human eye, appears white. There are basically two ways of doing this using single-color LEDs. Perhaps the obvious way is to combine three LEDs producing light of the primary colors (red, green, and blue) and combining these into white light (Figure 17.3). The human eye perceives this as white light, even though the spectrum is quite different from the broad continuum of sunlight shown in Figure 8.1.

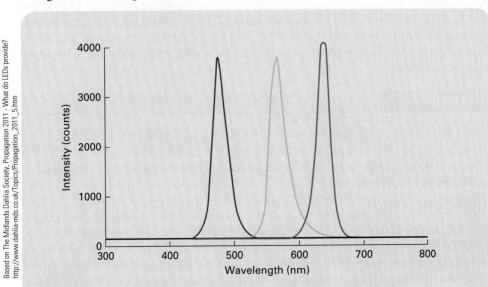

Figure 17.3: Spectrum of a three-color LED producing white light.

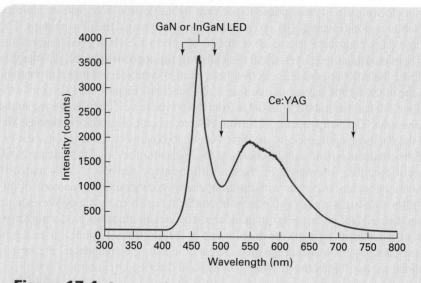

Figure 17.4: Spectrum of a blue GaN or InGaN LED with a cerium-doped yttrium aluminum garnet (YAG) phosphor coating.

An alternative, less expensive, and more commonly used method is to begin with a blue LED, which produces photons near the short-wavelength (highest-energy) end of the optical spectrum. The blue light is incident on a phosphor, which absorbs the blue light and re-emits it over a broad spectrum of lower energies (or longer wavelengths). The spectrum of this device contains the original spectrum of the blue LED and the broad re-emitted spectrum from the phosphor (Figure 17.4).

An example of the design of an LED streetlight fixture is shown in Figure 17.5. The finned geometry on the top side of the light fixture provides a large surface area to dissipate heat. This design of a heat sink is typical of that used on many electronic devices. An example of the effectiveness of LED streetlights is illustrated in Figure 17.6, where 250 W high-pressure sodium fixtures consuming 275 W total (lamp and ballast) were replaced by 200 W LED streetlights (LED Roadway Lighting Ltd. model SAT-96M) leading to a reduction in energy consumption and increased public safety due to improved lighting conditions.

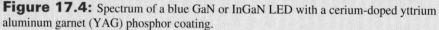

Example 17.2

A 1-km section of roadway has one lighting fixture every 40 m. If 200-W sodium vapor lamps are replaced with 80-W LED lamps, and the lights are on an average of 12 h per day. What is the annual savings for this section of road if electricity costs $0.10 per kWh?

Solution

The 1-km section of road has 1000 m/40m = 25 fixtures. The power savings are (200 W − 80 W) = 120W per fixture for a total of (25) × (120W) = 3 kW. During a 1-year period at 12 hours per day, this is

$$(3 \text{ kW}) \times (365 \text{ d/y}) \times (12 \text{ h/d}) = 13{,}140 \text{ kWh,}$$

or a value of (13140 kWh) × (0.10 $/kWh) = $1314.

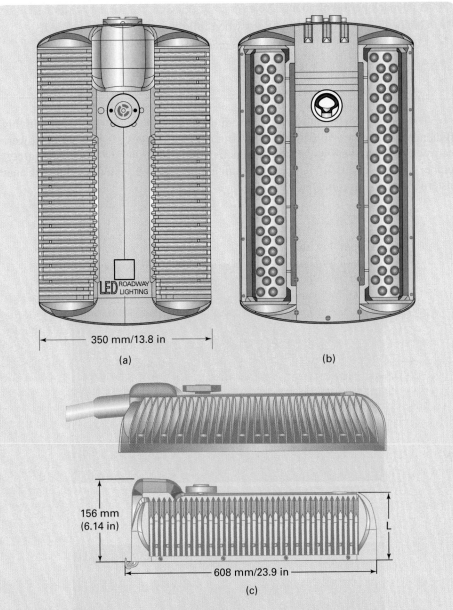

Figure 17.5: Top (a), bottom (b), and side (c) views of an 86-W LED streetlight containing 96 LEDs manufactured by LED Roadway Lighting.

Overall, LED streetlights have several clear advantages over traditional sodium vapor or metal halide streetlights:

- Low power consumption (about one-fifth that of traditional lights for the same light output)
- Long lamp lifetime (about 12–15 years, or about 3 times that of traditional lamps)

- A payback period of about 3 years
- Can be dimmed during conditions when full illumination is not needed
- Minimize wastage light. Light is directed to road surface
- Accurate color rendering
- Elimination of toxic materials (e.g., Pb or Hg) found in many traditional streetlights

Thus, the replacement of existing sodium or metal halide street lights with high-efficiency LED fixtures is a straightforward and cost-effective technique for lowering energy costs with a very reasonable payback period and for improving public safety. A number of installations are in place in municipalities across the United States, Canada, Eastern Europe, Australia, and Asia.

Figure 17.6: Conversion of streetlights on King William Street, Adelaide, Australia from (top) high-pressure sodium fixtures to (bottom) LED streetlights.

ENERGY EXTRA 17.1
LEED certification

The Leadership in Energy and Environmental Design (LEED) certification system was developed in 1998 by the United States Green Building Council (USGBC). This system provides third-party verification that a building or group of buildings achieves a certain level of excellence in environmental performance in areas such as energy and water efficiency, CO_2 emissions, and indoor environmental quality. A second version (LEED NCv2.2) was released in 2005 and a third version (LEED NCv3) in 2009. In 2003, the Canada Green Building Council received approval from the USGBC to implement its own version of the LEED system. Thirty countries have now created LEED certification systems.

Five categories of constructions can apply for LEED certification:

1. *Green Building Design and Construction* includes new constructions and some major renovations of existing structures, for example, schools (as shown in the figure), retail businesses, and hospitals.

2. *Green Interior Design and Construction* includes the interior design of commercial buildings.

3. *Green Building Operations and Maintenance* deals with the operations and maintenance systems of existing buildings.

4. *Green Neighbor Development* deals with the development of environmental excellence in neighborhood design.

5. *Green Home Design and Construction* deals with residential homes.

Buildings are judged in a number of areas of environmental leadership and are awarded points in each category. The total number of points that can be achieved is 100 plus 10 bonus points. LEED accreditation is awarded according to the total score, and buildings are certified at one of the following levels:

- *Platinum certification*: 80 points and above
- *Gold certification*: 60–79 points
- *Silver certification*: 50–59 points
- *Certification*: 40–49 points

Studies have indicated that LEED-certified buildings are indeed more energy efficient and that workers in the buildings have increase productivity due to improved environmental conditions such as air quality, ventilation, temperature, and lighting. LEED-certified buildings have also been found to achieve higher rents, higher occupancy rates, and increased sale prices compared to their nongreen counterparts.

© Britton Images Photography

Dr. David Suzuki Public School in Windsor, Ontario; the first LEED Platinum–certified Education Building in Canada.

Continued on page 434

Energy Extra 17.1 continued

A directory of LEED-certified building projects in the United States is available online at http://www.usgbc.org/LEED/Project/CertifiedProjectList.aspx. For Canada, a directory is available at http://www.cagbc.org/Content/NavigationMenu/Programs/LEED/ProjectProfilesandStats/default.htm.

Topic for Discussion

Locate LEED-certified buildings in your area. Observe the building from the outside, or, if it is a public building or if access can be arranged, view the interior. What features do you see that have contributed to its LEED certification?

17.6 Home Heating and Cooling

A system that controls the climate in a building is often referred to as an *HVAC (heating, ventilation, and air conditioning) system*. The importance and design of the various components of such a system depend largely on the climate in the region. A single-family home may be connected to a *district system* (such as a district heating system supplied with steam or hot water from a cogeneration facility or geothermal resource) through a heat exchanger, or it may have a self-contained system within the home. The latter provides the home owner the opportunity to make decisions concerning system design and operation that may have economic or environmental consequences. A home HVAC system may consist of individual distributed components, such as an electric heating system with baseboard heaters in the rooms or window air conditioners, or it may be a centralized facility, i.e., a furnace with a heat distribution system or central air conditioning. Although home HVAC systems can be quite varied in design, this discussion deals with some typical ways in which such systems operate, and how their efficiency can be optimized. We begin with systems that produce heat (i.e., furnaces) and then move on to systems that transfer heat into a building (i.e., a heat pump) and finally cooling (i.e., air conditioning) systems.

17.6a Furnace Efficiency

The term *furnace* is somewhat ambiguous. In North America, it is often used to refer to a centralized facility for producing heat for distribution to a building. It is also used, especially outside North America, to indicate an industrial furnace or kiln used for manufacturing processes. Sometimes the term furnace is used for a heater that distributes heat by circulating hot air, while the term *boiler* is used for a heater that distributes heat by circulating hot water or steam. Here we use a fairly generic definition of furnace as a centralized heater for providing heat to a building.

A furnace may produce heat by burning a combustible fuel or in some instances through resistance heating using an electric current. The fuel that is burned in a furnace may be natural gas, propane, oil, or in some cases an alternative fuel such as wood pellets. The efficient utilization of the energy content of the fuel may be viewed in two ways: (1) the efficient conversion of chemical energy into thermal energy and (2) the efficient distribution of the heat produced. The latter factor involves a control system to ensure that heat is provided when and where it is needed. A cost-effective means for doing this is with programmable thermostats that can provide heat at times and in rooms that are compatible with the occupants' needs.

The efficiency with which the chemical energy of the fuel can be converted into heat depends on the design and operational parameters of the furnace. Furnaces

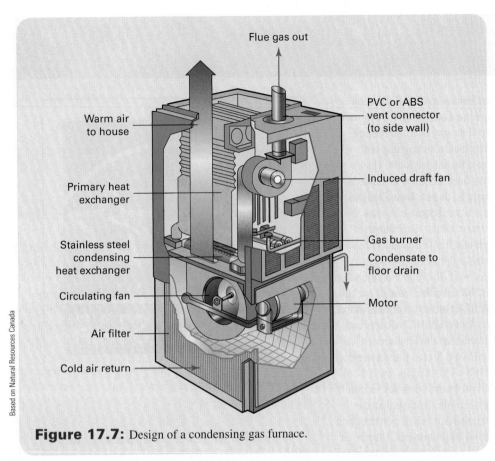

Based on Natural Resources Canada

Figure 17.7: Design of a condensing gas furnace.

generally fall into two categories: *noncondensing* and *condensing*. The combustion of a hydrocarbon produces a number of by-products, principal among these are CO_2 and H_2O (in the form of water vapor). It is desirable if the heat produced during combustion is transferred to the fluid (typically air, water, or steam) that distributes the heat throughout the building. However, in practice, some of the heat produced is carried away by the by-products or exhaust gas of the combustion. If the exhaust gases, sometimes called flue gases, carry away substantial thermal energy, then they need to be vented through a heat-resistant chimney. This is traditionally the design of a heating system. Modern, high-efficiency furnaces are condensing gas furnaces (Figure 17.7). These furnaces extract enough of the thermal energy from the exhaust gases for use in the building in order that these exhaust gases are sufficiently cool and the water vapor in the gases condenses into a liquid. In such cases, the exhaust gases can be at a temperature of around 50°C and can be vented by means of a pipe (which is often plastic) directed through a wall or the roof of the house. Typically, furnaces with efficiencies of greater than about 90% are condensing gas furnaces. Such furnaces with efficiencies as high as 98% are available to the home owner. Older furnaces may have efficiencies in the range of about 60%, and modern noncondensing furnaces typically have efficiencies of 80–85%. A home owner who is installing a furnace and is concerned with efficient utilization of fuel should consider a condensing gas furnace. ENERGY STAR qualified furnaces are almost always condensing gas furnaces and are an assurance of efficient operation.

ENERGY EXTRA 17.2
ENERGY STAR

In 1992, the United States Environmental Protection Agency (EPA) established the *ENERGY STAR* program to improve the efficiency of energy use and reduce greenhouse gases and to provide consumers with information about energy-efficient products. The program has subsequently been adopted by Canada, the European Union, Australia, Japan, New Zealand, and Taiwan. Consumers are most commonly aware of the ENERGY STAR program by the logo shown in the figure that appears on electronic devices (such as televisions and computers) and appliances (such as refrigerators and dishwashers). The logo indicates that the device has exceeded the specified level of energy efficiency in the ENERGY STAR testing program. In the United States, the criteria for ENERGY STAR approval are set for each product by either the Environmental Protection Agency or the United States Department of Energy.

The ENERGY STAR program, however, is much more diverse than just electronics and appliances. The ENERGY STAR rating also applies to construction materials, such as windows and furnaces. Upgrades to older homes can be made within the guidelines of the ENERGY STAR program, and newly constructed homes can earn the ENERGY STAR label if ENERGY STAR components and ENERGY STAR construction practices are utilized.

Commercial buildings such as hospitals, schools, and manufacturing plants can also be awarded ENERGY STAR approval. LEED and ENERGY STAR for buildings are complimentary approaches. ENERGY STAR rating deals more or less exclusively with energy consumption, whereas LEED takes a broader approach toward the environmental quality of the building. Buildings can achieve both LEED and ENERGY STAR certification. For existing buildings, the ENERGY STAR certification is a requirement for LEED approval.

U.S. Environmental Protection Agency

ENERGY STAR logo.

The ENERGY STAR program has promoted public awareness of the need for energy conservation and the corresponding benefits in greenhouse gas emission reduction. Consumer recognition of the ENERGY STAR logo has also motivated manufacturers to provide products that offer environmentally conscious options.

Topic for Discussion
Try to find similar appliances or electronic devices with and without the ENERGY STAR logo. Is a clear energy savings associated with the ENERGY STAR product? If so, make an estimate of the number of hours a typical user may use such a device and the annual electrical savings (in kilowatt-hours and dollars).

Heat produced by a furnace may be used to heat air, which is circulated through ductwork to various locations in the building. Usually, a fan or blower is used to circulate the hot air (hence the common terminology applied to this design, *forced hot air heat*), and separate ducts return cool air to the furnace. It is important that ductwork be well sealed against air leaks and be appropriately insulated. Inferior ductwork that passes through unheated or minimally heated portions of the house (e.g., basements or attics) wastes energy. A furnace (often called a boiler) that heats water uses circulating pumps to distribute heat to the rooms of a house. Like air-carrying ductwork, pipes carrying hot water for heating must be appropriately insulated.

Example 17.3

If fuel oil costs $1.05 per liter, and a house requires net heat of 8.7×10^{10} J per year, calculate the annual savings that a home owner could expect if an old 60% efficient system is replaced with a 94% efficient modern condensing gas furnace.

Solution

A 60% efficient furnace requires fuel with a total energy content of

$$\frac{8.7 \times 10^{10} \text{ J}}{0.6} = 1.45 \times 10^{11} \text{ J per year,}$$

whereas a 94% efficient would require fuel with a total energy content of

$$\frac{8.7 \times 10^{10} \text{ J}}{0.94} = 9.26 \times 10^{10} \text{ J per year,}$$

for an energy difference of

$$(1.45 \times 10^{11} \text{ J} - 9.26 \times 10^{10} \text{ J}) \text{ per year} = 5.24 \times 10^{10} \text{ J per year.}$$

The energy content of fuel oil (about the same as crude oil) is given in Appendix IV as 3.85×10^7 J/L, so this energy difference translates to

$$\frac{5.24 \times 10^{10} \text{ J per year}}{3.85 \times 10^7 \text{ J/L}} = 1361 \text{ L/y}$$

This has the monetary value of

$$(1361 \text{ L/y}) \times (\$1.05/\text{L}) = \$1429 \text{ per year.}$$

17.6b Heat Pumps

The basic physics of a heat pump was presented in Section 1.5. The heat pump's operating principle can be incorporated into the design of a system that can use mechanical input to transfer heat from a cold reservoir (the outside of a building) to a hot reservoir (the inside of a building). The amount of heat (energy) transferred from the cold reservoir to the hot reservoir, Q_h, is related to the amount of work done, W, as

$$Q_h = COP \cdot W, \tag{17.2}$$

where the Carnot coefficient of performance (*COP*) is defined as

$$COP = \frac{1}{1 - \dfrac{T_c}{T_h}}.$$ **(17.3)**

T_h and T_c are the temperatures of the hot and cold reservoirs, respectively. (Recall that the temperatures in this equation must be expressed on an absolute temperature scale such as Kelvin.) To maintain a high coefficient of performance, the extraction of heat from the cold reservoir should not significantly lower the temperature of that reservoir. Thus, it is necessary that the total heat capacity of the cold reservoir be much larger than the heat capacity of the structure being heated. This is analogous to the situation for a thermal generating station where it is important that depositing waste heat from the facility into the cold reservoir does not raise the temperature of the reservoir. Raising the temperature of the cold reservoir lowers the ideal Carnot efficiency. For a home heat pump system, the obvious choices for an appropriate cold reservoir are the air outside the house (*air source heat pump system*) or the earth (*ground source heat pump system*). Where it is available, a water-based cold reservoir (a lake for example) can also be used. Home heat pump systems are sometimes designed to provide heating during the cold weather and to function as an air conditioner for cooling during the warm weather.

A schematic illustrating the operation of an air source heat pump is shown in Figure 17.8. This system utilizes the air outside the building as the cold reservoir and transfers heat by means of a working fluid. It has components outside the house to absorb heat from the environment and components inside the house to deposit the heat inside the building. The principle of operation is as follows: Fluid in the liquid phase at low temperature and low pressure enters into coils outside the building, where it becomes

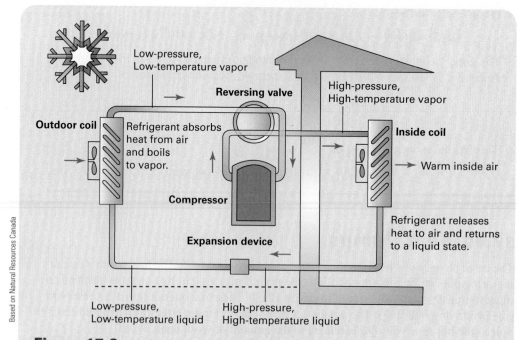

Figure 17.8: Schematic illustrating the operation of an air source heat pump.

Richard A. Dunlap

Figure 17.9: Exterior portion of a typical residential air source heat pump system.

a low-temperature, low-pressure vapor. This vapor is compressed by the compressor and becomes a high-temperature, high-pressure vapor and enters into the house. This vapor flows into coils inside the house, where it releases heat and becomes a high-temperature, high-pressure liquid. It then exits the building, where it passes through an expansion device to become a low-temperature, low-pressure vapor. It then flows into the cold outside the house and repeats the cycle. During this cycle, heat is extracted from the outside air and deposited inside the house. The heat released inside the house can be distributed via ductwork, as in the case of heat produced by a furnace. A typical exterior unit of an air source heat pump is illustrated in Figure 17.9.

Figure 17.10 shows the design of a ground source heat pump system. Water circulated through pipes in the ground transfers heat via a heat exchanger to a working fluid. The operation of the system is analogous to the air source system, and ultimately heat may be transferred to the air via a second heat exchanger for distribution within the house.

An analysis of the effectiveness of this heat transfer system is based on equations (17.2) and (17.3), along with a consideration of the heating requirements for the house, as presented in Chapter 8. Combining equations (17.2) and (17.3), the rate of heat transfer (or power) provided to the house can be expressed as

$$P = \frac{P_{\text{in}}}{1 - \dfrac{T_{\text{c}}}{T_{\text{h}}}}, \qquad \text{(17.4)}$$

where P_{in} is the power input into the heat pump. T_{c} is the outside temperature, and T_{h} is the inside temperature. From equation (8.5), the rate of heat loss from a building is

$$P = \frac{A(T_{\text{h}} - T_{\text{c}})}{R}, \qquad \text{(17.5)}$$

where A is the surface area, and R is the R-value. For an actual building, heat loss through the various components (walls, windows, etc.) can be combined to yield the total heat loss. To determine the effectiveness of a heat pump, the heat gain from equation (17.4) must be compared to the heat loss given by equation (17.5). The relationship

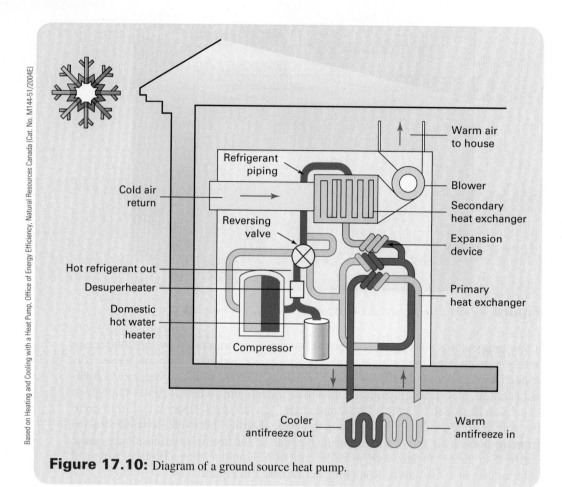

Figure 17.10: Diagram of a ground source heat pump.

Example 17.4

Calculate the power available from a heat pump when the input power is 4 kW and the inside and outside temperatures are 19°C and −3°C, respectively.

Solution

It is necessary to first express the temperatures in Kelvin:

$$T_c = -3°C + 273 = 270 \text{ K}$$

and

$$T_h = 19°C + 273 = 292 \text{ K}.$$

From equation (17.4), the power available is

$$P = \frac{4 \text{ kW}}{1 - \dfrac{270 \text{ K}}{292 \text{ K}}} = 53 \text{ kW}.$$

Example 17.5

A wall is 3 m by 6 m and has an R-value of 3.5 [in $(s \cdot m^2 \cdot °C)/J$]. A window in the wall with dimensions 1.0 m by 1.5 m has an R-value of 0.35 $(s \cdot m^2 \cdot °C)/J$. The inside and outside temperatures are 20°C and $-10°C$, respectively. Calculate the rate of heat loss through the wall in kW.

Solution

The power loss is given by equation (17.5):

$$P = \frac{A(T_h - T_c)}{R}.$$

For the wall with a window, the loss through the wall and the loss through the window can be combined as

$$P = \frac{A_{wall}(T_h - T_c)}{R_{wall}} + \frac{A_{window}(T_h - T_c)}{R_{window}}.$$

The window has an area of $(1.0 \text{ m} \times 1.5 \text{ m}) = 1.5 \text{ m}^2$ and the wall (without the window) has an area of $(3 \text{ m} \times 6 \text{ m} - 1.5 \text{ m}^2) = 16.5 \text{ m}^2$. Using the appropriate R-values for the two components and $(T_h - T_c) = (20 - (-10))°C = 30°C$ gives the total power loss as

$$P = \frac{(16.5 \text{ m}^2) \times (30°C)}{3.5(s \cdot m^2 \cdot °C/J)} + \frac{(1.5 \text{ m}^2) \times (30°C)}{0.35(s \cdot m^2 \cdot °C/J)}$$

$$= 141 \text{ W} + 129 \text{ W} = 270 \text{ W}$$

Interestingly, in this case the heat loss through the window is about the same as through the rest of the wall. This is not an uncommon situation for residential buildings.

between these two quantities depends on the design of the heat pump (i.e., the value of P_{in}) and the details of the building construction (i.e., the total areas and R-values of the building components). However, a general analysis of equation (17.4) shows that, as the value of T_c increases, the power increases, and it diverges as T_c approaches T_h (in this ideal treatment). An analysis of equation (17.5) shows that the power loss decreases linearly with increasing temperature and goes to zero as T_c approaches T_h. These characteristics are shown in Figure 17.11. The point (i.e., temperature at which the heat pump becomes useful as temperature is increased) is the balance point. This simple analysis shows that below a certain temperature a heat pump is not useful. There are therefore limits to the climate conditions where heat pumps are economical, as well as constraints on the need for additional contributions to the heating requirements for a home (Section 17.6d).

17.6c Air Conditioning

Centralized residential air conditioning systems are common in many locations with warmer climates. These systems generally consist of internal and external components. The general design and operation follow closely along the lines of an air source heat

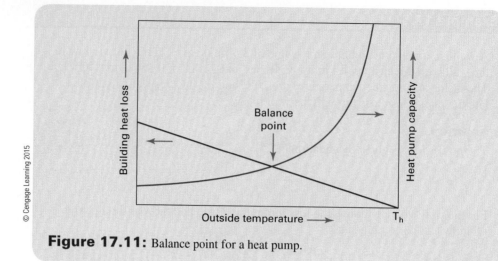

Figure 17.11: Balance point for a heat pump.

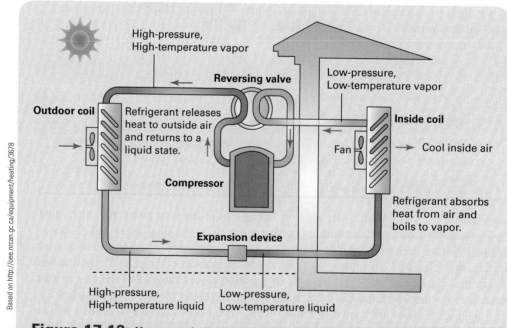

Figure 17.12: Heat pump in the cooling configuration (air conditioner). The flow of fluid is opposite to that of the heat pump in Figure 17.8.

pump, except that heat is transferred from inside the building to outside the building. Figure 17.12 shows a schematic of an air conditioning system. The high-pressure and low-pressure sides are interchanged compared with the heat pump system. This is accompanied by a reversal of the flow of the working fluid that results from changing the direction of the compressor and the expansion valve. A typical external component of a

Figure 17.13: External portion of a residential central air conditioning system.

residential central air conditioning is shown in Figure 17.13; the similarity with the heat pump shown in Figure 17.9 is obvious.

17.6d Integrated HVAC Systems

The design of an efficient residential HVAC depends substantially on the climatic conditions, the building design, and the needs of the occupants. In many regions, both heating and cooling are necessary, or are at least desirable, at different times during the year. The ways in which different approaches to residential HVAC systems can be integrated are very diverse. This section gives some common examples.

It is fairly straightforward to integrate forced hot air heating and central air conditioning (Figure 17.14). Heat from a furnace is provided to the home in cold weather and heat is extracted though a heat exchanger and transferred to the outside by the air conditioning system when cooling is required.

In cooler climates, heating is generally a priority, and many alternative heating approaches can be beneficial and economical. Some approaches include heat pumps, as described, and solar thermal heating (Chapter 8). Figure 17.15 shows a system for combining heat extracted from the environment using a heat pump and an oil-fired forced hot air furnace. The coefficient of performance of the heat pump decreases with decreasing temperature (Figure 17.11), and the heat loss from the building increases with decreasing temperature. Thus, as the outside temperature decreases, heating requirements increase, and the effectiveness of the heat pump diminishes. At some point, supplemental heating from another heating system (in this illustration, the oil-fired furnace) becomes necessary to maintain the interior temperature at the desired level. Another important factor that must be considered for efficient energy utilization is the

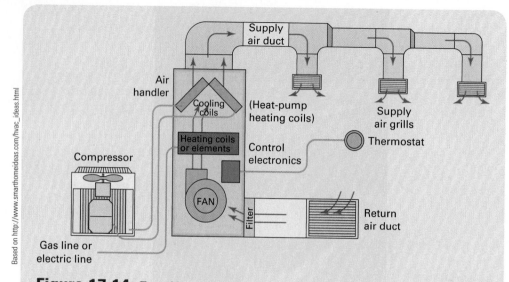

Based on http://www.smarthomeideas.com/hvac_ideas.html

Figure 17.14: Forced hot air heating system with integrated central air conditioning.

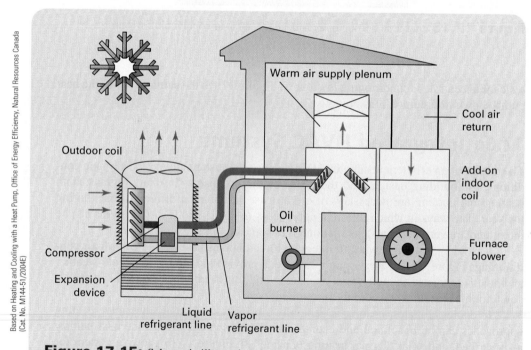

Based on Heating and Cooling with a Heat Pump, Office of Energy Efficiency, Natural Resources Canada (Cat. No. M144-51/2004E)

Figure 17.15: Schematic illustrating the operation of an air source heat pump in conjunction with an oil-fired furnace for residential heating.

coefficient of performance (*COP*) of the heat pump. The *COP* decreases with decreasing outside temperature, and if this approaches unity, then the energy required to run the heat pump, even in an ideal Carnot system, exceeds the energy gained. Thus, the effective control of the HVAC system is important for its most efficient operation.

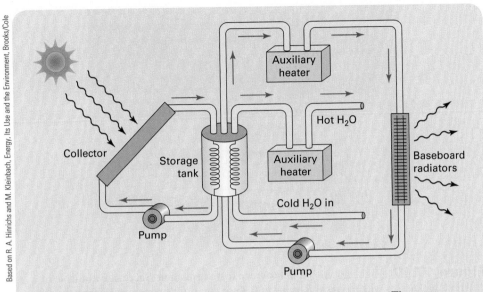

Figure 17.16: Design of an integrated solar space heating system. The system provides heating and hot water. Back-up heating is provided by an auxiliary system (e.g., fossil fuel-fired boiler).

The integration of a solar thermal heating system with a conventional fossil fuel–fired system (Figure 17.16) needs to deal with somewhat different requirements. Solar energy is available during the day and, as described in Chapter 8, must be stored during the night. This is often most convenient if water is used as the heating medium because it is also a convenient storage medium. Thus, Figure 17.16 shows a forced hot water heating system where heat is stored in a tank of water. Heat from the solar collector heats the water in the storage tank, but an auxiliary heating system, such as an oil- or natural gas–fired boiler provides additional heat as needed to maintain the proper storage water temperature, in accordance with the heating and hot water requirements of the home. Daily and seasonal variations in both the home energy requirements and the solar insolation need to be considered in the design of the system.

17.6e Minimizing Heat Loss: Insulation, Windows, and Air Leaks

Although it is important to efficiently produce energy for space heating, it is also important to utilize that energy effectively. Thus, it is necessary to minimize heat losses from a building. In Chapter 8, the basic physics of heat transfer through walls and windows was presented in order to determine the heating needs for a house. In this section, these ideas are expanded to determine the most effective ways of reducing heat losses in a building and conserving energy used for space heating. Actions that are effective in minimizing heat flow from the warm (heated) interior of a building to the cold exterior in the wintertime are also effective at minimizing heat flow from the warm outside to

Richard A. Dunlap

Figure 17.17: Rigid closed-cell polystyrene foam insulation. Sheets are available in different thicknesses with a typical R-value of 0.35 (s · m^2 · °C)/J per cm of thickness.

the cool (air conditioned) interior of a building in the summertime. In this section, we focus on three important aspects of limiting heat loss from a building: insulation, windows, and eliminating air leaks.

Insulation is a material that occupies the space around the interior of the building, for example, inside the exterior walls. Its effectiveness is based on its ability to limit air circulation within the space. A reflective layer such as aluminum foil can also be effective in reducing radiative heat transfer. Insulation comes in various forms:

- *Blankets:* A flexible material such as fiberglass wool that comes in a roll
- *Rigid insulation:* A fibrous or foam material that comes in sheets, such as closed-cell polystyrene foam, often referred to by its proprietary name, Styrofoam (Figure 17.17)
- *Foam insulation:* Sprayed on material that adheres to wall or roof surfaces, such as polyicynene or polyurethane
- *Loose fill:* Loose fiber pellets such as vermiculite or particles of fiberglass that are blown into wall or ceiling spaces using pneumatic equipment

The R-values per unit thickness are given in Table 17.1 for some typical insulating materials. The R-value of insulation depends on the properties of the insulation and is linearly proportional to thickness. The heat loss per unit area is, according to equation (17.5), inversely proportional to the insulation thickness. As discussed in Chapter 8, the R-values of interior and exterior walls need to be added to the R-values of the insulation to obtain the total R-value of the wall. From a practical standpoint, the values given in Table 17.1 should be used as guidelines for understanding the effects of insulation. However, in practice, building structure and construction techniques can often lead to actual R-values that differ substantially from those in the table. Insulation in walls fills the spaces between the wall studs; three factors related to the effectiveness of its installation are worthy of note.

1. The insulating properties of insulation are the result of its ability to minimize air transport. Compressing insulation such as fiberglass wool diminishes its

Table 17.1 Approximate *R*-values per unit thickness of some insulating materials.

material	R-value (s · m² · °C)/J per cm thickness
vermiculite loose fill	0.15
fiberglass loose fill	0.21
polyicynene spray foam	0.25
fiberglass wool	0.26
polystyrene foam rigid sheet	0.35
polyurethane spray foam	0.42
polyisocyanurate spray foam	0.45

© Cengage Learning 2015

insulating properties. Packing thicker insulation into a space than the space should accommodate reduces its effectiveness.

2. The insulation must fill the entire space between the studs. Air spaces between the edge of the insulation and the stud allow for heat loss.

3. The *R*-value for a wall is less than that of the insulation itself, because heat can be transported through the studs by conduction. This is referred to as a thermal bridge. It is a factor for wooden framed buildings but can be an even greater factor for metal framed buildings.

Considerations for insulation in attics are somewhat different because insulation can be laid or sprayed across ceiling joists, reducing concern for factors 2 and 3.

The construction of new buildings can accommodate insulation as appropriate for the climatological conditions in the area. The utilization of appropriate materials, as well as careful construction practices, optimizes the effectiveness of insulation.

In the case of upgrading insulation in existing buildings, access to spaces that need to be insulated may place restrictions on the available options. Adding additional insulation to existing insulation in attics is often fairly straightforward, although caution should be exercised to avoid compressing existing insulation under the weight of additional insulation and thereby reducing its effectiveness. Adding insulation to walls is somewhat more difficult because access to spaces that require insulation is often limited. In many cases, blown-in insulation is the only practical option. In addition to insulating wall and ceiling spaces, improving insulation on heating ducts and/or pipes carrying water for heating purposes through unheated spaces can be productive.

The economics of improving insulation for the purpose of reducing heat loss is often an important consideration. The payback period must make such improvements attractive for a homeowner. In general, a detailed analysis of the effects of improving insulation can be complex. However, Example 17.6 shows a very simple approach to understanding the results of adding insulation to an existing uninsulated building.

The significance of heat loss through windows has been emphasized in Example 17.5. The importance of ensuring that energy-efficient windows are utilized in new construction is obvious. The benefits of improving low-efficiency windows in existing buildings are also clear. Replacing older windows can be an effective means of reducing heating costs.

Example 17.6

Consider a simple house that is a cube 10 m on a side located near Portland, Oregon. It loses heat through the walls and ceiling but not through the floor. What is the heating requirement, in British thermal units per year, if the walls are uninsulated and have an average R-value of 1.0 (s · m² · °C)/J? Repeat this calculation for an insulated house with an R-value for the walls of 3.5 (s · m² · °C)/J. What is the annual savings in energy cost if heat costs $19.75 per GJ (typical of the price of oil or natural gas)?

Solution

From equation (17.5).

$$P = \frac{A(T_h - T_c)}{R}.$$

If A is in m², temperature is in °C, and the R-value is in (s · m² · °C)/J, then the power on the left-hand side is in W. If the right-hand side of the equation is multiplied by 86,400 s/d, then the left-hand side is in J/d. If the temperature on the right-hand side is expressed as the number of degree days per year, then the left-hand side gives the energy requirement in J/y. For Portland, Oregon, Chapter 8 gives the number of degree days per year as 2426 in °C. The surface area of the walls and roof is 5 × 10 m × 10 m = 500 m². Thus

$$\text{energy/year} = (500 \text{ m}^2) \times (2426°C \text{ d/y}) \times \frac{86,400 \text{ s/d}}{1.0 \text{ (s · m}^2 \text{ · °C/J)}} = 1.05 \times 10^{11} \text{ J/y.}$$

Repeating this calculation for an R-value of 3.5 gives

$$\text{energy/year} = (500 \text{ ft}^2) \times (2426°C \text{ d/y}) \times \frac{86,400 \text{ s/d}}{3.5 \text{ (s · m}^2 \text{ · °C/J)}} = 0.30 \times 10^{11} \text{ J/y.}$$

The energy savings are $(1.05 - 0.30) \times 10^{11}$ J/y = 7.5×10^{10} J/y for a monetary value of (75 GJ/y) × ($19.75/GJ) = $1481 per year.

The purpose of a window is to allow light into a house while preventing (as necessary) the transfer of heat between the inside and outside. The simplest window is a single pane of glass (referred to as a single-glazed window). A double-glazed window consists of two panes of glass separated by an air space, and a triple-glazed window consists of three panes of glass, and so on. The primary purpose of the air space between the panes of glass is to reduce heat transfer from convection by preventing the circulation of air that comes in contact with the glass panes. The design of a typical double-glazed window is shown in Figure 17.18.

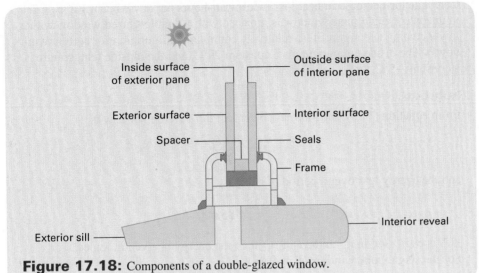

Figure 17.18: Components of a double-glazed window.

A typical single-glazed window has an *R*-value of about 0.18. The increase in *R*-value is approximately 0.18 per additional pane of glass. Thus, a basic double-glazed window (Figure 17.18) has *R*-value of about 0.36, and a triple-glazed window has an *R*-value of around 0.54. The *R*-value can be further increased by adding low-emissivity (low-e) coatings on the glass (see Energy Extra 8.1 for a description of the physics of optical coatings). Such a coating typically increases the *R*-value by about 0.18. Finally, the space between the panes of glass can be filled with a gas other than air. The most commonly used gas is argon. Argon has a higher viscosity and a lower thermal conductivity than air. The first property reduces heat transfer by convection because it reduces gas circulation in the space between the panes. The second property, obviously, reduces heat transfer by conduction. The net result is that replacing the air between the panes with argon increases the *R*-value by about 0.18.

From a practical standpoint, triple-glazed windows with argon fill and low-e coatings are about the practical limit for energy-efficient, high–*R*-value windows. These windows would have an *R*-value around 0.9 (0.54 for the triple glazing, 0.18 for the argon, and 0.18 for the coating). Vacuum glazed windows, where the gas is evacuated from the space between the panes have the highest *R*-values, up to around 2.2. However, these may not be practical for installations if wide temperature variations are expected because the vacuum seals may not be reliable under excessive thermal stress.

The actual net *R*-value for a window may be somewhat different (typically lower) than this simple approach would indicate. Heat transfer through the frame (Figure 17.18) can be important, particularly for windows with metal frames, and this additional heat transfer lowers the *R*-value. Nominally similar windows made by different manufacturers can have appreciably different actual performance.

The effectiveness of replacing lower *R*-value windows with higher *R*-value windows for conserving energy can be estimated on the basis of the physics of heat transfer as a function of *R*-value. The amount of energy that can be saved by upgrading windows is illustrated in Example 17.7, and the economics of this action are summarized in Example 17.8.

Example 17.7

Calculate the decrease in heat loss in kilowatt-hours per day when a 1.2 m × 2.1 m uncoated single-pane window is replaced with a double-glazed window with a low-e coating and argon fill (a relatively common configuration for current windows). The inside temperature is a constant 20°C, and the outside temperature is a constant −5°C.

Solution

From equation (17.5), the thermal power lost through the window is

$$P = \frac{A\Delta T}{R},$$

and the energy loss over a period of a day is

$$E = \frac{A\Delta T}{R} \times 86{,}400 \text{ s/d}.$$

If A is expressed in m^2 and ΔT in °C, then the energy is given in joules. For the single-pane window ($R = 0.18$),

$$E = (1.2 \text{ m} \times 2.1 \text{ m}) \times [20°C - (-5°C)] \times \frac{86{,}400 \text{ s/d}}{0.18 \text{ s} \cdot \text{m}^2 \cdot °C/J}$$
$$= 30.2 \text{ MJ/d}.$$

For the double-glazed window with low-e and argon ($R = 0.72$),

$$E = (1.2 \text{ m} \times 2.1 \text{ m}) \times (20°C - (-5°C)) \times \frac{86{,}400 \text{ s/d}}{0.72 \text{ s} \cdot \text{m}^2 \cdot °C/J}$$
$$= 7.6 \text{ MJ/d}.$$

The difference in energy loss is

$$30.2 \text{ MJ/d} - 7.6 \text{ MJ/d} = 22.6 \text{ MJ/d}.$$

Converting to kWh gives

$$\frac{22.6 \text{ MJ/d}}{3.6 \text{ MJ/kWh}} = 6.2 \text{ kWh/d}.$$

Example 17.8

Consider the situation in Example 17.7. If a house has 12 single-glazed windows that are replaced by low-e, argon-filled double-glazed windows, calculate the savings during 3 months of cold winter weather as typified in the example. Assume that heat is provided by electricity at a rate of $0.11 per kWh.

Solution

At the heat loss reduction given in Example 17.7, 6.2 kWh per day per window, the total savings in kilowatt-hours for 12 windows for 3 months are

$$(6.2 \text{ kWh/d}) \times (12) \times (90 \text{ d}) = 6696 \text{ kWh}.$$

At $0.11/kWh, the value of energy saved is

$$(6696 \text{ kWh}) \times (\$0.11/\text{kWh}) = \$737.$$

Heat loss through walls and windows is relatively straightforward to understand and quantify. Heat loss due to air leaks is due to convection and is more difficult to quantify. However, its importance is obvious; leave a window open in the wintertime in a cold climate, and the heat loss is apparent. Small spaces where air can leak into or out of a house can add up. These leaks include poor seals around windows and doors or between the house and the foundation (sill plate), but they also include places like dryer vents where outside air can get into the house. Figure 17.19 is an illustration of some of the common air leaks in a home. It is important, as in the case of considering insulation, to note that air leaks between the heated portion of the house and unheated spaces, such as attics, can be as significant as those to the outside. Because warm air is less dense than cold air, the warm air inside the house tends to leak out through openings in the higher portions of the house. These often lead into attic spaces (Figure 17.19). Lighting fixtures in the ceiling are a common source of air transport between the house and

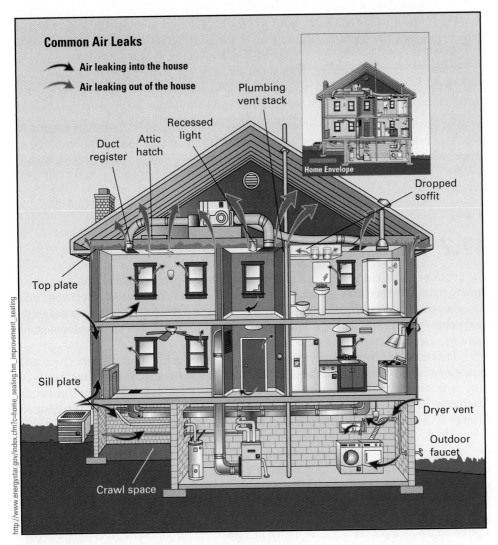

Figure 17.19: Common air leaks in a house.

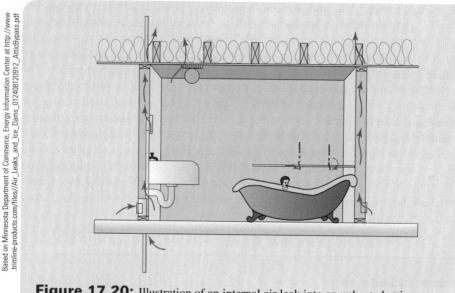

Based on Minnesota Department of Commerce, Energy Information Center at http://www.trimline-products.com/files/Air_Leaks_and_Ice_Dams_01240812091z_AtticBypass.pdf

Figure 17.20: Illustration of an internal air leak into an unheated attic space through electrical outlets and plumbing.

the attic. Also, air can leak into uninsulated interior walls through electrical outlets and switches or through spaces around plumbing and travel upward to the unheated spaces above, such as the attic (Figure 17.20). Thus sealing openings, even if they do not involve a direct connection to the outside, is important and can be accomplished with weather stripping or caulking as appropriate.

17.7 Residential Lighting

The efficient use of electricity for residential lighting involves personal actions such as turning lights off when they are not in use, as well as the analogous automated approach (i.e., the use of timers or photoelectric switches to control lighting). The most significant improvements in efficiency, however, are likely to come from the development of more efficient devices.

Traditionally, room lighting consisted of incandescent bulbs or, where space and geometry allowed, fluorescent tubes. Fluorescent tubes have the advantage that they produce more light output for the same electrical energy input, typically about 5 times as much light output, compared with incandescent bulbs. The amount of light is measured in units of *lumens* (lm). A lumen is a measure of the amount of light over the visible portion of the electromagnetic spectrum and accounts for the actual light intensity, as well as the variations in the sensitivity of the human eye, over that range of wavelengths. Thus, it is an accurate measure of the brightness of the light as perceived by a person. Different devices that produce light output from an electrical input yield different amounts of light for the same power input, depending on the mechanism used for producing the light, as well as the wavelength distribution of the light produced. The light output (in lumens) for a given electrical input (in watts) is shown in Table 17.2 for some common light sources. It is clear that fluorescent lamps are substantially more efficient at converting electrical

Table 17.2: Typical light output per electrical input in lumens per watt (lm/W) for some light-producing devices.	
light source	output/input (lm/W)
incandescent bulb	15
compact fluorescent lamp (CFL)	60–70
light-emitting diode (LED)	100–150

© Cengage Learning 2015

Richard A. Dunlap

Figure 17.21: Compact fluorescent lamps. Front left to right: 9-W spiral, 13-W spiral, 23-W spiral, 26-W spiral, and 9-W tube, with light output equivalent to incandescent bulb with ratings of 40 W, 60 W, 100 W, 120 W, and 40 W, respectively. Lamps shown have the standard North American Edison base.

energy into light than incandescent bulbs. The energy that does not become light is given off as heat, clearly incandescent lights produce more heat than light.

In recent years, compact fluorescent lamps (CFLs) have been developed. These are available in sizes that approximate the size of incandescent bulbs and are equipped with a standard base (either the Edison screw base, used in the United States and many other countries, or a bayonet base) so that they can be inserted into a standard lamp socket. Some examples of commercially available CFLs are shown in Figure 17.21. Two geometries are in common use: a spiral geometry, more common in North America, and a tube geometry, often used elsewhere. These have been promoted for their efficient production of light and their net positive contribution to the conservation of energy.

Example 17.9

A package for a 9-W CFL claims an expected lifetime of 8000 hours and a savings of $24.80 in energy costs. Confirm this claim.

Solution

A 9-W CFL has a light output equivalent to a 40-W incandescent bulb for a savings of 31 W. Over a lifetime of 8000 hours, this corresponds to a total energy savings of

$$(0.031 \text{ kW}) \times (8000 \text{ h}) = 248 \text{ kWh}.$$

If the cost of electricity $0.10 per kWh, then this corresponds to a savings of $24.80.

The principle of operation of the CFL is the same as for all fluorescent lamps. Electrons traveling through a gas interact with the atomic electrons of the gas atoms and cause them to be excited into higher energy levels; from there, they undergo transitions back to the ground state and thereby emit photons. Specifically, in a fluorescent lamp, the electrons are produced by the filaments at the ends of the tube, which are heated by electric current flowing through them. The filaments, referred to as cathodes because they emit electrons, are typically made of tungsten coated with a metallic oxide that readily gives off electrons. The tube contains mercury vapor at a very low pressure, and it is the electronic transitions in the mercury atoms that produce light. The photons that are produced by most electronic transitions in mercury are at fairly high energy and are in the ultraviolet portion of the electromagnetic spectrum (Figure 1.1) that is not visible to the human eye. To produce visible light, the ultraviolet photons are incident on a phosphor coating on the inside of the tube. The atoms in the phosphor coating absorb the ultraviolet photons and re-emit them at lower energies (in the visible portion of the spectrum) over a broad range of wavelengths. This is similar to how white light is produced by a blue LED (Section 17.5). The wavelength spectrum of the fluorescent lamp can be varied by using different combinations of phosphors. Classifications of fluorescent lamps such as cool white and daylight result from different combinations of phosphors.

Fluorescent lamps have an interesting electrical property. As the current flows through the mercury vapor, exciting electronic transitions in the gas atoms, the excited atoms actually make it easier for the current of electrons to flow. Thus as the current, I, increases, the resistance, R, decreases, making it easier for more current to flow. In the case where a constant voltage, V, is applied to the tube, the current increases in an uncontrolled manner according to Ohm's law:

$$I = \frac{V}{R}.$$

(17.6)

To avoid this situation, a ballast is utilized. A simple ballast is merely an inductor in series with the tube that limits the current flow in the AC circuit. CFLs typically use an integral electronic circuit built into the base (Figure 17.22) to fulfill this requirement.

Potential environmental hazards are associated with CFLs. Like other fluorescent lamps, CFLs contain mercury (Hg), which poses health risks. Government regulations (in the United States and most other locations) require that all mercury-containing products display the (Hg) symbol (Figure 17.23). Each CFL manufactured at present contains 1–5 mg of mercury. Many locations have initiatives in place to divert CFLs from disposal in, say, landfills where the mercury can disperse in the environment. However, this practice is far from universal, and in most places discarded CFLs are likely to contribute to environmental mercury release. Although this may appear to be an undesirable situation, the U.S. Environmental Protection Agency has estimated that the mercury content of all CFLs sold in the United States during a year is only about 0.1% of the total annual U.S. anthropogenic mercury emissions.

A simple calculation provides information about the energy and economic advantages of CFLs. The difference in light output to produce the same light level in a room is a measure of the energy savings that can result for the replacement of incandescent bulbs with compact fluorescent lamps. However, this is not a complete picture of how CFLs affect energy utilization. A CFL mostly produces light and a relatively small amount of heat. By comparison, about 10% of the electrical energy used by an incandescent bulb is converted into light; the remainder of about 90% is

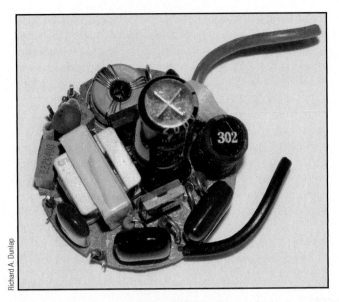

Richard A. Dunlap

Figure 17.22: Internal electronic ballast from the base of a 13-W spiral compact fluorescent lamp.

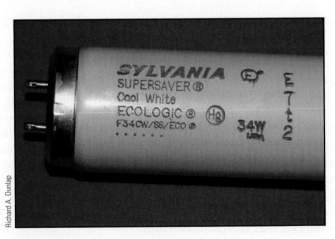

Richard A. Dunlap

Figure 17.23: The mercury symbol, Hg inside a circle, on a fluorescent lightbulb.

converted into heat. Thus, most of the energy goes into heat, which may be a desirable or an undesirable by-product of the light production, depending on the situation. For example, in, say, Minnesota in the wintertime, the excess heat is not wasted energy but contributes to heating the interior of the building, thus reducing the energy necessary for heat by that amount. In this case, the total energy saved by converting incandescent bulbs to CFLs is not as great as one would expect. On the other hand, if we consider the situation, say, in Alabama in the summertime, the excess heat produced by an incandescent bulb is really waste heat and would typically have to be eliminated from the building by the air conditioning system. Thus, in this case, the incandescent bulb not only produces less light (per watt electrical input) but also puts

Richard A. Dunlap

Figure 17.24: A 7-W LED lamp with Edison screw base.

additional energy demands on the cooling system, and the total energy saving for converting to CFLs is greater than expected.

As Table 17.2 indicates, compact fluorescent lamps are substantially more efficient at converting electrical energy into visible light than traditional incandescent bulbs. The table also shows that LEDs are even more efficient at producing light than CFLs. LEDs have become an important alternative to traditional lighting methods for roadway lighting. They have also recently become available as replacements for domestic room lighting. Due to their high efficiency, a 9-W LED lamp can provide as much light output as a 60-W incandescent bulb. An example of a commercially available LED lamp for room lighting is shown in Figure 17.24. The lifespan of a typical LED lamp exceeds that of a CFL; about 30,000 hours compared to about 8000 hours. Due to the substantially higher initial cost (at present), about a factor of 10 compared with CFLs, their use has not been as extensive, although prices have continued to decline, and this situation may change in the future.

17.8 Transportation

17.8a Fuel Economy

As illustrated in Figure 2.5, transportation accounts for nearly 30% of the energy use in the United States.and transportation accounts for over 70% of the energy obtained from petroleum. It is therefore an important concern to ensure that transportation technology makes efficient use of fuel. The problem of optimizing the energy efficiency of transportation is a complex one with many facets. One side of the problem deals with the prevalent mode of transportation. This varies considerably from one part of the world to another, from modes of transportation that are very efficient (i.e., bicycles) to moderately efficient (i.e., public transportation by bus or train) to the least efficient (personal automobiles). Globally, there has been a consistent shift from more efficient transportation modes to less efficient transportation modes with time. The move toward a transportation system that is dependent on automobiles occurred in the 1920s and 1930s in North America, in the 1950s and 1960s in much of Europe, and even later in many other parts of the developing world. Thus, a historical analysis of the use of energy for transportation is greatly influenced by economic, social, and cultural factors. Promoting more efficient use of transportation resources can help to conserve energy, and this might include encouraging consumers to drive less, ride bicycles, or take public transportation.

Another way of viewing the efficiency of transportation is to consider the trends in the efficiency of, say, the automobile over time. This approach deals primarily with the science and technology of transportation, rather than changes in human behavior and is the approach taken in the following discussion.

The fuel economy of a vehicle can be quantified in several different ways. The units used may be given as distance traveled per volume of fuel (commonly referred to as *fuel economy* or *fuel efficiency*) or as volume of fuel used per distance traveled (commonly referred to as *fuel consumption*). Fuel economy is commonly measured in kilometers per liter while fuel consumption is measured in liters per 100 kilometers (L/100 km). Different conventions are used in different countries around the world. There is an inverse relationship between fuel economy and fuel consumption given by

$$\frac{100}{(\text{L}/100 \text{ km})} = \text{km/L} \qquad (17.7)$$

This relationship is illustrated graphically in Figure 17.25.

In the 1960s and 1970s, it became increasingly apparent that the use of fossil fuels for transportation was a matter for concern. Concerns about the pollution that resulted from fossil fuel use (Chapter 4) were responsible for many governments' establishing

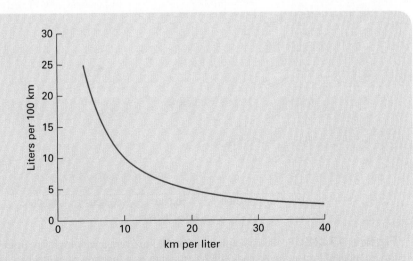

Figure 17.25: Relationship between fuel consumption measured in L/100 km and fuel efficiency measured in km/L.

policies to deal with emissions from vehicles. Concerns also arose over the long-term availability of fossil fuel resources, and this provided motivation to improve the fuel efficiency of automobiles. Government regulations in a number of countries specified average fuel efficiency that manufacturers needed to achieve for their vehicles in that country (in the United States, this is the Corporate Average Fuel Economy (CAFE), as discussed in Section 17.2. Usually vehicles were categorized as, say, passenger vehicles, light trucks, heavy trucks, and so on. Fulfilling fuel consumption regulations can be viewed in two ways: (1) preferentially marketing vehicles that meet fuel consumption guidelines or (2) improving vehicle technology to increase efficiency and hence reduce fuel consumption. The former approach depends on the needs and desires of the consumer and is a function of factors like economy and geography, whereas the latter deals primary with vehicle technology.

Per-capita GDP plays an important role in what vehicles can be most successfully marketed, and this, in turn, is directly related to fuel economy. Generally, more expensive vehicles burn more fuel. Figure 17.26 shows the fuel consumption for all passenger vehicles manufactured by Toyota and marketed in North America as a function of their cost. People who purchase more expensive vehicles use more fuel.

Geographic and cultural factors also influence vehicle choice. People who live in urban (rather than rural) environments or in regions of higher population density may chose smaller (and typically more fuel-efficient) vehicles. For example, in 2010 the average fuel consumption for passenger vehicles in North America was around 7.8 L/100 km, while for Japan and Europe Union it was around 5.5 L/100 km. While this is significant, it should be noted that the comparison of data from different countries can sometimes be misleading. The same vehicle may receive different fuel consumption ratings in different countries due to different testing procedures.

From a scientific and engineering standpoint, the development of new technologies that actually make an internal combustion engine vehicle more efficient in terms of converting the chemical energy of the fuel into mechanical energy is perhaps the

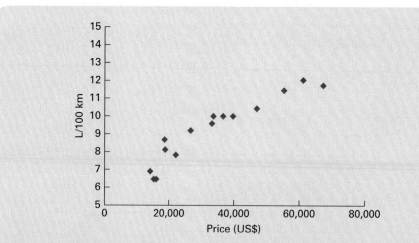

Figure 17.26: Fuel consumption in L/100 km average city/highway from U.S. Environmental Protection Agency (EPA) estimates as a function of model base price (in US$) for all nonhybrid passenger automobiles marketed in North America by Toyota (Scion/Toyota/Lexus) in 2012.

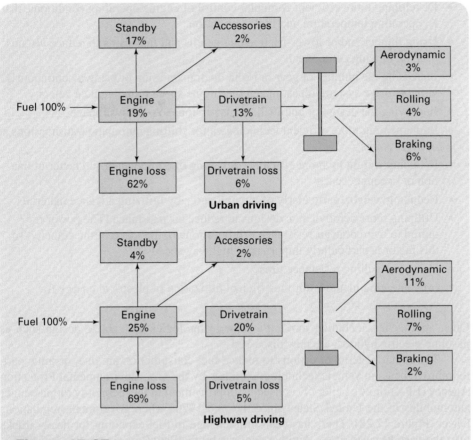

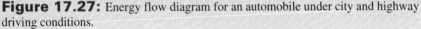

Figure 17.27: Energy flow diagram for an automobile under city and highway driving conditions.

most interesting. To understand where the most significant gains can be achieved, it is necessary to analyze where the energy losses actually occur. Based on studies by the U.S. Department of Energy, these losses are summarized for city and highway driving conditions in Figure 17.27. As illustrated in the figure, the typical overall efficiency for conversion of chemical energy of the fuel to energy delivered to the wheels for a gasoline-powered passenger vehicle for all driving conditions is about 17% (i.e., 13% for city and 20% for highway). This is an important reference point for comparison with other vehicle technologies, as discussed in the next two chapters.

Clearly, the greatest energy losses are from the engine. These losses are to some extent inevitable because of the basic thermodynamics of a heat engine. However, engine design and operation can optimize the efficiency of converting the chemical energy of the fuel into mechanical energy. The design of the drivetrain, control systems, and the overall geometry of the vehicle are some of the other factors that can make the vehicle more fuel efficient. Specifically, some of the design criteria that can be targeted are as follows:

- Improving cooling system efficiency (The use of efficient coolants can reduce the cooling system volume and lead to faster warmups and better temperature regulation.)

- Developing more efficient computer control of engine operating conditions, that is, operating temperature and fuel distribution
- Using thinner and/or lower-friction engine oils and lubricants to reduce viscous drag on moving components
- Increasing the number of gear ratios in the transmission or using a continuously variable drive system to best match the engine speed to the vehicle velocity
- Improving the design of automatic transmissions to reduce slippage
- Implementing more efficient technologies for shifting automatic transmissions at optimal points
- Reducing weight by more efficient packaging of components and better utilization of occupant space
- Reducing weight, particularly of moving parts, by utilizing advance materials
- Utilizing more aerodynamic designs to reduce air resistance (The power required to overcome air resistance is proportional to the cube of the velocity, so this factor is particularly important at highway speeds.)
- Using low rolling resistance tires
- Implementing methods for the efficient utilization of electrical energy for accessories

Other approaches, including moving to new technologies for vehicle design such as gasoline-electric hybrids, are now discussed.

The actual benefits of actions to reduce fuel consumption are due, in some part, to improvements in vehicle technology. Studies by the U.S. Environmental Protection Agency (EPA) show a generally consistent improvement in fuel economy for passenger automobiles in the United States since the mid-1970s when government regulations began (Figure 17.28). There has been little change in fuel economy for heavy trucks over the past 40 years or so. Another analysis by the Pew Environmental Group (Figure 17.29) showed results that suggest substantial improvement from the mid-1970s to the mid-1980s. There is much less improvement over the next 20 years or so, although there seems to be consistent improvement in fuel economy in the last five years.

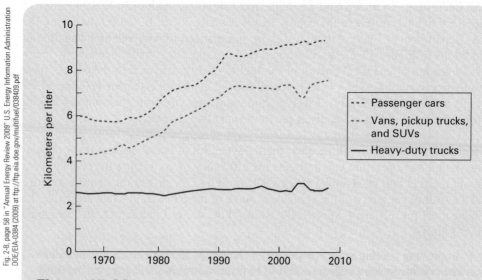

Fig. 2-8, page 58 in "Annual Energy Review 2009" U.S. Energy Information Administration DOE/EIA-0384 (2009) at ftp://ftp.eia.doe.gov/multifuel/038409.pdf

Figure 17.28: Fuel economy of vehicles in the United States, 1966–2008.

Based on http://www.pewenvironment.org/news-room/
fact-sheets/history-of-fuel-economy-one-
decade-of-innovation-two-decades-of-inaction-329037

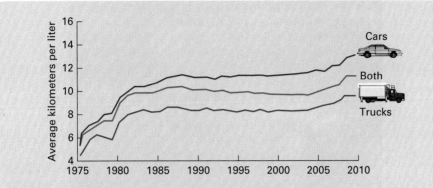

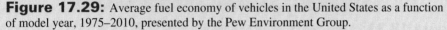

Figure 17.29: Average fuel economy of vehicles in the United States as a function of model year, 1975–2010, presented by the Pew Environment Group.

Recent and projected fuel economies in different countries worldwide is shown in Figure 17.30. The clear differences between Japan/European Union and the rest of the world are apparent in the figure.

In addition to the characteristics of a vehicle itself, the fuel economy that is actually realized also depends on drivers' habits. Energy-conserving driving behavior is promoted by many government energy policies (Section 17.2) and includes driving at moderate speeds on the highway, using cruise control, and minimizing idle time.

Based on http://www.pewclimate.org/federal/executive/vehicle-standards/fuel-economy-comparison

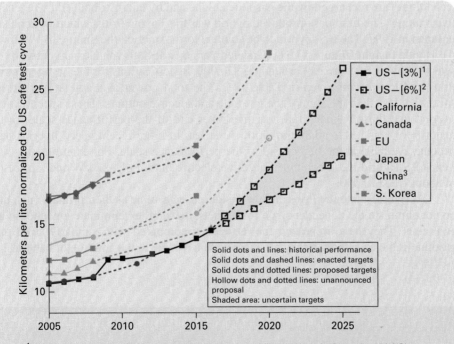

[1]Based on 3% annual fleet GHG emissions reduction between 2017 and 2025 in the September 30th NOI.
[2]Based on 6% annual fleet GHG emissions reduction between 2017 and 2025 in the September 30th NOI.
[3]China's target reflects gasoline fleet scenario. If including other fuel types, the target will be higher.

Figure 17.30: Average fuel economy for passenger vehicles in different countries. Historical data for 2005 to 2010 and projected data until 2025.

Richard A. Dunlap

Figure 17.31: A first-generation Honda Insight hybrid.

17.8b Hybrid Vehicles

Hybrid vehicles have become fairly common in recent years as an alternative to traditional gasoline-powered internal combustion engine vehicles. A hybrid vehicle contains both an internal combustion engine and batteries that power an electric motor and is a step toward moving to a nonfossil fuel transportation technology. A gasoline engine is used to power the vehicle, and batteries drive an electric motor to provide supplemental power. The batteries (see the next chapter for more details on battery technology) are recharged by the gasoline engine and by energy obtained by regenerative braking. The first mass-produced hybrid was the Toyota Prius, which went on sale in Japan in 1997. The first hybrid to be sold in North America (beginning in 1999) was the Honda Insight (Figure 17.31). Updated versions of both vehicles are currently on sale worldwide for a price of around US$20,000. Specifications for the first-generation Prius and Insight are given in Table 17.3. The total power available at any time is somewhat less than the sum of the gasoline and electric outputs. These specifications suggest that these vehicles fall near the lower end of the performance scale for the majority of vehicles available in North America. In more recent years, hybrids have become available at the higher end of the performance scale. For example, the 2011 Lexus LS Hybrid has a combined gasoline/electric output of 327 kW and a price tag of about US$110,000.

As hybrids, like the Prius and the Insight, use gasoline as their primary fuel, they do not eliminate the dependency on fossil fuels, nor do they eliminate greenhouse gas emissions. They have improved fuel efficiency compared with similarly sized pure gasoline vehicles, although they are more expensive due to the addition of batteries, an

© Cengage Learning 2015

Table 17.3: Power output of the first-generation Toyota Prius and Honda Insight.					
vehicle	years	mass kg	gasoline power kW	electric power kW	fuel consumption L/100 km
Toyota Prius	1997–2001	~1200	43	30	5.6
Honda Insight	1999–2006	838	52	9.7	3.7

Figure 17.32: Chevrolet Volt (series hybrid).

electric motor, and the associated systems. In a sense, they may be considered a transitional technology that helps to mitigate a problem (pollution, greenhouse gases, and dwindling fossil fuels) but cannot solve it.

A more recent development, the *series hybrid vehicle*, is exemplified by the Chevrolet Volt (Figure 17.32). This design contrasts the more common parallel hybrid technology and is sometimes called a *plug-in hybrid* because it can be recharged by plugging the vehicle into an external source of electricity. The Volt is basically an electric vehicle that uses an electric motor powered by Li-ion batteries to drive the wheels. A small gasoline engine starts only when the batteries need to be recharged. Power output from the electric motor is 111 kW and from the generator used to charge the batteries, 53 kW. A gasoline engine runs very efficiently (relatively speaking) if it runs at a constant speed and under constant load (typical efficiency of 30–35%) rather than having to respond to the fluctuations in driving conditions (typical efficiency of 15–20%). A fuel consumption of 1.6–4.7 L/100 km for the Volt is claimed. The range running on batteries is only about 65 km. However, the range using the gasoline engine to recharge the batteries is limited only by the size of the gasoline tank and is about 1000 km. The cost is around US$40,000. It remains to be seen whether this step toward a nonfossil fuel vehicle is a successful interim technology. If it is, it can provide an effective approach to conserving fossil fuel resources and reducing greenhouse gas emissions.

Although hybrids do not eliminate the need for fossil fuels, advances in battery technology and electrical components associated with the development of hybrid vehicles may have a direct benefit in the design of a new generation of battery electric vehicles and fuel cell vehicles, as described in Chapters 19 and 20.

17.9 **Summary**

Energy conservation is an essential component of a viable long-term plan to provide sustainable energy, and various aspects of its effective implementation are discussed in this chapter. Options for energy conservation are available on a variety of different scales, from the implementation of national energy policies by federal governments to what individuals can do in their own homes.

While scientific and technological advances can lead to improved conservation methods in the future, substantial technology already exists that can be implemented to aid in conservation efforts. As discussed in this chapter, the development of national and regional energy policies that include conservation are needed to provide guidelines for actions that promote the implementation of positive conservation measures. On a regional level, the most effective approaches to conservation are often related to factors such as climate, geography, and social traditions and economy. Regional and municipal governments, as well as public utilities, can do much to further conservation efforts in areas such as cogeneration, smart grid implementation, and street lighting.

Cogeneration refers to the direct use of excess heat from the thermal generation of electricity for space heating needs. This approach has been particularly effective for small generating facilities, which are appropriate for universities, large industrial plants, or isolated communities.

The implementation of a smart grid has been presented. The smart grid is an essential component for the efficient integration of renewable energy sources with tradition fossil fuel and nuclear generating facilities and for the effective regulation of electricity use in the community.

The principle of operation of LED lamps follows from the description of photovoltaic devices presented in Chapter 9. Whereas photovoltaic cells convert photons into a voltage, LEDs convert voltage into light. This chapter described the use of LEDs for municipal roadway lighting. Because municipal governments can pay as much as 40% of their budgets on street lighting, and LEDs lamps can provide a 60% savings in electricity consumption over traditional lighting, a net savings of up to a quarter of a municipality's budget can be realized by converting to a more energy-efficient lighting technology. Additional benefits include improved illumination and the elimination of the potentially toxic substances used in many traditional lamps.

This chapter also presented ways in which individuals can reduce home energy use by the use of efficient heating, ventilation, and air conditioning (HVAC) systems, the reduction of heat loss by efficient insulation, and the reduction of electricity consumption for lighting through the use of compact fluorescent lamps or LED lamps.

Finally, this chapter overviewed efforts to conserve energy associated with transportation. Government standards for vehicle fuel economy, present in most countries since the 1970s, have reduced the energy requirements of passenger vehicles and trucks and have provided a substantial reduction in the consumption of fossil fuels. This chapter also reviewed the viability of hybrid vehicles. Traditional parallel hybrid passenger cars offer improved fuel efficiency, but typically at the expense of increased capital cost. Series or plug-in hybrids, which have recently become available, offer increased fuel savings and are a step toward the widespread use of battery electric vehicle technology.

Problems

17.1 $R = 0.18$ windows in a house are replaced with $R = 0.52$ windows at a cost of $250 per m^2. Assume that the outside temperature is a constant 5°C [a reasonable approximation for a region corresponding to about 4300 degree days per year (°C)] that and heat costs $0.03 per MJ. How long will it take to recover the cost of the window replacement? (Do not include the cost recovery factor.)

17.2 The following table gives specifications for some 2012 automobiles sold in the United States.

automobile	average gasoline consumption (L/100 km)	base price (US$)
Honda Civic DX (automatic)	6.9	$ 16,605
Honda Civic Hybrid	5.3	$ 24,134
BMW 750i	13.1	$ 84,300
BMW Active Hybrid 750i	11.7	$ 97,000

If it is assumed that maintenance costs are the same for the gasoline and hybrid versions of the same vehicle, use the current price of gasoline in your area to determine how many miles would need to be driven for the better fuel economy of the hybrid to outweigh its higher purchase price. Repeat this calculation for the BMW.

17.3 Refer to Figure 2.5. What fraction of waste heat from electricity generation in the United States would be needed to satisfy all of the country's residential fossil fuel use (primarily for heating)?

17.4 A home owner in Maine replaces six 60-W incandescent bulbs in a family room with equivalent CFLs. Using the following information, estimate the net annual energy savings (in dollars) for lighting and heating. The lamps are on an average of 3.5 hours per day, and electricity costs $0.105/kWh. The home is heated with an oil furnace at an efficiency of 87%, and home heating fuel costs $0.74 per liter. Heat is required 191 days per year, and there is no air conditioning.

17.5 In a local store, find the price of a 60-W incandescent bulb and CFL and LED bulbs with an equivalent light output. Based on a use of 4 hours per day and an electricity cost of $0.11/kWh, calculate the payback period for each of these bulbs compared to the incandescent.

17.6 Compare the overall efficiency of heating a house with oil at 85% efficiency and heating a house with a heat pump with a coefficient of performance of 6 using electricity generated by a heat engine at 35% efficiency.

17.7 Consider the heat losses through the four exterior walls of a house. The house is 9.0 m × 12.0 m and the walls are 3.0 m high. There are 12 windows, each 1 m × 1.6 m. The walls are uninsulated and have an R-value of $R = 1.0$ and the windows are (uncoated) single pane. The home owner has the option of either upgrading the windows to (uncoated) triple pane or introducing insulation into the walls to increase their R-value to $R = 3.5$. Which action will provide the greatest benefit? In this problem ignore doors and heat losses through the floor and roof.

17.8 Consider typical passenger automobiles in the United States. Calculate the reduction in CO_2 emissions (in tonnes) for a 2008 vehicle compared to a 1966 vehicle during the lifetime of the vehicle (assumed to be 250,000 km).

Bibliography

B. Anderson and M. Riorden. *The New Solar Home Book*, 2nd ed. Brick House Publishing, Andover, MA (1996).

B. Anderson and M. Wells. *Passive Solar Energy: The Homeowner's Guide to Natural Heating and Cooling*, 2nd ed. Brick House Publishing, Andover, MA (1996).

A. K. Athienitis and M. Santamouris. *Thermal Analysis and Design of Passive Solar Buildings.* James & James, London (2002).

D. Chiras. *The Solar House: Passive Heating and Cooling.* Chelsea Green Publishing, White River Junction, VT (2002).

Natural Resources Canada. *Heating and Cooling with a Heat Pump.* Cat. No. M144-51/2004 (2004).

V. Quaschning. *Renewable Energy and Climate Change.* Wiley, Chichester (2010).

U.S. Department of Energy. *Insulation Fact Sheet.* DOE/CE-0180 (2008).

Energy Storage

Learning Objectives: After reading the material in Chapter 18, you should understand:

- The need for energy storage.
- Pumped hydroelectric storage and its use.
- The properties of compressed air.
- The use of compressed air for energy storage.
- Energy storage capabilities of flywheels and the relevance of materials properties.

- Properties of superconductors and the history of their development.
- The use of superconducting magnets for electrical storage.

18.1 Introduction

Storage is a major component of our energy use. Energy storage is necessary because it is sometimes inconvenient or impossible to produce energy when it is needed or where it is needed. The first instance may arise because of variations in demand compared with capacity (i.e., daily, weekly or seasonal fluctuations in electricity use) or inherently variable production methods (i.e., solar, wind, tidal, etc.). The second instance is most obvious in the case of energy needs for transportation. It must be realized that the storage of any form of energy often involves the conversion of one form of energy to another (and back again) and that any conversion is less than 100% efficient. So energy storage means energy loss, and this must be considered in the analysis of the viability of any particular method.

Chapter 8 presented some information about the storage of thermal energy. It is also important to consider other energy storage possibilities, particularly for the storage of electricity. Chapter 19 considers electrical storage using batteries, particularly with respect to electric vehicles, and Chapter 20 considers the possibility of using hydrogen as a method of storing energy. This chapter considers various other energy storage methods aimed at large-scale electricity storage for the grid.

18.2 Pumped Hydroelectric Power

The typical demand for electricity as a function of time throughout a week is shown in Figure 18.1. There are also longer-term seasonal variations. It is impractical to have generating facilities that can routinely provide power that can meet the peak demand. One approach is to have supplementary generating facilities that can be brought online quickly when the need arises. These might be combustion turbines used to supplement electricity generation by a coal-fired or nuclear power plant. Another approach is to store excess energy produced during periods of low demand and use this when demand exceeds capacity.

Pumped hydroelectric power is a well established technology that is in common use by public utilities to account for variations in electricity demand. It was first used extensively in the 1930s and remains the most economical and practical method for large-scale electricity storage. The general design of a pumped hydroelectric facility is shown in Figure 18.2. Excess electricity is used to pump water from the lower reservoir to the upper reservoir. In periods of high demand, the water is allowed to flow back into the lower reservoir to generate electricity by means of a turbine/generator. Most

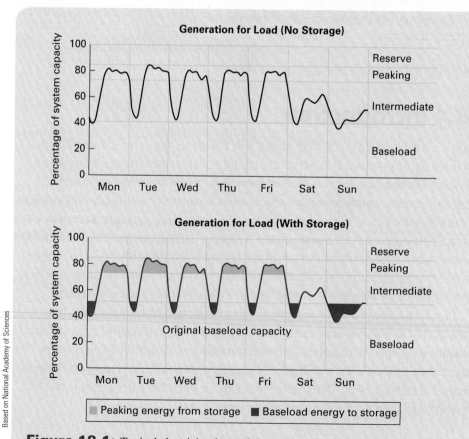

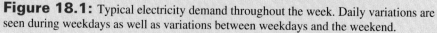

Figure 18.1: Typical electricity demand throughout the week. Daily variations are seen during weekdays as well as variations between weekdays and the weekend.

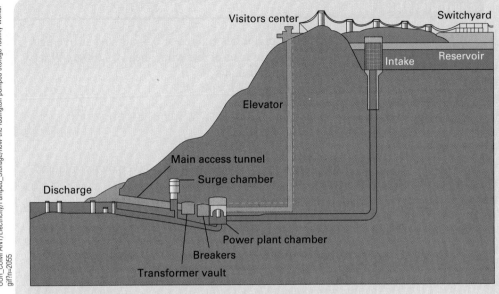

Figure 18.2: Diagram of pumped hydroelectric facility.

commonly, turbines such as a Francis turbine are used reversibly both as the pump and to drive the generator. The energy stored in water in the upper reservoir is given by

$$E = mgh, \tag{18.1}$$

where m is the mass of the water, g is the gravitational acceleration, and h is the average head. The power generated is

$$P = \frac{dE}{dt} = gh\frac{dm}{dt} = \rho gh\frac{dV}{dt}, \tag{18.2}$$

where ρ is the water density, and (dV/dt) is the flow rate (in volume of water per unit time). It might seem that there is a trade-off between head and flow rate and that the same power can be achieved in a high-flow low-head facility as in a low-flow, high-head facility. In practice high head is preferred because, to achieve a large enough flow in a low-head facility, the pipe may have to be of an impractical size. Also, to maintain power output for an extended period of time, the total volume of a low-head facility must be greater. Pumped hydroelectric facilities are therefore most practical in geographic locations that provide for sizable reservoirs that are relatively close together but furnish a sufficient head. Although this is most commonly achieved where only minor modifications of the natural geography are needed, it is also possible to utilize underground chambers (e.g., unused mines) for the lower reservoir. Although one might view pumped hydroelectric generation as related to normal hydroelectric generation, the geographic requirements are not necessarily compatible, and the electricity stored in a pumped hydroelectric facility need not be produced by any particular method or in particularly close proximity to the storage facility. In fact, pumped hydroelectric storage is most suitable as a means of electricity storage in conjunction with generating methods that can be naturally variable, such as solar or wind.

Example 18.1

A pumped hydroelectric storage reservoir has an area of 0.5 km^2 and an average depth of 10 m. The water flows to a lower reservoir with a head of 100 m. What is the total energy available in megawatt-hours?

Solution

The gravitational energy stored in the water is given by

$$E = mgh.$$

The total mass of the water stored in the upper reservoir is given by its volume and the density of water:

$$m = \rho V = (1000 \text{ kg/m}^3) \times (0.5 \times 10^6 \text{ m}^2) \times (10 \text{ m}) = 5 \times 10^9 \text{ kg}.$$

The energy stored is therefore

$$E = (5 \times 10^9 \text{ kg}) \times (9.8 \text{ m/s}^2) \times (100 \text{ m}) = 4.9 \times 10^{12} \text{ J}.$$

Converting to MWh gives

$$E = (4.9 \times 10^{12} \text{ J}) \times (2.78 \times 10^{-10} \text{ MWh/J}) = 1361 \text{ MWh}.$$

Pumped hydroelectric storage has fairly substantial infrastructure cost, but the maintenance and operating costs are quite low. Also, it is fairly low technology and quite reliable. It can be brought online quickly in response to demand variations. The lifetime of the storage is long, although not indefinite, because of natural seepage and evaporation. The efficiency is quite high, reaching about 80%. This is a result of the intrinsically high efficiency of converting the mechanical energy of falling water to electricity and the high efficiency of water pumps, although friction in the pipes decreases the efficiency by a small amount.

Example 18.2

For the pumped hydroelectric example in Example 18.1, how long could the energy stored in this facility provide electricity for a town of 50,000 people if the average power requirement for each person is 800 W$_e$? Assume a generator efficiency of 90%.

Solution

The town of 50,000 people uses a total of

$$(50,000) \times (800 \text{ W}_e) = 40 \text{ MW}_e.$$

At 90% efficiency the pumped hydroelectric facility can produce a total of

$$(1361 \text{ MWh}) \times (0.9) = 1225 \text{ MWh}_e.$$

The total time that the facility can supply the required electricity is

$$\frac{1225 \text{ MWh}_e}{40 \text{ MW}_e} = 30.6 \text{ h}.$$

Pumped hydroelectric is the most commonly used storage method for large-scale electricity storage for the grid. It is based on a well established, reliable technology and is reasonably efficient. There is the potential for large energy storage capacity and readily available high power output. Although infrastructure costs can be high, there is relatively little maintenance.

The largest pumped hydroelectric facility in the world is the station in Bath County, Virginia, constructed between 1977 and 1985 at a cost of US $1.6 billion. The facility consists of two reservoirs separated by a vertical distance of 385 m. The lower reservoir has an area of 2.25 km^2, and the upper reservoir has an area of 1.07 km^2. The water flows between the reservoirs through three tunnels, each approximately 6 km in length. Each tunnel splits into two 5-m diameter penstocks (for a total of six) before reaching the turbines. The map below shows the area around the facility and the locations of the upper and lower reservoirs. When demand exceeds the base load grid supply, water flows from the upper reservoir to the lower reservoir through six Francis-type turbines (at up to 850 m^3/s, generating a maximum of 2772 MW$_e$ for up to 11 hours. When electric demand is low, the generators function as electric motors, producing up to 480 MW to drive the turbines backward and pump water from the lower reservoir to the upper reservoir at a rate of 800 m^3/s.

A major advantage of pumped hydroelectric power to supplement base load generation compared to, say, bringing additional thermal generation online is response time. The Bath County station can bring 400 MW online just 6 minutes after start-up to add to base load capacity during periods of high demand.

Topic for Discussion

It may seem natural to think of pumped hydroelectric power in the context of normal hydroelectric power. However, pumped hydroelectric power is, in many ways, more relevant to other forms of electricity generation. Energy technologies such as wind, solar, tidal, and the like are the most variable in time, and most require an energy storage mechanism to best utilize their potential. Although it is not absolutely necessary for the facility that generates electricity and the facility that stores it to be in close proximity, it is convenient if they are. Discuss the ways in which the geographical requirements for various alternative energy technologies and the geographical requirements for pumped hydroelectricity are compatible or incompatible.

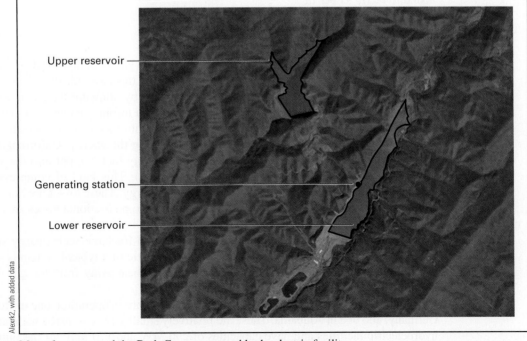

Upper reservoir

Generating station

Lower reservoir

Alexrk2, with added data

Map of area around the Bath County pumped hydroelectric facility.

Total pumped hydroelectricity capacity is in excess of 80 GW worldwide. This represents about 300 facilities that range from a few megawatts up to more than 2 GW. In the United States, pumped hydroelectricity capacity represents about 2.5% of base load generating capacity. In Europe, it is about 5.5%.

18.3 Compressed Air Energy Storage

Compressed air is another mechanism for storing energy. In one sense, it can be viewed much like pumped hydroelectric power, as a means for large-scale energy storage. It can also be viewed as a more portable power source that could compete in some ways with batteries or as a means of distributing power through pipelines that might be analogous to electricity. While current ideas concerning the use of compressed air fall primarily into the first two categories, compressed air was put to considerable use at the end of the 1800s as a means of distributing power. In the 1890s, Paris had more than 50 km of compressed air lines across the city that provided mechanical power for industrial applications, sewing machines, printing presses, and other machines.

For the purpose of storing energy in a compressed gas, it can be shown that (ideally) the energy required to change the pressure of a gas from P_i to P_f is

$$E = nRT \cdot \ln\left(\frac{P_f}{P_i}\right), \qquad (18.3)$$

where n is the number of moles of gas, R is the ideal gas constant, and T is the absolute temperature (in Kelvin). Analogously, this equation gives the ideal amount of energy that can be extracted by reducing the pressure of the gas from P_f to P_i. Within certain limits of temperature and pressure, the energy stored in compressed air can be approximated to be (in kJ per m^3)

$$E/V = 100 \cdot \ln\left(\frac{P_f}{P_i}\right). \qquad (18.4)$$

The derivation of these expressions assumes that the gas is compressed or expands isothermally, that is, without change in temperature. In this case (ideally), 100% of the energy input into compressing a gas can be extracted by allowing the gas to expand. Thus, a compressor can be used to compress gas, and a turbine can be used to convert the energy associated with the compressed gas back into mechanical or electrical energy, thereby using the gas as a mechanism for storing the energy. Unfortunately, in practice, when a gas is compressed, some of the energy that is input into the system goes into heating the gas (rather than compressing it). This kind of compression is called *adiabatic* and results in a loss of some of the energy. These losses are minimized if the gas is compressed and expands slowly, although some frictional losses are always associated with the compressor and turbine.

Compressed air is a reasonably practical method for large-scale energy storage that parallels pumped hydroelectric storage. A schematic of a typical system is shown in Figure 18.3. Heat exchangers are used to transfer heat away from the gas during compression and into the gas during expansion.

At present, two commercial facilities of this type are in operation: one in Huntorf, Germany, and one in Alabama. These are relatively small scale compared with pumped hydroelectric facilities and have capacities of 290 MW and 110 MW for the European

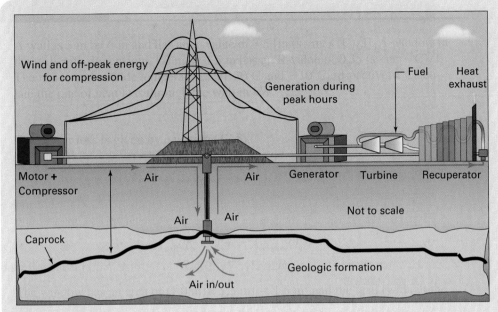

Based on Sandia National Laboratories

Figure 18.3: Schematic of typical large-scale compressed air energy storage system.

Example 18.3

Estimate the theoretical maximum energy that can be stored by compressed air in a $10 \times 10 \times 10 \ \text{m}^3$ chamber at a pressure of 5 MPa.

Solution

From equation (18.4),

$$\frac{E}{V} = 100 \cdot \ln\left(\frac{P_f}{P_i}\right).$$

A pressure of 5 MPa relative to atmospheric pressure is

$$\frac{5 \ \text{MPa}}{0.1013 \ \text{MPa}} = 49.3$$

So the energy stored per unit volume is

$$\frac{E}{V} = 100 \ln(49.3) = 390 \ \text{kJ/m}^3,$$

and for a volume of $10 \ \text{m} \times 10 \ \text{m} \times 10 \ \text{m} = 1000 \ \text{m}^3$ the total energy stored is $390 \ \text{kJ/m}^3 \times 10^3 \ \text{m}^3 = 390 \ \text{MJ}$.

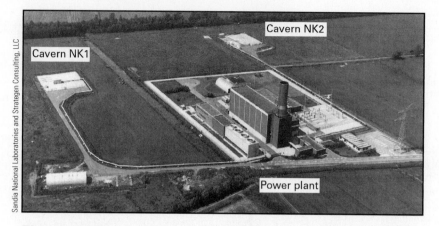

Figure 18.4: Huntorf compressed air energy storage facility.

and North American plants, respectively. Both facilities utilize natural underground caverns for gas storage.

The Huntorf facility was constructed in 1978 and has been in operation since then. An aerial photograph of this facility is shown in Figure 18.4. It utilizes two underground caverns with a total usable volume of 300,000 m^3 (equivalent to a cube about 70 m on a side). The caverns extend from 650 m underground at the top to 800 m at the bottom. The natural overburden of rock allows for a maximum operating pressure of 10^7 Pa. The construction of an above-ground storage tank of that volume and pressure capabilities would be impractical. The compressors require about 12 hours to fully compress the air within the caverns, and the turbines can produce up to 290 MW for up to 3 hours (total 870,000 kWh) as the air is released from the caverns.

The McIntosh facility in Alabama was constructed in 1991 and has improved efficiency over the Huntorf facility due to an improved heat recovery system. It uses a much larger underground cavern (about 5×10^6 m^3), and although it produces less power, it can supply power for up to 26 hours.

18.3a Implementation of Compressed Air Energy Storage

Although both operational compressed air energy storage facilities have performed very well for many years, the implementation of this approach to energy storage requires the availability of fairly specific geological resources. There are proposals for the more widespread application of compressed air energy storage using underwater reservoirs. Large underwater plastic bags could be used for gas storage at relatively low pressures. Open-bottomed containers can also be used to store energy associated with the displacement of water by compressed air inside the container. Underwater approaches to compressed air energy storage are most appropriate for deep lakes and oceans where the depth of the water provides the pressure needed to compress the gas.

On a much smaller scale, compressed air has been considered as a possible energy carrier for vehicles. Compressed air from a high-pressure cylinder can be used to run a piston or turbine engine for vehicle propulsion. This approach has been utilized on a limited basis for well over 100 years. It is also nonpolluting, as long as the air is compressed using energy that is produced by nonpolluting methods.

Theoretically, if 1 m³ of air is compressed into a cylinder with a volume of 5 L, it will have a pressure of 2×10^7 Pa and can store 530 kJ [according to equation (18.4)]. From a practical standpoint, only about half of this can actually be achieved as usable energy to propel a vehicle. A suitable cylinder made of advanced materials (i.e., Kevlar or carbon fiber) would weigh about 2 kg, and 1 m³ of air weighs about 1 kg. Thus a compressed air energy storage system has an energy density of about $(0.5 \times 530 \text{ kJ})/(2 \text{ kg} + 1 \text{ kg}) = 90$ kJ/kg. As a means of comparison with traditional internal combustion engines, the energy density of gasoline is 45 MJ/kg. At an efficiency of, say, 20% for the conversion of the chemical energy associated with gasoline to usable mechanical energy, gasoline, as an energy carrier, could provide about 10 MJ/kg of fuel (assuming the weight of the gas tank is relatively small). This comparison implies severe limitations for the range of a compressed air vehicle.

An analysis of the practicality of compressed air energy storage for transportation should also include a comparison with the more common alternative approach of using batteries to power an electric vehicle (Chapter 19). On the positive side, compressed air systems use straightforward technology involving nontoxic materials, and recharging the fuel supply in the vehicle can be done very quickly. On the negative side, the energy density that can be achieved in a compressed air system is comparable (at best) to rather low-tech batteries (i.e., lead acid) and is much less than that for currently available advanced batteries. Also, the safety of compressed air cylinders during accidents needs to be considered.

18.4 Flywheels

A *flywheel* is a mechanical device with a substantial moment of inertia. They have been used for many years as a method of evening out fluctuations in the rotational speed of a shaft that are caused by varying torque. This situation is common for, say, piston engines where the driving force (ignition of gasoline) is periodic. In recent years, interest has developed in using flywheels as a mechanism for the storage of energy.

Objects that are in motion have a kinetic energy associated with their movement given in terms of their mass, m, and velocity, v:

$$E = \frac{1}{2} mv^2. \tag{18.5}$$

Objects that are rotating have a kinetic energy associated with their rotational motion given by

$$E = \frac{1}{2} I\omega^2, \tag{18.6}$$

where I is the moment of inertia, and ω is the angular frequency of rotation. This is related to the frequency of rotation in revolutions per second, f, as

$$\omega = 2\pi f. \tag{18.7}$$

Table 18.1: Constants, k, from equation 18.5 for different geometries.

geometry	k
disk	1/2
ring	1
solid sphere	2/5
spherical shell	2/3

© Cengage Learning 2015

The moment of inertia is related to the mass and the geometry of the object. For objects with axial (cylindrical) symmetry, the moment of inertia is related to the mass and radius of the object as

$$I = kmr^2. \qquad (18.8)$$

The constant k is related to the details of the geometry and is given for some common simple shapes in Table 18.1. These equations can be combined to give

$$E = 2\pi^2 kmr^2 f^2. \qquad (18.9)$$

The rotational energy of an object is greatest when the rotational frequency, the mass, and the radius are maximized and the geometry is optimized to give the largest value of k. Table 18.1 shows that the most advantageous geometry for a flywheel is a ring. This ensures that the mass is distributed as far away from the axis of rotation as possible. This has been a common design for flywheels used in the past to even out the load on a rotating shaft (Figure 18.5).

Obviously from Example 18.4, a bicycle wheel would not make a very useful energy storage device, at least not at rotational frequencies that are typical for a bicycle wheel in normal use. We could, of course, rotate the wheel at a higher frequency, but there is a limit to how rapidly the wheel can rotate without sustaining damage. A more massive example might be a steel wheel on a freight train (which is approximately a solid disk with a typical mass of $m = 400$ kg and a typical diameter of $d = 0.85$ m (Example 1.3). The rotational energy content of such a wheel rotating at a frequency typical of a moving freight train would illuminate the 60-W bulb for about half an hour. A practical energy storage would require larger, more massive, and/or faster rotating flywheels.

Richard A. Dunlap

Figure 18.5: Steam engine with rim-loaded flywheel.

Example 18.4

Consider an ideal bicycle wheel, all of whose mass is located in a thin ring at the edge with a diameter of 0.7 m and a mass of 1.2 kg. If the wheel has an initial rotational frequency of 2 revolutions per second (equivalent to a bicycle traveling at a speed of about 4 m/s), calculate the length of time the rotational energy content of the wheel could illuminate a 60-W lightbulb (at 100% conversion efficiency).

Solution

From equation (18.9),

$$E = 2\pi^2 k m r^2 f^2.$$

Substitute the following values in the equation: $k = 1$ (as appropriate for a ring of mass given in Table 18.1), $m = 1.2$ kg, $r = 0.35$ m, and $f = 2$ s^{-1}. This gives

$$E = 2 \times (3.14)^2 \times (1) \times (1.2 \text{ kg}) \times (0.35 \text{ m})^2 \times (2 \text{ s}^{-1})^2 = 11.6 \text{ J}.$$

A 60-W lightbulb uses 60 J per second, so that 11.6 J will illuminate the bulb for

$$t = \frac{11.6 \text{ J}}{60 \text{ W}} \approx 0.2 \text{ s}.$$

From a practical standpoint, it is necessary to consider the strength of the material that the flywheel is made of. As the rotational speed of the wheel increases, the internal stresses that the wheel experiences also increase, and at some point the material fails. It turns out that optimizing k may result in unacceptable consequences in the strength of the flywheel because the spokes that support the heavy loaded rim may not be able to withstand the stresses at high rotational speeds. The mechanical properties of the flywheel material are therefore of fundamental importance for the design. A simple quantitative approach to flywheel design considers the stress at the rim of a rotating circular object. Here the stress is the greatest because the actual velocity is the greatest. The rim stress can be expressed as

$$\sigma = \rho r^2 \omega^2, \tag{18.10}$$

where ρ is the density of the material. This equation may be combined with equations (18.6) and (18.8) to give

$$E = \frac{1}{2} k m \left(\frac{\sigma}{\rho} \right). \tag{18.11}$$

The maximum amount of energy that is possible is given by this expression when the stress σ is the breaking stress for the material (i.e., the tensile strength). Thus, for a flywheel of a given mass, m, and geometry (resulting in a value of k), the energy storage capability is optimized by maximizing (σ/ρ). The values of this quantity are given for some possible flywheel materials in Table 18.2. It is clear from this table that the use of advanced materials enables flywheels to store more energy than traditional metal designs.

From the numbers in the table and equation (18.11), it can be calculated that a practical-sized flywheel of a few hundred kilograms could store a maximum of about a

Table 18.2: Densities, ρ, and tensile strengths, σ, and their ratios for some possible flywheel materials.

material	density (kg/m³)	tensile strength (N/m²)	σ/ρ (N · m/kg)
steel	7800	1.72×10^9	0.22×10^6
aluminum	2700	0.59×10^9	0.22×10^6
titanium	4500	1.22×10^9	0.27×10^6
fiber glass	2000	1.60×10^9	0.80×10^6
carbon fiber–reinforced polymer	1500	2.40×10^9	1.60×10^6

© Cengage Learning 2015

hundred megajoules of energy (assuming a safety factor to prevent failure) at rotational rates of 50,000 to 60,000 rpm. If a motor/generator system is used for inputting and extracting energy, the efficiency should be quite good (typically 80–90%), and the flywheel could be used to provide, say, a few tens of kilowatts of electricity for a few hours. Although this is considerably less than pumped hydroelectric or compressed air systems could provide, the flywheel system is, relatively speaking, quite compact and may be useful for some specific applications. A number of flywheels can be combined to increase total energy storage capabilities. One possible use for such a storage system would be in smart grid systems (Chapter 17), which incorporate energy resources that are variable on a fairly short timescale (for example, a photovoltaic system that is susceptible to variations due to variable cloudiness). In such a case, a flywheel energy storage system could even out grid fluctuations. Another possible use of a flywheel energy storage system is in a vehicle where energy can be input into the flywheel during regenerative braking and extracted when needed. For automobiles, the gyroscopic effect associated with the rotating

Example 18.5

For a solid, cylindrical, steel flywheel with diameter equal to length, calculate the maximum energy stored for a wheel diameter of 1 m.

Solution

From equation (18.11), the maximum energy is

$$E = \frac{1}{2} km \left(\frac{\sigma}{\rho} \right).$$

The mass of a 1-m diameter by 1-m long flywheel made of steel ($\rho = 7800$ kg/m³) is

$$m = \pi r^2 l \rho = 3.14 \times (0.5 \text{ m})^2 \times (1 \text{ m}) \times (7800 \text{ kg/m}^3) = 6120 \text{ kg}.$$

Using the value of $\sigma/\rho = 0.22 \times 10^6$ Nm/kg for steel (from Table 18.2) and a value of $k = 0.5$ (from Table 18.1) for a solid flywheel, the maximum energy is

$$E = (0.5) \times (0.5) \times (6120 \text{ kg}) \times (0.22 \times 10^6 \text{ Nm/kg}) = 337 \text{ MJ}.$$

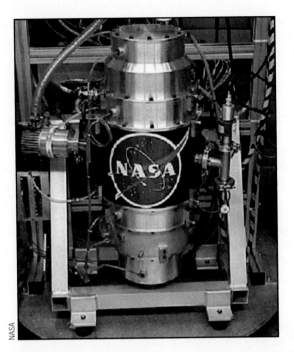

Figure 18.6: Small experimental high-speed flywheel constructed by NASA. The device is enclosed-in a vacuum chamber to minimize air friction. The total height of the cylindrical housing is about 60 cm.

flywheel may introduce problems, but experimental applications in trains have met with limited success. A small experimental high-speed flywheel is shown in Figure 18.6.

A final important consideration in the design of a flywheel system is the minimization of frictional losses because of the high rotational speeds involved. Friction can occur in the bearings that support the flywheel and with the surrounding air. Practical high-speed flywheels that use magnetic bearings and are contained in a vacuum enclosure typically experience energy losses of about 1% per hour.

18.5 Superconducting Magnetic Energy Storage (SMES)

Electrical energy is transmitted as a current through a conductor. In a normal conductor, some resistance to the flow of current is always present, resulting in the loss of some of the electrical energy and the dissipation of heat. In a superconductor, the resistance is zero, and no losses in the energy are associated with the flow of current. Thus electrical energy could be stored (in principle, without loss) in a coil of superconducting wire. Electric current could be injected into the coil and would circulate indefinitely until it was needed. Superconducting magnetic energy storage (SMES) is in relatively common use in research laboratories as a mechanism to even out fluctuations in the line voltage resulting from variations in demand or to act more like a temporary backup for brief power outages. It is particularly useful when a user requires a highly stable source of power. In this sense, the SMES is analogous to the surge tank associated with a hydroelectric facility rather than actual pumped hydroelectric energy storage.

The energy stored in a current circulating in a coil is given by

$$E = \frac{1}{2}LI^2,\qquad(18.12)$$

where L is the inductance of the coil, and I is the current. When L is measured in henries (H), and I is measured in amperes, then E is measured in joules. The inductance of a coil depends on its dimensions, the number of turns of wire, and the material in its core. For an air-core solenoid, a semiempirical relationship that gives a good approximation to the inductance is

$$L = frN^2,\qquad(18.13)$$

where r is the mean radius, and N is the total number of turns of wire. The factor f is a geometry factor that, for r measured in meters, has a value of

$$f = \frac{3.9 \times 10^{-5}}{\left[9 + \dfrac{10l}{r}\right]}\mathrm{J/(m \cdot A^2)},\qquad(18.14)$$

Here, l is the length of the solenoid in meters.

Example 18.6

Consider a superconducting solenoid 1 m in diameter (0.5 m radius) and 1 m in length, consisting of 5000 turns of wire. Calculate the energy stored in the coil for a current of 1000 A.

Solution

Combining equations (18.13) and (18.14) gives the inductance of the coil as

$$L = \frac{3.9 \times 10^{-5}\mathrm{J/(m \cdot A^2)}}{\left[9 + \dfrac{10l}{r}\right]}rN^2,$$

or for r and l in meters, the inductance in henries is

$$L = [3.9 \times 10^{-5}\ \mathrm{J/(m \cdot A^2)}] \times (0.5\ \mathrm{m}) \times \frac{5000^2}{9 + (10/0.5)} = 16.8\ \mathrm{H}.$$

Using equation (18.12), the energy is found to be

$$E = \frac{1}{2}LI^2 = (0.5) \times (16.8\ \mathrm{H}) \times (1000\ \mathrm{A})^2 = 8.4\ \mathrm{MJ}.$$

A superconducting coil, as described in Example 18.6, would be suitable for smoothing out power fluctuations for a single industrial or research user. An SMES facility that would compete with pumped hydroelectric storage for use by a public utility would need to be hundreds of meters or even kilometers in diameter.

A major difficulty in constructing a SMES facility is achieving the conditions that are necessary for making the coil superconducting. Superconducting materials exhibit normal electrical properties at high temperatures. If the temperature is lowered,

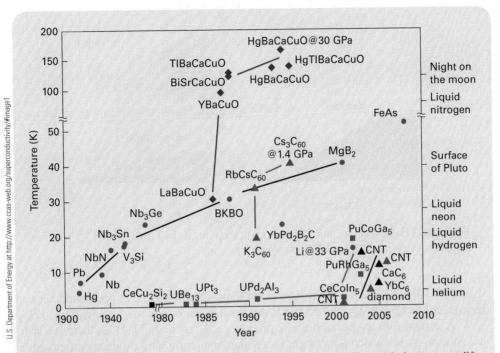

Figure 18.7: Progress in increasing T_C for superconductors. The symbols represent different classes of materials: carbon compounds (triangles), rare earth and actinide compounds (squares), perovskites (diamonds), and transition metal and alkali compounds (circles).

they will, at some temperature, T_C (the *critical temperature*), undergo a transition to a superconducting regime. In the superconducting regime, the electrical resistivity goes to zero, and from a practical standpoint, the materials can be utilized for various applications possibly including energy storage. This phenomenon was first observed in mercury, which becomes superconducting below 4.2 K, by Heike Kammerling-Onnes in 1911. Since then, efforts have been made to find materials with as high a superconducting transition temperature as possible. Figure 18.7 shows the highest observed superconducting transition temperature as a function of year. Materials have now been discovered with T_C over 150 K. The rapid increase in observed T_C in the mid-1980s corresponds to the discovery of the so-called *high-temperature superconductors* (HTSCs) by Karl Alexander Müller and J. Georg Bednorz.

Using a superconductor is not as simple as lowering its temperature below T_C. There is a close relationship between superconductivity and magnetism. In fact, the application of a magnetic field to a superconductor destroys its superconducting properties, making the material *normal* (from an electrical standpoint). In the simplest type of superconductors (Type I), a critical field at which the superconductivity suddenly and completely disappears can be defined. This critical field, H_C, is a function of temperature [Figure 18.8(a)]. A few superconducting materials (i.e., some elements) are Type I superconductors. However, most superconductors are Type II. As the magnetic field is increased in a Type II superconductor, a field is reached, H_{C1}, when portions of the sample become normal. However, other portions of the sample remain superconducting, and because an electric current takes the path of least resistance, the sample continues to carry current without resistance. As the field is increased, a point is reached, H_{C2}, when the entire sample becomes normal. At that point, the material does not conduct without

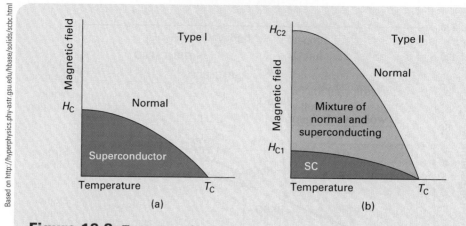

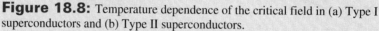

Figure 18.8: Temperature dependence of the critical field in (a) Type I superconductors and (b) Type II superconductors.

resistance [Figure 18.7(b)]. Thus, at a given temperature, a Type I superconductor can be used at a field up to H_C, and a Type II superconductor can be used at a field up to H_{C2}.

A current flowing through a wire produces a magnetic field. Thus, there is a critical current, j_C, for which the current in a superconducting wire produces a field equal to the critical field, thereby driving the material normal. Although there is clearly a relationship between H_C and j_C, it is not straightforward. However, it is clear that superconductors just below their critical temperature cannot carry very much current and still remain superconducting. From a practical standpoint, it is customary to use superconductors as far below T_C as is possible. The maximum amount of current that can be carried is related to the value of the critical field at zero temperature, $H_C(0)$ for a Type I and $H_{C2}(0)$ for a Type II. To determine the usefulness of a superconductor, it is important to know the value of T_C and the value of the critical field.

Some examples of the properties of superconductors are given in Table 18.3. The table shows that Type I superconductors have low values of T_C and low values of $H_C(0)$. This indicates that they are generally not very useful for carrying large amounts of current. However, some Type II superconductors (particularly the high-temperature superconductors) have fairly high values of T_C and $H_{C2}(0)$. Because j_C is not directly related to H_{C2}, it turns out that optimizing T_C and $H_{C2}(0)$ does not always optimize j_C. In this respect, YBCO, which was one of the earliest high-temperature superconductors to be discovered, seems to be in the lead.

Table 18.3: Properties of some superconducting materials. (T = tesla)

material	type	T_C (K)	$H_C(0)$ (T)	$H_{C2}(0)$ (T)
Sn	I	3.7	0.03	—
Pd	I	7.2	0.08	—
Nb	II	9.3	—	0.4
NbTi	II	10	—	12
Nb_3Sn	II	18	—	25
$YBa_2Cu_3O_7$ (YBCO)	II	93	—	168
$Bi_2Sr_2Ca_2Cu_3O_{10}$ (BSCCO)	II	110	—	200

While it may seem advantageous to utilize HTSC materials for the purpose of constructing SMES systems, the choice is not so clear. HTSC materials would minimize the refrigeration costs. However, for a large system, the increase in refrigeration costs between operating at, say, 80 K and operating at, say, 5 K, is not as substantial as it would be for a smaller system. In fact, even if HTSC materials are used, it may be advantageous to run the system at 5 K rather than at 80 K because the increase in critical field could outweigh the increase in refrigeration costs. However, as discussed in Chapter 2, HTSC materials are typically brittle and difficult to fabricate, so that capital costs and reliability may be more important factors in design viability than refrigeration costs.

SMES is not 100% efficient. Although the current circulates in the coil without loss, energy is required for refrigeration to maintain the coil at a temperature below the superconducting transition temperature of the wire. This is typically on the order of 0.1% of the energy stored in the coil per hour of storage. Also, losses are incurred when current is put into or taken out of the coil. Electric current carried in transmission lines is typically alternating current (AC), whereas the current stored in a superconducting magnet is direct current (DC). So to store electrical energy in a SMES and then use it, the AC must be converted to DC, and the DC must be converted back to AC for use. This AC-DC-AC conversion cycle is about 95% efficient and generally accounts for the major loss of energy in this system.

At present, small-scale SMES systems have been successful in providing grid stability (Figure 18.9). The world's largest (as of 2009) SMES system is at Sharp's Kameyama Plant in Japan and can provide up to 10 MW. Other similar or larger systems are in the testing stages. The practicality of large-scale systems that could compete with relatively low-tech options like pumped hydroelectric or compressed air has not been demonstrated.

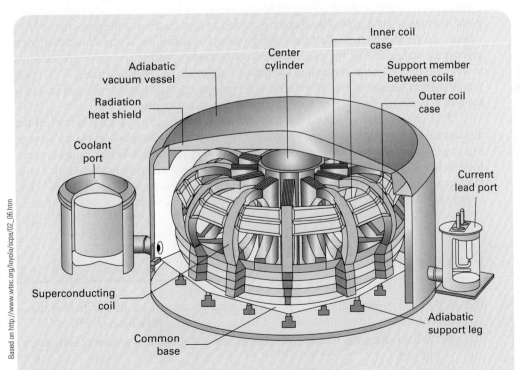

Based on http://www.wtec.org/loyola/scpa/02_06.htm

Figure 18.9: Design of a SMES system for grid stabilization.

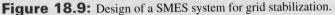

18.6 Summary

Methods for energy storage are required because energy is not always available when and/or where it is needed. The former case results largely from variations in energy production capabilities and in energy demand and is particularly relevant when implementing renewable energy technologies. The latter case arises primarily because energy produced in centralized facilities, such as thermal generating plants, requires an appropriate storage mechanism for portable applications, such as vehicles. Several possible technologies for energy storage, that is, specifically electrical energy storage, were discussed in this chapter. These technologies do not include batteries and hydrogen; these are presented in Chapters 19 and 20, respectively.

The chapter discussed pumped hydroelectric storage, which is in common commercial use for large-scale electricity storage for the grid. This is a straightforward method where excess electricity generated in times of low demand is used to pump water into an elevated reservoir. When demand exceeds capacity, the reservoir is drained through turbines at a lower elevation to generate additional electricity. Although capital construction costs can be high, this method is low maintenance and cost-effective over the long term.

Energy storage using compressed air is also discussed. Excess electricity can also be used to compress air, which can be used to drive turbines in times of higher demand. Two major commercial facilities of this type are in use, one in the United States and one in Germany. While this method of energy storage is potentially viable, requirements for an appropriate chamber are a major factor.

As pointed out in this chapter, other methods for large-scale energy storage, such as flywheels and superconducting magnet energy storage (SMES), are not at a technological point where they are viable. The chapter provided an introduction to the properties of flywheels and the factors that affect their energy storage capabilities. It was shown that, for a given mass and geometry, the maximum energy storage capability is directly proportional to the ratio of the tensile strength to the density (σ/ρ) of the flywheel material. Of commonly used materials, carbon fiber–reinforced polymer has the largest ratio and provides the best opportunity for high-density energy storage in a flywheel.

This chapter also reviewed the basics of superconductivity and the history of the development of superconducting materials. The physical principles of energy storage in a superconducting magnet have been presented. It is not clear whether future technological developments for flywheel and SMES devices will make them competitive with the more traditional large-scale energy storage methods. However, SMES, at least, has found viable applications as a means of stabilizing power sources on a smaller scale.

Problems

18.1 A pumped hydroelectric storage reservoir has an area of 1 km^2 and an average depth of 30 m. The water flows to a lower reservoir with a head of 150 m.

   (a) What is the total energy available in MWh$_e$? Assume a generator efficiency of 85%.

   (b) What is the average power available (in megawatts) if the upper drains through the turbines over a period of 12 hours?

18.2 The total volume of the upper reservoir of a pumped hydroelectric facility flows through the turbines over a period of 8 hours. The upper reservoir has an area of 300,000 m^2 and an average depth of 15 m.

(a) What is the average flow rate (in cubic meters per second) between the upper and lower reservoirs?

(b) If the water between the two reservoirs flows through a (circular) penstock with a diameter of 7.5 m, what is the average velocity of the water?

18.3 Design a steel flywheel-based electricity backup system to provide 1 day of typical electricity use for a single-family home (no electric heat). Assume that 70% of the maximum theoretical energy storage capacity is available and can be converted to electricity at 90% efficiency.

18.4 Consider the possibility of constructing a vehicle that operates on compressed air. A practical vehicle requires about 0.7 MJ/km, and a practical high-pressure storage tank might store gas at a pressure of 7.5 MPa. Derive a relationship between range (in kilometers) and storage tank volume (in cubic meters). If a maximum practical tank volume is 1 m^3, what is the range of the vehicle?

18.5 The world's largest pumped hydroelectric facility in Bath County, Virginia, has a total energy storage capacity of about 30 GWh$_e$. Assuming the same conversion efficiency for a compressed air storage facility (which is optimistic), calculate the dimensions of a cubic chamber that, at a pressure of 5 MPa, has the same energy storage capacity as the Bath County station.

18.6 Consider solid cylindrical flywheels with the diameter equal to the length. For flywheels made of steel, aluminum, and carbon-reinforced polymer, calculate the diameter of a wheel that has a maximum energy storage capability of 10 MJ.

18.7 Energy is extracted from a flywheel with an initial rotational frequency of f_0 in such a way that the frequency decreases linearly with time. Describe the power provided as a function of time as the flywheel slows.

18.8 For large-scale electric storage for the grid, a capacity in the order of 1 GWh or greater is desirable. Consider the possibility of creating an SMES system that is 0.5 km in diameter, consisting of a 2 m high coil with 20,000 turns of wire. What current is needed to store 1 GWh of energy?

Bibliography

G. J. Aubrecht II. *Energy: Physical, Environmental, and Social Impact*, 3rd ed. Pearson Prentice Hall, Upper Saddle River, NJ (2006).

J. A. Kraushaar and R. A. Ristinen. *Energy and Problems of a Technical Society*, 2nd ed. Wiley, New York (1993).

C. Ngô and J. B. Natowitz. *Our Energy Future: Resources, Alternatives and the Environment.* Wiley, New York (2009).

A. Ter-Gazarian. *Energy Storage for Power Systems.* Peter Peregrinus, Stevenage, United Kingdom (1994).

Battery Electric Vehicles (BEVs)

Learning Objectives: After reading the material in Chapter 19, you should understand:

- The properties and applications of different types of batteries.
- The use of the Ragone plot to illustrate energy-power relationships for energy storage mechanisms.
- Energy requirements for vehicle propulsion.

- The historical development of electric vehicles and reasons for changes in their popularity.
- The advantages, disadvantages, and economic viability of electric vehicles.
- The utilization of supercapacitors for energy storage and their significance in electric vehicle design.

19.1 Introduction

Transportation places special demands on a source of energy; specifically, the source of energy must be portable. For this reason, petroleum-based products have been the overwhelming choice for transportation purposes. This stems from the fact that petroleum products, such as gasoline and diesel fuel, have a very high energy density (i.e., many megajoules per kilogram). In addition, there is a highly developed technology (i.e., the internal combustion engine) for converting the chemical energy stored in petroleum products into the mechanical energy needed to propel a vehicle, and for now, at least, petroleum is readily available and comparatively inexpensive. Many alternative energy sources are not appropriate for direct application to transportation needs. A major exception, perhaps, is biofuels, which are a direct replacement for petroleum and which can, more or less, make direct use of the existing distribution infrastructure and vehicle technology. As discussed in Chapter 16, these have been particularly successful in some parts of the world (e.g., Brazil), but their implementation elsewhere is not necessarily straightforward. Other environmentally advantageous energy sources are not portable or suffer from very low energy density. The latter is the case, for example, with solar or wind energy. Solar vehicles (Figure 9.27) are really not viable, and wind is potentially suitable only for ocean transport (i.e., sailing ships). The implementation of alternative energy strategies for transportation, therefore, requires a storage mechanism with sufficiently high energy density. While some of the methods discussed in the previous chapter (e.g., compressed air or flywheels) may be utilized, the most likely energy storage possibilities for vehicle use are batteries and hydrogen. Both of these serve as convenient mechanisms for storing the electrical energy produced by most alternative methods and are discussed in this and the next chapter.

19.2 Battery Types

A battery is an electrochemical device for storing electrical energy as chemical energy. It consists of one or more cells. Each cell consists of two half cells connected in series. One half cell contains the *anode* (or negative electrode) and an electrolyte, while the other half cell contains the *cathode* (or positive electrode) and an electrolyte. The two electrolytes may be either solid or liquid and may be common to the two half cells, or they may be separated by a membrane that is porous to the passage of certain ions. When a battery is charged, active atoms in the anode are ionized. The ions, thus formed, travel to the cathode through the electrolyte, and the electrons that are liberated travel to the cathode through an external circuit. At the cathode, the ions and electrons are recombined in the so-called *redox* (reduction-oxidation) reaction. When the battery is discharged, the ions travel from the cathode back to the anode through the electrolyte, while electrons flow from the cathode back to the anode through the external circuit and in the process can do useful work—heat a resistor, produce a magnetic field, turn a motor, and so on.

The voltage difference between the electrodes of a battery depends on the nature of the chemical reactions in the cell. For example, zinc-carbon batteries (by the nature of the chemical reactions present) produce about 1.5 V, whereas NiCd batteries produce about 1.2 V. Several cells can be connected together in series to produce larger voltages. During discharge, an ideal battery produces a constant voltage up to the point where all of the excess electrons and positive ions at the cathode have traveled through the external circuit and the electrolyte, respectively, and have recombined at the anode. Such an ideal battery would have no internal resistance. In practice, batteries do have internal resistance, and this resistance increases during discharge. As a result, the voltage produced by the battery decreases during the discharge. The exact details of the voltage curve during discharge depends on the battery chemistry and internal design.

Batteries are in common use in our everyday lives. They are prevalent in electronic devices and vary considerably in their design and electrical storage capacity. On the small end of the scale, common batteries include those used in watches. On the large end of the scale, common batteries include automobile batteries. The current chapter deals primarily with the use of batteries in electric vehicles.

Generally, batteries can be classified as *primary* batteries (nonrechargeable) or *secondary* batteries (rechargeable). Primary batteries are generally small, relatively inexpensive, and suitable for devices that require relatively little power (e.g., a watch or a smoke detector) and/or utilize power only on an intermittent basis (e.g., a flashlight). These batteries are designed to be replaced when they become discharged, and most comply with battery industry standards for size and voltage characteristics. The most common types of primary batteries are primary lithium (Li) cells (e.g., button cells used for watches, calculators, etc.), zinc-carbon cells, or alkaline cells (used for flashlights, radios, toys, etc.). Some characteristics of standard size zinc-carbon and alkaline cells are given in Table 19.1.

Table 19.1: Some properties of standard zinc-carbon and alkaline cells. All are cylindrical and produce approximately 1.5 V.

size	diameter (cm)	length (cm)	volume (cm³)	Zn-C mass (g)	Zn-C energy (mWh)	Zn-C capacity (mAh)	alkaline mass (g)	alkaline energy (mWh)	alkaline capacity (mAh)
AAA	1.05	4.45	3.85	7	270	540	12	1875	1200
AA	1.35	5.05	7.23	14	560	1100	23	4275	2700
C	2.62	5.0	26.96	41	1600	3800	66	12,500	8000
D	3.42	6.15	56.50	90	3500	8000	148	30,750	12,000

© Cengage Learning 2015

It is clear from the table that the energy content of a battery depends on the battery chemistry and also on the size of the battery. The specific energy or energy density (i.e., energy per unit mass) depends on the battery chemistry, and for larger batteries (at least), amounts to about 36 Wh/kg for Zn-C batteries and 160 Wh/kg for alkaline batteries. To construct a viable battery electric vehicle (BEV), it is important to optimize the specific energy of the batteries in order to minimize the mass of the batteries. From a practical standpoint, primary batteries are not suitable for electric vehicles because they would have to be replaced at the end of their discharge. Secondary or rechargeable batteries are necessary for electric vehicles, and Table 19.2 gives the properties of some common secondary battery chemistries. To properly assess the suitability of certain chemistries, it is most relevant to consider factors such as the specific energy rather than the energy for a specific size battery.

ENERGY EXTRA 19.1
The Cost of electricity

Most of the electricity purchased for personal use comes from a public utility. This is the electrical energy is used for lights, televisions, household appliances, and so on, and it typically costs about $0.12 per kWh. It is, however, sometimes convenient to use batteries as a portable source of electricity for flashlights, cameras, portable electronics, remote controls, and the like. It is interesting to consider the monetary cost of electricity that is used from such a source. Primary (i.e., nonrechargeable) AAA batteries are in common use, and the cost of the electrical energy obtained from such a battery is really the same as the cost of the battery itself. This is because, once the battery has been discharged, it is disposed of (or preferably recycled) and is not recharged. Table 19.1 shows that the energy content of a Zn-C AAA battery is 270 mWh, or 2.7×10^{-4} kWh. If a package of four AAA cells is purchased at a discount store for

$1.00, then the cost of the electricity is $0.25/2.7 \times 10^{-4}$ kWh = $926 per kWh, or nearly 8000 times the cost charged by a public utility for the same amount of energy. To dry a typical load of laundry, a clothes dryer might use about 2.5 kWh of electricity. At a rate of $0.12 charged by a public utility, this amounts to a cost of about $0.30. If a clothes dryer ran on AAA batteries, the energy to dry a load of laundry would cost over $2000. While primary batteries are a convenient source of energy for portable electronic devices, the energy comes at a premium cost.

Topic for Discussion

Find prices at a local store for both Zn-C and alkaline AA and D cells. For the most objective comparison, consider the same name brand batteries in each category. In terms of economics, does size matter? Are alkaline batteries worth the extra cost?

Table 19.2: Properties of some secondary battery chemistries. Typical values are given. Specific energy is given in standard metric units (MJ/kg) and in traditional units often used for battery properties (Wh/kg), where 278 × (MJ/kg) = (Wh/kg). (NiMH = Nickel Metal Hydride)

chemistry	cell voltage (V)	specific energy (MJ/kg)	specific energy Wh/kg	specific power (W/kg)	self-discharge (%/month)	life (cycles)
Pb acid	2.1	0.13	36	100	4	600
Ni-Cd	1.2	0.22	56	150	20	1500
NiMH	1.2	0.28	78	800	20	1000
Li-ion	3.6	0.58	160	300	7	1200

© Cengage Learning 2015

The total energy contained in a battery (MJ) is related to the specific density (MJ/kg) and the mass of the battery (kg) by

$$\text{energy} = (\text{specific energy}) \times (\text{mass}).$$ **(19.1)**

Example 19.1

Calculate the mass of Pb-acid batteries needed to supply 1 kWh of energy. Repeat for Li-ion batteries.

Solution

From Table 19.2, the energy density of a Pb-acid battery is 36 Wh/kg. From equation (19.1), the mass of a battery is related to its specific energy and the total energy available by

$$\text{mass} = \frac{\text{energy}}{\text{specific energy}}.$$

Thus

$$\text{mass} = \frac{1 \text{ kWh}}{0.036 \text{ kWh/kg}} = 27.8 \text{ kg.}$$

For a Li-ion battery, the specific energy is 160 Wh/kg, so the mass is

$$\text{mass} = \frac{1 \text{ kWh}}{0.160 \text{ kWh/kg}} = 6.25 \text{ kg.}$$

For a given vehicle, the driving range of the vehicle is related (more or less) to the total energy available. By comparison with a conventional gasoline-powered vehicle, the energy content of the batteries is analogous to the volume of the gas tank. The rate at which energy can be extracted from the batteries is also important because this is analogous to the power available from the engine of a traditional vehicle and determines the vehicle performance (e.g., acceleration). Thus, in choosing a battery type for the design of an electric vehicle, it is important to maximize both the specific energy and the specific power or power density (i.e., the power per unit mass) as much as possible. The choice of a given battery chemistry and a given mass of batteries determines

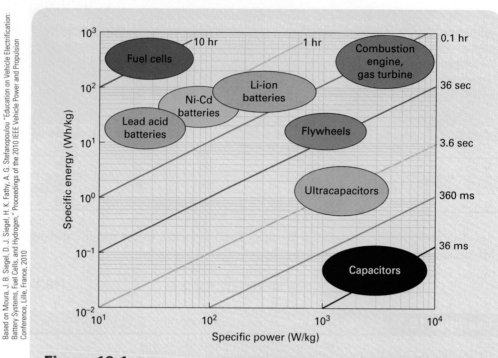

Based on Moura, J. B. Siegel, D. J. Siegel, H. K. Fathy, A. G. Stefanopoulou "Education on Vehicle Electrification: Battery Systems, Fuel Cells, and Hydrogen." Proceedings of the 2010 IEEE Vehicle Power and Propulsion Conference, Lille, France, 2010

Figure 19.1: Relationship of specific energy and specific power for secondary battery types (Ragone plot). Other energy storage methods relevant to vehicles are shown for comparison.

the total energy and power available and establishes the driving characteristics of the vehicle. In the next section, it will be shown how battery properties can put limitations on the vehicle's performance and range. Table 19.2 certainly suggests that, of the types listed, Li-ion batteries are the most desirable for electric vehicle use. The relationship of specific energy and specific power for major secondary battery chemistries are shown in Figure 19.1. This graph is referred to as a *Ragone plot* (pronounced ra-GOH-nee). It is also important to consider other factors such as safety, cost, and availability of materials. These considerations are now discussed.

19.3 BEV Requirements and Design

A comparison with traditional gasoline-powered vehicles is inevitable when considering the viability of an electric vehicle. Drivers are accustomed to the operating characteristics of these vehicles and will inevitably use them as a benchmark for comparison. Of course, all gasoline-powered vehicles do not have the same characteristics. Table 19.3 gives the properties of some typical classes of gasoline vehicles designed for the consumer market.

Table 19.3: Examples of some typical classes of consumer vehicles and nominal technical specifications (ca. 2012). The specified energy used is the primary energy content of the gasoline. The energy utilized at the wheels is this value multiplied by an efficiency of 0.17.

class	typical example	mass kg	power kW	power/ mass kW/kg	fuel consumption L/100 km	energy used MJ/km	range km	new cost (US$1000)
subcompact	Smartcar	820	53	0.065	6.2	2.2	530	13
compact	Honda Civic	1200	105	0.088	7.9	2.7	630	16
family	Toyota Camry V6	1575	200	0.13	10.7	3.7	650	25
luxury	Mercedes Benz S63	2120	390	0.18	16.9	5.9	530	130
sport	Lamborghini Murcielago	1670	475	0.28	23.6	8.2	420	350
compact SUV	Mazda Tribute	1520	130	0.085	9.8	3.4	630	23
full-sized SUV	Land Rover Range Rover	2590	230	0.09	15.7	5.5	660	78

Fuel consumption is the average for city and highway and is given in liters of gasoline per 100 km (L/100 km). Energy used per kilometer is based on the energy content of gasoline (34.8 MJ/L) by

$$0.348 \cdot (L/100 \text{ km}) = MJ/km. \qquad \textbf{(19.2)}$$

An overview of the automobile market suggests that vehicles like compact cars, family cars, and compact SUVs are viewed as the most practical and economical by the general public. Certainly from a practical standpoint, an electric vehicle that fits into this range of vehicles would be the most marketable. On the basis of the vehicle characteristics shown in Table 19.3, Table 19.4 gives the specifications of an ideal electric vehicle (in the context of market considerations for gasoline vehicles).

Current battery technologies dictate the properties of electric vehicles, and on the basis of the information in Table 19.2, the degree to which an actual electric vehicle can achieve the ideal design can be determined. The efficiency of an internal combustion vehicle is determined by the ratio of the energy provided to the wheels to the energy content of the gasoline in the tank. Typically this is only about 15–20%. Electric vehicles are, by comparison, quite efficient; typically 80–90% for the ratio of energy to the wheels to the energy stored in the batteries. Table 19.3 shows that a practical family vehicle might use about 3.5 MJ/km. The energy required to actually drive the wheels based on the efficiency (say 17%) of the gasoline engine is about 0.6 MJ/km. Thus the energy that has to be supplied by the batteries in an electric vehicle (based on an efficiency of 85%) is (0.6 MJ/km)/0.85 = 0.7 MJ/km. While this value could vary from about 0.38 MJ/km to about 1.4 MJ/km for different classes of vehicles (Table 19.3),

Table 19.4: Specifications of a benchmark family electric vehicle.	
property	goal
mass	1400 kg
power	150 kW
range	600 km
cost	US$20,000

© Cengage Learning 2015

the value of 0.7 MJ/km is used as a typical value in most of the examples in this text. If Pb-acid batteries, which contain 0.13 MJ/kg, are used, a range of 600 km would require batteries with a mass of

$$(600 \text{ km}) \times \frac{0.7 \text{ MJ/km}}{0.13 \text{ MJ/kg}} = 3230 \text{ kg}, \qquad (19.3)$$

which is well above the design goal. The available power provided by these batteries would be

$$(3230 \text{ kg}) \times (0.10 \text{ kW/kg}) \times (0.85) = 275 \text{ kW}. \qquad (19.4)$$

This substantially exceeds the power requirements. In fact, to satisfy the power needs would require only

$$\frac{150 \text{ kW}}{0.10 \text{ kW/kg} \times 0.85} = 1760 \text{ kg}, \qquad (19.5)$$

of batteries, still a substantial amount relative to expectations for total vehicle mass. However, this exercise clearly indicates that range may be a more difficult criterion to satisfy than power. While Pb-acid batteries may be acceptable for hybrid vehicles where the electric motor only supplements a traditional gasoline engine (Section 17.8), a purely electric vehicle requires batteries with a higher energy density.

Example 19.2

Calculate the mass of Li-ion batteries needed to power a vehicle with a range of 600 km and the corresponding maximum power available.

Solution

Using an energy requirement from the batteries of a typical electric vehicle of 0.7 MJ/km and a specific energy of 0.58 MJ/kg for a Li-ion battery gives the required mass:

$$(600 \text{ km}) \times \frac{0.7 \text{ MJ/km}}{0.58 \text{ MJ/kg}} = 724 \text{ kg}.$$

The maximum power available for 724 kg of Li-ion batteries is determined from the specific power of 300 W/kg:

$$(724 \text{ kg}) \times (0.3 \text{ kW/kg}) = 217 \text{ kW}.$$

Table 19.5: Characteristics of electric vehicles using Li-ion batteries.		
battery mass	power	range
100 kg	30 kW	80 km
300 kg	90 kW	250 km
725 kg	217 kW	600 km

Using the values of Li-ion battery parameters shown in Table 19.2, the performance characteristics calculated for Li-ion powered BEVs are shown in Table 19.5. There is active research in the development of new batteries with improved characteristics, and batteries with improved performance will certainly be available in the future. Calculations have considered the case for a battery mass that satisfies the requirements for power and range, and the table suggests that a reasonable compromise between power and range can be achieved by a battery-operated electric vehicle utilizing Li-ion batteries. It should be noted that the maximum power to the wheels is, in general, limited by the specifications of the electric motor and may be much less than the capabilities of the batteries and that, during normal driving, the power utilized is much less than the maximum available. In fact, a quick calculation would show that, at maximum power output, a Li-ion battery would be drained in less than half an hour.

Overall, a comparison of different energy sources for vehicles can be illustrated by a Ragone plot (Figure 19.2). For a particular energy source, the trade-off between energy and power is represented (more or less) by a diagonal line with a slope of −1,

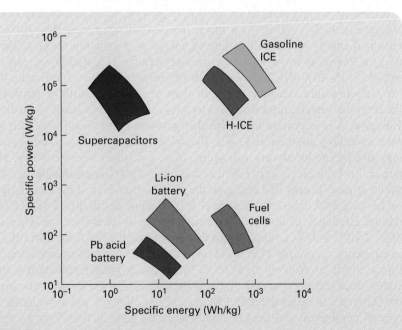

Figure 19.2: Power energy relationship (Ragone plot) for various types of batteries in comparison with other possible energy storage methods that might be used for transportation.

for example; the yellow area represents gasoline internal combustion engines (ICE). However, all known battery chemistries fall below the region for gasoline on the Ragone plot. The plot also illustrates some other energy storage mechanisms that have been discussed previously (e.g., flywheels) and some that will be discussed later in this chapter (supercapacitors) or in the next chapter [hydrogen internal combustion engines (H-ICE) and hydrogen fuel cells].

The production of a marketable electric vehicle requires the consideration of several important points.

First, there is a consideration of economics; Li-ion batteries are quite expensive. An electric vehicle competing with a $20,000 internal combustion vehicle might consist of a $15,000 basic vehicle with $25,000 of Li-ion batteries for a total new vehicle cost of about $40,000. Added to this is the fact that rechargeable batteries have a finite lifetime and that the $25,000 worth of batteries might need to be replaced in about 6 years.

Second is the question of recharging the batteries when they have depleted their charge. In the early days of electric vehicles, typical recharge times could be 8–12 hours or longer. Advanced Li-ion batteries can sometimes be recharged in 1–2 hours. This is still much longer than the time required to refill a gasoline tank (a couple of minutes). Vehicles that are utilized for short-range commuting and that can be recharged at the user's home overnight are more practical than those that might be used for long trips.

There is also the need to establish an infrastructure of recharging stations, just as an infrastructure of gasoline stations was developed to make the use of gasoline-powered vehicles practical.

Another point deals with the overall energy consumption of an electric vehicle. Electric vehicle manufacturers claim that the equivalent fuel consumption of an electric vehicle in comparison with gasoline-powered vehicles is very low. This feature is related to the fact that the conversion of energy stored in a battery to energy available at a vehicle's wheels is a much more efficient process than the conversion of chemical energy stored in gasoline to energy at a vehicle's wheels. This is because no heat engine is involved in the electric vehicle. If a traditional gasoline-powered vehicle is 15–20% efficient and an electric vehicle is 80–90% efficient, then the electric vehicle might consume the equivalent of about one-fifth of the fuel needed for the gasoline vehicle, or perhaps 1.5 to 2.0 L/100 km for a small to medium-sized car. However, it is important to realize that gasoline is produced very efficiently from a primary energy source (crude oil). Electricity that is used to charge batteries may or may not be produced very efficiently from a primary energy source. Hydroelectric power, for example, is very efficient and only slightly decreases the net efficiency between primary energy and energy at the wheels. However, if the electricity is produced by burning coal, then the efficiency of producing electricity is only about 40%, and the net efficiency of converting primary energy to energy at the wheels is typically $85\% \times 40\% = 35\%$, or about twice that of burning the fossil fuel directly. The increase in efficiency represents a net decrease of about 50% in carbon emissions, even if electricity comes exclusively from fossil fuels (and a greater reduction if the electricity comes from greener sources). An additional factor (in the favor of electric vehicles) is the fact that it is easier to control the release of pollutants and to sequester carbon from a small number of coal-fired electric generating stations than from a large number of individual vehicles.

A final factor that requires consideration is the environmental impact of manufacturing batteries and the availability of starting materials for battery production. Recall the discussion in Chapter 7 of the limitations of fusion power based on the availability

ENERGY EXTRA 19.2
Flow batteries

Flow batteries are a type of rechargeable battery in which electricity is produced by certain chemical reactions between liquid electrolytes. The first flow battery was constructed in the 1880s, although it was only in the 1970s that serious interest in this technology arose. The general features of a flow battery are illustrated in the following figure. Electrical energy can be used to charge the battery by causing an oxidation reaction in one electrolyte and a reduction reaction in the other. The two electrolytes are separated by an ion exchange membrane that allows the ions involved in the reaction to pass through. On discharge, the reverse reaction occurs, allowing electrical energy to be extracted from the cell. The electrolytes are stored in tanks external to the cell itself and are circulated through the cell, where they undergo chemical reactions by pumps.

Flow batteries have several interesting features that distinguish them from traditional rechargeable batteries. First, the power rating of the cell is a function of the size and geometry of the electrodes, and the energy storage capacity is limited only by the size of the electrolyte storage tank. This feature allows for the construction of flow batteries with very large energy storage capabilities and has made flow batteries an attractive possibility for large-scale electrical storage.

Another important feature of flow batteries is the ability to rapidly recharge them by merely pumping out depleted electrolytes (for later reprocessing) and pumping in freshly charged electrolytes. This feature would enable a flow battery electric vehicle to fill up much like filling up a conventional internal combustion engine vehicle at the gas station, thus eliminating the lengthy recharge stops that are a potential inconvenience for Li-ion battery vehicle users. A prototype scale vehicle (shown in the next page figure), operating on a flow battery, has been constructed

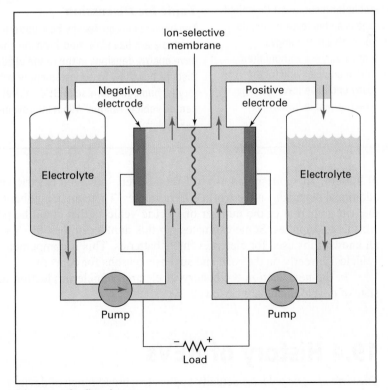

Diagram of a flow battery.

Continued on page 496

Energy Extra 19.2 continued

Courtesy of Robin Vanhaolst

Prototype vehicle running on a flow battery.

by researchers at the Fraunhofer Institute for Chemical Technology in Germany and is a step toward the possible future implementation of this technology.

Flow batteries are, however, not without their disadvantages. They are more complex and (at the present, at least) more costly and have lower energy densities than traditional rechargeable batteries.

Topic for Discussion

The highest energy density flow batteries currently available are based on zinc-bromine chemistry. These have energy densities in the range of 50–70 Wh/kg. Discuss how flow batteries compare with other battery technologies for use in BEVs. What are the expected vehicle ranges? What improvements would be desirable?

of lithium. The widespread use of Li-ion battery–powered electric vehicles would put additional demands on the world's Li resources. There are a number of uncertainties in the determination of the number of electric vehicles that could be produced from the world's Li supplies. Some estimates put this number in the 500–600 million range, if all known Li is used for electric vehicle batteries. This is comparable to the number of vehicles currently on the road and suggests reasons for concern.

In the next section the history of electric vehicle production and as the current state of development, are reviewed.

19.4 History of BEVs

The term *electric vehicles* actually applies to a broad range of devices that includes well established technologies such as electric trains. Battery electric vehicles (BEVs) are devices that carry their own energy source in the form of batteries, and this requirement

ENERGY EXTRA 19.3
Rechargeable sodium batteries

At present, the preferred battery choice for BEVs is Li-ion batteries because of their high specific power and high specific energy. Lithium is not the most desirable material for battery construction for at least two reasons: It is expensive, and, if BEVs become widespread, the world lithium resources may not be sufficient to satisfy the need for this element. In this case, dwindling lithium supplies would only exacerbate problems of high lithium cost.

It is of interest to consider other battery chemistries that may make use of chemical reactions similar to those in lithium batteries. Lithium is the lightest of the alkali metals. A glance at the periodic table shows that the next heaviest alkali metal is sodium. In fact, sodium exhibits much of the same chemical behavior as lithium. Sodium has two huge advantages over lithium: It is inexpensive, and its supply is virtually unlimited, being half of all the atoms in common salt.

The earliest dedicated research effort on sodium-based batteries was conduced in the 1960s by Ford Motor Company. This work dealt with batteries utilizing the reaction between sodium and sulfur during discharge of the cell:

$$2Na + 4S \rightarrow Na_2S_4.$$

Because this reaction is reversible, the cell can be re-charged. The design of a typical cell is illustrated in the figure and consists of a liquid sodium electrode and a liquid sulfur electrode separated by a solid electrolyte, through which sodium ions migrate to react with the sulfur. Na-S batteries have very high specific energy, and this feature would seem to make them attractive candidates for a number of portable power applications. However, Na-S batteries have two serious drawbacks: (1) The properties of the reaction and the behavior of sodium make it necessary to operate the cell at high temperature (typically around 350°C). (2) The sodium polysulfide formed during the discharge reaction is highly corrosive and represents a potential safety hazard.

Because of these features, Na-S batteries may be most suitable for large-scale stationary applications such as grid backup power storage rather than transportation use. Recent progress has been made in the development of lower temperature (<100°C) Na-S batteries where the electrodes remain solid. An

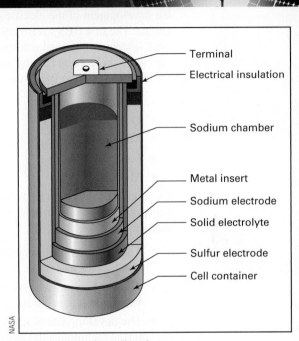

Construction of a sodium battery.

ion selective membrane that allows Na ions to pass is used in place of the solid electrolyte.

Sodium-ion batteries are another approach to using Na-based chemistry to produce rechargeable batteries that have been researched in recent years. Designs are similar to those used for Li-ion batteries and often use a solid carbon-based anode and a solid transition metal compound for the cathode. Major advantages of this design are the elimination of highly corrosive reaction by-products and the ability to operate the battery at or near room temperature. These features, along with a high specific energy, make Na-ion batteries an attractive possibility for vehicle use. Although research in this field is at its very early stages (compared with Li-ion battery research), Na-based batteries show promise as a viable future technology.

Topic for Discussion

If the world's supply of lithium were to be replaced by sodium (kilogram for kilogram) obtained from the oceans, estimate what fraction of the salt in the world's seawater would be utilized? Would this have undesirable environmental effects?

makes their technology much more complex. BEVs may include buses on one end of the size range and nonroad-going vehicles such as golf carts on the other. This discussion concentrates on vehicles for personal transportation (automobiles) that are suitable for road use. The history of BEVs predates that of gasoline-powered vehicles because the development of the electric motor predates the invention of the internal combustion engine. The first BEVs were constructed around 1830, but not until the 1880s were practical vehicles for human transportation available on the market. For some time, electric cars were more plentiful and popular than gasoline-powered automobiles. The reasons for this were that the BEV was

- Cleaner.
- Quieter.
- More reliable.
- Easier to start.
- More powerful.

The dominance of electric vehicles continued until around 1920, and by 1930 gasoline vehicles were in use almost exclusively. The reasons for the decline of the electric vehicle are:

- The invention of the electric starter in 1912 made starting gasoline vehicles much easier.
- The need for traveling longer distances made the limited range of the electric vehicle less convenient (particularly a factor in the United States).
- The development of mass production techniques (primarily a result of manufacturing techniques developed by Ford) for the manufacture of gasoline-powered vehicles made them much less expensive than they had been in the past
- Technological developments made gasoline vehicles more reliable than they had previously been.

It was not until the 1970s and 1980s, with diminishing oil supplies, the increasing reliance (in the United States) on foreign energy, and concerns over pollution and greenhouse gas emissions, that the electric vehicle was reconsidered.

In the early days of automobile development, most speed records were held by electric vehicles because of their greater power output compared with gasoline-powered vehicles. The first automobile to exceed 100 km/h was the Jamais Contente (Figure 19.3). On the more practical side, electric vehicles for urban use (Figure 19.4) were common in the early part of the last century. These typically had a range of 100–150 km and a top speed of around 30 km/h. This performance was sufficient to fulfill the needs of many drivers at that time. Some novel electric vehicles were produced during that era (Figure 19.5). The Bugatti Type 32 was a miniature electric sports car and was more of an expensive toy than an automobile.

In more recent years, the development of electric vehicles has been aimed at providing a clean replacement for gasoline vehicles. Perhaps the earliest of these recent vehicles that met with some success was the General Motors EV1 (Figure 19.6). The EV1 was produced from 1996 to 2003. Some of its specifications are given in Table 19.6. Another vehicle that was produced in fairly substantial numbers during about the same time was the Toyota Rav4 EV (Figure 19.7; specifications in Table 19.6). By comparison with gasoline vehicles, these cars tended to be heavy, were somewhat underpowered, and had a relatively small range. This, coupled with recharge times of up to 8 hours, made their use limited. Cost was a factor that was difficult to analyze. Many

Figure 19.3: The Jamais Contente (1899), designed and built by Camille Jenatzy.

Figure 19.4: A 1916 Detroit Electric vehicle.

Figure 19.5: A 1929 Bugatti Type 52 electric vehicle.

of the electric vehicles produced during this era were available to the general public, but not for sale; rather for lease only. In general, these vehicles were heavily subsidized and the EV1, for example, had a lease cost that was similar to a vehicle in the $30,000 to $40,000 range, whereas a detailed cost analysis suggested production costs that were closer to $80,000 per vehicle. General Motors destroyed virtually all EV1s after their leases expired.

Figure 19.6: The GM EV1 (produced from 1996–1999) being charged.

Figure 19.7: Toyota Rav4 EV produced 1997–2003.

Table 19.6: Specifications of the General Motors EV1 and the Toyota Rav4 EV. Because specifications (particularly of the EV1) changed over the years of its production, the information shown is for later year vehicles.

maker	General Motors	Toyota
model	EV1	Rav4 EV
years produced	1996–2003	1997–2003
number produced	1117	1249
battery type	NiMH	NiMH
mass	1320 kg	1565 kg
power output	102 kW	50 kW
top speed	128 km/h	126 km/h
range	260 km	160 km

Figure 19.8: The 2001–present Reva electric vehicle made in India.

As a result of battery properties and cost considerations, most recent electric vehicles have tended to fall into one of two categories: (1) relatively inexpensive, low-speed, short-range commuter vehicles or (2) expensive, high-power sport/luxury cars. In the first category, the Reva (Figure 19.8), built in India, has been the most successful worldwide and has been marketed in a number of countries (sometimes under different names; e.g. G-Wiz in England).

Example 19.3

Calculate the range of a 75 kW BEV utilizing NiMH batteries.

Solution

From Table 19.2, the specific power of a NiMH battery is 0.8 kW/kg. Thus, 75 kW requires

$$\frac{75\ \text{kW}}{0.8\ \text{kW/kg}} = 94\ \text{kg}$$

of batteries. Using the specific energy of 0.28 MJ/kg for NiMH from Table 19.2 gives a total energy available of

$$(94\ \text{kg}) \times (0.28\ \text{MJ/kg}) = 26\ \text{MJ}.$$

Using an estimated energy requirement of 0.7 MJ/km gives a total range of

$$\frac{26\ \text{MJ}}{0.7\ \text{MJ/km}} = 37\ \text{km}.$$

Figure 19.9: The Nissan Leaf.

Figure 19.10: Tesla Roadster.

Somewhat upscale from the Reva is the recently marketed Nissan Leaf (Figure 19.9).

The most notable example of the higher-end electric cars is the Tesla Roadster (Figure 19.10), made by a U.S. company, although the components are fairly international in nature, and many of the (nonelectric) components of the car are manufactured in collaboration with Lotus in the United Kingdom. The large amount of power available from batteries is an attractive feature in the design of a sports car. A comparison of the properties of the Reva, Leaf, and Tesla Roadster is given in Table 19.7.

At perhaps the far end of the range of practicality is the Eliica (*electric lithium ion car*, Figure 19.11). This is a prototype electric vehicle built in Japan. As seen in the figure, it has eight wheels. The electric motors produce 480 kW, and the vehicle has

Table 19.7: Properties of three recent model battery electric vehicles (* = electronically limited).

vehicle	mass (kg)	power (kW)	top speed (km/h)	range (km)	cost (US$)	charge time (h)	battery type
Reva	665	13	80	80	13,000	8	Pb-acid
Nissan Leaf	1521	80	145	117	33,000	7	Li-ion
Tesla Roadster	1235	185	210*	390	109,000	3.5	Li-ion

a top speed of 370 km/h. The projected sales price (if it is ever produced for sale) is around US$255,000.

The interest of the general public in electric vehicles is illustrated in Figure 19.12, where the number of electric cars in the United States is plotted as a function of year. These trends suggest that major changes in the battery electric vehicle market are needed to increase the number of these vehicles on the roads. Specifically, in the North American market, the development of an electric vehicle that is competitive with cars such as a gasoline-powered Toyota Corolla or Honda Accord, is necessary. As explained,

Figure 19.11: The Eliica prototype electric vehicle built in Japan.

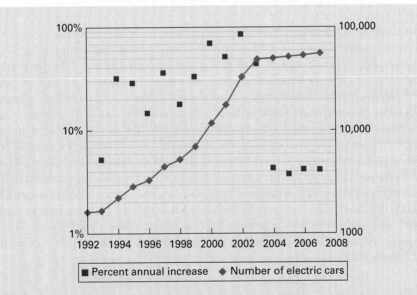

Figure 19.12: Number of electric cars in the United States since 1992.

some concerns need to be addressed in electric vehicle design and production in order to make a marketable vehicle:

- Range
- Charge time
- Cost
- Battery replacement cost

Figure 19.13: The 2011 Tesla Model S electric vehicle: (a) exterior and (b) trunk (the floor folds up to add two rear-facing seats for children).

There are, of course, some trade-offs in dealing with these factors. The public may be willing to accept a vehicle with a smaller range if recharge times are 10–15 minutes rather than several hours. Similarly, long recharge times may be more tolerable if the range is, say, 600–700 km, as most people rarely drive more than that in a day. Initial cost and battery replacement cost are important factors for most potential buyers. Although technological developments may lower these costs, a consideration of current and projected gasoline prices is always a major component of cost analysis. In general, the development of new batteries with lower cost, greater capacity, shorter recharge times, and greater capacity retention (cycleability) would be the most significant contribution to electric vehicle development. Also, the consumer's environmental consciousness is a factor that may have some influence on automobile choice.

A possible step toward a battery electric vehicle that is compatible with usual North American expectations for a family vehicle is the Tesla Model S [Figure 19.13(a)], first available for sale in the United States in 2012. With seating for 7 (five adults in front and rear seats and two children in seats that face backward and fold down into the trunk floor [Figure 19.13(b)], a range of up to 426 km, and a price tag of between about US$60,000 to US$100,00, this vehicle begins to fill in the gap between battery electric commuter cars and performance sports cars. (*Note:* The stated range stated is for the higher-priced models with larger battery packs, while the less expensive models have smaller battery packs and corresponding smaller ranges.)

19.5 Supercapacitors

A mechanism for storing electrical energy that may be of relevance to battery electric vehicles, as well as series hybrids (Chapter 17) and fuel cell vehicles (Chapter 20), is the *supercapacitor* (sometimes referred to as an *ultracapacitor*). The position of the supercapacitor on a Ragone plot is shown in Figures 19.1 and 19.2. The figures show clearly that the supercapacitor stores relatively little energy (compared to batteries, for example) but that it can release this energy very quickly. Thus, this is a low-energy, high-power device. Its ability to release energy quickly also means that it can

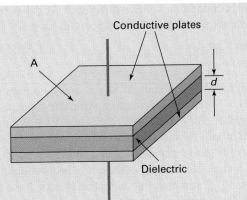

Figure 19.14: Schematic diagram of a parallel plate capacitor.

store energy quickly. The timescale for energy storage and release is illustrated on the Ragone plot in Figure 19.1. In principle, a supercapacitor is really just a normal capacitor that can store more energy and release more power. Thus, we begin with a discussion of the operation of capacitors in general.

Unlike a battery, which stores electrical energy using chemical reactions, a capacitor stores electrical energy in an electric field by separating positive and negative charges. The simple parallel plate capacitor shown in Figure 19.14 consists of two conductive plates of area A separated by an insulating dielectric of thickness d. The capacitance of this device is

$$C = \frac{\varepsilon_r \varepsilon_0 A}{d}, \tag{19.6}$$

where ε_r is the *relative static permittivity* (sometimes in the past this has been referred to as the *dielectric constant*), and ε_0 is the permittivity of vacuum. In SI units, d is measured in meters, A is measured in square meters, ε_r is dimensionless, and ε_0 is 8.854×10^{-12} F/m (F = farad = J/V^2).

Example 19.4

Calculate the capacitance (in Farads) of a parallel plate capacitor utilizing a dielectric with a relative static permeability of $\varepsilon_r = 1$ and with an area of 10 cm $\times$ 10 cm and a distance between the plates of 0.1 mm.

Solution
Using all dimensions in meters in Equation (19.6), the capacitance is found to be

$$C = \frac{\varepsilon_r \varepsilon_0 A}{d} = \frac{(1) \times (8.854 \times 10^{-12}\,\text{F/m}) \times (0.1\,\text{m}) \times (0.1\,\text{m})}{0.0001\,\text{m}} = 8.85 \times 10^{-10}\,\text{F}.$$

inductiveload

If the capacitor is charged, a voltage difference, V, exists between the two plates. The energy stored in the capacitor under these conditions is

$$E = \frac{1}{2} CV^2. \tag{19.7}$$

It is easily seen that, for a capacitance in farads and a voltage in volts, the energy is given in joules. If a 1-F capacitor (compare with the capacitance calculated in Example 19.4) is charged to 10 V, then this represents 50 J of energy; this is enough to illuminate a 60-W lightbulb for just under a second, not a significant factor for electric vehicle use. The energy storage capabilities can be improved by increasing either V or C. Increasing V has a clear advantage because E is proportional to V^2. However, there is a limit for voltage, V_{max}, at which the dielectric breaks down and starts conducting, thereby shorting out the capacitor. The maximum voltage is related to the breakdown electric field of the dielectric material, E_b:

$$V_{max} = dE_b, \tag{19.8}$$

where E_b is in V/m. C can be increased by increasing either ε_r or A or by decreasing d. The simplest approach is merely to make the capacitor larger. However, there are certain limits to this approach if the capacitor is to be practical, and it is generally most desirable to maximize the energy per unit volume (or mass). Decreasing d beyond some limit is somewhat counterproductive. Although it increases C [equation (19.6)], it also decreases V_{max} [equation (19.8)]. An analysis of these equations shows that the maximum energy storage capacity is

$$E_{max} = \frac{1}{2} CV_{max}^2 = \frac{1}{2} \frac{\varepsilon_r \varepsilon_0 A}{d} (E_b d)^2 = \frac{1}{2} A d \varepsilon_r \varepsilon_0 E_b^2. \tag{19.9}$$

The volume of the capacitor is Ad, and its mass is given in terms of its density, ρ, as $m = Ad\rho$. Most dielectric materials have densities around that of water. Thus, the maximum energy density per unit mass (in J/kg) from equation (19.9) is

$$\left(\frac{E}{m}\right)_{max} = \frac{1}{2} \left(\frac{\varepsilon_r \varepsilon_0}{\rho}\right) E_b^2. \tag{19.10}$$

Traditionally, capacitors have often been made by rolling up parallel plates with an insulating material between the layers. This does not increase the energy density, but it makes the geometry more compact. To actually maximize the energy density, it is important to consider ε_r and E_b. Although there is often a trade-off between these factors, maximizing E_b is obviously the more important.

Supercapacitors approach the energy density problem from two directions: maximizing E_b and reconsidering the geometry. Polyethylene terephthalate (PET) is the most commonly used dielectric. This material has values of $\varepsilon_r = 3$ and $E_b = 7.1 \times 10^8$ V/m and increases the capacitance by about a factor of four over traditional materials. It is electrode geometry, however, where supercapacitors benefit the most. As shown in Figure 19.15, nanoporous carbon is used as the electrode material. This material has numerous pores into which the dielectric penetrates, providing a substantially larger surface area onto which to accumulate charge than the macroscopic area of the surface. The improvement over normal capacitors is illustrated in Figure 19.1.

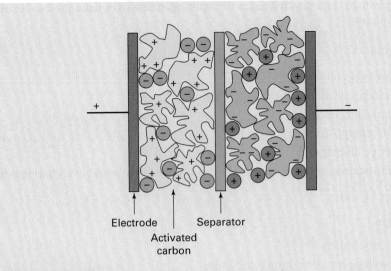

Electrode | Separator

Activated
carbon

Figure 19.15: Schematic diagram of a supercapacitor.

Example 19.5

Using a specific energy for a supercapacitor of 2 Wh/kg (Figure 19.1), calculate the range of a vehicle utilizing 1000 kg of supercapacitors.

Solution

The total energy content of 1000 kg of (fully charged) supercapacitors is

$$(2 \text{ Wh/kg}) \times 1000 \text{ kg} = 2 \text{ kWh}.$$

Converting to MJ gives

$$2 \text{ kWh} \times 3.6 \text{ MJ/kWh} = 7.2 \text{ MJ}.$$

Using a typical energy requirement for a vehicle of 0.7 MJ/km gives

$$\frac{7.2 \text{ MJ}}{0.7 \text{ MJ/km}} = 10.3 \text{ km}.$$

Even given the improvement of supercapacitors over normal capacitors, Figure 19.1 shows that the energy density is very low compared to batteries. An electric vehicle built by NASA using 1000 kg of supercapacitors rather than batteries had a range of about 7 km, consistent with the preceding example calculation. Clearly, this would suggest that it is unlikely that supercapacitors would compete with batteries as a sole source of energy for an electric vehicle. However, the place where supercapacitors excel is in power density. As Figure 19.1 shows, they can provide ten times or more the amount of power per unit mass as a Li-ion battery.

The advantages and disadvantages of supercapacitors over batteries can be summarized as follows:

Advantages

- High power density
- High voltages available from a single unit
- No chemical reactions
- Up to 10^6 charge/discharge cycle life
- Fast charge/discharge
- No sophisticated charge/discharge electronics needed to avoid overcharging or damage

Disadvantages

- Low energy density
- Linear voltage decrease during discharge
- High self-discharge rate compared with batteries

These properties make them suitable for specific applications, including:

- Additional power for a BEV when there is a demand for excess power
- Short-term power storage for regenerative braking
- Short-term power applications such as starter motors, where they could replace conventional batteries

Also, nonvehicle applications can include short-term backup power for electronic devices.

19.6 Summary

Battery electric vehicles provide several very attractive features as alternatives to petroleum internal combustion engine vehicles. The chapter began with an overview of battery technology and showed that batteries are highly efficient storage devices for electrical energy and, combined with the high efficiency of an electric motor, provide a very energy-efficient vehicle. The Ragone plot has been introduced as a means of understanding the energy-power relationship for energy storage mechanisms. While the energy density of batteries is less than that of fossil fuels, Li-ion batteries, in particular, offer a good compromise between high specific energy and high specific power.

Electric vehicles themselves do not emit greenhouse gases during use, although an analysis of the net carbon footprint must take into account the source of the electricity used to charge the batteries. Even in the worst case (from a carbon standpoint) where electricity is generated from coal, the overall primary-energy-source-to-wheel efficiency of the BEV is very good, perhaps 30–35%, or about twice that for a conventional internal combustion engine. An added bonus is that, even if fossil fuels are used to generate electricity, it is much easier to sequester carbon from a few large coal-fired generating stations than from a very large number of individual vehicles.

The chapter reviewed the history of the development of the BEV and has shown that this was a popular technology in the early development of the passenger vehicle. In recent years, interest in BEVs has once again increased due to concerns over the future availability of petroleum-based fuels and their effects on the environment.

The major drawbacks of BEVs that utilize current battery technology are the range and recharge time. While current vehicle characteristics are suitable for typical urban vehicle use and for driving habits in some parts of the world, they would put constraints on much of the current vehicle use in North America.

The chapter showed that improved BEV characteristics require advances in battery technology. Increasing specific capacity will increase range without increasing battery weight. Improved cycleability will increase battery life and reduce replacement costs. Other factors that affect the viability of BEVs include vehicle cost and availability of resources. As long as internal combustion engine vehicles are (relatively) inexpensive and oil is readily available, the market for BEVs is greatly influenced by economic factors. Also, the availability and cost of lithium associated with the large-scale replacement of fossil fuel-powered vehicles with BEVs are uncertain.

The development of new battery chemistries based on other elements is attractive. There is active research in the development of sodium-based batteries. Sodium shares many of the chemical properties of lithium (because of similarities in their electronic structure) but is inexpensive, and its availability is virtually unlimited (as a component of salt in the oceans).

This chapter has shown that the use of supercapacitors may help to improve some aspects of vehicle performance. However, because power is generally less of a concern for BEVs than range, supercapacitors do not resolve the principal concerns related to BEVs.

Problems

19.1 What are the ranges of typical (family size) battery electric vehicles that utilize 250 kg of batteries if the batteries are (a) Pb-acid, (b) NiMH, and (c) Li-ion?

19.2 A solar photovoltaic installation generates an average of 20 MW of power during a 12-hour period of sunlight and no power during 12 hours of night. A Pb-acid battery system is designed to store one day of solar energy for use at night, and during periods of less than full sunlight. What is the mass of the battery system?

19.3 What is the range of a family-sized vehicle utilizing 100 kg of supercapacitors as an energy storage mechanism? What is the maximum power output (in watts)?

19.4 A family vehicle uses 100 kg of Li-ion batteries in conjunction with 200 kg of supercapacitors for energy storage. What is the maximum power output, and what is its range? Would it be advantageous to increase or decrease the relative mass of the batteries compared to the supercapacitors in order to improve vehicle characteristics?

19.5 A typical Pb-acid automobile battery weighs 20 kg. Compare the amount of energy it stores when fully charged to the energy content of 1 L of gasoline.

19.6 **(a)** Consider a Pb-acid battery backup system for a residence. Using information in Chapter 2, estimate the mass of batteries needed to provide the average electric power for a home (exclusive of heating).

(b) Using the energy density of Pb-acid batteries, how long could backup power be provided at this level?

19.7 A gasoline-powered automobile burns 7.0 L of fuel per 100 km while traveling on the highway at a constant velocity. If the engine is 19% efficient, how much power is actually being used to move the vehicle?

19.8 Assume that the batteries in an electric vehicle are designed to provide the same energy to the vehicle's wheels as 75 L of fuel in the tank of a gasoline-powered vehicle. The batteries are charged by connecting them to a DC charger that can supply 30 A at 220 V (this is typical of the AC circuit used to provide electricity for a kitchen stove if differences between DC and AC power are not considered). How long will it take to charge the batteries? Be sure to take the relative efficiencies of gasoline-powered vehicles and BEVs into account.

Bibliography

J. O. Besenhard (Ed.). *Handbook of Battery Materials*. Wiley VCH, Weinheim, Germany (1999).

D. R. Blackmore and A. Thomas. *Fuel Economy of the Gasoline Engine*. Macmillan, London (1977).

H. Braess and U. Seiffert (Eds.). *Handbook of Automotive Engineering*. Society of Automotive Engineers, Warrendale, PA (2004).

J. A. Kraushaar and R. A. Ristinen. *Energy and Problems of a Technical Society*, 2nd ed. Wiley, New York (1993).

J. MacKenzie. *The Keys to the Car: Electric and Hydrogen Vehicles for the 21st Century*. World Resources Institute, Washington, DC (1994).

D. A. J. Rand, R. Woods, and R. M. Dell. *Batteries for Electric Vehicles*. Research Studies Press, Somerset, United Kingdom (1998).

R. Stone and J. Ball. *Automotive Engineering Fundamentals*. Society of Automotive Engineers, Warrendale, PA (2004).

F. M. Vanek and L. D. Albright. *Energy Systems Engineering: Evaluation and Implementation*. McGraw Hill, New York (2008).

Hydrogen

Learning Objectives: After reading the material in Chapter 20, you should understand:

- The basic properties of hydrogen as a gas and as a liquid.
- Methods for hydrogen production.
- The factors that must be considered when storing gaseous or liquid hydrogen.
- The use of hydrogen as a fuel in internal combustion engines.

- The properties and types of fuel cells.
- The design of fuel cell vehicles.
- The viability of hydrogen as a fuel and efficiency considerations for its use.

20.1 Introduction

In previous chapters, two possible technologies for future nonfossil fuel transportation have been discussed: biofuels and battery electric vehicles. Both of these have been the subject of intense development in recent years, and both have seen commercial availability in North America in the form of flex fuel vehicles and electric vehicles. However, neither of these technologies is without its own drawbacks. An alternative transportation technology that has also been promoted is hydrogen-powered vehicles.

Hydrogen, like a battery, is an energy storage mechanism. It is not a primary energy source, like oil or coal, because no significant sources of naturally occurring hydrogen exist in nature; although hydrogen generation schemes based on organic processes may be viewed somewhat like biofuels. Hydrogen may be produced from electricity through electrolysis and can be used to provide thermal or mechanical energy by burning, or by conversion back into electricity by means of a fuel cell. The oxidation of hydrogen yields water as its only by-product (although nitrogen compounds can be created from atmospheric nitrogen in any combustion process that occurs at an elevated temperature). Like all methods of producing usable energy, the use of hydrogen must be viewed in detail, from production to end use, in order to fully understand its viability in a future energy economy.

20.2 Properties of Hydrogen

Energy is produced by combining the hydrogen with oxygen to produce water through the process

$$2H_2 + O_2 \rightarrow 2H_2O. \tag{20.1}$$

The energy content of hydrogen, that is, the energy gain during oxidation, is quite substantial compared with other fuels (Figure 20.1). Although this figure suggests the advantages of hydrogen as a fuel, there are some important considerations for its use. Firstly, hydrogen is a gas at standard temperature and pressure (STP = room temperature and 101 kPa) and occupies a substantial volume. Table 20.1 shows a comparison between liquid gasoline and gaseous hydrogen at STP. Clearly, the volume occupied by hydrogen at STP, as a result of its low density (0.0899 kg/m^3), is a drawback that would necessitate some means of reducing the volume of the fuel. This is true even for stationary applications, but is particularly the case for transportation use. The hydrogen equivalent of 60 L of gasoline (typical for an automobile gasoline tank) at STP would be about 200 m^3, or a cube 6 m on a side. One option would be to compress hydrogen gas while keeping it at room temperature. Another option would be to liquefy hydrogen to make a liquid fuel and to maintain it at low temperature. The energy density of liquid hydrogen is 8520 MJ per m^3; this is less than gasoline by about a factor of four. Although the energy density per kilogram is greater for hydrogen, its density is less than that of gasoline (71 kg/m^3 vs. 720 kg/m^3). The possibilities for storing hydrogen in a convenient form are discussed further in Section 20.4. The various mechanisms for producing hydrogen are considered first.

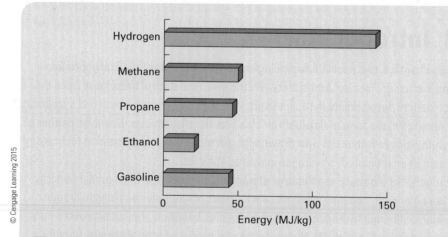

© Cengage Learning 2015

Figure 20.1: Energy content (per kg) of various fuels.

© Cengage Learning 2015

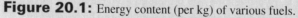

fuel	energy per kg (MJ)	energy per m^3 (MJ)
gasoline	44.5	34,800
hydrogen	142	11.8

Table 20.1: Specific and volumetric energy densities of gasoline and hydrogen at STP.

20.3 Hydrogen Production Methods

There are four basic methods for producing hydrogen:

1. Electrolysis
2. Thermal decomposition of water
3. Chemical reactions
4. Biological processes

The basic features of each of these are now discussed.

20.3a Electrolysis

Electrolysis is the best known method of producing hydrogen. It is the separation of the oxygen and hydrogen atoms of water molecules using electricity. A basic electrolysis cell is shown in Figure 20.2. The overall reaction is

$$2H_2O + energy \rightarrow O_2 + 2H_2, \qquad (20.2)$$

where the energy is supplied in the form of electrical energy. It is convenient to consider the two half reactions that take place at the electrodes in the cell. Hydrogen is formed by decomposition of water at the cathode (negative electrode) by the reaction

$$2H_2O \rightarrow H_2 + 2OH^-, \qquad (20.3)$$

where the negative charge is supplied by an electron coming from the power supply. The negatively charged OH^- ion is repelled by the cathode and travels through the water to the anode where it undergoes the reaction

$$2OH^- \rightarrow (1/2)O_2 + H_2O. \qquad (20.4)$$

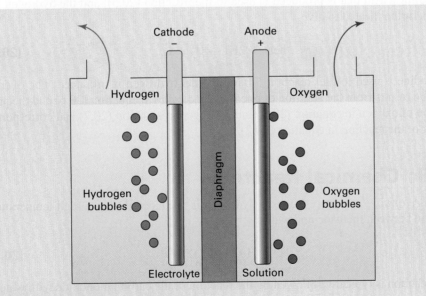

Based on http://www.greencarcongress.com/images/standard_electrolysis.png

Figure 20.2: Basic electrolysis cell.

The negative charge travels back through the external circuit, in the form of an electron, to the power supply. The net result of this process is half of the reaction given in equation (20.2). The electrical conductivity of the water is increased with charged ions produced by dissolving an ionic compound such as H_2SO_4, KOH, or the like.

Electrical energy from the power supply is stored as chemical energy in the hydrogen. Another way in which this process can be viewed is to consider that the energy supplied by the circuit is used to reduce (that is the opposite of oxidize) the water. Energy stored in the hydrogen can be recovered by oxidizing the hydrogen either in a combustion process (Section 20.5) or in a fuel cell (Section 20.6). It is important to realize that this process, like all energy conversion processes, is not 100% efficient. In fact, it is typically about 70% efficient, and this must be considered in an evaluation of any process that involves the storage of energy in the form of hydrogen.

20.3b Thermal Decomposition of Water

The thermal decomposition of water corresponds to the reaction given in equation (20.2), except that the energy is supplied in the form of thermal energy rather than electrical energy. Unfortunately, this process is not as simple as it may seem because the temperatures required to dissociate water are unrealistically high. It is, however, possible to use a catalyst to make the process easier. One common approach is the sulfur-iodine thermochemical cycle. This begins with heating sulfuric acid, which causes the reaction

$$\text{heat} + H_2SO_4 \rightarrow (1/2)O_2 + SO_2 + H_2O. \tag{20.5}$$

The sulfur dioxide is combined with iodine and water and heated further to yield

$$\text{heat} + SO_2 + I_2 + 2H_2O \rightarrow 2HI + H_2SO_4. \tag{20.6}$$

The sulfuric acid that is produced is used to fuel reaction (20.5), and the hydrogen iodide is further heated to give

$$\text{heat} + 2HI \rightarrow H_2 + I_2. \tag{20.7}$$

The iodine is used to fuel reaction (20.6), and net hydrogen is produced. The overall reaction results from the addition of one water molecule in equation (20.6) and the production of oxygen in equation (20.5) and hydrogen in equation (20.7) and corresponds to half of the reaction in equation (20.2).

20.3c Chemical Reactions

Steam reforming is a process where hydrogen is produced during a high-temperature reaction between methane and water:

$$CH_4 + H_2O \rightarrow CO + 3H_2. \tag{20.8}$$

This reaction is typically followed by the reaction of the carbon monoxide with water:

$$CO + H_2O \rightarrow CO_2 + H_2. \tag{20.9}$$

ENERGY EXTRA 20.1
Alternative hydrogen production methods

One interesting approach to the production of hydrogen, sometimes referred to as solar hydrogen, involves the possibility of producing hydrogen directly from the reaction of certain materials with sunlight. These materials may be organic materials that produce hydrogen as a result of photosynthetic processes, or they may be semiconducting materials that produce hydrogen by direct electrolysis with surrounding water molecules. One might envision a simple approach where sunlight is incident on a piece of silicon immersed in water, and electron-hole (e^--h^+) pairs are formed by the interaction of photons with the charges in the material (Chapter 9). These electron-hole pairs can then flow through the water, generating hydrogen and oxygen. The reaction is

$$2h^+ + H_2O \rightarrow 2H^+ + \tfrac{1}{2}O_2$$

and

$$2e^- + 2H^+ \rightarrow H_2.$$

There are several problems with this simple approach:

- The electron-hole pairs tend to recombine in the silicon rather than interact with the water molecules.

- The efficiency for converting photons into hydrogen is very low.
- The silicon is chemically reactive with water and degrades due to corrosion.

Peidong Yang at the University of California, Berkeley, has developed materials that are effective at dealing with these difficulties. By creating nano-sized wires of silicon with a coating of semiconducting TiO_2 on their surface (see the photograph), he has been able to resolve many of these issues. TiO_2 is much less reactive with water than Si, and the coating protects the Si from adverse chemical reactions with water. The junction between the two semiconducting materials helps to prevent electron-hole recombination. Finally, the large surface area of the nanowires presents a large cross section for absorption of photons and interaction with the water.

These materials have shown promise for the direct conversion of sunlight into hydrogen, which can then be used for creating heat by combustion or electricity by means of a fuel cell. It is hoped that this approach can help to alleviate problems related to the inefficiencies and expense of the traditional hydrogen production route.

Lawrence Berkeley National Laboratory

Highly dense array of 20-μm long nanowires made from TiO_2-coated Si.

Continued on page 516

Energy Extra 20.1 continued

Topic for Discussion

The preceding description is only one of quite a few possible mechanisms for obtaining hydrogen by methods other than the electrolysis of water. Coal is not really pure carbon, but it contains various other elements, including hydrogen. It is possible to extract the hydrogen from coal using catalytic reactions. One possibility is the use of sulfur to react with the carbon and hydrogen in the coal:

$$4CH_{0.5} + S \rightarrow 4C + H_2S.$$

The hydrogen sulfide can then be decomposed to yield hydrogen,

$$H_2S \rightarrow H_2 + S,$$

and the sulfur can be recovered for further use. It has been suggested that soft coal may contain enough hydrogen that the hydrogen can be extracted for use as a carbon-free fuel and the carbon may be left sequestered in the coal and not burned. Discuss the merits of this approach.

Other chemical reactions can also release hydrogen. Another common reaction used for hydrogen production is the high-temperature reaction of water with carbon:

$$H_2O + C \rightarrow H_2 + CO. \tag{20.10}$$

This can be followed by the reaction in equation (20.9). These chemical reactions can be an important source of hydrogen for industrial purposes. However, they may not suitable for producing hydrogen in the context of an environmentally conscientious energy technology because they produce CO_2.

20.3d Biological Processes

These are analogous to the production of hydrocarbons from biological processes, that is, the production of biofuels. In some cases, biological processes can release hydrogen without associated carbon. Research into these possibilities is in the relatively early stages.

20.4 Storage and Transportation of Hydrogen

Once produced, hydrogen must be transported to its place of distribution and/or use, and it must be stored in an appropriate form. Because of the large volume occupied by hydrogen gas at STP, the hydrogen must be stored in a more compact form. There are three possible options for this;

1. Compressed hydrogen gas (CHG)
2. Liquid hydrogen (LH_2)
3. Metal hydrides.

To understand storage as CHG, it is important to look at the properties of hydrogen under pressure. Figure 20.3 shows the relationship between density and pressure for hydrogen.

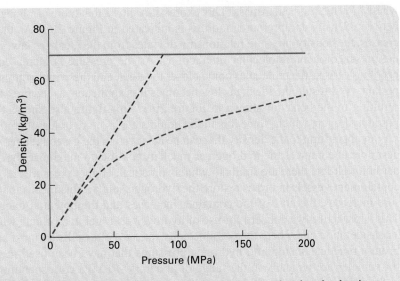

Figure 20.3: Density-pressure relationship for hydrogen gas showing the density of liquid hydrogen (red line) and the ideal gas law relationship (blue line). The green line shows the density of liquid hydrogen.

Example 20.1

A 1 m³ tank of hydrogen at 100 MPa is used to generate electricity by combustion in a heat engine that drives an electric generator with an overall efficiency of 15%. What is the total electrical energy available in kilowatt-hours?

Solution

From Figure 20.3, the density of hydrogen gas at 100 MPa is about 40 kg/m³. From Table 20.1, the energy content of hydrogen is 142 MJ/kg. Thus 1 m³ of hydrogen at 100 MPa will yield

$$(40 \text{ kg}) \times (142 \text{ MJ/kg}) \times (0.278 \text{ kWh/MJ}) = 1580 \text{ kWh}.$$

At an effiiency of 15%, the total electrical energy generated is

$$(1580 \text{ kWh}) \times (0.15) = 237 \text{ kWh}.$$

At relatively low pressures, the density-pressure curve follows the ideal gas relationship reasonably well (i.e.):

$$PV = nRT. \tag{20.11}$$

Here P is pressure, V is volume, n is the number of moles of gas, R is the universal gas constant ($R = 8.315 \text{ J} \cdot \text{K}^{-1} \cdot \text{mol}^{-1}$), and T is the absolute temperature. This expression can be rewritten in terms of the density, ρ:

$$\rho = \frac{MP}{RT}, \tag{20.12}$$

where M is the molecular mass. This expression shows that there is a linear relationship between density and pressure, and this is indicated in Figure 20.3. As the pressure increases, the density-pressure relationship deviates from the ideal gas law, which ignores the presence of intermolecular interactions. As the pressure is increased and the gas becomes denser, the molecules come closer together, and these interactions become important. As a result, the ideal gas law becomes less applicable as the pressure increases, resulting in the behavior shown in Figure 20.3. This figure also shows the density of liquid hydrogen, which, since it is in the liquid state, is more or less incompressible and sets an upper limit to the density that can be achieved in the gaseous state. Although it is desirable to compress the hydrogen as much as possible to maximize the energy stored per unit volume, there are limits to what is practical. Higher pressures require stronger containment vessels in order to store the hydrogen safely. This means that, as the pressure increases, the cost of the container increases and the potential dangers increase. This is particularly a concern for use of hydrogen as a fuel in a vehicle where weight is a consideration and where the possible consequences of a collision must be considered. From a practical standpoint, hydrogen pressures that have been used in storage tanks for vehicles are typically in the range of about 35–70 MPa corresponding to densities in the range of about 20–35 kg/m^3, or about one-third to one-half the density of liquid hydrogen. A factor to consider for this and any other method for storing hydrogen is its efficiency; that is, how much energy is needed to store the hydrogen? In the case of CHG, the energy needed is that required to run a compressor to compress the gas. This energy is lost and is not recovered when the hydrogen is utilized and typically amounts to about 10% of the energy content of the hydrogen, making this process about 90% efficient.

Example 20.2

If hydrogen behaves as an ideal gas, calculate the room temperature density at a pressure of 50 MPa.

Solution

We use equation (20.12), $\rho = \dfrac{MP}{RT}$, where the molecular mass for hydrogen is 2 g/mol, $R = 8.315$ J·K^{-1}·mol^{-1}, and $T = 293$ K. Note that in basic SI units the joule is kg·m^2·s^{-2}, and the pascal is Pa = kg·m^{-1}·s^{-2}. Thus equation (20.12) may be written as

$$\rho = \frac{(0.002 \text{ kg·mol}^{-1}) \times (50 \times 10^6 \text{ kg·m}^{-1}\cdot\text{s}^{-2})}{(8.315 \text{ kg·m}^2\cdot\text{s}^{-2}\cdot\text{K}^{-1}\cdot\text{mol}^{-1}) \times (293 \text{ K})}$$

$$= 41 \text{ kg/m}^3,$$

consistent with the value shown for the straight (ideal gas) line in Figure 20.3 for a pressure of about 50 MPa.

From Figure 20.3, it is obvious that the energy density (MJ/m^3) of liquid hydrogen is greater than for CHG at any pressure. At atmospheric pressure, the boiling point of hydrogen is 20.3 K. Gaseous hydrogen must be cooled down below this temperature in order to convert it into a liquid. Two important factors need to be considered: how

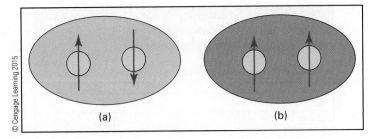

© Cengage Learning 2015

(a)   (b)

Figure 20.4: (a) Parahydrogen (ground state) and (b) orthohydrogen (excited state), showing the alignment of proton spins.

much energy is needed to liquefy the hydrogen, and how can it be contained to prevent it (as much as possible) from boiling away. Typically, the energy required for liquefaction compared to the energy content of the hydrogen is about 30%, making the process about 70% efficient.

The effective and efficient liquefaction of hydrogen is somewhat more complex than merely lowering the temperature. The hydrogen molecule consists of two hydrogen atoms bonded together. Each hydrogen atom consists of a nucleus (a proton) and an electron. In the hydrogen molecule, the magnetic moments (spins) of the protons of the two atoms interact and can align either antiparallel or parallel (Figure 20.4). These two spin arrangements are referred to as parahydrogen and orthohydrogen, respectively. The ortho form is at a higher energy than the para form and can be considered an excited state of the parahydrogen molecule ground state. At room temperature, thermal energy causes many of the parahydrogen molecules to acquire enough energy to become orthohydrogen. The equilibrium state at room temperature corresponds to about 25% parahydrogen and 75% orthohydrogen. As the temperature is lowered, the equilibrium parahydrogen fraction increases at the expense of the orthohydrogen fraction. In the liquid state, the temperature is low enough that the equilibrium concentration of orthohydrogen is very small. However, if hydrogen is cooled and liquefied fairly quickly, then the liquid contains excess orthohydrogen. This slowly converts to parahydrogen by flipping the spin of one of the protons. Since the ortho–to-para conversion is an exothermic process (i.e., a transition from an excited state to the ground state), heat is released, causing some of the liquid hydrogen to boil. In practice, catalysts such as iron oxide or carbon, which convert orthohydrogen to parahydrogen, need to be used during the cooling process to minimize the formation of a nonequilibrium mixture of ortho and parahydrogen.

Even given these considerations, liquid hydrogen must be effectively insulated from its surroundings to reduce heat transfer into the liquid, causing it to boil. Specially designed containers have been developed for this purpose. The loss rate depends on the quality of the insulation but depends on another important factor; the size of the container. As the heat capacity of a liquid depends on its volume (or mass) and the heat transfer depends on the surface area, the loss rate depends on the surface-to-volume ratio of the container, which is a function of the container volume. Some examples of loss rates (in percent per day) for hydrogen are given in Table 20.2. Although containers at distribution or storage facilities can be quite large, the fuel tank in a vehicle is limited in size. This means that loss of fuel may be a factor if the vehicle is unused for any period of time.

Table 20.2: Loss rates for typical liquid hydrogen storage tanks of different volumes.

volume (m^3)	loss rate (%/day)
0.1	2
50	0.4
20,000	0.06

© Cengage Learning 2015

A final point concerns the transfer of liquid hydrogen. Transfer tubes must be well sealed to avoid contact between the liquid hydrogen and air. This is because air (consisting primarily of nitrogen and oxygen) has a higher boiling point than hydrogen, and contact between the two causes air to condense and contaminate the hydrogen.

The final method of storing hydrogen is in the form of metal hydrides. This method relies on reactions of hydrogen with certain metals of the form

$$2M + H_2 \rightarrow MH_2. \tag{20.13}$$

This reaction is exothermic and releases excess heat. One of the commonly used metals for this purpose is titanium, which forms the hydride TiH_2.

Example 20.3

Calculate the mass of hydrogen that can be stored in 1 m^3 of titanium. *Note:* The density of titanium is 4510 kg/m^3

Solution

In the TiH_2 phase, two-thirds of the atoms are hydrogen. Using the approximate atomic masses of 1 g/mol for hydrogen and 48 g/mol for titanium, the ratio of the weight of hydrogen to titanium is

$$\frac{2 \text{ g/mol}}{48 \text{ g/mol}} = 0.042.$$

Thus, 1 m^3 or 4510 kg of Ti contains

$$(4510 \text{ kg}) \times (0.042) = 188 \text{ kg}$$

of hydrogen.

A comparison of hydrogen storage capabilities of a reasonably sized fuel tank for vehicle use (e.g., 100 L, or 0.1 m^3) is given for CHG, LH$_2$, and metal hydrides in Table 20.3. A comparison with gasoline is also shown.

While the total energy stored in 0.1 m^3 of TiH$_2$ is attractive (in comparison with hydrogen stored by other methods) and begins to approach that of gasoline, there are two major difficulties with this approach. First, the mass of the titanium is quite large and adds substantially to the total mass of the fuel system. Secondly, it is not a simple matter to form the metal hydrides and to subsequently release the hydrogen for use. The process in equation (20.13) is exothermic, and this means that energy is required to remove the hydrogen from the metal hydride. Appropriate technology must be developed to enable effective charging and discharging of the metal hydride.

Table 20.3: Storage capabilities for a 0.1-m³ volume for hydrogen in different forms and a comparison with gasoline. Total masses include the mass of a suitable storage container.

fuel	fuel mass (kg)	total mass, typical (kg)	energy per volume (MJ/0.1m³)	energy per mass (MJ/kg)
CHG (35 MPa)	2.0	100	280	2.8
CHG (70 MPa)	3.5	150	500	3.3
LH$_2$	7.2	100	1000	10
TiH$_2$	18	450	2550	5.7
gasoline	72	85	3500	41

The transport of hydrogen also presents some challenges. In some ways, this can benefit from the methods and infrastructure associated with the use of natural gas. Hydrogen can be transported across land in pipelines or as CHG in tanker trucks. It may also be transported in appropriately insulated trucks in the form of LH$_2$. Across oceans, LH$_2$ ships are the most logical method. It may seem unclear why the transportation of hydrogen across oceans may be of interest because it is an energy storage medium that can be produced from electricity on the continent where it is to be used. However, hydrogen is a potential energy storage mechanism for transporting energy that is produced far offshore to users on land. This may be most relevant to possible future energy technologies such as OTEC, ocean currents, and possibly waves (if not close enough to shore for power transmission cables). Of course, the question of efficiency of energy conversions is very important, especially given the already marginal (at best) efficiency of methods like OTEC.

20.5 Hydrogen Internal Combustion Vehicles

The world's first automobile is shown in Figure 20.5. This vehicle utilized an internal combustion engine (ICE) that ran on hydrogen, and, although not practical for general use, it demonstrated the operational principle of a motorized vehicle. Other vehicle technologies (e.g., gasoline and even battery) dominated the development of the automobile for many years. In recent years, however, there has been renewed interest in the use of hydrogen as a fuel. The remainder of this chapter is devoted to a consideration of hydrogen as a fuel, primarily for transportation. However, possible uses extend beyond the transportation sector and some of these are discussed here as well.

Current technologies for the use of hydrogen as a transportation fuel fall into two categories: internal combustion engines (discussed in this section) and fuel cells (discussed in the next section).

Hydrogen ICE vehicles are based on production gasoline-powered vehicles. The gasoline internal combustion engine will run on hydrogen with relatively minor modifications. The most significant differences between gasoline and hydrogen ICE vehicles are in the fuel supply system. Thus, the manufacturing infrastructure for the production of hydrogen ICE vehicles is largely in place. The use of hydrogen as a fuel in an ICE

Based on http://www.quantium.plus.com/derivaz/isaac/rivaz.jpg

Figure 20.5: The first automobile was built in 1807 by Francois Isaac de Rivaz of Switzerland. It utilized compressed hydrogen gas as a fuel and used an electric spark produced by means of a battery for ignition.

has clear advantages. Energy is produced by the reaction in equation (20.1) with the by-product being water. No carbon emissions are produced because there is no carbon in the fuel. Some NO_x compounds are produced becasue this inevitably occurs when anything is burned in air, but these emissions are relatively minor. The efficiency of a hydrogen ICE is similar to that of a gasoline ICE and is ultimately limited by the theoretical Carnot efficiency. For hydrogen ICEs developed for automobiles, this means an efficiency of about 15–20%. According to Table 20.3, it is clear that a major problem with hydrogen vehicles is carrying enough hydrogen in a compact enough space to provide a usable range. For this reason, at present, virtually all hydrogen vehicles that are under serious development are hybrids, that is, vehicles that utilize two energy sources. This is true both of hydrogen ICE vehicles, as well as hydrogen fuel cell vehicles (discussed in the next section). In the case of hydrogen ICE vehicles, the logical alternative fuel is gasoline because the engine can be designed as a two-fuel engine running on either hydrogen or gasoline, and fuel systems delivering both fuels can be incorporated into the vehicle.

Two companies that have been particularly active in pursuing hydrogen ICEs are BMW and Mazda. BMW's vehicle may be close to limited commercialization, and Mazda's vehicle is well along in its development. The BMW Hydrogen 7 (Figure 20.6) is a luxury vehicle derived from the 760Li gasoline-powered BMW that uses a 6.0-L V-12 internal combustion engine modified to run on either hydrogen or gasoline. It is perhaps unique in that it uses liquid hydrogen (LH_2) rather than compressed hydrogen gas (CHG) as a fuel. This feature has required the development of a specialized

Table 20.4: Specifications of two hydrogen internal combustion engine (ICE) vehicles.

manufacturer	model	fuels	hydrogen range [km]	gasoline range [km]
BMW	Hydrogen 7	LH$_2$/gasoline	200	480
Mazda	RX-8 RE	CHG/gasoline	100	530

© Cengage Learning 2015

© ZUMA Press, Inc./Alamy

Figure 20.6: BMW Hydrogen 7 powered by a two-fuel internal combustion engine that runs on gasoline or hydrogen (LH$_2$).

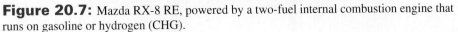

© BG Motorsports/Alamy

Figure 20.7: Mazda RX-8 RE, powered by a two-fuel internal combustion engine that runs on gasoline or hydrogen (CHG).

insulated hydrogen fuel tank with a suitable filling mechanism. Some of the specifications of this vehicle are given in Table 20.4.

Mazda has modified the rotary engine RX-8 for use as a two-fuel vehicle running on either hydrogen or gasoline. They claim that the details of the combustion cycle in the rotary (Wankel) engine make it particularly suited to burning hydrogen. The Mazda RX-8 RE (Figure 20.7) uses compressed hydrogen gas as a fuel. This is a common feature of virtually all (except the BMW) hydrogen-powered vehicles and has required the development of safe and light high-pressure gas cylinders. Some of the specifications of the Mazda RX-8 RE are shown in Table 20.4.

Clearly, from the information in the table, both vehicles are primarily gasoline-powered vehicles with supplementary hydrogen power. This fact is indicated by the relative ranges of the vehicles on these two fuel sources and is a direct consequence of the difficulty in designing an internal combustion vehicle with sufficient usable range utilizing hydrogen as a sole fuel source. Hydrogen may be best utilized as a vehicle fuel in fuel cells, as discussed in the next section.

20.6 Fuel Cells

A fuel cell is a device that catalyzes the reaction given in equation (20.1) and that is thereby able to produce energy (Figure 20.8). The reaction at the anode is

$$H_2 \rightarrow 2H^+ + 2e^-. \tag{20.14}$$

This corresponds to the ionization of two hydrogen atoms. The liberated electrons travel through an external circuit, where they can do work, and the hydrogen ions travel into the electrolyte. At the cathode, the two electrons combine with an oxygen atom according to the reaction

$$2e^- + (1/2)O_2 \rightarrow O^{--}, \tag{20.15}$$

and release an oxygen atom with two excess electrons into the electrolyte. This oxygen atom combines with the hydrogen from the anode to form water:

$$2H^+ + O^{--} \rightarrow H_2O. \tag{20.16}$$

In most cases, it is suitable to use air to provide the oxygen at the cathode.

The concept of the fuel cell was first proposed by the German physicist C. F. Schönbein in 1838. The first operational fuel cell was constructed in 1843 by the Welsh scientist W. R. Grove; however, it was not until the 1950s that practical fuel cells were developed. In 1959, the farm equipment company Allis-Chalmers constructed a tractor that utilized a 15-kW fuel cell. Over the years, a variety of different types of fuel

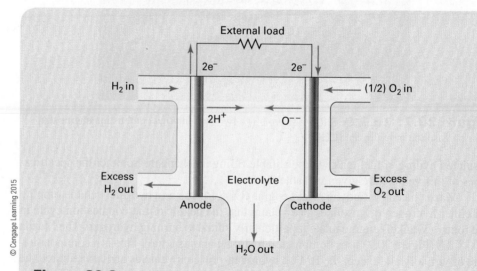

Figure 20.8: Schematic diagram of a generic hydrogen fuel cell.

Table 20.5: Properties of the common varieties of fuel cells.

type	electrolyte	operating temperature (°C)	power density (kW/m²)	typical power output (kW)	lifetime (10³ h)	efficiency (%)
phosphoric acid	H_3PO_4	~200	0.2	< 200	40	40
alkaline	KOH	−40 to 60	0.25	0.3 to 12	20	70
molten carbonate	K_2CO_3 or Na_2CO_3	~800	0.15	< 2000	40	60
solid oxide	ZrO_2	600–1000	0.3	100	40	60
solid polymer	PEM	60–80	0.5	50–250	40	80

cells have been developed, and this section reviews the major types, their potential applications, and their advantages and disadvantages. Table 20.5 gives properties of the common varieties of fuel cells. These differ primarily in the electrolyte used. Most individual fuel cells produce an output voltage of about 0.86 V. Combinations of cells are arranged in series and/or parallel to provide the voltage and current needed for a particular application. The different types of fuel cells are discussed below.

20.6a Phosphoric Acid Fuel Cells

The earliest fuel cells produced in the 1800s were similar in design to today's phosphoric acid fuel cells. They tend to be somewhat expensive, as are most fuel cells, because they use platinum as the catalyst. Recent research on this and other types of fuel cells is aimed at developing new materials with the aim of reducing their cost. Phosphoric acid fuel cells are available commercially and have been used for a number of applications. Because of the relatively low power density, these devices tend to be fairly large and heavy. For this reason, they are most applicable for stationary applications, such as backup electrical power for hospitals, police stations, and the like. One advantage of phosphoric acid fuel cell is that they are fairly insensitive to impurities in the hydrogen fuel. For this reason, hydrogen can be produced by less expensive processes (than electrolysis). Like many fuel cells, phosphoric acid fuel cells operate at an elevated temperature. This has certain disadvantages because it is necessary to utilize some energy to heat the cell in order to warm it up before it can be used. This feature may be most disadvantageous for applications, such as automobiles, where quick-starting capabilities are desirable. However, one advantage of the elevated temperature operation of this and some other fuel cells is that the water produced is actually steam. In some applications, the energy content of this steam may be utilized either directly as a source of heat or for the production of additional electricity by traditional generation methods. If the heat content of the steam is fully utilized, the overall efficiency of a phosphoric acid fuel cell can be as high as 85%. This is referred to a cogeneration, analogous to the discussion in Chapter 17.

20.6b Alkaline Fuel Cells

Alkaline fuel cells have been utilized by NASA as a source of energy for space vehicles. Although the need for a platinum catalyst makes these fuel cells expensive, cost is a secondary consideration for these applications. Also, these fuel cells are best suited to relatively low-power applications, which, again, is acceptable for spacecraft use. One of the most attractive features of these fuel cells is that the water produced is of sufficiently high purity that it is drinkable. Thus, the fuel cell provides a source of both electricity and drinking water for the astronauts. The ability to operate over a wide range of low temperatures is also an advantage for space applications.

20.6c Molten Carbonate Fuel Cells

As seen in the table, these fuel cells can be quite large and can provide considerable power. For this reason, they may be quite suitable for stationary backup power applications. An additional advantage (as well as a disadvantage) is that they can utilize hydrogen from gaseous fossil fuels. This means that some natural fuels (i.e., fossil fuel products such as methane) can be used, as well as hydrogen produced by cheap-and-dirty methods. However, it also means that the by-products of the cell are hydrocarbons as well as water. Thus, these fuel cells contribute to the release of greenhouse gases. Because they operate at elevated temperatures, cogeneration can be used, and efficiency can be increased to about 85%.

20.6d Solid Oxide Fuel Cells

The operating characteristics of the solid oxide fuel cell are, in many respects, similar to those of the molten carbonate fuel cell. They also can utilize hydrocarbon-based sources of hydrogen, such as natural gas or methane. They tend to be fairly expensive (per kilowatt-hour of output).

20.6e Solid Polymer Fuel Cells

These fuel cells use a polymer electrolyte membrane (PEM) and are often referred to as polymer electrolyte membrane fuel cells. They are probably the best choice for automobiles and are used in most prototype and limited-production fuel cell vehicles. The reasons for this include the fact that they can produce the power necessary for vehicle propulsion, and they have a high power density, so they are (comparatively) fairly light.

Other features that make them attractive for vehicle use include the facts that they can operate at low temperatures, that they can respond quickly to a varying load, and that they have a high efficiency. One disadvantage is that they require fairly high-purity hydrogen (but as a result produce virtually no pollution or greenhouse gases). Another interesting feature of the PEM fuel cell is that it is reversible; that is, when hydrogen and oxygen are supplied as a fuel, it produces electricity and water. When water and electricity are supplied, then the system functions as an electrolysis cell and produces hydrogen and oxygen. This feature makes them attractive for remote power systems when they can produce and store electricity reversibly. This ability may be appropriate for use in conjunction with solar photovoltaic or wind-generated electricity. Ballard Power Systems in British Columbia, Canada, has been one of the leaders in the development of PEM fuel cells for vehicles and other uses. They provide cells for a number of vehicle manufacturers, although some vehicle manufacturers utilize in-house–produced cells. An example of a fuel cell installed in a prototype vehicle is shown in Figure 20.9.

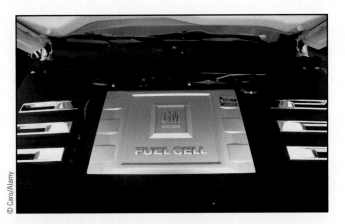

Figure 20.9: A fuel cell installed in the engine compartment of a prototype passenger vehicle.

20.7 Fuel Cell Vehicles

Many major automobile manufacturers have programs for the development of fuel cell vehicles. The electricity produced by the fuel cell is used to power the vehicle using an electric motor. These vehicles are virtually all hybrid hydrogen/electric vehicles. Because the electric drive system is in the vehicle, it is convenient to utilize a supplementary electric storage mechanism to increase the vehicle's range and efficiency. This storage mechanism may consist of batteries or supercapacitors. The batteries or supercapacitors may be charged externally and/or charged through regenerative braking. Many recent fuel cell vehicles are built on existing gasoline vehicle platforms, and all utilize compressed hydrogen gas as a storage mechanism. Some examples of fuel cell vehicles that utilize PEM fuel cells are shown in Figures 20.10 through 20.12. Specifications of these vehicles are given in Table 20.6.

Figure 20.10: The Ford Focus FCV showing the fuel cell system in the engine compartment. NiMH batteries are used as a secondary energy source.

Figure 20.11: The Honda FCX Clarity utilizes Li-ion batteries as a secondary energy source.

Figure 20.12: The Kia Borrego utilizes supercapacitors as a secondary energy source.

Table 20.6: Specifications of some hydrogen fuel cell vehicles.

make	model	hydrogen storage	mass [kg]	power at wheels [kW]	supplementary energy storage	range [km]
Ford	Focus FCV	25 MPa CHG	1727	65	NiMH battery	280
Honda	FCX Clarity	34.5 MPa CHG	1625	95	Li-ion battery	430
Kia	Borrego FCEV	69 MPa CHG	2250	109	supercapacitor	680

© Cengage Learning 2015

20.8 Hydrogen: Present and Future

Table 20.6 allows for an analysis of the characteristics of fuel cell vehicles. A comparison with traditional gasoline vehicles (Table 19.3) and battery electric vehicles (Tables 19.6 and 19.7) is informative. For a comparison with gasoline vehicles, it is probably most appropriate to compare the Honda FCX Clarity with the Honda Civic and to compare the Kia Borrego FCEV with the Land Rover Range Rover. The Clarity has less range and is heavier than the Civic, while the Kia has much less power than the Land Rover. Fuel cell vehicles typically have greater range and similar or more power compared with past electric vehicles, like the GM EV1 or the Toyota Rav4 EV. A comparison with current or probable future battery electric vehicles, such as the Tesla Roadster or Model S (Chapter 19), suggests that the ranges are similar, but pure electric vehicles win in a power comparison. A major selling point in favor of hydrogen vehicles is the time for refueling and general claims of the nonpolluting character of hydrogen as a fuel. Progress in the production of hydrogen-powered vehicles is perhaps evidenced by several manufacturers' marketing, on a limited basis, of hydrogen ICE or fuel cell vehicles and a number who have plans for marketing them in the relatively near future.

Nevertheless, hydrogen for transportation is a new technology and there are a number of concerns that need to be addressed. Principal among these are safety, infrastructure, and cost.

From a safety standpoint, hydrogen's reputation is greatly influenced (in a negative way) by the Hindenburg disaster (Figure 20.13). Recent studies have shown that it was, in fact, a combustible coating on the airship's exterior that was primarily responsible for the disaster and not the hydrogen fuel itself. Hydrogen, however, is a flammable substance, and safety is a concern, as it is for gasoline. In the case of gasoline, appropriate safety standards have been developed, and the public perception is that the risks of using gasoline as a vehicle fuel are acceptable.

U.S. Navy

Figure 20.13: The German Zeppelin Hindenburg burned on May 6, 1937 in Lakehurst, New Jersey. Of the 97 people on board, 35 people and one person on the ground died in the incident.

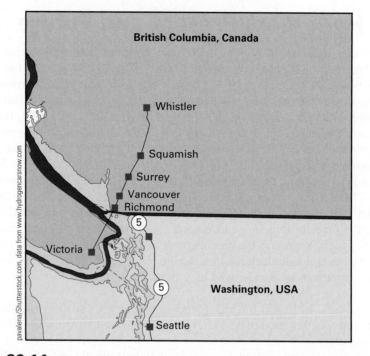

Figure 20.14: Hydrogen fueling station locations in British Columbia and the U.S. Northwest.

The infrastructure for the use of hydrogen as a transportation fuel involves the establishment of hydrogen stations to supplement or eventually replace gasoline and diesel stations. In fact, hydrogen and gasoline filling facilities could be integrated into common stations in the same way that many gas stations currently sell propane. Movement in this direction is already taking place and hydrogen fueling stations are currently located in the United States (led by California), Canada, Australia, Hong Kong, Japan, South Korea, Singapore, Taiwan, and at least ten European countries (led by Germany). The locations of some hydrogen fueling stations worldwide are shown on the maps in Figures 20.14 and 20.15. A typical station is shown in Figure 20.16.

The cost of developing new technologies tends to be high, although in the long term, technical developments and demand are expected to bring marketable products within the reach of the general public. A cost assessment for hydrogen vehicles is difficult, as it has been (and probably still is) for battery electric vehicles. Honda is in the early stages of marketing the FCX Clarity. The lease price for this vehicle is around US$600 per month. This far exceeds the lease price of a gasoline-powered Honda Civic.

This discussion may suggest that hydrogen ICE vehicles and hydrogen fuel cell vehicles are a technology that may provide environmentally friendly transportation and that could compete with battery electric vehicles (and in some ways may be more attractive). However, in May 2009, the U.S. Department of Energy proposed the discontinuation of government support for the development of hydrogen vehicles. The relative efficiency of various transportation technologies (discussed in the next section) is important in understanding why this proposal has been made.

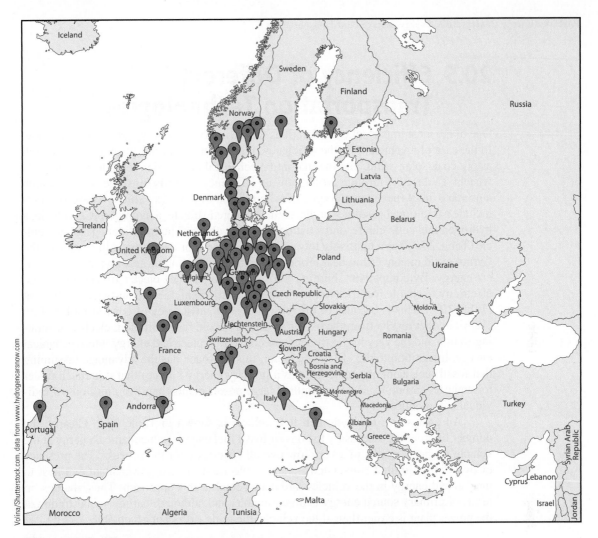

Figure 20.15: Hydrogen fueling station locations in Europe.

Figure 20.16: Hydrogen fueling station. This facility in Reykjavík, Iceland opened in 2003 and was the world's first public hydrogen fueling station.

20.9 Efficiency of Different Transportation Technologies

One of the most important considerations for the use of any energy source is efficiency. In the case of a vehicle, it is the efficiency of converting a primary source energy such as oil, coal, solar, wind, and the like into mechanical energy delivered to the vehicle's wheels. Tables 20.7 through 20.10 show an efficiency analysis for some ways in which a primary fossil fuel energy source may be utilized to power a vehicle. These analyses include a traditional gasoline- or diesel fuel-powered vehicle, a hydrogen ICE vehicle, a hydrogen fuel cell vehicle, and a battery electric vehicle. The traditional gasoline vehicle (Table 20.7) has a relatively low efficiency because of the intrinsic limitation of the Carnot cycle in converting heat produced by the combustion of gasoline into mechanical energy. By comparison, burning a fossil fuel to produce electricity and using that electricity to charge batteries to run a battery electric vehicle is about twice as efficient (Table 20.8). This is because of the reasonably high efficiency of converting heat into electricity that results from the use of an effective cold reservoir at an electric generating station and the very high efficiency of converting electrical energy into mechanical energy. In addition to the efficiency factor, this approach has the advantage (assuming that fossil fuels continue to be used as a primary energy source) that it is much easier to sequester carbon from a few hundred large power plants than from millions of individual automobiles.

An analysis of the hydrogen ICE vehicle is shown in Table 20.9. Clearly, the number of processes necessary to convert fossil fuel energy to mechanical energy at the vehicle's wheels results in a very low overall efficiency for the process. The presence of two Carnot limited conversions—heat to electricity in the power plant and heat to mechanical energy in the vehicle—are detrimental to this approach. Thus, the cost in terms of primary source energy is unreasonable, and, although carbon sequestration at the power plant is easier than at the vehicle, the trade-off is not acceptable.

Based on data from R. A. Dunlap, *Energy and Environment Research* "A simple and objective carbon footprint analysis for alternative transportation technologies" 3 (2013): 33–39.

Table 20.7: Efficiency analysis for gasoline-powered internal combustion engine vehicle showing net efficiency for conversion of primary energy (gasoline) to mechanical energy delivered to the vehicle's wheels.

process	efficiency (%)
fossil fuel → mechanical energy	17
net efficiency	17

Based on data from R. A. Dunlap, *Energy and Environment Research* "A simple and objective carbon footprint analysis for alternative transportation technologies" 3 (2013): 33–39.

Table 20.8: Efficiency analysis for battery electric vehicle showing net efficiency for conversion of primary energy (oil of coal) to mechanical energy delivered to the vehicle's wheels.

process	efficiency (%)
fossil fuel → electricity	40
electricity → mechanical energy	85
net efficiency	34

Based on data from R. A. Dunlap, *Energy and Environment Research* "A simple and objective carbon footprint analysis for alternative transportation technologies" 3 (2013): 33–39.

Table 20.9: Efficiency analysis for hydrogen-powered internal combustion engine vehicle showing net efficiency for conversion of primary energy (oil or coal) to mechanical energy delivered to the vehicle's wheels.

process	efficiency (%)
fossil fuel → electricity	40
electricity → hydrogen gas	70
hydrogen gas → CHG or LH_2	80
CHG or LH_2 → mechanical energy	17
net efficiency	4

Based on data from R. A. Dunlap, *Energy and Environment Research* "A simple and objective carbon footprint analysis for alternative transportation technologies" 3 (2013): 33–39.

Table 20.10: Efficiency analysis for hydrogen fuel cell-powered vehicle showing net efficiency for conversion of primary energy (oil or coal) to mechanical energy delivered to the vehicle's wheels.

process	efficiency (%)
fossil fuel → electricity	40
electricity → hydrogen gas	70
hydrogen gas → CHG	80
CHG → electricity	70
electricity → mechanical energy	90
net efficiency	14

An analysis of the fuel cell vehicle (Table 20.10) shows that this is a somewhat better approach. However, the overall efficiency is barely comparable to that of the gasoline ICE vehicle and is less than half that of the battery electric vehicle.

Thus, if we continue to use fossil fuels as a primary energy source, the development of battery electric vehicles would seem to be a better approach than fuel cell or hydrogen ICE vehicles. This approach has the advantage of reducing carbon emissions by a factor of two, making carbon sequestration easier and allowing for the use of relatively plentiful coal for electricity generation.

If more environmentally friendly primary sources of energy (e.g., wind, solar, etc.) are used and the concern for carbon emissions is eliminated, then the comparison of efficiencies for battery electric vehicles and fuel cell or hydrogen ICE vehicles still shows that the net energy to the wheels, compared to the total primary energy input, favors batteries. This feature results from the fundamental fact that it is much more efficient to get energy into and out of batteries than into and out of hydrogen, and this basic principle is unlikely to change in the foreseeable future. However, the production of hydrogen by biological mechanisms may be an approach that can help to make hydrogen more attractive.

With respect to this type of assessment, *Technology Review* (March–April 2007) concluded (www.technologyreview.com/Energy/18301/):

In the context of the overall energy economy, a car like the BMW Hydrogen 7 would probably produce far more carbon dioxide emissions than gasoline-powered cars available today. And changing this calculation would take multiple breakthroughs—which study after study has predicted will take decades, if they arrive at all. In fact, the Hydrogen 7 and its hydrogen-fuel-cell cousins are, in many ways, simply flashy distractions produced by automakers who should be taking stronger immediate action to reduce the greenhouse-gas emissions of their cars.

ENERGY EXTRA 20.2
Carbon footprint analysis for transportation technologies

The efficiencies that have been derived for different transportation technologies in this chapter certainly provide some guidance for the assessment of the environmental impact of these approaches. However, this analysis can be taken one step further by actually calculating the mass of carbon dioxide emitted per kilometer traveled. A fairly straightforward approach to calculating this quantity for the comparison of different types of vehicles is described here.[1] The amount of CO_2 emitted per km traveled is expressed as

$$\frac{kg(CO_2)}{km} = \frac{kg(CO_2)}{(MJ)_p} \times \frac{(MJ)_p}{(MJ)_w} \times \frac{(MJ)_w}{km},$$

where p is the primary energy, and w is the energy delivered to the vehicle's wheels. The first term on the right-hand side of the equation represents the amount of CO_2 generated per megajoule of primary energy consumed. The second term is the inverse of the efficiency from primary energy to wheel energy, as calculated in this chapter. The final term on the right-hand side is the average energy to the wheels needed to move the vehicle 1 km. This analysis does not consider the carbon footprint of nonfossil fuel–generated electricity or a life cycle analysis of vehicle materials. It also does not consider the possibility of sequestrating carbon from fossil fuel–generating stations. Because these effects tend to average out somewhat, this analysis is a reasonable comparison of different vehicle technologies.

For a gasoline internal combustion engine- (ICE-) powered vehicle, gasoline is very close to a primary energy source, whereas for battery electric vehicles (BEVs) or hydrogen-powered vehicles the

primary energy sources are first used to produce electricity. Thus $kg(CO_2)/(MJ)_p$ depends on how electricity is produced. A rough breakdown of current electricity production in the United States follows:

fuel	% electricity
coal	44.9
natural gas	23.4
petroleum	1.0
nonfossil	30.7

© Cengage Learning 2015

The CO_2 emission per MJ for different fossil fuels is obtained from a simple analysis of their energy content and is shown in the following table. The CO_2 emissions from nuclear, hydroelectric, and alternative energy sources is considered to be zero.

fossil fuel	kg (CO_2)/MJ
coal (~pure carbon)	0.11
natural gas (methane, CH_4)	0.055
heavy hydrocarbons (>6 C/molecule)	0.069

Based on data from R. A. Dunlap, Energy and Environment Research "A simple and objective carbon footprint analysis for alternative transportation technologies" 3 (2013): 33–39.

A weighted average of these values according to how electricity is produced in the United States is 0.063 kg$(CO_2)/(MJ)_p$. For a typical family vehicle, the energy at the wheels per kilometer traveled was shown in the last chapter to be about 0.6 $(MJ)_w$/km. Combining all of this information, the CO_2 emission per kilometer is obtained.

technology	kg $(CO_2)/(MJ)_p$	$(MJ)_p/(MJ)_w$	$(MJ)_w$/km	kg (CO_2)/km
gasoline ICE	0.069	5.9	0.6	0.24
H_2 ICE	0.063	25	0.6	0.94
H_2 fuel cell	0.063	7.1	0.6	0.27
BEV	0.063	2.9	0.6	0.11

© Cengage Learning 2015

Continued on page 535

Energy Extra 20.2 continued

This analysis shows that the net environmental effect of BEVs is quite positive (compared with gasoline vehicles), whereas fuel cell vehicles are about neutral. Hydrogen ICE vehicles are seen to have a very negative environmental effect. Incremental improvements to the results for nonfossil fuel vehicles can be obtained by increasing energy conversion efficiencies (i.e., decreasing $(MJ)_p/(MJ)_w$) and by increasing vehicle efficiencies (i.e., decreasing $(MJ)_w/km$). However, it is through the shift in electricity-generating technology away from fossil fuels [i.e., the reduction of $kg(CO_2)/(MJ)_p$] that the most substantial improvements can be made. However, H_2 ICE vehicles would require that at least 85% of electricity would come from nonfossil fuel sources in order to be an environment improvement over gasoline vehicles (although from a financial standpoint this would not be an attractive alternative). A reduced value of $kg(CO_2)/(MJ)_p$ already exists, and puts alternative transportation technologies in a better light compared to gasoline-powered vehicles, in countries like France and Canada, where a large fraction of electricity comes from nonfossil fuel sources.

[1]R. A. Dunlap, *Energy and Environment Research* **3** (2013): 33.

Topic for Discussion

Although the efficiency of energy use is an important consideration, it is also important to understand the financial aspects of any technological development and how results can be marketed. Consider, as an example, the fuel cost to the driver (in dollars per megajoule delivered to the wheels of a vehicle) for a gasoline-powered BMW 740Li. For comparison, estimate the cost (in U.S. dollars per megajoule) for a fossil-fuel-free BMW Hydrogen 7, running on hydrogen produced by the electrolysis of water using wind-generated electricity. Discuss the economic viability of this technology.

Based on R. A. Dunlap, *Energy and Environment Research* "A simple and objective carbon footprint analysis for alternative transportation technologies" **3** (2013): 33–39.

The U.S. Department of Energy (DOE) is continuing research into hydrogen fuel cells for stationary use. These systems may be used for backup power or as an energy storage mechanism in remote locations. The criteria for these applications are different, in many ways, from those that are appropriate for vehicles used by the general public. Portability is typically not a concern, and cost may be viewed as a secondary factor compared with convenience and reliability.

20.10 Summary

Hydrogen is not a primary energy source but merely an energy storage mechanism. Energy is required to produce hydrogen (typically from water), and this energy is recovered (at least in part) when the hydrogen is recombined with oxygen, either by combustion or in a fuel cell. This chapter reviewed the properties of hydrogen and the pros and cons of using it as a storage method.

Hydrogen is a gas at standard temperature and pressure. Its heat of combustion is 142 MJ/kg, compared with about 45 MJ/kg for gasoline. Unfortunately, its density is very low, leading to a very low energy-per-unit volume. It can be liquefied by cooling to about 20 K or compressed to increase its density. Even so, its energy per unit volume is less than one-third that of gasoline. This chapter reviews the common methods of producing hydrogen.

Hydrogen is released in a variety of chemical reactions and by the decomposition of water into hydrogen and oxygen. Some reactions that yield hydrogen involve the release of CO_2 and are not desirable from an environmental standpoint. The electrolysis of water is a convenient and common method of hydrogen production that has no direct adverse environmental effects.

As described in this chapter, hydrogen is an attractive storage mechanism for energy for transportation use because its use to produce heat by combustion or to produce electricity by means of a fuel cell, yields no greenhouse gases. As noted, however, its low density requires that it be liquefied or compressed to high pressures for portable use. Liquid hydrogen must be stored in an insulated tank to minimize evaporation (although this cannot be eliminated entirely). Compressed hydrogen gas requires storage in sufficiently strong cylinders to withstand the high pressure. Either approach involves careful consideration in vehicle design.

This chapter has provided an overview of the operation of a fuel cell. This device combines hydrogen fuel with oxygen from the atmosphere to produce water and electricity. The chapter summarized the different types of fuel cells and showed that polymer electrolyte membrane fuel cells are probably the most suitable for vehicle use.

A description of a variety of prototype hydrogen ICE and fuel cell vehicles was presented in the chapter. The ICE vehicles are modified gasoline ICE vehicles and are typically hybrid vehicles because they carry gasoline to supplement the somewhat limited hydrogen range that results from the low energy-per-unit volume of hydrogen.

Fuel cell vehicles are also typically hybrid vehicles because they incorporate batteries or supercapacitors to increase the range and store electrical energy generated by regenerative braking.

A full analysis of hydrogen as a fuel must include how the hydrogen is produced and the overall efficiency of the energy production process from primary energy source to usable end-product energy. This has been considered in detail in this chapter. Clearly, the route by which hydrogen is produced, stored, and utilized is an important factor in determining the efficiency, as well as its environmental impact. If hydrogen is produced by chemical methods, the carbon footprint of this process is important in determining the overall greenhouse gas emissions. If hydrogen is produced by electrolysis, then the source of the electricity used for the process is important in determining the carbon emissions. The low efficiency for transportation using hydrogen ICE technology makes this approach of questionable economic viability, and any carbon emissions during the process become accentuated. Fuel cell technology is more promising and warrants further consideration as a future approach to transportation. The development of new technologies for the production of hydrogen would be an important step in making this technology competitive with battery electric vehicles and improvements in fuel cell efficiency and cost could lead the way to commercial utilization.

Problems

20.1 Hydrogen gas is burned to provide heat for a typical North American home (see Chapter 2 for heating requirements). If the hydrogen is stored in a spherical tank at a pressure of 80 MPa, what would be the diameter of the tank needed to supply the average monthly heating requirement? Assume a typical furnace efficiency (Chapter 17).

20.2 Hydrogen gas is burned according to the reaction

$$2H_2 + O_2 \rightarrow 2H_2O$$

to provide heat for a typical North American home (see Chapter 2 for heating requirements). What would be the average daily water production in liters? Assume a typical furnace efficiency (Chapter 17).

20.3 If hydrogen followed an ideal gas law at all pressures, calculate the pressure at which the density of CHG would exceed that of LH_2 ($71 \ kg/m^3$).

20.4 Assume that the cost of hydrogen per unit energy is the same as gasoline ($0.03/MJ). What is the loss (in dollars per day) due to evaporation from a 100-L LH_2 fuel tank?

20.5 **(a)** Given a fuel cell efficiency of 75% and a required energy to the wheels of 0.6 MJ/km, calculate the mass of hydrogen fuel required to give a vehicle a range of 300 km.
   (b) Calculate the fuel volume if hydrogen is stored in the form of (i) liquid hydrogen, (ii) CHG at 50 MPa, and (iii) TiH_2.

20.6 A solid oxide fuel cell operates at 55% efficiency using methane as a fuel. Calculate the mass (in grams) of water and carbon dioxide produced for every kilowatt-hour electrical of output. Note the heat of combustion of methane is 50 MJ/kg (Chapter 3).

20.7 Calculate the percentage of improvement in the energy density of a CHG fuel tank if the pressure is increased from 35 to 70 MPa.

20.8 For stationary applications, mass and volume considerations for a fuel source are much less important than they are for use in transportation. It therefore may be the most reasonable to sacrifice compactness to minimize requirements for expensive advanced technologies. Consider a plan to provide power to a small city with a population of 100,000, using hydrogen fuel cells with an efficiency of 60%. The average net per-capita power consumption is 6 kW for electricity, heat, and power for an electric vehicle.
   (a) Calculate the volume of hydrogen needed to provide the city's energy for one month if the hydrogen is stored at a (relatively low) pressure of 1 MPa.
   (b) If the city has a land area of 100 km^2, what fraction of this area would need to be devoted to hydrogen storage if the storage tanks were vertical cylinders 3 m in diameter and 5 m high and were packed tightly together in a square array?

Bibliography

A. I. Appleby and F. R. Foulkes. *Fuel Cell Handbook.* Krieger, Malabar (1993).

G. J. Aubrecht II. *Energy: Physical, Environmental, and Social Impact,* 3rd ed. Pearson Prentice Hall, Upper Saddle River, NJ (2006).

D. R. Blackmore and A. Thomas. *Fuel Economy of the Gasoline Engine.* Macmillan, London (1977).

B. K. Hodge. *Alternative Energy Systems and Applications.* Wiley, Hoboken, NJ (2010).

P. Hoffmann. *Tomorrow's Energy: Hydrogen, Fuel Cells, and the Prospects for a Cleaner Planet.* MIT Press, Cambridge, MA (2001).

T. Koppel. *Powering the Future: The Ballard Fuel Cell and the Race to Change the World,* Wiley, New York (1999).

K. Kordesch and G. Simader. *Fuel Cells and Their Applications.* VCH Publishers, New York (1996).

P. Kruger. *Alternative Energy Resources: The Quest for Sustainable Energy.* Wiley, Hoboken (2006).

J. Larminie and J. Dicks. *Fuel Cell Systems Explained.* Wiley, Chichester, United Kingdom (2000).

J. MacKenzie. *The Keys to the Car: Electric and Hydrogen Vehicles for the 21st Century.* World Resources Institute, Washington, DC (1994).

C. Ngô and J. B. Natowitz. *Our Energy Future: Resources, Alternatives and the Environment.* Wiley, Hoboken, NJ (2009).

R. O'Hayre, S. W. Cha, W. Colella, and F. B. Prinz. *Fuel Cell Fundamentals.* Wiley, Hoboken, NJ (2006).

V. Quaschning. *Renewable Energy and Climate Change.* Wiley, West Sussex, United Kingdom (2010).

J. J. Romm. *The Hype About Hydrogen: Fact and Fiction in the Race to Save the Climate.* Island Press, Washington, DC (2005)

D. Sperling and J. Cannon (Eds.). *The Hydrogen Energy Transition: Moving Toward the Post-Petroleum Age.* Elsevier, Amsterdam (2004).

C. Spiegel. *Designing and Building Fuel Cells.* McGraw-Hill, New York (2007).

S. Srinivasan. *Fuel Cells.* Springer, New York (2006).

PART VI

The Future

Photovoltaic collectors with pumped hydroelectric
storage in Geeshacht, Schleswig-Holstein, Germany

© Jörg Müller/Alamy

Future Prospects and Research and Design Projects

21.1 Introduction

Humanity has been dependent on fossil fuels to fulfill its energy needs for well over a century. The future of fossil fuel use is ultimately limited by the availability of oil, natural gas, coal, and other petroleum resources. Traditional oil resources are likely to provide energy for only a few more decades, and coal and natural gas produced by fracking and some alternative petroleum resources, such as oil sands, may last somewhat longer. There is, however, concern that the continued use of these resources will cause irreparable environmental damage. In addition to the clear relationship between greenhouse gas emissions and the quantity of energy produced by the combustion of fossil fuels, other adverse environmental consequences are associated with fossil fuel production and use, particularly for the many approaches to enhanced fuel recovery, such as shale oil extraction and natural gas well fracking. Thus, there are obvious environmental incentives to alternative energy development. Understanding future energy scenarios requires an understanding of future energy needs and possibilities for future energy production.

This concluding chapter provides an overview of the key points covered in this text in terms of possible technological approaches to future energy production, as well as methods for the efficient conservation, storage, and distribution of energy. The projects in this chapter provide an opportunity to take a broad look at various technologies and compare the advantages and disadvantages of different approaches in particular circumstances. In many cases, additional resources from the Internet, texts, and other sources are needed to provide the data necessary to undertake the project.

21.2 Approaches to Future Energy Production

Our energy needs over the years have increased, and this trend is expected to continue for some period into the future. Much work has been done on trying to predict future energy use, because knowing our future needs is beneficial in understanding which energy technologies are most viable for development. Figure 2.13 shows one prediction of future energy needs for the next couple of decades, and indicates a reasonably consistent increase on this timescale. However, predictions for the distant future can be quite variable, and some suggest that future energy use will actually decrease as a result of factors that include improved efficiency of equipment and increased awareness of energy conservation.

Predictions that extend very far into the future become progressively less certain because the effects of assumptions become compounded. Perhaps the most systematic recent approach to future energy prediction was presented in a report from the Intergovernmental Panel on Climate Change (IPCC) that was jointly established by the World Meteorological Organization (WMO) and the United Nations Environment Programme (UNEP). This study involved the prediction of future world energy use until the year 2100 on the basis of 40 different scenarios. Scenario assumptions dealt with factors such as predicted world population, economic growth, technological development, and the degree of cooperative energy management. Figure 21.1 show the extremes of these predictions over the next century or so. The huge range of predictions for the year 2100 is apparent in the figure. In the best-case scenario, energy use in 2100 will be roughly the same as it is today. In the worst case, energy use will be about 5 times what it is today. The best case results from limited population growth, development of new energy-efficient technologies, and rapid economic growth. The worst case results from continuous population growth, poor economy, and lack of social equality.

The distribution of energy use from the different available sources must be considered in the context of overall energy needs, as suggested by Figure 21.1. Predictions concerning the role of various energy sources in future energy production are perhaps even more uncertain than predictions for total energy needs. Certainly, the distribution

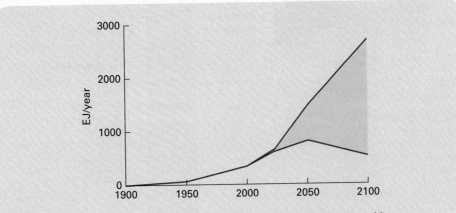

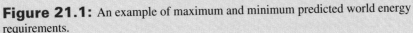

Figure 21.1: An example of maximum and minimum predicted world energy requirements.

of energy use can be no more accurate than the prediction of the total. This uncertainty may also be a feature of predictions of the relative proportions of the various contributions to energy use. Certain energy options—wind, for example, are not unlimited and may satisfy most of our energy needs if those needs do not exceed the availability. If the need is higher than the availability, the distribution of energy sources needs to be reevaluated. In any case, the evaluation of possible future energy economies is an interesting exercise that sheds some light on the viability of the options discussed in this text, but they should not be taken too literally.

Predictions for the future must, of course, start with what exists today. A typical current breakdown of energy sources (in the United States) is illustrated in Figure 21.2. It is inevitable that, at some point in the future, fossil fuel use will decrease. This decrease must be at the expense of the increased use of nuclear and/or renewable energy sources. Predictions concerning nuclear power that were made around 1970 were overly optimistic, and predictions concerning the use of OTEC have certainly not materialized so far. In one case, political factors played an important role in the development of nuclear energy, and in the other case technological advances fell short of expectations. The significant implementation of alternative energy sources comes with a substantial economic cost, but the failure to implement these changes comes with a substantial environmental cost. There is a reluctance to change things that we are accustomed to, and, up to some point, there is a financial incentive to avoid the development of a new infrastructure.

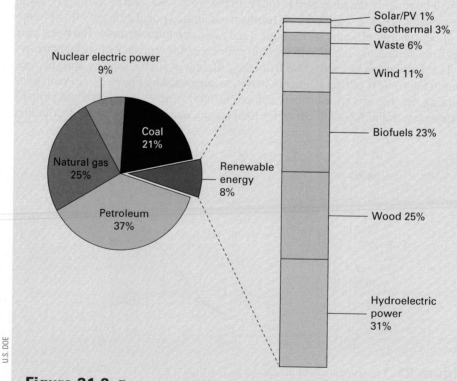

U.S. DOE

Figure 21.2: Energy sources in the United States in 2009.

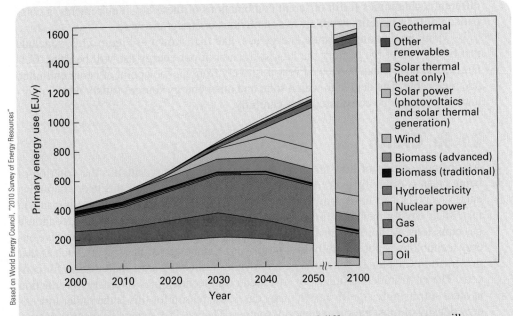

Based on World Energy Council, "2010 Survey of Energy Resources"

Figure 21.3: One prediction of how the utilization of different energy sources will develop over the next century.

The most reasonable options for a particular region are a function of a number of factors, including geography and climate. An appropriate approach to alternative energy in the U.S. Midwest would be quite different than it would be in Brazil or Scotland. Many regions have some commonality in resources; sunlight and wind are available in varying amounts almost everywhere. However, other regions may have more unique resources available, such as hydroelectric, tidal, or wave energy. Some energy options may not be mutually exclusive but may complement each other, as rows of photovoltaic collectors between the turbines of a wind farm.

It is likely that a combination of approaches will be taken in the foreseeable future and that our energy economy will be a mix of technologies. The exact nature of our energy economy is a complex matter that will develop over time. One prediction for a future energy economy is from the World Energy Council, shown in Figure 21.3. This prediction has some interesting features. The total energy use, 1600 EJ per year, falls almost exactly in the middle of the range of predictions from Figure 21.1. In this prediction, less than 20% of our energy comes from fossil fuels. The majority (more than 60%) comes from solar. Implementation of such a future energy mix will require ambitious efforts to develop new solar technologies, as well as a dedicated approach to freeing ourselves from the cheap and easy route of burning more coal for the next few hundred years.

21.3 Key Considerations

As pointed out in the introduction to Part IV of the text, an alternative energy technology must ideally be *c*lean, *u*nlimited, *r*enewable, *v*ersatile, and *e*conomical. It is clear, at this time at least, that there is no perfect solution to our future energy needs. Although different technologies fulfill different combinations of these key requirements, a comparison of their benefits is not always so obvious or straightforward.

Many predictions of future energy use, like that shown in Figure 21.3, conclude that by the end of the century the largest fraction of our energy use will be solar. Certainly solar energy can be viewed very positively from the viewpoint of being unlimited and renewable. The degree to which solar and other energy sources satisfy other important criteria must be considered very carefully.

21.3a Clean

An objective rating of how an alternative energy technology satisfies the criterion of being clean is often difficult. As has been discussed in Part IV of this text, alternative energy sources can have a measurable or even significant environmental impact. In fact, such environmental influences inevitably include some quantity of greenhouse gas emissions. A thorough analysis of the environmental effects of each alternative energy technology is not generally straightforward. Quantitative assessments, such as that presented in Chapter 16 for biofuels, often yield results that are a major cause for concern. The environmental impact of so-called clean technologies for energy production is often surprisingly significant. In many cases, the reason for this is the rather low energy density of most alternative energy sources. The need for extensive manufacturing, transportation, maintenance, and related activities contributes to a net environmental impact per unit energy that can be greater than for traditional fossil fuel technologies. The analysis of this impact follows along the same lines as the risk assessment for energy sources presented in Chapter 6.

PROJECT 21.1
CO$_2$ sequestration

Consider the possibility of sequestering CO$_2$ from fossil fuel generating stations in depleted natural gas wells. Begin by summarizing the current production of electricity in the United States. Calculate the volume of CO$_2$ produced annually from the combustion of coal and natural gas in the United States for electric generation. Compare this volume to the volume of natural gas produced from wells in the United States and the volume of CO$_2$ from electric generating stations currently sequestered in this country.

(a) Are efforts to sequester CO$_2$ appropriate to the magnitude of the environmental problem of CO$_2$ emissions?

(b) In this electric generating scenario, is the availability of natural gas wells compatible with the needs for CO$_2$ sequestration?

(c) Would the volume of depleted oil wells in the United States contribute in a substantial way to CO$_2$ sequestration capacity?

(d) Investigate the potential technical difficulties in transporting CO$_2$ to the necessary locations for injection into depleted wells. Make a rough analysis of the distribution of fossil fuel–fired plants in the United States and the distribution of wells. It might be simplest to concentrate on coal-fired plants and natural gas wells. Make an estimate of the infrastructure that would be needed to sequester CO$_2$ by this method.

21.3b Unlimited

The power that is available from alternative energy resources depends on the nature and extent of the resources, and on the existence of a viable technology to utilize these energy sources. In some instances, such as with wind energy, a suitable technology exists that enables us to effectively make use of the resource, although this, in itself, does not ensure that development of the resource is viable economically and environmentally. In other cases, such as with OTEC or salinity gradient energy, further technological advances are clearly needed before an energy resource can be exploited in a practical way. Table 21.1 presents the amount of power available from different alternative energies on the basis of existing technologies or technologies that might conceivably be available in the foreseeable future. It is important to realize that the availability of power from a particular resource is limited by the physical availability of the resource itself, as well as by the availability of the technology to make use of the energy resource. This situation might, for example, arise as a result of the limits to the world's supplies of indium for the production of photovoltaic cells or limits to the supplies of the rare earth elements that are used for magnets in wind turbine generators.

The values in Table 21.1 must be viewed in the context of future energy needs as discussed in Section 21.2. As illustrated in Figure 21.1, annual energy requirements in the range of 1000 to 2000 EJ are reasonable expectations over the next century or so. This requirement amounts to an average power consumption in the range of 30–60 TW. Predictions like those shown in Figure 21.3 can be viewed in the context of the resources given in Table 21.1. The table also shows that solar energy exceeds (by far) the total energy requirements of society and that this is the only alternative energy resource (nuclear not included) that can fulfill all of the world's energy needs. Even the utilization of a fairly small fraction of available solar energy at a relatively low efficiency is sufficient to satisfy humanity's requirements. Biomass and wind can, at least in principle, make a major contribution to possible future energy mixes. Other resources may be locally significant but can contribute a relatively small fraction of total energy needs on a global scale. Thus, on a relevant scale, only solar energy may be considered to be unlimited.

Table 21.1: Power available from different renewable energy sources. Power listed is that which is technologically feasible and economically viable on the basis of technical and scientific capabilities at present or in the foreseeable future.

energy source	power available (TW)
solar	>1000
biomass	~6
wind	~4
tidal/waves/currents	~2
hydroelectric	~1
geothermal	<1
OTEC/salinity gradients	<1

Many alternative energy technologies utilize materials (or particularly elements) with limited availability. While some cases have been made quite apparent in the text, such as lithium for rechargeable batteries or indium for photovoltaic cells, other cases are not so apparent. A somewhat controversial article by David Cohen ["Earth audit," *New Scientist* **194** (2007): 34], overviews the concerns for future resource availability. Although not specifically discussed in this article, the rare earth elements require careful consideration in the context of energy production and utilization, and, as their name implies, these elements are not common. Rare earth–containing materials are of importance largely because of their magnetic properties, and they form a critical component in many alternative energy technologies. Modern wind turbines and battery electric vehicles, for example, depend heavily on magnets based on rare earth elements (mostly neodymium). In fact, Nd-Fe-B is the industry standard for high-energy permanent magnets.

Research the ways in which rare earths are essential for the development of alternative energy technologies. Concentrate particularly on wind turbines and electric vehicles, but do not overlook other possible needs for these materials. How many kilograms of rare earth elements, such as Nd, are needed (for example) for a 1-MW wind turbine or a family-sized BEV? If all the world's energy needs were satisfied by electricity generated by conventional (nonsuperconducting) wind turbines and all passenger vehicles were BEVs, would enough rare earth elements be available? A good place to start researching the world's supplies of various elements is the website http://minerals.usgs.gov/minerals/pubs/commodity/ and the links therein.

21.3c Renewable

All of the alternative energy sources listed in Table 21.1 are basically manifestations of solar energy, except for geothermal and tidal energy. Because the longevity of solar energy is, for all practical purposes, infinite, we would anticipate that the longevity of these energy resources would also be infinite. This is only partly true for two reasons: (1) Changes that affect our ability to exploit the resource and (2) the need for nonrenewable resources to convert primary energy into a usable form can both limit our ability to utilize a resource. In the first case, this might be the reduction in the viability of hydroelectric energy from a particular location due to sedimentation (Chapter 11). In the second case, this might necessitate the recycling of rare materials (Section 21.3b). For other possible energy technologies, these two factors may also be limitations. Some examples are the limited longevity of geothermal energy that results from the fact that thermal energy from geothermal deposits is typically extracted faster than it is replenished (Chapter 15) and limits to the resources of lithium used to breed fuel for a fusion reactor (Chapter 7).

Lithium may be a crucial element in future energy technologies. Summarize the ways in which lithium may be of importance. Discuss the availability of lithium, the requirements for its use in different energy technologies, the longevity of the world's lithium resources for different applications, and the possibility of recycling lithium from these technologies. Specifically contrast the need for lithium for current battery technology and the potential future need for fusion power.

21.3d Versatile

The goal of most alternative energy technologies is the generation of electricity. Although electricity is suitable for many of our energy needs, its utilization for transportation requires careful consideration. In fact, transportation places particular challenges on technology because it requires an energy storage mechanism that must, at a minimum, be portable and have a sufficiently high energy density. Added to this difficulty is the fact that the route from primary energy source to mechanical energy at a vehicle's wheels often involves a number of energy conversions that add up to an unacceptably low overall efficiency. To meet the needs of a suitable transportation technology, the method of energy production, as well as energy storage, must be versatile enough to provide acceptable overall efficiency. From a practical standpoint, this basically means that in the context of current technology nonfossil fuel–based transportation will rely on either biofuels or involve some electricity storage mechanism.

Biofuels have been successful in Brazil but are unlikely to be an effective approach in more temperate countries unless new technologies are developed for utilizing the cellulose component of plant matter. Electric vehicles require an electricity storage mechanism, and at present this means either batteries or hydrogen. The efficiency analysis in Chapter 20 clearly favors batteries, and research into new battery chemistries has progressed in recent years. Lithium resources are somewhat limited. If fusion energy becomes a reality, then fusion reactors will compete with batteries for the world's lithium supply. From Chapter 7, the estimated Li resources worldwide are about 1.7×10^{10} kg. Based on the discussion in Chapter 19, a rough estimate of the amount of Li required for the Li-ion cells in a practical vehicle would be about 10 kg. Thus, the amount of Li available would provide batteries for about 1.7×10^9 vehicles, although improvement in vehicle efficiency and changes in our perception of vehicle requirements may increase this number. The current automobile population worldwide is about

---

PROJECT 21.4
The use of methane

The United States has substantial natural gas (methane) resources (compared with its oil resources). Natural gas has been promoted as a clean alternative to coal for producing electricity in thermal power plants. Consider the following options for use of methane in future transportation technologies:

- Combustion of methane in traditional ICE vehicles

- The production of electricity by thermal generation by methane combustion and the use of electricity in battery-operated vehicles

- The production of electricity by thermal generation using methane combustion and the subsequent use of electricity to produce hydrogen by electrolysis for fuel cell vehicles

- The production of electricity by thermal generation by methane combustion and the use of electricity to produce hydrogen through electrolysis for hydrogen ICE vehicles

- The direct use of methane in solid oxide fuel cell vehicles

In each case, consider the environmental impact of the approach and its economy in terms of overall efficiency.

PROJECT 21.5
Dimethyl ether

Natural gas (primarily methane) is a fairly plentiful fossil fuel (Chapter 3), that has found a number of applications, e.g. electricity generation, home heating, and the like. A major factor that limits methane use for some applications (e.g., transportation) is the fact that it is a gas and therefore occupies a large volume. Because of its low boiling point, it is inconvenient to liquefy methane. Petroleum gas, which is about 85% propane combined with other gaseous hydrocarbons, has a much higher boiling point than methane and can be readily liquefied at ambient temperature under a modest pressure ($\sim$ 1.5 MPa) to produce liquefied petroleum gas (LPG). This product has advantages over methane for some applications.

There has been significant recent interest in the use of dimethyl ether (DME) as a fuel (by combustion) because it has properties similar to propane and can be produced from methane.

Investigate the use of DME as a fuel, including the following points:

1. What are the properties of DME in comparison with methane and propane?

2. What are the reactions and techniques by which DME can be produced from methane?

3. What are the energetics of this production method? [Specifically, what is the net energy input or output (ideally) when methane is converted into DME?]

4. What are the advantages (and disadvantages) of DME compared with methane and propane, and what applications would benefit most from this approach?

5. If one considers the details of the process of producing DME from methane and the energy produced by combustion, are there environmental advantages to producing DME rather than merely using methane in terms of $kg(CO_2)$ per MJ of energy produced?

6×10^8 vehicles, and that number is expected to double over the next few decades, as personal motor vehicles become more common in developing countries. Thus, the viability of this approach as an alternative transportation technology may basically be a question of resource availability. The long-term sustainability of a BEV-based transportation system must consider this type of analysis and emphasizes the importance of understanding the global uses of Li and of effective Li recycling. Lithium-sulfur batteries are being developed (Figure 21.4) that have superior energy storage capabilities. Sodium shares common features with lithium in terms of its electronic structure and its subsequent chemical properties. Research into sodium-based batteries is underway and shows promise for batteries that rely on a plentiful element (contained in the salt of the oceans).

21.3e Economical

The economics of various approaches to energy production involve a number of factors. These include the cost of producing, operating, and maintaining the necessary infrastructure, as well as the overall efficiency of converting primary energy into a usable form. Alternative energy technologies are, for the most part, energy conversion technologies, with electricity, in many cases (although not all), as the end usable energy

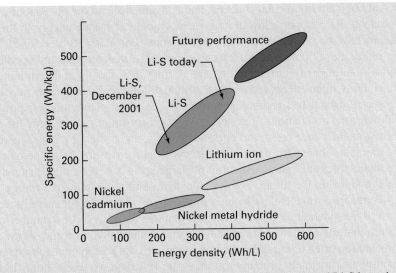

Figure 21.4: Ragone plot showing performance of newly developed Li-S batteries.

Based on Thom Mason, Oak Ridge National Laboratory

product. All alternative energy technologies must compete with fossil fuels. While coal, oil, and natural gas are readily available and (relatively) inexpensive, a major factor in promoting new energies is cost. It is also necessary to consider availability and environmental impact, as well as the availability of materials needed to utilize the resource. Table 21.2 gives an estimate of the cost of producing electricity from some renewable and nonrenewable energy resources. Accurate estimates are very difficult to make, so these should be interpreted as only rough figures. From a financial standpoint, Table 21.2 shows that coal is the least expensive method of producing electricity. Most of the renewables are similar to nuclear and/or oil, although the uncertainty of these numbers at present makes it somewhat difficult to distinguish among most of them. Unfortunately, solar (photovoltaics) is clearly more expensive than other electricity production methods.

Table 21.2: Current cost of electricity from various sources.

energy resource	cost (US$/MJ)	cost (US$/kWh)
solar (photovoltaic)	0.084	0.30
tidal/wave	0.028	0.10
geothermal	0.022	0.08
nuclear (fission)	0.018	0.065
wind	0.017	0.06
oil	0.017	0.06
hydroelectric	0.014	0.05
natural gas	0.011	0.04
coal	0.007	0.025

Although environmental concerns and convenience can offset economic factors to some extent, the cost per kilowatt-hour or megajoule of usable energy to the end user must somehow be competitive with other options. For a facility developed by, say, a public utility, development costs, infrastructure costs, maintenance and operating costs, as well as the longevity of the facility, are all important factors in determining the economic viability of an energy technology. Although long payback periods may be acceptable for well established technologies such as coal-fired plants, nuclear power plants, or hydroelectric installations, new technologies carry more risk in terms of long-term reliability, maintenance costs, and operational costs, and a shorter payback period is certainly desirable. There may also be a tendency to delay investing in technologies, such as photovoltaics, pressure-retarded osmosis, or large-scale battery storage, that may become more economical in the future. For individuals investing in alternative energy resources for residential use, financial gains on a timescale of few years is generally required. In any case, a consideration of energy options requires an analysis of a number of factors. A careful analysis of the environmental effects of alternative energies shows that their implementation often has undesirable and sometimes not so apparent consequences. The availability and longevity of many resources are less than we would hope. In fact, only solar energy can clearly supply all of our energy needs and is renewable indefinitely, but, at present, it is substantially more expensive than many other alternative options.

21.4 Overview of Future Energy Technologies

The need to develop energy technologies that are not based on fossil fuel combustion is clear. For this discussion, we will divide these technologies into two basic categories: nuclear energy and renewable energy. In the first case, nuclear energy, at least in the foreseeable future, refers to fission energy. In the second case, a number of options have been discussed in Part IV of this text. A brief summary of some of the available energy options is now presented.

21.4a Nuclear Energy

Traditional thermal neutron reactors are a well established technology that contributes significantly to our present energy needs. The availability of ^{235}U resources is unlikely to extend very far into the future. However, the utilization of the energy content of non-fissile ^{238}U by more extensive fuel reprocessing and/or the use of fast breeder reactors will make this a viable option for quite some time. These possibilities are reasonably well established technologically. The development of ^{232}Th as an energy source can extend these possibilities even further and has been promoted as an environmentally more attractive option. Fusion is unlikely to be technologically feasible in the foreseeable future.

21.4b Solar Energy

Solar energy can contribute to our energy needs in two ways: solar thermal and photovoltaic. Solar thermal energy is best suited to residential heating and can contribute to a small extent to our future energy use. It is based on a well established and very mature technology. While there may be small technological advances that improve efficiency, no huge breakthrough is on the horizon for the use of solar thermal energy. Wide-scale use of solar thermal energy use, such as solar thermal electricity generation, is probably not a technology that can compete with other alternatives.

Solar photovoltaic electricity generation has received much interest in recent years, and there has been substantial research into improving the technology in this area. On the one hand, photovoltaics have been around for many years, while on the other hand, it is a fairly immature technology, and substantial developments are still likely. The low efficiency of solar photovoltaics is often seen as its major difficulty, and this is certainly an important consideration that has prompted much research into developing new photovoltaic materials with higher efficiency. The practical problem with photovoltaics is its relatively low efficiency combined with its relatively high cost. The significance of developing new materials must be weighed in the context of their cost and also their availability if photovoltaics are to fulfill a significant fraction of the world's energy needs. Also, resources such as the elements indium, cadmium, tellurium, and the like, which could play an important role in the development of high-efficiency photovoltaics, are limited. While it would be desirable to have photovoltaic efficiencies of 40%, we should consider whether 15–20% efficiency would be acceptable if the required materials are more plentiful and less expensive. Certainly, this situation would require more land area to fulfill our energy needs, but total land area may not be the most serious limiting factor for the development of this technology. At the moment, the limiting factor seems to be cost, and, in the future, the limiting factor may be material availability. Lower efficiency may be a viable alternative if devices are inexpensive and utilize more common materials.

The availability of solar energy varies according to location around the earth. Clearly, latitude is significant, and locations far from the equator are less likely to benefit from solar energy. Also, local variations that depend on geography and climate influence the viability of this resource.

PROJECT 21.6
Future photovoltaic applications

Evaluate the possibility of developing a sufficient solar energy infrastructure during the remainder of the twenty-first century to fulfill the predictions shown in Figure 21.3. You will need to estimate from the figure the solar energy per year (that is, W_e) as a function of year from now until 2100. What is the area of photovoltaic arrays needed to meet this requirement? On the basis of these needs, discuss the following questions:

(a) What is the most reasonable photovoltaic material based on current technology?

(b) What is the mass of materials (particularly critical elements) needed per year to satisfy this development? How does this compare with current photovoltaic cell production? Comment on the need for the development of additional photovoltaic manufacturing infrastructure.

(c) What is the anticipated lifetime of a facility, and will cells have to be replaced during the time period you are considering? How does this affect the quantity of cells that needs to be produced per year?

(d) Assuming that future developments increase the efficiency of photovoltaics to nearly the theoretical limit, how will this influence your analysis?

(f) If new photovoltaic technologies are developed from current research directions, are certain types of cells less desirable in the context of economics or resource availability?

PROJECT 21.7
Use of wind energy

Wind power is based on a well established technology, its net cost per kWh_e is competitive with other renewable sources, and it has minimal environmental impact (compared to other technologies). A convenient aspect of wind power is that it can be implemented on a vast range of scales, ranging from residential wind turbines of a few kW_e capacity to wind farms of close to a GW_e capacity. Not surprisingly, therefore, the utilization of wind energy has grown substantially in recent years. Five countries that have been very active in developing wind power in recent years are China, Denmark, Germany, India, and the United States. These countries have quite different energy needs and serve as good examples of how wind power can be implemented in different parts of the world.

(a) Locate information about wind energy use in these five countries. What fraction of the total electricity is produced from wind?

(b) Estimate the capacity factor for each country, realizing that the total energy produced per year

(in Wh_e) is the rated capacity (in W_e) times the number of hours per year times the capacity factor.

(c) What fraction of its total electricity needs could each country expect to obtain from wind, assuming that 5% of its land area was utilized? Consider, for example, the diameter of a 2-MW_e rated wind turbine and the optimal spacing of turbines.

(d) For each of the five countries prepare a brief report discussing the following points:

- How serious has the effort been thus far to develop wind power to meet the countries' electrical needs?
- Can a much greater capacity be expected from further development, and what fraction of total electrical needs would be reasonable to expect from wind?
- How important is the development of offshore wind farms?

21.4c Wind Energy

Wind energy is based on a well established technology that is unlikely to see any major changes in the foreseeable future. Wind energy is relatively inexpensive, its use has increased significantly in recent years, and it is likely to see continued increases in the future. Its possible use is widespread, although local geographic and climatic considerations are important. The major difficulty that limits the use of wind energy is the total power available worldwide. While wind is likely to be a factor in our future energy mix, it cannot satisfy all our energy needs.

21.4d Hydroelectric Energy

Hydroelectric energy, like wind, is based on a mature technology that has been utilized extensively in the past. There are relatively few opportunities for the expansion of large-scale hydroelectric facilities, particularly in North America. There are more possibilities for run-of-the-river hydroelectric development. Like wind energy, hydroelectricity will be a small factor in worldwide energy production. However, two considerations should be addressed: environmental impact and longevity. As discussed in Chapter 11, greenhouse gas emissions may be a factor for high head hydroelectric facilities, particularly in tropical areas, and many resources may not be renewable indefinitely due to sedimentation.

PROJECT 21.8
Environmental effects of hydroelectric energy

Consider, as quantitatively as possible, the environmental effects of a moderately large high head hydroelectric facility with a capacity of about 1 GW_e compared with a typical coal-fired generating station of the same capacity. As a basis of comparison, calculate the CO_2 emissions per year for the coal station operating at full capacity. For the hydroelectric station determine the following:

(a) The typical area of a reservoir needed for such a facility. Look up specifications of similarly sized hydroelectric facilities to get an idea of reservoir size.

(b) If the facility is located in a temperate region with average forestation, calculate the loss of carbon sequestration capabilities that results from the replacement of forested areas with the reservoir.

(c) How would the calculated effect differ in tropical regions? Also, find information about methane production from reservoirs in tropical regions. What is the greenhouse gas effect of this methane production compared with the loss of CO_2 sequestration capabilities? How does the overall environmental impact of a hydroelectric facility in a temperate or in a tropical region compare with a fossil fuel plant of similar capacity?

21.4e Tidal and Wave Energy

Energy from the movement of the oceans has been exploited in a few locations. Technology is developing in this area but is based on well established scientific principles. Tidal energy has been harnessed in France and Canada, and it is possible that other locations may be developed in the future. The availability of locations where this source of energy may be utilized is fairly limited, and it is likely to be a local factor in certain areas rather than a substantial component of global energy use. Waves (and possibly ocean currents) are somewhat more widespread, but again the utilization of wave energy is likely to be only a small factor in the energy production for certain countries.

21.4f OTEC and Salinity Gradients

Ocean energy from gradients in temperature and salinity may be utilized. OTEC development has fallen well below one-time expectations. The low efficiency and technological difficulties associated with developing this resource are likely to limit its use for some time. Salinity gradient energy is a field that is still in the early stages of development, and time will tell whether it may be a small factor in our future energy mix.

21.4g Geothermal Energy

Geothermal energy has been developed extensively and relies, for the most part, on mature technologies. It is an important factor in energy production in specific geographical locations where appropriate underground resources are available. New technologies may be possible that will better utilize this resource, although the total power available is limited. Its longevity is perhaps questionable if thermal energy is extracted from a deposit more rapidly that it is replenished.

PROJECT 21.9
Comparison of different water based energy technologies

A comparison of the advantages of various alternative energy technologies is complex. It must include an analysis of factors such as economics, resources availability, and environmental impact, many of which are difficult to quantify. A simple (and probably overly naïve) approach might be to look at energy produced per unit of resource. We should not try to overinterpret the meaning of such results because the analysis does not consider environmental and other factors, but this type of analysis can provide some insight into the relative desirability of different approaches. Consider, for example, the ways in which energy can be extracted from water, where we can define the quantitative measure of megajoules of energy per cubic meter of water. For the following water-based energy technologies, determine MJ/m^3 as indicated in the description.

- *Hydroelectric:* This technology converts the potential energy of the water into electricity. Calculate MJ/m^3 (for a freshwater density of 1000 kg/m^3) for a head of 100 m (typical for a high head hydroelectric facility) and a conversion efficiency of 90%.
- *Wave/tidal (current):* This technology coverts the kinetic energy of moving water into electricity. Assuming a seawater density of 1025 kg/m^3, a velocity of 5 m/s, and a conversion efficiency of 35%, calculate MJ/m^3.

- *OTEC:* This technology converts thermal energy in the oceans into electricity. For typical (tropical) conditions, calculate the ideal Carnot efficiency for conversion from thermal to mechanical energy by means of a heat engine and a conversion efficiency of 90% from mechanical to electrical. Calculate MJ/m^3.
- *Osmotic energy:* This method converts chemical energy in the seawater into electricity. Assume a seawater molarity of 500 mol/m^3. Calculate the effective head based on the osmotic energy, and assume that this can be converted into electricity by means of a turbine/generator with an efficiency of 80%

Tabulate your results for each of these technologies and comment on the results. Which is the most efficient method of extracting electrical energy from water?

Does this method of assessment provide a reasonable view of which technologies are likely to be viable? To answer the more complex question of viability, consider an assessment of the following factors:

- Infrastructure needed per unit energy produced
- Operating costs and difficulties
- Energy distribution considerations
- Overall efficiency of prime energy conversion to electricity for the user

PROJECT 21.10
Hydrogen as an energy storage mechanism for OTEC

Research the following energy scenario:

A 10-MW$_e$ OTEC facility is located 500 km offshore in a very desirable ocean thermal environment. Electricity produced at the facility must be transported to shore. It is proposed that hydrogen will be produced by electrolysis and that the hydrogen will be liquefied and transported to shore by ship, where it will be converted back into electricity using fuel cell technology.

To be environmentally friendly, the ship will utilize liquid hydrogen for propellant in a hydrogen ICE. Prepare a report summarizing the expectations for the overall economics and energy efficiency of this approach. Provide recommendations for investing in a program to develop this technology, and compare its viability to traditional alternative energy technologies, such as terrestrial-based wind energy.

PROJECT 21.11
A comparison of alternative energy resources

A major river runs through a productive agricultural region. The region has good insolation, good wind condition, and a reasonable elevation change. It is planned to develop alternative energy for the region using one of the following methods:

- Construct a nuclear power plant and use the river as a heat sink.
- Construct a 100-km² wind farm.
- Construct a 100-km² photovoltaic array.
- Convert 100 km² of food crops to corn for use for ethanol production.
- Dam the river and create a 100-km² reservoir for a hydroelectric generating station.

Discuss the pros and cons of each of these plans in terms of their energy productivity (i.e., in terms of energy per land area) and environmental impact. Be as quantitative as possible, and include an analysis of the following factors:

- Infrastructure cost, estimate cost per net MW_e capacity
- Anticipated longevity of resource
- Operational cost and net cost per kWh_e
- Environmental impact of the facility in terms of effects of infrastructure construction and environmental impact of operation.

21.4h Biomass Energy

The viability of biomass energy depends on the efficiency, that is, energy in (at the moment mostly from fossil fuels) vs. energy out of converting biomatter into a usable fuel. Bioenergy could be a substantial component of future energy production, but, using present technology, its efficient use is limited to locations with suitable climate conditions (i.e., tropical regions). The development of efficient processes for using cellulose as a source for ethanol production would make biofuels more universally useful. Municipal waste is a minor factor in energy production. The use of municipal waste for energy production, however, is potentially beneficial from an environmental standpoint because any environmentally benign energy gain from waste material is preferable to most other methods of disposal.

21.5 Efficient Energy Utilization

While the efficient and economical production of energy from environmentally conscientious sources is a necessary step for a sustainable future, the efficient and effective use of this energy is also essential. The factors discussed in this text can be categorized in three ways:

1. Conservation of energy
2. Efficient distribution of energy
3. Storage of energy

PROJECT 21.12
Analysis of an integrated home heating system

A homeowner plans to install an integrated central home heating system consisting of solar thermal panels on the roof plus a natural gas-fired furnace that distributes heat through forced hot air ducts. The house is one story with a footprint of 15 m × 10 m and is located in a reasonably sunny location at a latitude of 40°N with one wall facing south. There are 3200 (°C) degree days per year. The heating requirements of the house are typical of those discussed in Chapter 8. For simplicity, consider the hot water supply system to be independent of the heating system. Assume that you work for an alternative heating consulting company that has been assigned the task of developing such a heating system. Prepare a proposal for this system. At a minimum, your proposal should show a diagram of the system illustrating how the various components would be integrated. Specify the parameters for the system, such as the area of the solar collectors, the size of an appropriate energy storage system, energy output requirements for the natural gas furnace, and any other relevant specifications. The home owner is particularly interested in obtaining an estimate of the amount of natural gas (in cubic meters at STP) that will be used per year.

21.5a Conservation

While developing new energy technologies is crucial for a sustainable future, making the best use of energy that is produced helps to alleviate the demands on energy production. Thus, conservation is an important part of maintaining an adequate energy supply for the future and an essential part of dealing with environmental concerns that result from energy production and use. Because virtually all energy production and use have some environmental impact, reducing energy use reduces this impact.

As seen in Chapter 17, energy conservation can be approached at various levels, from national government energy policies and international cooperative efforts, to regional policies, to business and industrial practices, and finally to the actions of individuals. Energy conservation is an ideal example of a situation where individuals can be proactive in addressing future energy issues and where simple actions can have an important cumulative effect on world energy use. Although there is often reluctance to change effective practices and there are always economic concerns that need to be addressed, many conservation actions are effective at improving the overall comfort in our lives and have financial benefits on a quite reasonable timescale. Improved environmental quality and long-term energy security are added benefits of our conservation efforts.

21.5b Distribution

The large-scale distribution of electricity is a major component of our energy use. The electric grid connects energy production facilities with users. In the past, electric generation was by means of fossil fuel-fired plants, nuclear power plants, and hydroelectric generating stations. Both fossil fuel and nuclear power stations have output that is controlled by operators. The timescale for changes to the output in response to changes in demand can be fairly short, as for combustion turbines, or much longer, as for coal-fired thermal or nuclear facilities. Apart from fairly predictable seasonal fluctuations, hydroelectric generating station output is readily controlled by the operator. Thus electricity supplied to the grid could be controlled to meet demands.

PROJECT 21.13
Analysis of an integrated smart grid system

A region with a population of 320,000 and an average annual electricity requirement of 33 MWh$_e$ per capita would like to develop a self-contained electrical system with no substantial imports of electricity. It has been decided to develop a system consisting of a coal-fired thermal generator to cover base load capacity, a wind farm to provide additional power for peak periods, and a pumped hydroelectric storage system to store excess energy generated from wind during nonpeak periods. A smart grid control system will manage the operation of the system. The region has good wind resources, a substantial water supply from a river, and variable terrain elevation. Prepare a proposal for such a system specifying the minimum requirements for each component. Make any reasonable assumptions that are needed about efficiencies, capacity factors, and natural resources availability. Use Figure 18.1 as a guideline for typical power requirement variations. As a minimum, consider the following points:

- *Coal generating facility:* What is the proposed capacity in MW$_e$? How much coal on, say, a weekly basis needs to be used? What is the estimated cooling water requirement?

- *Wind farm:* What is the required capacity? Be sure to include a reasonable estimate of capacity factor. How many 1-MW$_e$ capacity turbines would be needed? What would be the land area required?

- *Pumped hydroelectric storage:* What total energy storage capacity (GWh) would be required? What would be reasonable dimensions (area, average depth) to achieve this? Propose minimum head and flow rates that would be compatible with the requirements.

- *Smart grid:* Describe how a smart grid system might integrate the components of the system to most efficiently supply the community's electricity needs.

The implementation in recent years of alternative energy sources, such as wind, solar, and tidal (for example), has made the efficient distribution of electricity more complex. The need for storage mechanisms that results from the variations in output from most alternative generating methods, combined with the desire to utilize resources more efficiently by better matching supply and demand characteristics, has led to the development of smart grid technologies. These technologies have been implemented in several locations worldwide, and development in this area will certainly continue in the future. This approach fulfills the need to integrate base load supply, from sources such as nuclear energy, with fluctuating sources such as solar or wind and appropriate energy storage mechanisms, such as batteries or pumped hydroelectric, in order to most efficiently make use of the available resources. This approach, combined with real-time monitoring (or even control) of consumer energy use, will provide the most effective means of energy distribution.

21.5c **Storage**

Energy storage is an essential component of many viable energy systems because energy cannot always be produced where and/or when it is needed. This is particularly the case for alternative energy sources. The nature of the most suitable energy storage technology for a particular application depends on a number of factors:

1. *The form of the initial and final energy:* Energy storage often involves the storage of electrical energy with the need for electrical energy as a final product. This might be, for instance, the storage of electrical energy produced by wind

turbines for later use on the electric grid when it is needed. In this case, pumped hydroelectric storage is a common approach. Another common situation is thermal-to-thermal energy storage. This might involve storing thermal energy from solar collectors for later use for domestic heating or hot water. In this situation, a tank of water or a container of rocks or sand might be appropriate, depending on the nature of the heat distribution system.

2. *Size and/or weight restrictions:* The size of an energy storage unit is most important for energy storage systems that need to be portable. This situation is most relevant to transportation where vehicle size and weight need to be within reasonable limits. Batteries for BEVs or hydrogen for fuel cell vehicles are the most common approaches to energy storage for transportation applications.

3. *Total energy storage capacity:* In cases (i.e., vehicles) where size is an important factor, high energy density is a requirement. Systems that store energy via chemical processes are the most suitable. These include batteries or hydrogen, and, although these mechanisms do not typically achieve the energy storage capacity of liquid fossil fuels, they are potentially adequate. Systems that store mechanical energy (e.g., compressed air or pumped hydroelectric) have lower energy densities and are most appropriate where large quantities of energy need to be stored but size limitations are not a factor.

4. *Maximum power available:* The rate at which energy can be extracted from an energy storage system is generally an important consideration. For transportation applications, sufficient power is needed to provide acceptable vehicle performance. In the case of BEVs or fuel cell vehicles, this is generally less of a limitation than the total energy storage capacity. Supplementary power from, say, supercapacitors is an option when needed. Large-scale energy storage systems (for example, for connection to the grid) need to meet the demands of the users. System design, such as head, flow rate, and so on, for a pumped hydroelectric facility, needs to address these requirements.

PROJECT 21.14
Residential energy storage

A home owner in a rural area would like to store enough energy to satisfy the electrical needs (without electric heating) for a period of 1 week in the event of an extended power outage (see Chapter 2 for typical electricity requirements for a single-family home in North America). This is a common situation that is typically satisfied by a gasoline (or sometimes diesel or propane) generator. In this case, the home owner would like to consider nonfossil fuel alternatives for storing energy. Some options are:

- An ethanol-powered ICE generator.
- A CHG-fueled fuel cell.

- Pb-acid batteries.
- Li-ion batteries.
- A flywheel/generator system.
- A compressed air-powered generator.

Discuss the design of each of these systems. Consider, particularly, the quantity of fuel (if any) that would be required to supply 1 week of electricity, the space requirements for the system, and the actual environmental advantages over a gasoline-powered generator.

PROJECT 21.15
Batteries and supercapacitors for transportation use

A 1500-kg fuel cell vehicle requires an average of 0.7 MJ/km at the wheels. Hydrogen fuel is stored in a 0.1-m^3 CHG tank at 35 MPa, and electricity is produced by a 100-kW PEM fuel cell. It is desired to increase both power and range by adding an energy storage system, consisting of either Li-ion batteries or supercapacitors or both, which does not exceed 5% of the total vehicle weight. When the batteries and/or supercapacitors are fully charged externally, a 30% increase in both power to the wheels and range is required.

(a) Will either batteries or supercapacitors alone fulfill the requirements?

(b) If not, will some combination of batteries and supercapacitors suffice?

(c) If a viable design is feasible, what is the minimum increase in vehicle weight required? If a viable design is not feasible, specify the design of a battery/supercapacitor system that comes closest to satisfying the requirements and stays within the weight limitations.

21.6 Conclusions

Our future energy choices will be determined by a consideration of a number of factors:

- Science
- Technology
- Environment
- Economics
- Politics

The basic science of many aspects of energy production still needs to be understood. For example, this would include understanding the physics of new materials for photovoltaic cells and batteries. Technologies need to be developed that make use of basic scientific principles—for example, establishing efficient methods for photovoltaic cell production or designing new devices for converting ocean wave motion into electricity. How energy production methods interact with our environment needs to be fully understood. It is clear from this discussion that so-called environmentally friendly or green energy production methods can have more of an environmental impact than might be initially perceived. Certainly, energy must be affordable to those who use it, and new technologies must, at some point, compete economically with traditional technologies; otherwise their development will be hindered. Political agendas often determine energy policy, and public perception of new energy technologies, such as hybrid vehicles or hydrogen vehicles, may not always be accurate. An informed public always makes the best decisions.

There are no obvious immediate solutions for our energy needs. Deciding the best route to take and developing the technology to follow that route will be a challenge and will require dedication and persistence. The effort required to deal with this challenge is enormous. In the Preface, the immensity of this task was suggested—one new major energy production facility every day for the next 50 years. A worldwide effort of this magnitude, with a clear vision of the development of facilities for sustainable and environmentally conscious energy is needed to secure our energy needs for the future.

Bibliography

J. Fanchi. *Energy: Technology and Directions for the Future.* Elsevier, Amsterdam (2004).

R. Heinberg. *Power Down: Options and Actions for a Post-Carbon World.* New Society, Gabriola Island, Canada (2004).

R. Hinrichs and M. Kleinbach. *Energy: Its Use and the Environment*, 5th ed. Brooks-Cole, Belmont, CA (2012).

T. B. Johansson, H. Kelly, A. Reddy, and R. Williams (Eds.). *Renewable Energy: Sources for Fuels and Electricity.* Island Press, Washington, DC (1992).

D. J. C. MacKay. *Sustainable Energy—Without the Hot Air.* UIT Cambridge, Cambridge, MA (2009).

V. Smil. *Energy at the Crossroads—Global Perspectives and Uncertainties.* MIT Press, Cambridge, MA (2003)

B. Sorensen. *Renewable Energy: Its Physics, Engineering, Environmental Impacts Economy and Planning Aspects*, 2nd ed. Academic Press, London. (2002).

Special Report on Emission Scenarios—Summary for Policymakers. WMO and UNEP, Geneva (2001), available at http/www.ipcc.ch/pub/SPM_SRES.pdf

C. Starr, M. F. Searl, and S. Alpert. "Energy sources: A realistic outlook." *Science* **256**, (1992): 981.

J. W. Tester, D. O. Wood, and N. A. Ferrari (Eds.). *Energy and the Environment in the 21st Century* Cambridge. MIT Press, Cambridge, MA (1991).

J. Tester, E. Drake, M. Driscoll, M. W. Golay, and W. A. Peters. *Sustainable Energy: Choosing Among Options.* MIT Press, Cambridge, MA (2006).

J. R. Wilson and G. Burgh. *Energizing Our Future: Rational Choices for the 21st Century.* Wiley, Hoboken, NJ (2008).

R. Wolfson. *Energy, Environment, and Climate*, 2nd ed. W.W. Norton, New York (2011).

World Energy Council. *2010 Survey of Energy Resources,* available online at http://www.worldenergy.org/documents/ser_2010_report_1.pdf

APPENDICES

Powers of Ten

number	prefix	abbreviation
10^{18}	exa	E
10^{15}	peta	P
10^{12}	tera	T
10^{9}	giga	G
10^{6}	mega	M
10^{3}	kilo	k
10^{-3}	milli	m
10^{-6}	micro	μ
10^{-9}	nano	n
10^{-12}	pico	p
10^{-15}	femto	f
10^{-18}	atto	a

Physical Constants

quantity	symbol	value	units
alpha particle binding energy	B_α	28.296	MeV
alpha particle mass	m_α	4.00150618 3727.409	u MeV/c^2
atomic mass unit	u	$1.6605402 \times 10^{-27}$ 931.494	kg MeV/c^2
Avogadro's number	N_A	6.0221367×10^{23}	$mole^{-1}$
Boltzmann's constant	k_B	$1.3806488 \times 10^{-23}$ 8.6173324×10^{-5}	$J \cdot K^{-1}$ $eV \cdot K^{-1}$
Coulomb constant	$1/4\pi\varepsilon_0$ $e^2/4\pi\varepsilon_0$	8.987551×10^9 1.439976	$N \cdot m^2 \cdot C^2$ $MeV \cdot fm$
deuteron binding energy	B_d	2.224	MeV
deuteron mass	m_d	2.013553214 1875.628	u MeV/c^2
electron mass	m_e	$5.48579903 \times 10^{-4}$ 0.5109988	u MeV/c^2
electron volt	eV	$1.60217733 \times 10^{-19}$	J
electronic charge	e	$1.60217733 \times 10^{-19}$	C
gas constant	R	8.314510	$J \cdot K^{-1} \cdot mol^{-1}$
gravitational acceleration	g	9.80665	m/s^2
gravitational constant	G	6.67259×10^{-11}	$N \cdot m^2 \cdot kg^{-2}$
neutron mass	m_n	1.008664904 939.56531	u MeV/c^2
permittivity of vacuum	ε_0	8.854×10^{-12}	F/m
Planck's constant	h $\hbar$ hc $\hbar c$	$6.6260755 \times 10^{-34}$ 4.13570×10^{-15} $1.05457266 \times 10^{-34}$ 6.58217×10^{-16} 1.240×10^3 1.973×10^2	$J \cdot s$ $eV \cdot s$ $J \cdot s$ $eV \cdot s$ $MeV \cdot fm$ $MeV \cdot fm$
proton mass	m_p	1.007276470 938.2723	u MeV/c^2
speed of light	c	2.99792458×10^8	$m \cdot s^{-1}$
Stefan-Boltzmann constant	σ	5.67051×10^{-8}	$W \cdot m^{-2} \cdot K^{-4}$

Miscellaneous Conversion Factors

To convert from the unit on the left to the units on the right, multiply the value in left-hand units by the conversion factor. To convert from the units on the right to the units on the left, divide the value in right-hand units by the conversion factor.

convert from (convert to)	multiply by (divide by)	convert to (convert from)
astronomical units (au)	1.49×10^{11}	meters (m)
barrels (bbl)	158.97	liters (L)
becquerels (Bq)	1.0	decays per second (s^{-1})
electron volts (eV)	1.6×10^{-19}	joules (J)
hectares (ha)	10^4	square meters (m^2)
kilowatt hours (kWh)	3.6×10^6	joules (J)
million electron volts (eV)	1.6×10^{-13}	joules (J)
years (y)	3.15×10^7	seconds (s)

Energy Content of Fuels

fuel	quantity	equivalence in joules
crude oil	liter	3.85×10^7
	barrel (bbl = 158.97 L)	6.12×10^9
gasoline	liter	3.48×10^7
bituminous coal	kilogram	3.10×10^7
natural gas (STP)	cubic meter	3.85×10^7
^{235}U (fission)	gram	8.28×10^{10}
deuterium (fusion)	gram	2.38×10^{11}
wood (maple, 20% water)	kilogram	1.4×10^7
ethanol	liter	2.35×10^7
municipal waste	kilogram	1.0×10^7

The Elements

element	symbol	atomic number	element	symbol	atomic number
actinium	Ac	89	gold	Au	79
aluminum	Al	13	hafnium	Hf	72
americium	Am	95	helium	He	2
antimony	Sb	51	holmium	Ho	67
argon	Ar	18	hydrogen	H	1
arsenic	As	33	indium	In	49
astatine	At	85	iodine	I	53
barium	Ba	56	iridium	Ir	77
berkelium	Bk	97	iron	Fe	26
beryllium	Be	4	krypton	Kr	36
bismuth	Bi	83	lanthanum	La	57
boron	B	5	lawrencium	Lr	103
bromine	Br	35	lead	Pb	82
cadmium	Cd	48	lithium	Li	3
cesium	Cs	55	lutetium	Lu	71
calcium	Ca	20	magnesium	Mg	12
californium	Cf	98	manganese	Mn	25
carbon	C	6	mendelevium	Md	101
cerium	Ce	58	mercury	Hg	80
chlorine	Cl	17	molybdenum	Mo	42
chromium	Cr	24	neodymium	Nd	60
cobalt	Co	27	neon	Ne	10
copper	Cu	29	neptunium	Np	93
curium	Cm	96	nickel	Ni	28
dysprosium	Dy	66	niobium	Nb	41
einsteinium	Es	99	nitrogen	N	7
erbium	Er	68	nobelium	No	102
europium	Eu	63	osmium	Os	76
fermium	Fm	100	oxygen	O	8
fluorine	F	9	palladium	Pd	46
francium	Fr	87	phosphorus	P	15
gadolinium	Gd	64	platinum	Pt	78
gallium	Ga	31	plutonium	Pu	94
germanium	Ge	32	polonium	Po	84

element	symbol	atomic number	element	symbol	atomic number
potassium	K	19	tantalum	Ta	73
praseodymium	Pr	59	technetium	Tc	43
promethium	Pm	61	tellurium	Te	52
protactinium	Pa	91	terbium	Tb	65
radium	Ra	88	thallium	Tl	81
radon	Rn	86	thorium	Th	90
rhenium	Re	75	thulium	Tm	69
rhodium	Rh	45	tin	Sn	50
rubidium	Rb	37	titanium	Ti	22
ruthenium	Ru	44	tungsten	W	74
samarium	Sm	62	uranium	U	92
scandium	Sc	21	vanadium	V	23
selenium	Se	34	xenon	Xe	54
silicon	Si	14	ytterbium	Yb	70
silver	Ag	47	yttrium	Y	39
sodium	Na	11	zinc	Zn	30
strontium	Sr	38	zirconium	Zr	40
sulfur	S	16			

Table of Acronyms

AC	alternating current
ACEA	Association des Constructeurs European d'Automobile
AEEFM	Assessment of Energy Efficiency Finance Mechanism
ALR	adiabatic lapse rate
ASDEX	Axially Symmetric Divertor EXperiment
BEE	Bureau of Energy Efficiency
BEV	battery electric vehicle
BWR	boiling water reactor
CAFE	Corporate Average Fuel Economy
CANDU	Canadian Deuterium Uranium (Reactor)
CFL	compact fluorescent lamp
CHG	compressed hydrogen gas
CHP	combined heat and power
CIGS	copper-indium-gallium-selenide
CRF	capital recovery factor
CURVE	clean, unlimited, renewable, versatile, economic
DC	direct current
DOE	Department of Energy
DSSC	dye-sensitized solar cell
ec	electron capture
EIA	Energy Information Administration
Eliica	electric lithium ion car
EPA	Environmental Protection Agency
EU	European Union
FBR	fast breeder reactor
FCP	Fuel Consumption Program
GDP	gross domestic product
GNP	gross national product
HC	hydrocarbon
H-ICE	hydrogen internal combustion engine
HTSC	high-temperature superconductor
HVAC	heating, ventilation, and air conditioning
IAEA	International Atomic Energy Agency
ICE	internal combustion engine
IEE	Intelligent Energy Europe
INES	International Nuclear and Radiological Event Scale
IPEEC	International Partnership for Energy Efficiency Cooperation
IPEEI	Improving Policies Through Energy Efficiency Indicators
IR	infrared
ITER	International Thermonuclear Reactor

LED	light-emitting diode
LEED	Leadership in Energy and Environmental Design
LH$_2$	liquid hydrogen-2
LMFBR	liquid metal fast breeder reactor
LNG	liquid natural gas
LOCA	loss of coolant accident
MCT	marine current turbine
NASA	National Aeronautics and Space Administration
NCAR	National Center for Atmospheric Research
NELHA	National Energy Laboratory of Hawaii Authority
NIMBY	not in my backyard
NOAA	National Oceanic and Atmospheric Administration
NREL	National Renewable Energy Laboratory
OECD	Organisation for Economic Co-operation and Development
OTEC	ocean thermal energy conversion
OWC	oscillating water column
PCRA	Petroleum Construction Research Association
PCV	positive crankcase ventilation
PEM	polymer electrolyte membrane
PET	polyethylene terephthalate
PRO	pressure-retarded osmosis
PTO	power takeoff
PV	photovoltaic
PWR	pressurized water reactor
RBMK	Reactor Bolshoi Moschnosti Kanalynyi (Russian water-cooled graphite moderated reactor)
RDF	refuse-derived fuel
RED	reverse electrodialysis
rpm	revolutions per minute
SAVE	specific actions for vigorous energy efficiency
SBN	sustainable building network
SEGS	solar energy generating station
SET	Strategic Energy Technology (Program)
SI	System Internationale
SMES	superconducting magnetic energy storage
STEP	Solar Total Energy Project
STP	standard temperature and pressure
SVO	straight vegetable oil
TBL	triple bottom line
TEPCO	Tokyo Electric Power COmpany
TMI	Three Mile Island
TVA	Tennessee Valley Authority
US$	U.S. dollars
USDOE	United States Department of Energy
USEIA	United States Energy Information Administration
UV	ultraviolet
WEACT	Worldwide Energy Efficiency Action Through Capacity Building and Training
ZNE	zero net energy (building)

I

PRINCIPAL UNITS USED IN MECHANICS

Quantity	International System (SI)			U.S. Customary System (USCS)		
	Unit	Symbol	Formula	Unit	Symbol	Formula
Acceleration (angular)	radian per second squared		rad/s^2	radian per second squared		rad/s^2
Acceleration (linear)	meter per second squared		m/s^2	foot per second squared		ft/s^2
Area	square meter		m^2	square foot		ft^2
Density (mass) (Specific mass)	kilogram per cubic meter		kg/m^3	slug per cubic foot		slug/ft^3
Density (weight) (Specific weight)	newton per cubic meter		N/m^3	pound per cubic foot	pcf	lb/ft^3
Energy; work	joule	J	N·m	foot-pound		ft-lb
Force	newton	N	kg·m/s^2	pound	lb	(base unit)
Force per unit length (Intensity of force)	newton per meter		N/m	pound per foot		lb/ft
Frequency	hertz	Hz	s^{-1}	hertz	Hz	s^{-1}
Length	meter	m	(base unit)	foot	ft	(base unit)
Mass	kilogram	kg	(base unit)	slug		$\text{lb-s}^2\text{/ft}$
Moment of a force; torque	newton meter		N·m	pound-foot		lb-ft
Moment of inertia (area)	meter to fourth power		m^4	inch to fourth power		in.^4
Moment of inertia (mass)	kilogram meter squared		kg·m^2	slug foot squared		slug-ft^2
Power	watt	W	J/s (N·m/s)	foot-pound per second		ft-lb/s
Pressure	pascal	Pa	N/m^2	pound per square foot	psf	lb/ft^2
Section modulus	meter to third power		m^3	inch to third power		in.^3
Stress	pascal	Pa	N/m^2	pound per square inch	psi	lb/in.^2
Time	second	s	(base unit)	second	s	(base unit)
Velocity (angular)	radian per second		rad/s	radian per second		rad/s
Velocity (linear)	meter per second		m/s	foot per second	fps	ft/s
Volume (liquids)	liter	L	10^{-3} m^3	gallon	gal.	231 in.^3
Volume (solids)	cubic meter		m^3	cubic foot	cf	ft^3

The Ghost Ship of Diamond Shoals

Outer Banks

In years past, mariners dreaded navigating the waters surrounding North Carolina's Outer Banks, a succession of skinny islands stretching roughly 175 miles from Virginia southward to Cape Lookout. They dubbed the area "Graveyard of the Atlantic" because over six hundred ships had sunk there. Mariners relied heavily on lighthouse keepers, lifesaving stations, and the Coast Guard crew to keep them afloat or rescue them when the sea tried to take them. These men knew that how well they performed their duties meant the difference between life and death.

When the crew at the Hatteras Inlet Coast Guard Station was awakened by an alarm on January 31, 1921, they moved quickly and efficiently. The surfman on lookout duty was letting them know that he had spotted a ship in distress. The lookouts at Big Kinnakeet, Creeds Hill, and Cape Hatteras Coast Guard Stations also noted the ship off in the distance. Men representing all four stations were sent in two rescue boats. The huge, five-masted schooner appeared to be stuck on the outer tip of Diamond Shoals, and was slowly sinking. The rescuers noted that no distress signal had been given and that there weren't any signs of life

The ghost ship *Carroll A. Deering* before its mysterious shipwreck.
Courtesy of the Outer Banks History Center

aboard the vessel. They could get only within a quarter mile of the ship, due to the high surf and rough swells. The men circled it several times, calling out for a response from someone aboard, but the rescuers received no answer. Finally the men returned to the station and sent in the report.

A quick investigation revealed that the craft was the *Carroll A. Deering*, a three-year-old ship launched in Maine. The owner, G.G. Deering Company, was notified, and it authorized the Coast Guard to go aboard and find out what was wrong. Coast Guard cutters *Seminole* (stationed in Wilmington, North Carolina) and *Manning* (stationed in Norfolk, Virginia) as well as the wrecking tug *Rescue* (stationed in Norfolk, Virginia) were dispatched to the shipwreck, but it was four days before the men were able to board because of the bad weather. The ship had suffered quite a bit of damage during that time. Once the officials were able to search the ship, they made some eerie discoveries.

Everything seemed to be in place except some key items. The

> *Carroll A. Deering*, named after the owner's son, was the last schooner ever built. It was launched on April 4, 1919.

ship's log and navigational instruments were missing. Nothing seemed to indicate that there had been a problem or reason for a quick departure, except that dinner had been served in the dining hall and it didn't look like anything had been eaten. Cooking pots remained on the stove of the galley, but the fire that kept the food warm had long expired. Everything was cold to the touch. The only living thing the search party found was a cat.

Someone called out that the yawl boat (lifeboat) was missing, and all the men scrambled up to the top deck. They discovered that the anchors were also missing. The cables that connected the steering wheel to the rudder had been cut. It would have taken an ax to cut through the thick rope cables. Why would anyone do that? There was also a gash on the outside of the ship near where the ladder had been positioned.

The continuous beating the vessel was taking from the combination of the wind and the pounding waves was driving the ship aground at a rapid rate. The rescuers scooped up the gray cat and made a hasty retreat. When the ship's condition and the strange circumstances were reported to Coast Guard headquarters, a major search was initiated all along the coast. Every inch of water, including tiny bays and inlets, was searched. No further wreckage of any kind was discovered. Despite an investigation that lasted many days, nothing else from the ship was ever found. No one known to be aboard the *Carroll M. Deering*, which has been nicknamed the "ghost ship," was ever heard from again.

It was discovered that the skipper of the ship, Captain F. Merritt, had gotten very sick en route from Newport News to Rio de Janeiro. He had put in at Lewes, Delaware, and was taken to a hospital. Captain W.M. Wormell was brought aboard to assume Captain Merritt's duties. The vessel delivered the cargo (coal) to Rio de Janeiro and then sailed on to Barbados. Upon arrival, the captain learned there was no cargo to pick up at Barbados. The ship was to return to Newport News. The investigation proved that the *Carroll M. Deering* left Bridgetown, a port in Barbados, on

Captain F. Merritt, a World War I hero with a citation for bravery, was chosen to command the ship. He became ill at Delaware Breakwater and was replaced by Captain W. M. Wormell of Portland, Maine.

January 9, 1921. Aboard were the captain, the first mate, and eight members of the crew. A friend or acquaintance of Captain Wormell gave testimony during the investigation that the skipper had told him he anticipated trouble. Wormell also told him that his first mate, Officer McClellan, was no good. While in port, the crew got rip-roaring drunk and McClellan was arrested, but the captain arranged his release and they set out.

Carroll M. Deering was logged as passing the Cape Fear Lightship, which was positioned at Wilmington, on January 23, 1921. No problems or oddities were noted. Six nights later the Cape Lookout Lightship sighted the ship. It is strange that it would have taken the vessel six days to go just seventy miles north.

Even more strange is the brief conversation that occurred between the watch officer of the Cape Lookout Lightship and the *Carroll M. Deering*. Someone on the *Deering* shouted through a megaphone that the captain had requested other ships steer clear of the vessel because she had lost her anchors during a storm. The lightship tender thought this was odd because no storms had been reported. Members of the crew were spotted on deck, although they didn't appear to be busy. The ship seemed to be making good time, yet nothing more was seen or heard of the *Carroll M. Deering* until the Hatteras Coast Guard spotted the foundering ship on January 31, 1921.

As if all of this were not enough of a mystery, it was discovered that a passenger boarded in South America. However, the investigation didn't uncover the name of the secret passenger. The ship was only permitted to carry ten men, as per its registration under U.S. maritime regulations, so that would probably explain why the eleventh passenger wasn't documented.

The shipwreck was added to navigational charts as a possible hazard to mariners. Despite the best efforts of the U.S. Navy, the Justice Department, and the Commerce Department, what hap-

During the same time as the *Carroll A. Deering* incident, the steamer *Hewitt* disappeared without a trace along the Atlantic Coast. Some believe that both vessels were taken by Russian pirates. Author Edward Rowe Snow included this possibility in his publication, *Mysteries and Adventures Along the Atlantic Coast*: "When this theory was investigated, it was proved that the American representative of the *Washington Post* knew of several vessels entering the port of Vladivostok under the command of Russian crews. The names of the ships in every case had been obliterated. Eight ships had disappeared at the same time that the *Deering* was wrecked, including the *Entine, Florina, Svartskag, Lorringa* and *Hewitt*. The romantic solution is that 20th Century pirates captured and killed the crew with intention of taking the ship, which in the meantime foundered on the Outer Diamond. The *Carroll A. Deering* incident is considered by many to be the greatest mystery of the seas during the first half of the 20th Century."

pened could not be discovered. Many speculated it must have had something to do with piracy, but there were no signs of an altercation and the ship's captain appeared to be above reproach. His daughter, Miss Wormell, took part in the investigation. Above all, she didn't want her father's name forever linked to piracy or negligence.

A break in the mystery came one day when a bottle with a message washed ashore at Cape Hatteras and was found by a local resident. The note said that pirates had attacked the *Carrol M. Deering*, and those who survived the attack were put into lifeboats without oars or provisions. The government investigated the note, despite the fact that they considered it ludicrous that pirates would seize a ship at Hatteras in 1921. The inquiry resulted in a confession by the man who wrote the note and put it in a bottle, and then claimed to have found it on the shore.

In the end, the government concluded that the crew left the sinking ship on a lifeboat and must have been lost at sea while trying to escape. That sounds plausible except that no trace of the yawl or the men was ever found; not even a lifejacket or an article

Some of the theories advanced over the years include mutiny, piracy, murder, and an escape boat lost at sea with all men drowned. Many believe the escape boat theory is the most plausible. The thinking is that the vessel became disabled due to storms and it was anticipated she wouldn't make it, so the crew jumped out of the ship before it hit the sandbar. This makes the most sense because it explains why no signs of drowned men were ever found in the area: They had abandoned the ship long before Diamond Shoals. The only flaw in this theory is that it does not explain why a meal was prepared and served but not touched. If it was decided it was best to abandon the ship, a meal would never have been prepared, or it would have been at least partially eaten. Furthermore, no explanatory letter or distress signal seems to have been attempted.

of clothing or a piece of the boat—nothing. Miss Wormell still contends that the crew encountered foul play, although there is no outstanding evidence to support that claim. Ultimately, the government has stopped looking for answers and says that the mystery is only another that the sea provides so frequently.

The owners of the sunken ship hired Merritt Chapman Wrecking Corporation to salvage some of the items, which were sold at an auction. Later, a nor'easter scattered the remaining hulk all over the shoreline. Locals grabbed up usable wreckage. Remnants from the shipwreck helped build many houses along the Outer Banks. The stern sank and is part of the Graveyard of the Atlantic. The Coast Guard used explosives to blow up the remaining wreckage to keep it from being a navigational hazard.

Although the physical evidence of the mysterious ship is long gone, legend has it that strange sounds are sometimes heard when storms blow through Hatteras in February. Some say it's just the wind, but others believe it may be the spirits of the crew of the *Carroll M. Deering* still trying to tell us what happened to them.

The Graveyard of the Atlantic Museum, with a projected opening date of 2003, will be the world's only museum devoted solely to the shipwrecks and maritime history of the Outer Banks. This 19,000-square-foot-museum and gift shop at the Hatteras–Ocracoke Ferry Terminal and U.S. Coast Guard Base will house paraphernalia from over one thousand ships wrecked along North Carolina's coast, including the *Carroll A. Deering*. Exhibits will include a logbook from the Civil War with descriptions of actual battles, as well as old Coast Guard and Lifesaving Station uniforms and medals. For more information call (252) 986-2995 or visit www.graveyardoftheatlantic.com.

The Chicamacomico Lifesaving Station was one of the first of many lifesaving stations placed along the Outer Banks. This group station and outbuildings are considered to be one of the most complete U.S. Lifesaving Service/Coast Guard Station complexes on the Atlantic Coast. The original building, constructed in 1874, was turned into a boathouse when the new shingle-style station was built in 1911. Both are included in the large complex. Part of the 30,318 acres that comprise Cape Hatteras National Seashore, the station is located on NC 12 in Rodanthe, about twenty-five miles south of Nags Head. For more information, call (252) 987-2401.

This is a report filed by Coast Guard Station Surfman C.P. Brady of Coast Guard Station No. 183 on the *Carroll A. Deering*.

Courtesy of the Outer Banks History Center.

TREASURY DEPARTMENT,
U. S. COAST GUARD.
Form 2625.
F. C., May 1-18.

No. 2

REPORT OF ASSISTANCE RENDERED

Coast Guard Station No. 183   District No. 7

Jan. 31 and Feb. 4, 19 21
(Date of casualty.)

1. Nationality, name, and rig	American schooner, Carroll A. Deering
2. Hail port and gross tonnage	Bath Me. 2,114.
3. Where from and where bound	Not known
4. Number of days out	Not known
5. Number of crew (including master)	Not known
6. Number of passengers	Not known
7. Estimated value of vessel	Not known
8. Estimated value of cargo	No cargo
9. Nature of cargo	No cargo
10. Estimated damage to vessel	Total loss
11. Estimated damage to cargo	No cargo
12. Name and address of { Master	Not known
{ Owner	G.G.Deering Co., Bath, me.
13. Supposed cause of DISASTER.	Thick weather
14. Nature of—stranded, collided, etc.	Aground
15. Time of day or night, and date	Discovered at 6:30 a.m. january 31, 1921
16. Exact location—direction and distance from the station	N.W.Point of S.W.Diamond shoal,9 miles S.S and Surfman Andrew Gray.
17. Discovered by whom, and time of discovery	6:30pm, Jan. 31 1921 by surfman C.P.Brady
18. Direction and force of wind. WEATHER.	S.W. 5.
19. State of sea and tide	Rough sea, Strong tide.

20. Temperature and weather symbols .... Tem:50. M.

ASSISTANCE RENDERED.
21. Time of starting to scene .... 7:30 a.m.

22. Time of arrival at scene .... 11:30 a.m.

23. Time of launching boat .... 10:00 a.m.

24. Number of shots fired .... None

25. If any shots were unsuccessful, state briefly cause, each case .... None

26. What members of the crew did not participate in the operations, and why? .... F.M.Miller absent sick, W.R.Midgett on liberty, L.G.Hooper on leave, C.D.Burrus absent sick.

27. Number and names of persons lost .... Not known

28. Number of persons resuscitated from apparent death by drowning or exposure to cold .... None

29. Number of bodies found, and disposition made of same .... None

30. Number of persons sheltered at station, how long, and number of meals furnished .... None

31. Accidents to crew of station .... None

32. Damage to boats or apparatus .... None

33. If anything occurred to interfere with operations, state fully the nature and cause .... Nothing occurred

34. At what time (give date, hour, and minute) and in what manner was cutter notified? .... 1:30 p.m. (Via) Cape Hatteras Radio station

35. Was vessel assisted in danger of damage or loss? .... No

36. Were the lives of persons on board imperiled? .... No persons aboard.

77

Haunted Houses

Slocumb-Sandford House

Fayetteville

She's been called everything from the "Lady in Black" to the "Woman on the Stairs," but who is this benevolent spirit that frequents this old house located at the intersection of Halliday and Dick Streets? It could be the ghost of Margaret Halliday, who married John William Sandford and moved into the dwelling in 1830. It is said that she absolutely loved her home and kept it in perfect order. Everything was always clean and neat and in place. Some believe she comes back to make sure the house remains in as good condition as she kept it.

Still others are convinced it is none other than the fiancée of a man who passed away in the house. During the War Between the States, the house was occupied by Union soldiers. It was reportedly used as an underground passage to the Cape Fear River. A man who was possibly a Confederate soldier or sympathizer was captured by Union soldiers and left in the basement to die. The young lady he was betrothed to never got over her loss, and many believe her spirit lingers in the dwelling because that's where she feels closest to her beloved. Those who have seen the spirit agree that it is an attractive young woman garbed in a black dress, floating on the stairs.

The house was built around 1800 and serves as a true compliment to antebellum architecture. Some of its outstanding features include a beautiful ballroom and a handmade marble mantel. In addition to being haunted, the house was reportedly a bank during the 1820s. It is also rumored that General LaFayette may have stayed in the mansion. The dwelling now belongs to the Woman's Club, which bought it in 1945.

Thompson House

The driver didn't see John Thompson until it was too late to avoid hitting the pedestrian. An ambulance was called and the injured man was taken to Baptist Hospital, where he was diagnosed with a fractured skull. The poor man never regained consciousness.

After his death, Mrs. Thompson began renting out rooms to supplement her income. Tenants have always complained of hearing a swinging gate, even when the gate is not moving, and footsteps coming up the walk when no one is there. They also say they have heard the basement door opening and soon afterwards the sound of running water. This was especially unnerving to one tenant, Ann Bobbit, since she had never been able to get the basement faucet to work. Her husband didn't believe her accounts until he experienced the same thing. The couple went to talk to Mrs. Thompson to try to get out of their lease, but Mrs. Thompson told them not to worry, that it was just her late husband!

"In 1937 he was killed while on his way home from work. He always walked home and he was hit by a car one night. John always entered the yard through the swinging gate. He came in the house by way of the basement and washed his hands down there before coming upstairs for supper. I think what you're hearing is John trying to get back home. He's harmless," she assured them.

The couple did stay and continued to hear the noises from time to time. The dwelling is now owned by Jeanette Wells, who also rents out rooms. It's been a long time since anyone has reported any strange events at the old Thompson house. The swinging gate and walkway Mr. Thompson always used are now gone.

Burgwin-Wright House

Wilmington

Built around 1770 by merchant and planter John Burgwin, the house represents Georgian-style architecture. The land the house is built on was purchased in 1744 so a jail could be constructed on it. Wilmington's abandoned jail serves as an outstanding stone foundation for the structure. A secret tunnel extends from the house down to the Cape Fear River. While its existence has been rumored for decades, the remains of the tunnel were discovered just a few years ago. The house itself also has an intriguing history. It was occupied in 1781 by Lord Cornwallis, who used it as his headquarters just after his defeat in the Battle of Guilford Court House. In fact, Cornwallis kept his prisoners in the former jail until he surrendered at Yorktown.

Joshua Grainger Wright bought the house in 1799. It remained a residence until 1937, when the National Society of the Colonial Dames of America in the State of North Carolina bought the house. The women opened the house to tourists in 1950. It is considered a great example of a colonial gentleman's town residence. The house has double porches on both sides and a formal garden. Inside, visitors can see its exquisite eighteenth- and nineteenth-century furnishings. They may even witness a ghost since the property is reportedly haunted, some say by the ghosts of prisoners whom Cornwallis stashed in the dungeon or former jail.

According to docent Ardell Tiller, something unexplainable occurs every day in a small bedroom upstairs. Tiller closes the little door to the Blue Bedroom each night before leaving, and every morning she finds it open. She says no one claims to have opened the door, and it closes tightly, so it couldn't just blow open. Tiller says that is the only strange thing that has happened since she

European traders and settlers came to Wilmington because of its location and resources. The city, incorporated in 1739, has become one of North Carolina's biggest tourist attractions and one of America's biggest districts on the National Register of Historic Places. Television and movie producers often use the city's two-hundred-block historic district and nearby Wrightsville Beach.

became a docent a couple of years ago, but she has heard the story of the spinning wheel.

Some years ago there was a spinning wheel on the hearth of a room upstairs. One day when a tour reached that room, the docent began her talk on the room and its furnishings. Suddenly, the spinning wheel began turning. The tour guide was so shocked she stopped talking and walked over to the wheel. It suddenly stopped. The woman explained to the participants that the spinning wheel had been stuck in place until now. Efforts to loosen it had been unsuccessful and the women had ceased trying because they didn't want to damage the old wheel. If that wasn't strange enough, another examination of the wheel showed that it was stuck in place again. The brief, rapid turning of the wheel could never be explained. The spinning wheel is no longer displayed in that room at the Burgwin-Wright House.

Whichard House

Located on Fifth Street, this circa-1940s dwelling was once the home of a man who owned several different kinds of games, including a few pinball machines. All the games required a nickel to play and the old man kept the proceeds in a big bucket in a room on the second floor. It was hard for him to get up and down the stairs, so he lived downstairs and used the upstairs as a storage area. Once a week, the gentleman took the week's collections upstairs. He spent hours counting the nickels, and after he filled a bucket, he started filling another. The old man moved them around by dragging the containers across the floor.

Later, the first floor of the house was used as a rental property. The second floor remained unoccupied. Tenants complained of hearing dragging and clinking noises overhead. One couple even called the police, who investigated but found nothing. When the noises persisted on a weekly basis, the frightened pair called the landlord. To their surprise, he laughed and told them not to worry, that it was only his grandfather counting and dragging his buckets of nickels.

Hannah Clark House

New Bern

Its most notorious resident was Captain Edward B. Tinker. One time when returning to New Bern from Baltimore, he crossed the Pamlico Sound and encountered a big storm. The deck was closed off and the crew scrambled to make repairs to equipment and keep the ship afloat, which was taking a powerful beating from the nor'easter. The vessel pitched and hurtled, and many aboard began to wonder if they were going to survive. Part of the cargo, which included gold, was moved from the cargo hold to the lifeboats. This was often done during storms, in the best interest of the ship. When the worst of the squall had passed, the precious metal was loaded back onto the ship. During this process, it was discovered that a significant amount of the gold was missing. The captain said it had probably gone overboard during the hasty transfer or while it was in the little lifeboats. He asked the crew to sign sworn statements corroborating his account. All agreed, except a cabin boy named Edward.

Captain Tinker asked this lad to crew for him on the trip from New Bern to his plantation on Brice's Creek. Shortly before reaching his home, the captain told the youth he had decided to do a little duck hunting. He just needed the boy to go disturb the ducks from their resting place in the marsh. It was understood that when they flew out, he would shoot the wild fowl. Instead, Tinker shot Edward! He tied rocks to the boy's body to weigh it down and then dumped it into the river.

But the rocks came loose and the corpse floated to shore, where it was later found. Tinker was arrested for the boy's murder. The old seaman bribed a guard and escaped to Philadelphia. The captain was found and brought back to New Bern for trial.

85

Another gentleman who had been on the boat and saw what happened testified against Tinker. This resulted in the hanging of Captain Tinker in 1811 for the murder of his cabin boy.

Built around 1800–1820, the house on Craven Street, known as the Hannah Clark House, was demolished in 1935. After it was torn down, workmen found a skeleton buried in a grave under the house. Long before this discovery, residents of the house swore they heard voices in empty rooms and saw rocking chairs move without a breeze or any other explanation.

The John Wright Stanly House was relocated to this site and bizarre incidents continued to occur. The cook was shot by her ex-husband, who then turned the gun on himself. The bodies were found side by side on the kitchen floor. Some months later, the lady of the house, Mrs. Farrow, and her mother, Mrs. Morris, were in the kitchen discussing the matter. As they talked, they glanced at the rocking chair in the corner of the kitchen, remembering how their former cook had loved sitting in it. Suddenly, the chair moved as if someone were in it. With a rapid motion, the chair rocked back and forth. As the women watched, the chair stopped rocking just as quickly as it had begun, and the room suddenly felt very cold.

Mrs. Farrow also reported hearing water running in the kitchen sink on two or three occasions. When she checked it, the faucet was always turned off but there was water in the sink. The sound of slippered feet crossing the room has been heard, but when someone checked to see if anyone was in the kitchen, no one was discovered. Most disturbing of all were the sightings. Men were seen in different rooms of the house for a moment or two before they vanished.

These strange men were seen even before the house was moved. Neighbors reported ghost sightings and sounds at night. These specters are believed to be descendants of Governor Richard Dobbs Spaight Sr., who lived in the John Wright Stanly House. In 1802, Congressman John Stanly, son of John Wright Stanly, killed Spaight in a political duel in New Bern.

Horace Williams House

Chapel Hill

Chapel Hill has many historic homes, several of which are reportedly haunted. This particular dwelling, originally an 1840s' farmhouse, is at 610 E. Rosemary Street. When Horace Williams bought it, it had an octagonal-shaped addition as well as a parlor and front porch that had been added in the 1880s. Williams was a distinguished professor of philosophy at the University of North Carolina at Chapel Hill. He died of natural causes in 1940 and willed the house to the school with one stipulation: that it be left as is. Through the years, the house has been home to many tenants.

Catherine Berryhill Williams, Special Events Coordinator for University of North Carolina President Molly Broad, was one such inhabitant. For four years, she and her family lived in the house. Because of strange incidents that occurred in the house when Mrs. Williams was a child, none of her friends wanted to come over and play, Williams recalls during an interview by the *Daily Tar Heel*. Though Williams claims she doesn't believe in ghosts, she did admit that she and her family felt a presence in the house.

Additionally, things were often found moved in the house and no one admitted to moving them. "In the evening, my mother would bank the fire before she went to bed and she would place the fire utensils on one side of the fireplace," Williams said. "And in the morning they would be on the other side of the fireplace."

Williams said her sister claimed to have had several conversations with Horace Williams' ghost. "My sister said he would just sit on the edge of her bed and talk with her," Williams said. "It was not a scary presence. It wasn't threatening. But we all felt it."

The Horace Williams House is part of the historic district and is now home to the Chapel Hill Preservation Society. The school

87

still owns the house, but the Preservation Society leases it. Society Director Catherine Frank says that she has not witnessed any unusual occurrences but that her predecessor claimed that a metal container that was kept on the mantel was often found in a different spot from where it was placed. She says they do have trouble periodically with one of the lights, but Frank believes that it has more to do with the old wiring than with ghosts. The caretaker for the property was interviewed a couple of years ago on a local radio station, and he revealed that there is a rocking chair that sometimes rocks for no apparent reason.

Some famous haunted places include Gettysburg, *Queen Mary*, the Tower of London (England), the Alamo (Texas), Hotel Del Coronado (California), Equinox (Vermont), and the White House. The home of the president of the United States is reportedly haunted by Abraham Lincoln. It's generally believed that his ghost appears when America is in turmoil or crisis. While her husband was president, Eleanor Roosevelt allegedly communicated with Lincoln through séances. Some maids reported seeing a ghost during Harry Truman's administration. Rosalyn Carter refused to discuss it, and Jacqueline Kennedy told the media that she "took great comfort in it."

Reed House

In 1892, Samuel Harrison Reed purchased eighteen acres and built an enormous home, which contains ten fireplaces, sixty-two windows, and a turret. Samuel Reed and his wife had nine children. Sadly, five of the children didn't live long enough to become adults. His wife also died prematurely at age fifty. Less than a year later, Samuel Reed passed away. It's thought that they all died at home. The four remaining children were sent to live with relatives and the house was rented out and eventually sold.

The estate had many owners through the years. Ultimately, it was abandoned for roughly ten years, which resulted in significant deterioration. In 1972, the city decided to condemn it. But Marge Turcot, president of the Preservation Society of Asheville and Buncombe County, decided the mansion was worth saving and undertook the task. She and her husband bought the run-down house and began restoring it.

The first night they stayed there, the Turcots heard loud footsteps on the back stairs. When Marge checked on her children, she found them all fast asleep. From that night on, the family often heard unexplainable footsteps on the stairs. The sounds of a pool game being played in the billiards room have been heard, but no one is found when a family member investigates. The Turcots also claim to have seen bedroom doors open and close on their own.

The renovated property became a bed and breakfast in 1985 and has accommodated many guests. Some claim to have seen and heard many unexplainable things: books leaping off the bookshelves, heavy footsteps, the sounds of children playing when no children *were* playing, strange sounds coming from the attic, and lights turning on or off on their own. Family members and some

The website www.hauntedhouse.com details haunted houses, forests, barns, and other buildings or places across the country that are open to the public at certain times of the year. These are not real haunted sites but places that have been decorated and made to be fun and somewhat scary for kids and adults. Admission costs, directions, and hours of operation are included. This website welcomes information on annual haunted places and happenings. The site also has over 150 stories that have been submitted by those who claim to have had scary experiences.

guests say they have felt a presence in their bedrooms.

Many believe the only explanation is that the spirits of the Reed family, except for the four who survived, remain in the structure. The Victorian mansion is listed on the National Register of Historic Places and has also been declared a local historic property by the Historic Resources Commission.

North Carolina "Ghost Watch" Contest Winners

Announcing a Statewide Ghost Hunt

North Carolina writer Terrance Zepke is seeking unusual ghost stories to include in her next book, *The Best Ghost Tales of North Carolina*. "I got the idea to hold a contest after hearing so many wonderful stories from people who attended book talks and signings for my previous book, *Ghosts of the Carolina Coasts*. This time, I want to focus solely on North Carolina ghosts, and not just coastal ones. I am looking for little-known or previously unpublished tales."

Submissions should be between 500–1000 words, and documentation (such as photographs or letters) is gladly accepted. Please note as much information as possible, such as names, addresses, and directions (if it is a haunted dwelling or site).

The grand prize winner will receive $100 and a credit line. The first place winner will be awarded $50, plus a credit line. All other accepted entries will receive $25 and a credit line. The *Greensboro News & Record* will publish some of the entries Halloween week. Submissions will be returned if a SASE is enclosed.

Send to: "Ghost Watch" c/o Terrance Zepke, P.O. Box 4881, Greensboro, NC 27404. Be sure to include your name and phone number with area code.

Announcement used for the "Ghost Watch" contest.

I would like to thank the media and participants who responded to my solicitation for ghost stories. The overwhelming interest confirmed my feeling that everyone appreciates a good tale about things that aren't always scientifically accountable or readily rationalized. I received submissions from all over North Carolina and even a few from neighboring states. I was impressed with how many people were willing to share their family history and personal experiences. I am glad to share the best responses on the following pages.

—Terrance Zepke

Grand Prize Winner

Gentle Spirit

Elizabeth W. Blythe, Greensboro

Nestled in the western part of North Carolina is the tiny town of Dillsboro. It boasts the Jarrett House, an old hotel that serves wonderful home-cooked meals and rents a few rooms to overnight guests.

Several years ago, my husband and I decided to spend the night there. We enjoyed a delicious meal and the place's quaint atmosphere. A pleasant breeze enveloped us as we sat in rocking chairs on the front porch, the perfect ending to a very busy day.

Our room on the second floor had the look of a bygone era, with oversized, antique furnishings and fluttering lace curtains. Positioned away from each other were two beds, double and single.

We went to sleep in the double bed. But well into the night, I awakened abruptly. Someone was touching my head. Then I felt a hand on my brow. With a very gentle touch, it patted my bangs and the top of my head again. Thinking it was my husband, I turned over to respond, but he wasn't next to me. Startled, I looked around the dimly lit room and saw him sound asleep in the

single bed next to the bathroom.

After waking him, we talked. He explained that he had gotten restless during the night and decided to sleep in the other bed so I would not be disturbed.

Next morning at breakfast I decided to ask our waiter if the hotel was haunted. He said he had only worked there a short time, but he would ask some of the kitchen help about it. He returned to tell us that there had been reports, but the staff was told not to discuss it, especially with guests. Then the waiter requested we not divulge his comments regarding the matter.

However, as we checked out of the hotel that morning, I decided to ask the cashier if any unusual happenings had ever occurred there. A bit surprised by my question, she first hesitated but then told us that there had been similar incidents reported through the years. She also said that some of the housekeeping staff had, at various times, sensed something unusual, particularly on the second floor, and it was thought to be a presence—that of a lady, perhaps.

The rumor was, the cashier said, that folks thought the presence might be Mrs. Jarrett, who had a reputation for tucking her children in at bedtime and caressing them gently, especially if they were ill.

First Place

Pot of Gold

Rob Kincaid, Greensboro

For many years, my grandmother has told me the story of her cousin, who was frightened beyond belief by "haints" he stumbled upon in the woods on a dark mountain night in rural North Carolina. The following is my recollection of that story.

Cousin Bobby Turbyfield was a young man who lived on the banks of the Elk River in eastern Avery County. He was, by all accounts, normal, with a fondness for the moving picture shows in town.

Mountain life was hard, filled with work and very little else. Farm boys were used to breaking a sweat well before the sun came up. Mountain life was also dull, and a young man of twenty needed an outlet. In the town of Elk Park, that outlet was the fifty-seat movie theater beside Brinkley's Hardware.

Bobby was fully versed in the trouble a boy could encounter after dark. His parents and older relatives had brought him up with the understanding that "no good could come to a body adder dark" and every attempt should be made to get back home. In the mountains, darkness falls like a thick, dark, woolen blanket and at the time was accentuated by the lack of electric lighting. There was very little to break the blackness.

One cool Friday night in late October, Bobby headed home after dark from the movie theater. He took a shortcut through the woods, a well-worn path dense with low-growing brush and still-leafy trees—a path he had walked many times in an effort to shorten the three-mile journey home. Bobby's stride kept pace with his fast-beating heart.

As he came over a small hill and headed down into a valley, he saw what appeared to be a light glowing in the distance. As he approached what he assumed would be local campers, he slowed his rapid pace to a less noisy, step-by-single-step progression.

When he got a clear view of the scene, he stopped cold. Two people, a man and woman, dressed in simple black and white clothes and surrounded by a heavy, cool mist, stood in the light. The man wore a black hat, full coat, knee britches, white knee socks, and black shoes with bright buckles. The woman wore a black, full-length dress, white cotton shirt, and a large bonnet that partially obscured her face. The dim yellow light coming from a lantern hanging in a nearby tree cast their faces in deep shadows, making them look old and wrinkled. They definitely weren't from around those parts.

As he watched the ghostlike figures, his idea of walking closer and speaking to them quickly changed. They carefully moved an old black iron pot that apparently was very heavy. The pot had come out of a huge hole surrounded by dead grass, making Bobby think a wagon-size boulder nearby had been covering its hiding place. While he stood there in frightened silence, the two figures maneuvered the pot to a spot close by and set it down. As they did so, its contents came into clear view: gold coins filled the pot to the brim.

The sight of the coins, given the hard times of the region, made Bobby even more uneasy. What had he stumbled onto? Would the people he was spying on appreciate his knowledge of their actions? Silently, he climbed back up the hill and headed home in another direction.

He could hardly contain his fear and amazement. Who were these people, and how had they moved a boulder the size of a wagon? Why were they dressed as they were, and where had all that gold come from?

When he got home, he told his parents what he'd seen and near-

ly collapsed from exhaustion and fright. But his parents were skeptical of the story and figured it was an excuse for being late.

In the morning, his curiosity still at a peak, Bobby persuaded his father to go with him into the woods to find the strangers. A heavy pot like the one they were moving would be difficult to handle and should have kept them in the area past dawn, he reasoned.

As Bobby and his father walked, Bobby started to feel both excitement and fright at the idea of a face-to-face encounter with the two mysterious people and their gold. He and his father slowed their pace as they neared the site to avoid making noise.

To the surprise and disappointment of both father and son, the pilgrims were gone, along with the pot and its contents. What was left, however, was a hole in the ground and a clear mark that the pot had made. The grass around that spot was pale and lifeless, like that found under large objects.

The story of this late-night encounter and the impression of the pot that had been left in the ground would dazzle the local residents for years to come. The spot remained visible for many years, drawing curious visitors.

My grandmother told this story to my brother and me, telling us to stay within sight at all times and get home before dark. She had no explanation for this event but said she had seen the giant rock that had been moved and the mark the pot had made in the ground.

The mountain folk decided that Bobby had stumbled upon the ghosts of early settlers who had gotten lost on the long, perilous journey out west. After they met with some unknown fate, they returned from the afterlife to retrieve their savings, buried forever in the wilderness of the North Carolina mountains.

Second Place

The Shadow Man

Richard Morton, Oakboro

I have had one experience that I classify as a ghost story, for lack of any other explanation. This incident happened in September 1969 in Stanly County, near the community of Big Lick. At the time, Big Lick was nothing more than a crossroads with a gas station on one corner. I was a junior at UNC Charlotte, majoring in civil engineering, and I was living at home with my parents, James and Imogene Morton.

One evening, my girlfriend, Gail Smith, and her friend Donna Lowder dropped by to see me. This was unusual because we usually called each other every day but did not see each other during the week since she lived in Albemarle. They arrived late in the day, maybe at 6:30 or so, and it was nearly dark.

Because my parents' house is small and they were both home, we decided to ride around a little while in Gail's Corvair. As the darkness grew, I decided to try and spook Gail and Donna a little by driving down the dirt road in front of my parents' house. The road is maybe two miles long and had only three houses then.

There is a long, deserted stretch where the road descends down a hill to a small creek, crossed by a little one-lane bridge without side rails, then rises steeply. I had always heard as a child that someone had committed suicide near this bridge years before, but I didn't really know any of the specifics.

As we neared the bridge, I started slowing down, preparing to tell them this story. By this time it was dark, and I had the headlights on. As we neared the bridge I could see what I thought was a man walking up the hill on the other side. He looked like a shadow man, as he had no real substance and you could see through him, but was still recognizable as a person.

Donna and I were looking at this thing, but Gail was looking down, fiddling with the cigarette lighter. I had just started to say, "What the heck is that?," when all of a sudden as if he realized we were watching, the shadow man condensed into a ball of white

102

light about the size of a basketball. I was hoping it would hover awhile so we could observe it, but it shot off to our left. Only two or three seconds had elapsed from the time I spotted the man until he disappeared.

I wasn't scared as much as fascinated, but Donna started screaming and I took off just to settle her nerves. Gail missed the whole thing and was wondering what in the world was going on. Donna was really upset, so we went back home and related the story to my parents. Donna's frightened demeanor made my parents believe us, but there really wasn't much they could do other than listen. It made the hair stand up on the back of your neck— you know you're looking at something that is not explainable in our cause-and-effect world.

I've been down that same road other times at night but have never seen anything strange again and have never heard anyone else report anything. The area looks pretty much the same now as it did then, but there is a subdivision going up nearby, and the state is going to pave the road soon. Without Donna's reaction, I might have tried to explain it away, but I know what I saw, and her description was identical to mine.

Third Place

Blackbeard's Tale

Robert F. Bell, Ridgeway, VA

(This is a story about the author's ancestors, written for his grandchildren.)

There are quiet times along the Pamlico Sound of North Carolina when the breeze is southerly, the night skies are dark, and the fishing boats are in port. Sound travels far across the water, and some say there is a persistent noise that can be heard only from three to four in the morning. The haunting sound is of oars "thunking" in wooden locks and of an angry wail for vengeance. Legend reports the sighting of a ghostlike figure standing astern a small wooden longboat while gaunt sailors row up and down the northern shore of Pamlico Sound from Swan Quarter to Bath. I have seen the vision and heard the cry on more than one occasion, and I was close enough once to make out the call. In an Elizabethan cockney accent, the tall, grizzled specter shouts out the name "Bell" and waves a mighty arm as if inviting us all to face his furious vengeance.

The Colonial Records of North Carolina are clear about the facts of the infamous pirate Blackbeard's death on Ocracoke Island in 1718; and the Virginia Colonial Records concur and even shed further light on the occasions leading up to that fateful battle. Blackbeard had a large and comfortable home in Bath, right next door to the residence of Governor Hyde. He was thought of as a patriot and, at worst, a privateer, but he wholesaled much of his ill-gotten goods to the merchants of the area and entertained widely. Somehow, Edward Thatch, alias Edward Teach, alias Blackbeard the Pirate, had gained tolerance (if not respectability) in Bath despite his heinous behavior at sea and his horrific appearance and evil glare.

Legend has it that Blackbeard sought the hand of Hyde's beautiful daughter in marriage, but she spurned his offer, claiming to love another gentleman. That unfortunate young gentleman promptly disappeared and was never seen again, except that his severed hands were sent in a package to the beautiful daughter of Governor Hyde.

The major player in the demise of Blackbeard was our forefather William Bell, a respected merchant and treasurer of Currituck, precinct of Hyde County. His family were early settlers inland at Bell Bay, about fifteen miles east of Bath. He had a loving wife and one young son, William Jr., whom he worshipped. In those days, there were no roads, and the only means of transportation was by boat. William Bell sold and traded his goods to other planters and settlers up and down Pamlico Sound from his wide, flat-bottomed sailboat (called a piragua).

On the night of September 14, 1717, William Bell was tied up at Chester's Landing, a few miles west of Bell Bay. Within the small cabin of their piragua were Bell, his young son, William Jr., and a young Indian companion. They were on their usual trading route circuit. Sometime between the hours of three and four o'clock that morning, they were stirred by the sound of an approaching longboat and boisterous laughter from the oarsmen. When the tall and swarthy skipper of the longboat stepped aboard Bell's piragua, he shouted for a drink of brandy. Still groggy from sleep, William Bell demanded, "Who are you, sir, and where do you come from?"—To which the dark intruder snarled, "My name is Thatch, and I come straight from Hell; and I shall soon take you thence if you don't fetch me some brandy."

It was Blackbeard himself, no doubt about it—drunk and in a very black mood. An argument ensued and Blackbeard called to his crew for his sword. The argument turned violent. Blackbeard's sword was broken in the fray, and Bell was beaten severely. Their piragua was towed out into the Sound, stripped of all its goods and cash, and set adrift without sails or oars. The longboat was rowed west with the loot, which included sixty-six pounds sterling and a few shillings, and half a keg of coveted brandy. Curiously, Blackbeard also stole a prized family possession: a small, silver cup, uniquely made.

Later that morning, Bell, his son, and the Indian lad were able to make it to shore. William set about to end the career of Blackbeard, whatever it took. He wrote to Alexander Spotswood, who was lieutenant governor of Virginia, and asked for help. He also gave specific information on where Blackbeard's ship, *Adventure*, was located. William did not request help from North Carolina Governor Hyde because it was common knowledge that Blackbeard was paying Hyde handsomely for amnesty and protection.

Spotswood was delighted with the request. The wealthy merchants of Virginia, constantly harassed and robbed by Blackbeard, had demanded action from the Crown. At last, he had a request to tread on the sovereign territory of North Carolina for the worthy cause of justice. The lieutenant governor dispatched Lt. Robert Maynard and two sloops with orders to arrest or kill Blackbeard and his men. Bell's directions were given to Maynard, and upon arrival at Ocracoke Island (where Blackbeard was hiding) on November 22, 1718, he found the *Adventure* at anchor, as expected, at Teach's Hole, the only deep-water anchorage on the southwestern corner of the island. A fierce battle ensued and resulted in hand-to-hand combat with swords. Blackbeard was a giant of a figure, some say seven feet tall, with long black hair over much of his body. He roared with anger and fought savagely but in the end was felled with over a hundred saber cuts. The few remaining crew were captured and taken to Williamsburg, Virginia, to be tried for piracy.

Maynard severed Blackbeard's head and threw the body overboard. Legend has it that the body of Blackbeard continued thrashing, even without a head, and swam around Maynard's ship three times before sinking to the bottom of Teach's Hole. The grizzled head was affixed to the end of the bowsprit of Maynard's ship for all to see as they set sail back to Virginia.

And among the curiosities found aboard Blackbeard's ship was a unique silver cup that could only have been that of William Bell. There are also many unanswered questions. Bell probably had a

history of trading with Blackbeard and his men. All of the other merchants of Pamlico Sound did, and there is no reason to think that Bell had not also profited richly in this trade with the infamous pirate. Bell even knew the precise location of the *Adventure*. Why then was Bell the one to blow the whistle? Was it just a matter of reaching the limit of tolerance for intimidation and bullying? Was this the time when something had to be done? Or were there other, more sinister motives?

Bell was a respected citizen, and Blackbeard certainly needed a cohort with whom he felt he could entrust his booty. We must remember that no one has ever found the very extensive loot of gold, silver, and jewels that Blackbeard amassed during his years of piracy. Although Blackbeard favored the luxury of the town of Bath, with its comfortable homes, it was unlikely he would hide his loot near his domicile. His anchorage was off Ocracoke Island, and he would have had to sail his longboats right past Bell Bay before the approach to Bath. What if he made regular deposits with his then-trusted friend, William Bell, before proceeding on to the respectable Bath? Would simple greed not explain why Bell was anxious to see a fatal end to Blackbeard? Knowing Blackbeard would never be taken alive, and knowing that Spotswood of Virginia was the only person who could complete the necessary tasks, Bell could have cunningly plotted the embezzlement of Blackbeard's millions. Could the untold millions still be buried somewhere within the hundreds of acres of land surrounding Bell Bay?

We can imagine the fury and anger that possessed Blackbeard when he realized the identity and treachery of his betrayer, and how he cursed the name of our beloved forefather and all his progeny. Perhaps, that would also explain reports that the head of Blackbeard, affixed to the bowsprit of Maynard's ship, silently mouthed one word over and over: "Bell . . . Bell . . . Bell." It would also explain why to this day we, the Bells, experience the specter of a seven-foot-tall image standing astern an ancient longboat as it is rowed along the shoreline of Pamlico Sound, and why we alone can hear the wailing and wrathful challenge to our family during the hours of three or four A.M. on very dark and quiet nights in these backwaters of eastern North Carolina.

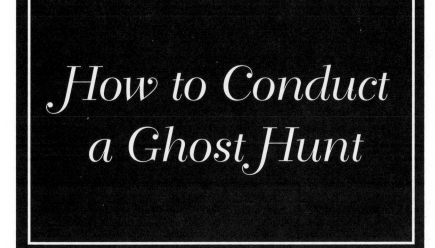

How to Conduct a Ghost Hunt

How to Conduct a Ghost Hunt

An investigation of this sort begins by going to a place you believe may be haunted and then using scientific techniques to prove or disprove the existence of a ghost or ghosts at that particular location.

I recommend that you tag along on a ghost hunt with an experienced amateur or a professional organization before attempting your own investigation. There is such a group in North Carolina, the Triangle Paranormal Research Society (TPRS). According to its website, TPRS was formed in 1997 "so that people in the Triangle area of North Carolina, who have an interest in learning more about ghosts, can meet, discuss the nature of ghosts, and investigate possible hauntings. The basic belief of TPRS is that ghosts (as they have manifested themselves) do exist. However, we feel that the nature of these entities, and the characteristics of their interaction with humans, have not been fully investigated. We intend to research the nature of ghosts, and do field work to gain personal experience and documented evidence."

If you feel you are ready to head up your own group, do some preliminary work before the actual investigation. Check out the area in advance. Make sure you can find the place and that you will be able to gain access. You do not want to trespass on property that is roped off, fenced in, or clearly marked "No Trespassing." Find the best parking spot and look for obstacles you could trip over in the dark, such as stumps or holes. Take note of good spots in which to carry on the investigation—in other words, where you want to be positioned when you come back for the real investigation.

Bring someone with you. You should never go alone in case you encounter a problem, such as car trouble or getting sick or injured. Dress appropriately. Wear comfortable shoes and bring a jacket, if necessary. Don't drink alcoholic beverages before or during the study as you need to have your wits about you. Don't talk, smoke, or wear cologne or perfume because spirits often emit sounds and scents to get attention. If you're wearing perfume or doing any of these other activities, you could jeopardize detection.

Since the best "psychic" hours are late at night and into the wee morning hours, make sure your vehicle has a full tank of gas and has been serviced recently. That means that the battery is good, all tires are in good shape and there is a spare in the trunk, headlights are operating properly, and antifreeze or coolant is sufficient. It's a good idea to bring along a cell phone, in case of emergency.

Do your homework. Before heading out to cemeteries, schools, theaters, battlefields, churches, inns, houses, former plantation homes, lighthouses, cottages, bridges, or other possible haunted sites, talk to long-time residents, former or current owners, and historians. For information on property ownership, go to the courthouse. Get permission from the owner to be on his property. It might also be a good idea to notify local authorities so they don't show up during your investigation and interfere with it.

Go to the local newspaper office and library and look for articles about the edifice or destination. Did any significant events occur there, such as a suicide, murder, or battle? Read books by local authors. If possible, verify whether the haunting is urban legend or if there have been actual sightings or corroborating events. Explore whether there are logical explanations for incidents, such as recent construction or animals (squirrels, mice, cats, etc.). Find out if the spirit is benevolent, mischievous, or downright mean and dangerous. Learn in which spot or direction you're most likely to encounter it.

What should you bring on such an investigation? A basic but

According to the Society for Paranormal Investigation Research and Information (SPIRIT), there are three ways that ghosts show up in photos:
- as mist (foglike discoloration)
- as vortex (crescent-shaped pale areas that are usually white)
- as orbs (these round pale areas resembling globes or balls of light are the most common)

necessary tool is a **35-mm camera** with **high-speed film**, such as 400, 800 or 1600 ASA. Since built-in flashes have an average range of twelve feet or less, a more powerful mounted flash is highly recommended. When dropping the film off for developing, indicate on the instruction box of the envelope that you want all photos developed and returned, regardless of quality. Also, make sure the batteries have plenty of life left. Bring extra batteries, even if the ones in the camera are new or almost new. Clean the lens thoroughly with a soft cloth.

A **video camera** is even better, if you have one or access to one. It should be able to record in low light. A tripod is optional. If possible, use both a camera and camcorder.

Never leave home without a **flashlight**, stocked with **new batteries**. Note that batteries run down faster in cold weather. Some professional ghost hunters recommend bringing extra batteries in case of "interference" from the spirits. It is also a good idea to bring extra flashlights. Avoid tiny flashlights, like pocket-size or pen lights. You need something that emits a bright beam.

You should take notes, possibly even make basic drawings, including as much detail as possible of what you saw and where, so a **small notepad** that will fit into your pocket and a couple of **pens** are essential. You may later input this information into a succinct record, such as the "Ghost Hunt Log" (see the sample log on pages 114–115).

A **tape recorder** with a **microphone** and **new batteries** is a good tool. There are voice-activated tape recorders on the market nowadays.

Some optional tools for the serious sleuth include:

A **digital or electronic thermometer** or **thermal scanner** used to detect cold spots.

A **motion detector** used to sense movements by unseen objects or forces. There are battery-operated ones for under $20.

An **EMF detector** reads an area's electromagnetic fields. It is commonly believed that high readings are indicative of spirits. It is a good idea to take a reading when doing the initial investigation preparation at the sight. Also, take note of things that might interfere with a normal reading, such as power lines. Some ghost hunters say a compass can be used instead of this costly device because it works the same way, in principle. The needle will move wildly or shake if something is interfering with a normal reading.

While talking is not encouraged on ghost hunts, it is a good idea to be able to communicate with whoever is accompanying you, especially if you are out of sight of one another. Cheap **walkie-talkies** can be purchased for $10–$50 at most drugstores, electronics outlets, and toy stores.

All equipment should be cleaned and calibrated beforehand to ensure accurate readings.

Happy hunting!

GHOST HUNT LOG

Date _____   Time _____

Investigator _____   Location _____

Weather _____

Other Investigators Present _____

_____

_____

Equipment:

Camera _____   **Video Camera** _____

Film Speed and Type _____   VHS _____ VHS-C _____

Brand _____   8 mm _____ Digital _____

Exposures _____   Other _____

                               Length _____

Tape Recorder _____   **Thermometer** _____

Microcassette _____   Standard _____

Standard Cassette _____   Electronic _____

Length _____   Infrared _____

EMF _____   Night Vision _____

Phenomena witnessed by investigator:

Time _____   Captured on Film _____

Phenomenon

_____

_____

_____

_____

_____

Time _____   Captured on Film _____
Phenomenon

_____
_____
_____
_____
_____

Other Comments

_____
_____
_____

Roles of film used _____
Audiotapes used _____
Videotapes used _____
Number of psychic photos _____
Phenomena captured on film _____
Summation

_____
_____
_____
_____
_____
_____
_____
_____

Investigator's Initials _____

Resources

Note: The following websites might be useful to those who are interested in learning more about this subject, but neither the author nor the publisher endorses them or the links they may offer.

The Triangle Paranormal Research Society (TPRS) has a website that explains its mission and details professional equipment and publications that are available for loan to members. **www.pagesz.net/~petrone/tprs.htm**

The Society for Paranormal Investigation, Research and Information (SPIRIT) has a website that includes numerous ghost stories. **www.ghosthunter.org**

Throughout history, almost every culture has acknowledged ghosts. The Egyptians called them *khus*. The Romans labeled the good spirits as *lares* and the bad ones as *lemures*. The Japanese call them *shojos*, while the Irish refer to them as *tash*. In India, there are several different classifications of ghosts, from *bautas* to *mumiais*. Some Native Americans call them *spooks*.

At one time, Icelandic law sanctioned judicial refuge against bothersome spirits. Victims were permitted to resurrect spirits before the court to put a restraining order against offending ghosts!

The International Ghost Hunters Society has a website that is full of photos, latest news and conferences, an on-line store, plus a home-study course on how to become a ghost hunter. Sign up for a free electronic newsletter that is full of ghost stories and updates. **www.ghostweb.com**

The ShadowLands is a useful website that contains many ghost stories submitted by researchers and visitors alike, as well as photos, articles, a listing of haunted places, and a training and tips page. It also has an interesting feature, "Find a Grave," which lists brief biographies and burial sites for famous people such as William Shakespeare and Marilyn Monroe. **www.theshadowlands.net**

The Spectre's website discloses ghost-hunting tips, current investigations, and research help. **www.geocities.com/coghosthunters**

Ghost America has a unique feature on its website. It offers a state-by-state listing of paranormal research groups, ghost investigation resources, tips, and stories. **www.ghostamerica.com**

In 1927, Dr. J. B. Rhine and his wife, Dr. Louisa Rhine, came to Duke University to study psychic phenomena with Psychology Department Chairman William McDougall. For a while, Duke had a parapsychology lab. Rhine later founded the Rhine Research Center Institute for Parapsychology to guarantee that his research progressed as he envisioned it should. The center is a "non-profit research and education organization established to explore unusual types of experiences that suggest capabilities as yet unrecognized in the domain of human personality, and to investigate

Some well-known movies featuring ghosts include *Ghost Story*, *Casper*, *The Canterville Ghost*, *Ghost Busters*, and *The Frighteners*.

those capabilities thoroughly by exact scientific methods." The center continues the impressive research and education missions set forth by the program Duke established. The center offers a Community Education Program and Summer Study Program, which consists of introductory classes and workshops for those who wish to gain a better understanding of parapsychological research. It also has a reference library and bookstore. The Rhine Research Center is located at 402 North Buchanan Boulevard (adjoining the east campus of Duke University), Durham, NC 27701. Telephone: (919) 688-8241.

For more information on North Carolina destinations, contact the North Carolina Travel and Tourism Division Department of Commerce at 430 N. Salisbury Street, Raleigh, NC 27611. Call 800-VISITNC or (919) 733-4171, or visit **www.visitnc.com.**

Index

Index

Index

Index

If you enjoyed reading this book, here are some other books from Pineapple Press on related topics. For a complete catalog, write to Pineapple Press, P.O. Box 3889, Sarasota, FL 34230 or call 1-800-PINEAPL (746-3275). Or visit our website at www.pineapplepress.com.

Ghosts of St. Augustine by Dave Lapham. The unique and often turbulent history of America's oldest city is told in twenty-four spooky stories that cover four hundred years' worth of ghosts. ISBN 1-56164-123-5 (pb)

Ghosts of the Carolina Coasts by Terrance Zepke. Taken from real-life occurrences and Carolina Lowcountry lore, these thirty-two spine-tingling ghost stories take place in prominent historic structures of the region. ISBN 1-56164-175-8 (pb)

Haunt Hunter's Guide to Florida by Joyce Elson Moore. Discover the general history and "haunt" history of numerous sites around the state where ghosts reside. ISBN 1-56164-150-2 (pb)

Haunting Sunshine by Jack Powell. Take a wild ride though the shadows of the Sunshine State in this collection of deliciously creepy stories of ghosts in the theatres, churches, and historic places of Florida. ISBN 1-56164-220-7 (pb)

Historic Homes of Florida by Laura Stewart and Susanne Hupp. Seventy-four notable dwellings throughout the state—all open to the public—tell the human side of history. Each is illustrated by H. Patrick Reed or Nan E. Wilson. ISBN 1-56164-085-9 (pb)

Houses of St. Augustine by David Nolan. A history of the city told through its buildings, from the earliest coquina structures, through the colonial and Victorian times, to the modern era. Color photographs and original watercolors. ISBN 1-56164-069-7 (hb); 1-56164-075-1 (pb)

Lighthouses of the Carolinas by Terrance Zepke. Eighteen lighthouses aid mariners traveling the coasts of North and South Carolina. Here is the story of each, from origin to current status, along with visiting information and photographs. ISBN 1-56164-148-0 (pb)

Mystery in the Sunshine State edited by Stuart Kaminsky. An enticing selection of Florida mystery fare from some of Florida's most notable writers. ISBN 1-56164-185-5 (pb)

Oldest Ghosts by Karen Harvey. In St. Augustine—the oldest settlement in the New World—spirits tap tourists on the shoulders, and the sightings are as intriguing as the city's history. ISBN 1-56164-222-3 (pb)

Pirates of the Carolinas by Terrance Zepke. Meet thirteen of the most fascinating buccaneers in the history of piracy. Gain insight into the personalities and lives of these sea marauders, who are all connected to the Carolinas. ISBN 1-56164205-3 (pb)

The Return by Mark T. Mustian. From Miami to the streets of Brazil, ex-priest Michael Mason follows a trail that could prove the second coming of Christ. A thriller of a page turner by a first-time author. ISBN 1-56164-190-1 (hb)